Emerging Systems Approaches in Information Technologies:
Concepts, Theories, and Applications

David Paradice
Florida State University, USA

INFORMATION SCIENCE REFERENCE

Hershey · New York

Director of Editorial Content:	Kristin Klinger
Senior Managing Editor:	Jamie Snavely
Assistant Managing Editor:	Michael Brehm
Publishing Assistant:	Sean Woznicki
Typesetter:	Michael Brehm, Michael Killian
Cover Design:	Lisa Tosheff
Printed at:	Yurchak Printing Inc.

Published in the United States of America by
Information Science Reference (an imprint of IGI Global)
701 E. Chocolate Avenue
Hershey PA 17033
Tel: 717-533-8845
Fax: 717-533-8661
E-mail: cust@igi-global.com
Web site: http://www.igi-global.com/reference

Library of Congress Cataloging-in-Publication Data

Emerging systems approaches in information technologies concepts : theories
and applications / David Paradice, editor.
     p. cm.
  Includes bibliographical references and index.
  Summary: "This book presents findings utilizing the incorporation of the
systems approach into fields such as systems engineering, computer science,
and software engineering"--Provided by publisher.
  ISBN 978-1-60566-976-2 (hardcover) -- ISBN 978-1-60566-977-9 (ebook) 1.
Information technology. 2. Systems engineering. 3. Software engineering. I.
Paradice, David B.
  T58.5.E523 2010
  620.001'171--dc22
                              2009032053

British Cataloguing in Publication Data
A Cataloguing in Publication record for this book is available from the British Library.

Editorial Advisory Board

Table of Contents

Detailed Table of Contents

    Manuel Mora, Universidad Autónoma de Aguascalientes, México
    Ovsei Gelman, CCADET, Universidad Nacional Autónoma de México, México
    Moti Frank, HIT - Holon Institute of Technology, Israel
    David B. Paradice, Florida State University, USA
    Francisco Cervantes, Universidad Nacional Autónoma de México
    Guisseppi A. Forgionne, University of Maryland, Baltimore County, USA

An accelerated scientific, engineering, and industrial progress in information technologies has fostered the deployment of Complex Information Technology (highly dependent) Organizational Systems (CITOS). The benefits have been so strong that CITOS have proliferated in a variety of large and midsized organizations to support various generic intra-organizational processes and inter-organizational activities. But their systems engineering, management, and research complexity have been substantially raised in the last decade, and the CITOS realization is presenting new technical, organizational, management, and research challenges. This chapter uses a conceptual research method to review the engineering, management, and research complexity issues raised for CITOS, and develop the rationality of the following propositions: P1: a plausible response to cope with CITOS is an interdisciplinary engineering and management body of knowledge; and P2: such a realization is plausible through the incorporation of foundations, principles, methods, tools, and best practices from the systems approach by way of systems engineering and software engineering disciplines. Discussion of first benefits, critical barriers, and effectiveness measures to reach this academic proposal are presented.

Chapter 2

    Frank Stowell, University of Portsmouth, UK

This chapter raises questions about the nature of Information systems, the way that they are designed and developed and suggest areas that IS researchers may wish to investigate. A concern is raised about the way we think about the domain of information systems and a suggestion made that rather than think of it in terms of the mnemonic IS, with IS association with IT, it should be thought of in terms of as systems of information. This suggestion is made as a means of highlighting considerations developers have to take into account which go beyond those of technology alone. As a means of instigating this proposition four questions are raised in the chapter which are intended to stimulate further information systems research. These questions are about the nature of IS, design Methods, the underpinning philosophy and finally, IS failure.

Chapter 3

    James Courtney, University of Central Florida, USA
    Yasmin Merali, Warwick Business School, UK
    David Paradice, Florida State University, USA
    Eleanor Wynn, Intel Corporation Information Technology, USA

This chapter addresses complexity in information systems. It defines how complexity can be used to inform information systems research, and how some individuals and organizations are using notions of complexity. Some organizations are dealing with technical and physical infrastructure complexity, as well as the application of complexity in specific areas such as supply chain management and network management. Their approaches can be used to address more general organizational issues. The concepts and ideas in this chapter are relevant to the integration of complexity into information systems research. However, the ideas and concepts in this chapter are not a litmus test for complexity. This chapter intends to provide a starting point for information systems researchers to push the boundaries of our understanding of complexity. The chapter also contains a number of suggested research questions that could be pursued in this area.

Chapter 4

    Miroljub Kljajić, University of Maribor, Slovenia
    John V. Farr, Stevens Institute of Technology, USA

The interrelationship between Information Systems (IS), Systems Engineering (SE), and Information Systems Development (ISD) is discussed from past, present, and future perspectives. While SE is relatively a well-established discipline based upon an interdisciplinary approach to enable the realization of successful systems, ISD has evolved to a variant of SE applied mainly for the development of IS. Given the growth in complexity, need for enterprise wide solutions, and the cost and schedule overruns that have be common place for modern software centric systems, well-established tools, techniques, and processes are needed for the development of good IS. Similarities and differences of methodology as well as their evolution and perspectives are also presented herein. A positive trend was found in the evolution of research methodology and published material in SE and its use in IS.

This chapter provides possible directions for the wider application of the systems approach to information systems development. Potential improvement of software development practices is linked by some leading experts to the application of more systemic ideas. However, the current state of the practice in software engineering and information systems development shows the urgent need for improvement through greater application of systems thinking.

The aim of this chapter is to outline some of the key themes that the author believes are important, first, in applying the systems approach to produce high quality IS research in general and, second, to consider more specifically some of the questions and debates that are of interest within the philosophy of IS and of the systems approach. Four themes are identified: being systemic, being critical and realist, being pluralist in approach, and having a concern for truth and knowledge.

This chapter investigates the complex nature of information in information systems (IS). Based on the systems thinking framework, this study argues that information in IS is a system in its own right. A conceptual model of information-as-system is built on the systems thinking perspective adopted from Gharajedaghi's holistic thinking rooted from Ackoff systems approach, which is developed through Peirce's semiotics with the validity support of Metcalfe and Powell's perspective of information perception, Mingers and Brocklesby's schema of situational actions, Toulmin's theory of argumentation and Ulrich's theory of systems boundary. The proposed model of information-as-systems is described in terms of triads—on the structure, function, and process, all interdependent—in a context of information-as-system in IS.

Money-information processes are important determinants of the homeostasis of modern exchange-based societies. While general principles of cybernetics may be approached from an abstract, holistic perspective, this chapter approaches certain aspects of social cybernetics from a concrete, internal perspective. From the chosen perspective, it is argued that the irreducible unit (the holon) of modern economic systems is the exchange. Highly complex economies are combinations of specific and observable exchanges. That complexity is facilitated by the introduction of money-information markers to temporally separate the reciprocating transfers of trades.

This chapter describes a complex adaptive systems (CAS)-based enterprise knowledge-sharing (KnS) model. The CAS-based enterprise KnS model consists of a CAS-based KnS framework and a multi-agent simulation model. Enterprise knowledge sharing is modeled as the emergent behavior of knowledge workers interacting with the KnS environment and other knowledge workers. The CAS-based enterprise KnS model is developed to aid knowledge management (KM) leadership and other KnS researchers in gaining an enhanced understanding of KnS behavior and its influences. A premise of this research is that a better understanding of KnS influences can result in enhanced decision-making of KnS interventions that can result in improvements in KnS behavior.

The increasing design, manufacturing, and provision complexity of high-quality, cost-efficient and trustworthy products and services has demanded the exchange of best organizational practices in worldwide organizations. While that such a realization has been available to organizations via models and standards of processes, the myriad of them and their heavy conceptual density has obscured their comprehension and practitioners are confused in their correct organizational selection, evaluation, and deployment tasks. Thus, with the ultimate aim to improve the task understanding of such schemes by reducing its business process understanding complexity, in this chapter the authors use a conceptual systemic model of a generic business organization derived from the theory of systems to describe and compare two main models (CMMI/SE/SwE, 2002; ITIL V.3, 2007) and four main standards (ISO/IEC 15288, 2002; ISO/IEC 12207, 1995; ISO/IEC 15504, 2005; ISO/IEC 20000, 2006) of processes. Description and comparison are realized through a mapping of them onto the systemic model.

A formal conceptualization of the original concept of system and related concepts—from the original systems approach movement—can facilitate the understanding of information systems (IS). This chapter develops a critique integrative of the main IS research paradigms and frameworks reported in the IS literature using a systems approach. The effort seeks to reduce or dissolve some current research conflicts on the foci and the underlying paradigms of the IS discipline.

As organizations strive to expand system capabilities through the development of system-of-systems (SoS) architectures, they want to know "how much effort" and "how long" to implement the SoS. In order to answer these questions, it is important to first understand the types of activities performed in SoS architecture development and integration and how these vary across different SoS implementations. This chapter provides results of research conducted to determine types of SoS lead system integrator (LSI) activities and how these differ from the more traditional system engineering activities described in Electronic Industries Alliance (EIA) 632 ("Processes for Engineering a System"). This research further analyzed effort and schedule issues on "very large" SoS programs to more clearly identify and profile the types of activities performed by the typical LSI and to determine organizational characteristics that significantly impact overall success and productivity of the LSI effort. The results of this effort have been captured in a reduced-parameter version of the constructive SoS integration cost model (COSOSIMO) that estimates LSI SoS engineering (SoSE) effort.

The work system method was developed iteratively with the overarching goal of helping business professionals understand IT-reliant systems in organizations. It uses general systems concepts selectively, and sometimes implicitly. For example, a work system has a boundary, but its inputs are treated implicitly rather than explicitly. This chapter asks whether the further development of the work system method might benefit from integrating general systems concepts more completely. After summarizing aspects of the work system method, it dissects some of the underlying ideas and questions how thoroughly even basic systems concepts are applied. It also asks whether and how additional systems concepts might be incorporated beneficially. The inquiry about how to use additional system ideas is of potential interest to people who study systems in general and information systems in particular because it deals with bridging the gap between highly abstract concepts and practical applications.

This chapter provides an overview of perspectives associated with information and knowledge resource management in systems engineering and systems management in accomplishing enterprise resource planning for enhanced innovation and productivity. Accordingly, the authors discuss economic concepts involving information and knowledge, and the important role of network effects and path dependencies in influencing enterprise transformation through enterprise resource planning.

Current developments in information systems (IS) evaluation emphasize stakeholder participation in order to ensure adequate and beneficial IS investments. It is now common to consider evaluation as a subjective process of interpretation(s), in which people's appreciations are taken into account to guide evaluations. However, the context of power relations in which evaluation takes place, as well as their ethical implications, has not been given full attention. In this chapter, ideas of critical systems thinking and Michel Foucault's work on power and ethics are used to define a critical systems view of power to support IS evaluation. This chapter proposes a system of inquiry into power with two main areas: 1) Deployment of evaluation via power relations and 2) Dealing with ethics. The first element addresses how evaluation becomes possible. The second one goes in-depth into how evaluation can proceed as being informed by ethical reflection. The chapter suggests that inquiry into these relationships should contribute to extend current views on power in IS evaluation practice, and to reflect on the ethics of those involved in the process.

This chapter combines disruptive innovation strategy (DIS) theory with the system dynamics (SD) modeling method. It presents a simulation model of the hard-disk (HD) maker population overshoot and collapse dynamics, showing that DIS can crucially affect the dynamics of the IT industry. Data from the HD maker industry help calibrate the parameters of the SD model and replicate the HD makers□ overshoot and collapse dynamics, which DIS allegedly caused from 1973 through 1993. SD model analysis entails articulating exactly how the structure of feedback relations among variables in a system determines its performance through time. The HD maker population model analysis shows that, over five distinct time phases, four different feedback loops might have been most prominent in generating the HD maker population dynamics. The chapter shows the benefits of using SD modeling software, such as iThink®, and SD model analysis software, such as Digest®. The latter helps detect exactly how changes in loop polarity and prominence determine system performance through time. Strategic scenarios computed with the model also show the relevance of using SD for information system management and research in areas where dynamic complexity rules.

This chapter uses soft systems methodology and complexity modeling to build an evaluation approach of a data warehouse implemented in a leading European financial institution. This approach consists in building a theoretical model to be used as a purposeful observation lens, producing a clear picture of the problematic situation under study and aimed at providing knowledge to prescribe corrective actions.

This chapter is concerned with methodological issues. In particular, it addresses the question of how is it possible to align the design of management information systems with the structure of an organization. The method proposed is built upon the Cybersin method developed by Stafford Beer (1975) and Raul Espejo (1992). The chapter shows a way to intersect three complementary organizational fields: management information systems, management control systems, and organizational learning when studied from a systemic perspective; in this case from the point of view of management cybernetics (Beer 1959, 1979, 1981, 1985).

This chapter seeks to address the dearth of practical examples of research in the area by proposing that critical realism be adopted as the underlying research philosophy for enterprise systems evaluation. The authors address some of the implications of adopting such an approach by discussing the evaluation and implementation of a number of automated performance measurement systems (APMS). Such systems are a recent evolution within the context of enterprise information systems. They collect operational data from integrated systems to generate values for key performance indicators, which are delivered directly to senior management. The creation and delivery of these data are fully automated, precluding manual intervention by middle or line management. Whilst these systems appear to be a logical progression in the exploitation of the available rich, real-time data, the statistics for APMS projects are disappointing. An understanding of the reasons is elusive and little researched. The authors describe how critical realism can provide a useful "underlabourer" for such research, by "clearing the ground a little ... removing some of the rubbish that lies in the way of knowledge" (Locke, 1894, p. 14). The implications of such an underlabouring role are investigated. Whilst the research is still underway, the chapter indicates how a critical realist foundation is assisting the research process.

Preface

ABSTRACT

The papers published in the first volume of the *International Journal of Information Technology and the Systems Approach* (IJITSA) and a special issue of *Information Resource Management Journal* (IRMJ) reveals that most of the papers describing research that is grounded in a systems approach are theoretical and conceptual. The papers are typically grounded in a constructionist epistemology. This outcome is influenced somewhat by the need to publish area-specific position papers in a new journal to establish the types of research sought be the editorial review boards. However, the analysis also identifies the opportunity for objectivist, empirical work and shows that a wide range of possible research purposes may be explored with ample opportunity to make significant contributions.

INTRODUCTION

Organizations have been facing increasingly turbulent environments for over half of a century. Fifty years ago the post-world-war era ushered in the foundation of the modern business world as nations around the globe rebuilt their economies. Forty years ago organizations operated in a world of turbulent social change marked by the emergence of new consumer markets and a host of technological innovations. Thirty years ago the global economy as we now know it began to take shape with oil producing countries influencing the energy markets that allow organizations to operate. Twenty years ago the political super power landscape changed, small computing devices became common, and the seeds of a mobile, computing-based population were sown. Ten years ago global commerce moved boldly into a digital environment.

Today, we witness the global interconnectedness of the world's manufacturing, banking, and commerce systems. Fluctuations in the price of the fuel required to bring goods to market impact the cost of goods. The prices of these goods impact the ability of consumers to purchase them. As spending slows, organizations adjust by cutting back on production or services, slowing the growth of the economy. Jobs become scarce and wages start to lag the cost of living. Consumers borrow more or purchase on credit to maintain a standard of living; organizations borrow to finance operations in anticipation of greater markets later. As debt increases, credit lines grow tighter. Eventually, defaults occur and a cascade of failures ripples through the system producing ever larger impacts. In today's world, the system is truly global. Now more than ever, we realize that only through truly systemic thinking do we have any chance of managing the complexity of the world around us to any successful result, regardless of how we define success.

The systems approach is an approach that maintains a holistic view of a problem while supporting a focused investigation on one or several aspects of the problem. It is a robust approach; one that can support multi-disciplinary and / or inter-disciplinary methods. It is an integrative approach.

Over the past year, Manuel Mora of Autonomous University of Aguascalientes (Mexico) has sought to provide outlets for research that was grounded in a systems approach. The *International Journal of Information Systems and the Systems Approach* (IJITSA) is an international, refereed journal supporting publication of papers that address the foundations, challenges, opportunities, problems, trends, and solutions encountered by both scholars and practitioners in the field of information systems as they are perceived from the perspective of the systems approach. IJITSA emphasizes a systemic worldview of managerial, organizational and engineering interaction, which is often reflected or implemented in modern complex information systems and information technologies. Articles published in IJITSA focus on information systems and often also include tenets of software engineering, systems engineering, complexity and philosophy. Applied and theoretical research papers are welcome.

Similarly, Professor Mora organized a special issue of *Information Resources Management Journal* (IRMJ) on the systems approach. The IRMJ is a refereed, international journal providing wide coverage of issues in information systems field. It is especially receptive of applied research. This book summarizes the work in these volumes to offer interested persons in both the research and practitioner areas access to the high-quality research in information systems and related disciplines that is grounded in a systems approach.

IJITSA'S GOALS AND MISSION

The *International Journal of Information Technologies and Systems Approach* (IJITSA) is a refereed, international journal on applied and theoretical research, aimed at providing coverage of the foundations, challenges, opportunities, problems, trends, and solutions encountered by both scholars and practitioners in the field of information systems when applying the systems approach to important issues. IJITSA was created to disseminate and to promote discussion of high quality research results on information systems. A long run goal of IJITSA is to facilitate the application of the systems approach to the systems field, thus obtaining a better understanding of the complexity inherent in the field. The current publication rate is two issues per year.

The articles published in IJITSA are organized into several subareas and a particular paper's home area is determined by the paper's primary orientation. These subareas are information systems, software engineering, systems engineering, complex systems, and philosophy of systems. However, IJITSA emphasizes a systemic worldview of modern complex information systems and information technologies. Therefore, the interaction of managerial, organizational and engineering facets that characterize complex situations is particularly emphasized in IJITSA.

A "systems approach" can have different meaning to different researchers, probably depending on their training and philosophy. The editors of IJITSA therefore focus on the rigor of submitted papers in applying a systems approach (broadly defined) to theoretical, empirical, modeling, engineering or behavioral studies in order to explore, describe, explain, predict, design, control, evaluate, interpret, intervene and/or develop organizational systems where information systems are the primary objects of study.

A FRAMEWORK FOR ANALYSIS & ASSESSMENT

This book provides an opportunity to assess the start that has been made in publishing research in the systems approach area. To that end, each paper published in the first volume of IJITSA and the IRMJ special issue was categorized on several dimensions. The dimensions were defined in prospectus for the journal's creation. They also appear in the descriptive information available online. Table 1 lists the papers.

The first evaluation of each paper was with respect to the overall basis of the paper. The overall basis of the paper was identified as either theoretical or empirical. There are, of course, many ways that a paper may be theoretical or empirical. Theoretical research papers may be further divided into several categories. Theoretical position papers are papers that study the whole discipline or a vast topic within the discipline, with a long-term and strategic perspective. These papers analyze the contributions, achievements and challenges of the topic, and may use theoretical or empirical arguments. Theoretical literature review papers are papers that report the state of the art of a topic. Formal theoretical papers are papers that report the development and/or utilization of a theoretical construct, framework, model, architecture or methodology. Finally, theoretical modeling papers are papers that report the development of a model. The model may be evaluated using simulation.

Empirical research papers may be engineering-oriented or behaviorally-oriented. Engineering papers are papers that report the design and/or building of a model or system, which is evaluated in an empirical manner. The empirical test may be an analysis of data to determine the fit of a model to reality or a pilot test evaluation of a system. Behaviorally-oriented papers are papers that report survey-based, case study or action research studies, where the behavior of human beings is the main focus of the system. These studies may occur in a laboratory or in a more ethnographic form in organizational settings.

The second evaluation of each paper was with respect to the epistemology of the paper. The epistemology of the paper indicates the philosophical underpinnings of the work. The categories considered are objectivism, constructionism, and critical inquiry. Crotty (1998) provides an excellent treatment of epistemologies and theoretical perspectives, upon which we draw. Research from an objectivism epistemology is rooted in the notion that truth and meaning reside in the objects of study, independently of any consciousness. It is typified by positivist and post-positivist studies. Constructionism, on the other hand, is the view that knowledge and reality is contingent upon human practices, being constructed from interaction between humans and the world they study. We currently combine both socially-constructed reality and individually-constructed reality (i.e., subjectivist research) under the constructionism umbrella. Research rooted in critical inquiry is research that challenges conventions. For the purposes of IJITSA, critical inquiry need not be pursued from an advocacy or emancipatory perspective, as is often the case in other social science research (although such a stance is not precluded by any means). Regardless of the epistemological approach taken, the philosophy is expected to be integrated or discussed from the perspective of the systems approach with the long-term aim of obtaining a unified view of theory of systems and the object of study.

The third evaluation of each paper was with respect to the systemic research methodology or approach employed. Conceptual papers apply a systems approach theory to bring order to a set of related concepts. Formal mathematical papers apply mathematical rigor to the analysis. Papers that employ systems simulation may use discrete, multi-agent or hybrid modeling. Papers that study feedbacks and information flows to gain a better understanding of the system over time typically draw on methods of systems dynamics modeling. Soft systems methodologies form the basis of another category of papers. These papers typically examine difficult problems with many opposing, and often conflicting, points of view. Action research papers recognize the influence the researcher may have on the system when s/he is

Table 1. Papers published using a systems approach

Paper	Title	Authors
1	Toward an Interdisciplinary Engineering and Management of Complex IT-Intensive Organizational Systems: A Systems View	M. Mora, O. Gelman, M. Frank, D. Paradice, F. Cervantes, G. Forgionne
2	Do We Mean Information Systems of Systems of Information	F. Stowell
3	On the Study of Complexity in Information Systems	J. Courtney, Y. Merali, D. Paradice, E. Wynn
4	Importance of Systems Engineering in the Development of Information Systems	M. Kljajić, J. Farr
5	Towards A Wider Application of the Systems Approach in Information Systems and Software Engineering	D. Petkov, D. Edgar-Nevill, R. Madachy, R. O'Connor
6	Pluralism, Realism, and Truth: The Keys to Knowledge and Information Systems Research	J. Mingers
7	Information-As-System in Information Systems: A Systems Thinking Perspective	T. Nguyen, H. Vo
8	An Analysis of the Imbursement of Currency in a Debt-Based Money-Information System	G. Swanson
9	A Complex Adaptive Systems-Based Enterprise Knowledge Sharing Model	C. Small, A. Sage
10	A Conceptual Descriptive-Comparative Study of Models and Processes in SE, SwE, and IT Disciplines Using the Theory of Systems	M. Mora, O. Gelman, R. O'Connor, F. Alvarez, J. Macías-Lúevano
11	Integrating the Fragmented Pieces of IS Research Paradigms and Frameworks: A Systems Approach	M. Mora, O. Gelman, G. Forgionne, D. Petkov, J. Cano
12	System-of-Systems Cost Estimation: Analysis of Lead System Integrator Engineering Activities	J. Lane, B. Boehm
13	Could the Work System Method Embrace Systems Concepts More Fully?	S. Alter
14	Information and Knowledge Perspectives in Systems Engineering and Management for Innovation and Productivity through Enterprise Resource Planning	S. Stephenson, A. Sage
15	A Critical Systems View of Power-Ethics Interactions in Information Systems Evaluation	J. Córdoba
16	Information Technology Industry Dynamics: Impact of Disruptive Innovation Strategy	N. Georgantzas, E. Katsamakas
17	Using a Systems Thinking Perspective to Construct and Apply an Evaluation Approach of Technology-Based Information Systems	H. Kefi
18	The Distribution of a Management Control System in an Organization	A. Reyes
19	Making a Case for Critical Realism: Examining the Implementation of Automated Performance Management Systems	P. Dobson, J. Myles, P. Jackson

engaged in the research of the system. Critical systems research combines a traditional systems approach with participatory methods so that the complex issues being examined may be reconciled from different viewpoints as the problem examination process unfolds. Finally, a study may be multi-disciplinary or inter-disciplinary. However, it can be argued that almost every paper published in these issues is multi-disciplinary or inter-disciplinary, so that category was not considered in the analysis below.

The fourth evaluation of each paper was with respect to its purpose. This was often the most difficult category to assess, because most studies of complex phenomena cannot be categorized on a single purpose. However, a wide range of possible purposes exists and it is beneficial for our purposes here to place each paper in one primary category to start. Scholastic papers are papers that organize disperse, fragmented and ignored theoretical knowledge to achieve a better understanding of the phenomenon of interest. Exploratory papers generate potentially useful insights for new situations. For the purposes of this effort, descriptive papers identify sets of variables and their measurement scales that best describes the profile of the information systems field. Predictive papers find plausible relations between variables. The relationships may be non-linear and include feedback. There may also be multiple predictors and multiple effects described. Explanatory papers describe cause and effect relationships, which may also be multiple in nature. Design and control papers design and possibly evaluate conceptual or physical artifacts used to control information systems, either as a whole or in part. Evaluative papers evaluate system actions such as politics, programs, or projects, among others, in an information systems context. Instrumental papers develop and validate a conceptual or physical instrument to measure a construct that is argued to be useful in better understanding information systems. Interpretation papers develop and test (or simply apply) a model in a complex event or situation. Often, there is conflict in these situations due to the different perspectives taken. Interpretation papers seek to achieve a mutual understanding and interpretation of the event or situation. Interventionist papers design, implement and verify human actions taken to ameliorate a problematic situation related to an information systems context. Developmental papers design, build and evaluate a physical artifact to exhibit the proof of a new information systems concept, either in whole or in part. Methodological papers propose a new methodological research process based on the systems approach. These papers rely on logical argumentation or proofs of the method's applicability.

Most of the papers examined can be argued to achieve multiple purposes. In spite of this realization, most of the papers were placed in only one category, which was selected based on my perception of the primary goal or achievement of the paper. However, other readers and the authors of the papers may rightfully disagree with my assessment, but my goal is to provide a "big picture" view of our current situation. I encourage future researchers to fill in the (in some case, missing) details with future work.

IJITSA also publishes interviews of internationally known information systems scholars who have published research in the areas covered by IJITSA. These interviews were not included in the analysis which follows.

ANALYSIS OF THE PUBLISHED PAPERS

Table 2 contains the results of the effort to analyze the papers on the categories described above. The horizontal lines in the table separate the issues of publication. The first row contains data for the papers published in IJITSA volume one, issue one. The second row contains data for the papers published in IJITSA volume 1, issue two. The third row contains data for the papers published in IRMJ volume twenty, number 1.

When considering the basis of the papers, the analysis reflects an overwhelming bias toward theoretical work. The first papers in IJITSA were intended to outline the positions of the senior editorial review board members in order to provide guidance to potential researchers who are interested in pursuing work in one of the subareas (information systems, software engineering, system engineering, complex systems, or philosophical issues) of the journal. Given the complexity that can be inherent in a systems approach to any issue, it is not surprising that a majority of the papers do not incorporate empirical arguments.

Table 2. Analysis of papers published using a systems approach

Basis		Epistemology			Methodology						
Theoretical	Empirical	Objectivism	Constructionism	Critical Inquiry	Conceptual	Formal Mathematical	Systems Simulation	Systems Dynamics	Soft Systems	Action Research	Critical Systems
1				1	1						
2			2		2						
3			3		3						
4	4		4		4						
5			5		5						
6			6		6						
7			7		7						
8			8				8				
9			9					9			
10			10		10						
11			11		11						
12			12		12						
13			13		13						
14			14		14						
15				15							15
	16	16						16			
	17		17						17		
18			18		18						
19				19	19						

Purpose											
Scholastic	Exploratory	Descriptive	Predictive	Explanatory	Design & Control	Evaluative	Instrumental	Interpretive	Intervention	Developmental	Methodological
1											
						2					
3											
		4									
						5					
6											
	7										
		8		8							
											9
10						10					
11											
					12						
	13										
					14						
											15
											16
						17					
18											18
19											19

However, a few papers do include empirical analysis. Paper number 4 ("Importance of Systems Engineering in the Development of Information Systems") uses data to confirm some of the approach described in the paper, so it was given both a theoretical and an empirical rating. Paper 16 ("Information Technology Industry Dynamics: Impact of Disruptive Innovation Strategy") tested a system dynamics model using simulation. Paper 17 ("Using a Systems Thinking Perspective to Construct and Apply an Evaluation Approach of Technology-Based Information Systems") tested a model of information systems / information technology using a systems thinking perspective with empirical data from interviews and questionnaires. Notably, this paper also confronts the positivist versus constructivist / interpretivist research dilemma directly, settling on the systems approach as a pragmatic manner for achieving a purposeful result. This paper was also given both a theoretical and an empirical rating.

With respect to epistemology, and given the bias toward a theoretical basis for the papers published, it also should not be surprising that a constructionism epistemology is used in most of the papers. Paper 16 ("Information Technology Industry Dynamics: Impact of Disruptive Innovation Strategy") was classified as a paper based on an objectivism epistemology because it reflects a fairly standard post-positivist approach to research.

Several papers were classified as based on a critical inquiry epistemology. Paper 1 ("Toward an Interdisciplinary Engineering and Management of Complex IT-Intensive Organizational Systems: A Systems View"), paper 15 ("A Critical Systems View of Power-Ethics Interactions in Information Systems Evaluation"), and paper 19 ("Making a Case for Critical Realism: Examining the Implementation of Automated Performance Management Systems") were placed in this category. As noted above, in this analysis the focus on the critical inquiry characteristic of challenging the status quo took precedence over the existence of an advocacy or emancipatory goal.

Paper 1 ("Toward an Interdisciplinary Engineering and Management of Complex IT-Intensive Organizational Systems: A Systems View"), the position paper for the journal, challenges readers to consider the benefits and advantages of using interdisciplinary concepts "to improve and reposition the information systems discipline to accommodate the emergence of" complex information technology intensive organizational systems. Paper 15 ("A Critical Systems View of Power-Ethics Interactions in Information Systems Evaluation") uses the ideas of critical systems thinking and Foucault's work on power and ethics to examine information systems evaluation processes. Paper 19 ("Making a Case for Critical Realism: Examining the Implementation of Automated Performance Management Systems") proposes that "critical realism be adopted as the underlying research philosophy" in the evaluation of enterprise systems.

Within the methodology section of Table 2, most of the papers were placed in the conceptual methodology category. This is consistent with the nature of the position papers that were published. Many of these papers are intended to demonstrate how an application of systems approach theory can bring order to the primary topic of the paper. For example, in paper 5 ("Towards a Wider Application of the Systems Approach in Information Systems and Software Engineering") the authors propose "directions for future research and practical work" that result from applying systems thinking to the fields of information systems and software engineering.

Paper 8 ("An Analysis of the Imbursement of Currency in a Debt-Based Money-Information System") is the only paper that was placed in the formal mathematical method category. The paper takes an analytical approach to examine money-information exchanges. It also takes an internal perspective to examine certain aspects of social cybernetics.

Paper 9 ("A Complex Adaptive Systems-Based Enterprise Knowledge Sharing Model") uses systems simulation to test a complex adaptive systems-based enterprise knowledge sharing model. The research found that the methodology can provide knowledge management executives with a better understanding of knowledge sharing behavior and influences. As noted above, paper 16 ("Information Technology Industry Dynamics: Impact of Disruptive Innovation Strategy") is a paper that also uses simulation to test a model. However, the primary focus of this paper was determined to be the systems dynamics aspect, so it was placed in the systems dynamics methodology category. It was the only paper placed in the systems dynamics category.

Paper 17 ("Using a Systems Thinking Perspective to Construct and Apply an Evaluation Approach of Technology-Based Information Systems") is a soft systems methodology paper. Soft systems methodology is combined with complexity modeling to build an evaluation approach of a data warehouse. As noted earlier, the systems approach adopted in this work was selected for the pragmatic fashion that it would support a purposeful outcome.

The final paper considered in the methodology section of the grid is paper 15 ("A Critical Systems View of Power-Ethics Interactions in Information Systems Evaluation"), which was mentioned earlier as a paper using a critical inquiry epistemology. The paper takes the position that information systems evaluation does not provide enough guidance to practitioners on how to act in relation to power as an issue that affects any action for improvement.

Moving next to the Purpose section of Table 2, we see that eight of the nineteen papers are classified as having a scholastic purpose. Papers 1 ("towards an Interdisciplinary Engineering and Management of Complex IT-Intensive Organizational Systems: A Systems View"), 3 ("On the Study of Complexity in Information Systems"), and 6 ("Pluralism, Realism, and Truth: The Keys to Knowledge in Information Systems Research") are all editorial position papers in the first issue. Paper 10 ("A Conceptual Descriptive-Comparative Study of Models and Standards of Processes in SE, SwE, and IT Disciplines Using the Theory of Systems") analyzes three fields in a single comparison paper with the objective of reducing the complexity inherent in business process schemes.

Three of the Information Resources Management Journal papers are scholastic papers. Paper 11 ("Integrating the Fragmented Pieces of IS Research Paradigms and Frameworks: A Systems Approach") applies the systems approach to information system research paradigms and frameworks. Paper 18 ("The Distribution of a Management Control System in an Organization") builds upon the Cybersin method to align systems and organizational structure. Paper 19 ("Making the Case for Critical Realism: Examining the Implementation of Automated Performance Management Systems") leverages a philosophical stance, so it is considered scholastic on that merit. (Papers 18 and 19 are also methodological papers, discussed below.)

Paper 7 ("Information-As-System in Information Systems: A Systems Thinking Perspective") has been categorized as an exploratory paper. It examines information as a system in its own right, thus proposing a new way of conceptualizing information.

The descriptive papers in this analysis are paper 4 ("The Role of Systems Engineering in the Development of Information Systems") and paper 8 ("An Analysis of the Imbursement of Currency in a Debt-Based Money-Information System"). These papers contain models that describe relationships identified by the authors. Paper 8 is also categorized as an explanatory paper. Through the use of analytical descriptions typically found in economic analysis, the author describes money-information exchanges and derives implications for the design of information systems.

Only two papers were categorized as design and control papers. Paper 12 ("System-of-Systems Cost Estimation: Analysis of Lead Systems Integrator Engineering Activities") examines the activities of architecture development and integration in order to answer questions related to time and effort needed to achieve desired results in systems implementations. Paper 14 ("Information and Knowledge Perspectives in Systems Engineering and Management for Innovation and Productivity through Enterprise Resource

Planning") focuses on enhancing innovation, productivity, and knowledge management through a better understanding of network effects and path dependencies in enterprises.

The only evaluative paper is paper 17 ("Using a Systems Thinking Perspective to Construct and Apply and Evaluation Approach of Technology-Based Information Systems"). As mentioned earlier, this paper uses a soft systems approach to build an evaluative mechanism of a data warehouse implementation.

The final category containing entries in Table 2 is the methodological purpose category. Five papers are listed there. Paper 9 ("A Complex Adaptive Systems-Based Enterprise Knowledge Sharing Model") and paper 16 ("Information Technology Industry Dynamics: Impact of Disruptive Innovation Strategy") use simulation. Paper 15 ("A Critical Systems View of Power-Ethics Interactions in Information Systems Evaluation") and paper 19 ("Making a Case for Critical Realism: Examining the Implementation of Automated Performance Management Systems") rely on a critical thinking methodology. As noted above, paper 18 ("The Distribution of a Management Control System in an Organization") builds upon the Cybersin method.

OPPORTUNITIES FOR FUTURE RESEARCH

The analysis indicates there are many, many opportunities for publishing new work grounded in the systems approach. Epistemologically, objectivism is almost nonexistent as a knowledge perspective in this analysis. Given that outcome, it is not surprising that only three of the nineteen papers published in these issues were empirical papers. Researchers may be reluctant to pursue empirical work in systems areas, due to a perceived increase in the complexity of modeling systems as compared to modeling components of systems (i.e., subsystems). Indeed, a major criticism of non-systems research is that it is necessarily reductionist and loses much of the richness that a more complete systems-oriented description more naturally captures. However, structural equation modeling has much to offer in the analysis of systems and researchers are encouraged to investigate the appropriate application of that type of approach where possible. In cases where structural equation modeling is not appropriate, researchers should explore the use of multivariate analyses. In some cases, nonparametric approaches may be applicable to describe differences in distributions of outcomes that do not meet the assumptions of parametric analysis methods.

On the other hand, the use of critical inquiry-like epistemology is very encouraging. Critical inquiry is a defining characteristic of scientific approaches (Popper 2000). Through critical inquiry we can surface assumptions in the descriptions and models of problem situations to better understand the nature of the complexity inherent in them. Critical inquiry is often pursued from an advocacy perspective with a goal that change will occur in the system (typically, a social / political system). As information systems researchers, we may be able to improve the application of the systems approach through a more aggressive advocacy!

All of the methodological categories except the conceptual category are in need of development. The rigor of formal mathematical approaches should be pursued where possible, including when it can only be applied to a portion of a system being analyzed. In such a case, a mixed-method approach (Creswell 2003) can be utilized to present an analysis that is richer than one which is based only on a qualitative approach or only on an empirical approach. Systems simulation, systems dynamics, and the use of the soft systems methodology can be leveraged to add to the body of work in these categories. This volume contains excellent examples of critical inquiry, which can be used to inform a critical systems methodology approach in research.

The absence of action research methods-based papers in this analysis is a call for researchers to get out of their offices and into the world. Information systems issues are inherently socio-technical issues, especially when aspects of decision making come into play. The role of systems in complex decision making environments such as public policy making, healthcare, financial fund administration, enterprise management and many other areas depends on understanding how the human element comes into play. Action research must be executed carefully, as the researcher cannot avoid being part of the system under investigation. However, guidelines exist for minimizing the researcher's influence on the study results so there is no need to hesitate on methodological grounds.

An increase in action research papers could lead to an increase in papers with an interventionist purpose. Research involving the design and implementation of systems that also incorporate an analysis of human actions taken to mitigate a problematic situation would fall into the interventionist category.

Actually, there is great opportunity for papers in all of the purposeful categories used in this analysis. As might be expected given a lack of empirically-based papers, predictive papers that find plausible relations between variables have not been published. Instrumental papers that develop and validate an instrument to measure a construct are also lacking. Interpretation papers that develop and test (or simply apply) a model in a complex event or situation are also needed.

Systems design, development, implementation, and assessment were core activities in the early years of information systems research. Developmental papers that describe these activities related to the design, implementation, and testing of an artifact to prove systems approach to an information systems concept, either in whole or in part, certainly would be welcomed.

CONCLUSION

As noted at the outset, this analysis paints a picture using a broad brush. The papers considered here are truly too complex to be accurately described on the discrete dimensions that exist in our categories. Yet, I believe we have established a few reasonable characterizations of the work that has been published. Much of it is theoretical and conceptual in nature. The underlying epistemology is constructionist, a term that is used in this work to include subjective perspectives that are either individually or socially constructed. The primary purposes of the work are to educate readers as to what could be studied, evaluate a few approaches, and outline several potential methodologies. There is a small but undeniable collection of papers that reflect a critical inquiry approach. I feel this is a valuable contribution and I hope to see more papers develop along that line of inquiry.

There are many opportunities for researchers to fill in gaps in our existing publication scheme. A greater use of objectivist epistemology (as typically underlies positivist and post-positivist approaches) will be likely to bring more empirically-based papers into the community. Action research is needed and increasing that approach can drive academically rigorous research into the world of relevant application, a need that is often identified inside and outside of academe.

REFERENCES

Creswell, J.W. (2003). *Research design: Qualitative, quantitative, and mixed methods approaches* (2nd edition). Thousand Oaks, CA: Sage Publications, Inc.

Crotty, M. (1998). *The foundations of social research*. Thousand Oaks, CA: Sage Publications, Inc.

Popper, K. (2000). *Conjectures and refutations* (5th edition). New York: Routledge.

Chapter 1
Toward an Interdisciplinary Engineering and Management of Complex IT–Intensive Organizational Systems:
A Systems View

Manuel Mora
Universidad Autónoma de Aguascalientes, México

Ovsei Gelman
CCADET, Universidad Nacional Autónoma de México, México

Moti Frank
HIT - Holon Institute of Technology, Israel

David B. Paradice
Florida State University, USA

Francisco Cervantes
Universidad Nacional Autónoma de México

Guisseppi A. Forgionne
University of Maryland, Baltimore County, USA

ABSTRACT

An accelerated scientific, engineering, and industrial progress in information technologies has fostered the deployment of Complex Information Technology (highly dependent) Organizational Systems (CITOS). The benefits have been so strong that CITOS have proliferated in a variety of large and midsized organizations to support various generic intra-organizational processes and inter-organizational activities. But their systems engineering, management, and research complexity have been substantially raised in the last decade, and the CITOS realization is presenting new technical, organizational, management, and research challenges. In this article, we use a conceptual research method to review the engineering, management, and research complexity issues raised for CITOS, and develop the rationality of the following propositions: **P1***: a plausible response to cope with CITOS is an interdisciplinary engineering and management body of knowledge; and* **P2***: such a realization is plausible through the incorporation of foundations, principles, methods, tools, and best practices from the systems approach by way of systems engineering and software engineering disciplines. Discussion of first benefits, critical barriers, and effectiveness measures to reach this academic proposal are presented.*

Businesses no longer merely depend on information systems. In an increasing number of enterprises, the systems are the business. (R. Hunter & M. Blosch, Gartner Group, 2003)

INTRODUCTION

An accelerated scientific, engineering, and industrial progress in information technologies and its convergence with communications technologies (the ICT concept) has fostered the deployment of Complex Information Technology (highly dependent) Organizational Systems (CITOS) in the last decade. The CITOS concept subsumes the well-known constructs of *mission-critical systems, large-scale information systems, enterprise information systems*, and *inter-organizational information systems*. Generic instances of CITOS are worldwide credit card systems, brokerage financial systems, military defense systems, large ERPs, governmental tax payment systems, and worldwide e-commerce and B2B supply-chain systems in automotive and publishing industries.

Empirical evidence, such as (a) the raising of the ICT budget (measured as a percentage of sales) to 5%-9% in the 2000s (Prewitt & Cosgrove, 2006); (b) the growing of world ICT trade from 8% in 1995 to 10% in 2001 with a 4% annual growth rate (OECD, 2004); (c) the IT commoditization or democratization phenomenon being more affordable the ICT infrastructure in midsized firms in the 1990s (Carr, 2003); (d) the maturing of the myriad of ICT in the last decade (e.g., mobile computing, wireless networks, Web services, grid computing, and virtualization services); (e) the new ways for performing business-oriented operational, tactical, and strategic organizational duties through ICT (e.g., workflow systems, business process management, and service-oriented management); (f) the several tangible and intangible organizational benefits from intra-organizational processes (as in Porter's value-chain activities) and inter-organizational activities (supplier-customer value chains, B2B, and e-government initiatives) leveraged by CITOS; and (g) the thousands of US dollars lost due to availability, continuity, and capacity failures in ICT services (van Bon, Pieper, & van deer Veen, 2006) because of an hour of system downtime. These factors and others show that CITOS are relevant for business and government organizations (as well as for nonprofit organizations).

Such systems are characterized by having (1) many heterogeneous ICT (client and server hardware, operating systems, middleware, network and telecommunication equipment, and business systems applications) (2) a large variety of specialized human resources for their engineering, management, and operation; (3) a worldwide scope; (4) geographically distributed operational and managerial users; (5) core business processes supported; (6) a huge financial budget for organizational deployment; and (7) a critical interdependence on ICT. Thus, these can be correctly labeled as "complex systems" (comprised of a large variety of components and inter-relationships in multiple scales generating unexpected emergent behaviors).

According to a systemic definition, the emergent properties from a system cannot be attributed to the individual actions performed from parts. Rather the interactions among people, machines, applications, procedures, data, policies, and the organizational setting and organizational environment are responsible for their coproduction. Consequently, and because of its raised engineering and management complexity, a holistic study of human and machine component inter-relationships and of its environment is needed when the efficiency and effectiveness of CITOS are considered. It could be expected that the current IS body of knowledge (IS BoK) addresses such issues. Nevertheless, because of the ICT technological progress, combined with the extended capabilities of CITOS services demanded in large organizations (Dougmore, 2006), have raised significantly the systems engineering, management, and research complexity for users, managers, engineers, and researchers, our premise is such a current IS BoK is insufficient and an extended (interdisciplinary) IS BoK is required.

According to the international expert in complex systems Bar-Yam (2003b), to design organizational complex systems, we must recognize that "the networked information system that is being developed, serves as part of the human socio-economic-technological system. Various parts of this system that include human beings and information systems, and the system as a whole, is a functional system" (p. 17). Also "the recognition that human beings and information technology are working together as an integrated system" (p. 25) should be an imperative consideration for its design. From a service oriented management and engineering perspective (Chesbrough & Spohrer, 2006), Dermikan and Goul (2006) suggest a similar finding:

A transdisciplinary education program needs to be developed by utilizing organizational sociology, law, services marketing, business strategy and operations, accounting and finance, information technology, and industrial and computer engineering to provide the knowledge necessary to equip new graduates to lead this culture change. (p. 12)

Because the current typical IS graduate curriculum lacks most systems approach foundations and contains few, if any, truly systems perspective courses (e.g., most IS development methods do not use a well-defined systems perspective (Alter, 2007; Avison, Wood-Harper, Vidgen, & Wood, 1998; Checkland & Holwell, 1995)), we hypothesize (**H1**) that a deep "system" view has been scarcely deployed in the graduate IS curriculum. We also hypothesize (**H2**) that an IS holistic view is better able than a partial view to cope with the new technical and organizational complexity that large organizations show. Feigenbaum[1] (1968) foresaw similar ideas by identifying a partial using of the systems approach in the IS discipline. His perspective, a well-defined systemic view where man-machine systems are combined (with new roles such as CSD (chief systems designer), CSMO (chief systems manager officer), CBPO (chief business process officer), and CBAO (chief business architect officer)) and their inter-relationships with their environment are considered, offers an initial effort toward a holistic view of the firm and the information systems deployed.

To test **H1** and **H2** is the long-term aim of the research stream fostered by IJITSA. In this article, we focus on a more limited but still useful purpose: to develop the rationality of two conceptual propositions relevant for the progress of the information systems discipline, and outline plausible courses of action. **Proposition P1 (the interdisciplinary IS BoK proposition)** argues that an interdisciplinary body of knowledge is called for in information systems because of the rising systems engineering, management, and research complexity of the emergent complex information technology-intensive organizational systems (CITOS). **Proposition P2 (the systems approach foundation proposition)** poses that **P1** is plausible through incorporating foundations, principles, methods, tools, and best practices from the systems approach by way of systems engineering and software engineering disciplines.

To support **P1** and **P2**, we use a conceptual research approach (Glass, Ramesh, & Vessey, 2004) with a descriptive and evaluative purpose. A similar scheme is used by Goul, Henderson, and Tonge (1992) in the domain of artificial intelligence (AI). The units of study are abstract elements (BoK of SE, BoK of SwE, and BoK of IS). This article is structured as follows: first, we identify the emergent engineering, managerial, and research challenges raised by CITOS. Second, we compare the underlying foundations (core definitions, disciplines of reference, teaching themes, and research methods) of information systems and of the two most emergent related disciplines (software engineering and systems engineering) that are considered essential to address such rising complexity in CITOS, and report the knowledge gaps. Third, we discuss its benefits, the hard and soft barriers to deployment, and its effectiveness measures.

THE INCREASING ENGINEERING, MANAGEMENT/BEHAVIORAL, AND RESEARCH COMPLEXITY DEMANDED BY CITOS

Conventional systems are characterized by being architecturally and functionally cohesive (they have low heterogeneity, low dispersion, low autonomy of their parts, low functional variety, and manageable functional scalability) and for being highly predictable. The classic software engineering (SE), software engineering (SwE), and IS disciplines have largely provided the adequate knowledge to design, build, and deploy conventional systems efficiently in organizations. When these systems (originally planned as conventional well-controlled physical entities with core software and hardware components), are combined with intensive human-activity systems and telecommunications components from multiples sources, as is characteristic of CITOS, the systems exhibit characteristics of complex systems: *many components, rich interactions and loose coupling among the components, the system evolves, system characteristics emerge over time, and the system pursues a mixture of component goals and system goals* (Frank, 2001; Jackson, 1991; Keating, Rogers, Unal, Dryer, Sousa-Poza, Safford, et al., 2003). Thus, an engineering complexity (manifested as many alternative designs, components, assembly procedures, equipments, tools/languages, and standards available for their realization and operation) and a behavioral and management complexity (manifested as unexpected interactions and emergent behaviors during their project management and deployment phases that might lead to critical failures and the user demand for enhanced system capabilities and functionalities) are introduced in such systems.

The Engineering Complexity of CITOS

The *complex systems* concept has long been present in the systems approach. Von Bertalanffy (1972) reports that "modern technology and society have become so complex that traditional branches of technology are no longer sufficient; approaches of a holistic or systems, or generalist and interdisciplinary, nature became necessary" (p. 420). The software engineering (Glass, 1998) and systems engineering (Bar-Yam, 2003a, 2003b) domains have also been concerned with the engineering complexity of *complex systems* (Shenhar & Bonen, 1997).

A main effect of the engineering complexity of systems is a "system failure." In the SE domain, failures occur during the operation of

the system but also those taking place during the various development stages. A system failure, as distinguished from a local component-based failure, is a failure expressed in degrading system performance. The symptom of failure is seen or measured but its source is not clear. A "system problem" refers to another scenario. An example of a system problem is a user expressing dissatisfaction with the system because of unanticipated environmental changes. System problems and failures, in a most basic systemic perspective, are not single events but a messy system of problems (Ackoff, 1981). Sources of system problems come from the conflicts raised from the interactions between the system, subsystems, and suprasystem to reach their objectives and from changes in such objectives (Ackoff, 1976).

New concepts have been developed in the SE and SwE domains to cope with engineering complexity. In the case of SE, a few of the concepts are *system of systems* (SoS) (Keating et al., 2003), *federation of systems* (FoS) (Sage & Cuppan, 2001), and *complex system* (Mage & de Weck, 2004). A SoS exists if (1) its component systems have well-substantiated purposes even if detached; (2) its component systems are managed for their own purposes; (3) its component systems, functionalities, and behaviors can be added or removed during its use; and (4) it shows emergent behavior not achievable by the component systems acting independently (at least one emergent property must be present to be considered a system). When SoS are human-activity intensive, these become a FoS (Sage & Cuppan, 2001). SoS and FoS are comprised of component systems that individually provide user-oriented functionalities and for the whole system. Each SoS and FoS are implicitly complex *systems* but not the converse. Because of the ambiguity and uncertainty in SoS, the strong interaction of the SoS and its context, and the limitations for deploying partial solutions the classic SE for single-complex systems must be updated (Keating et al., 2003). A true *systemic worldview* (our philosophy), conceptualized as "a

way to thinking, deciding, acting, and interpreting what is done and how it is done" (p. 44), as well as an action-research orientation that links theory building and theory testing, are the main updates suggested. Because SoS and FoS architectures are found in CITOS, these concepts cannot be omitted in an interdisciplinary IS BoK.

Complex systems can be classified to avoid inadequate deployment of engineering and management processes. Shenhar and Bonen (1997) derived a 4x3 matrix of instances of systems based on uncertain technological and system scope dimensions. Mage and de Weck (2004) developed a more detailed classification (also based on the Theory of Systems) of a 5x4 matrix of operators (transformation/process, distribution/transport, store/house, trade/exchange, and regulate/control) and operands (matter, energy, information, and value). Natural, noncomplex artificial, and complex (artificial) engineering systems are also differentiated by the authors. Noncomplex artificial systems have either technical or social complexity. Complex engineering systems have both. Similar to other studies, a complex system is defined as a system "with numerous components and interconnections, interactions or interdependencies that are difficult to describe, understand, predict, manage, design, and/or change" (p. 2). The engineering (and management) of CITOS can be based on these classifications. Other studies have also complemented such concepts to update the classic SE view (Calvano & John, 2004; Cleary, 2005; Franke, 2001).

In the software engineering discipline, concepts such as software-intensive systems (Andriole & Freeman, 1993; Boehm, 2000), sociotechnical software-intensive systems (Sommerville, 1998), and software-intensive systems of systems (Boehm, 2006; Boehm & Lane, 2006) are examples of research efforts for addressing the engineering complexity issue associated with CITOS from an information systems perspective.

For Andriole and Freeman (1993), the best engineering strategy to address engineering complexity involves unifying the SE and SwE disciplines. These authors argue (1993) that:

Our working premise is simple: software-intensive systems (regardless of their application domains) are among the most important, yet hardest to create and maintain artifacts of modern society. Thirty years ago, there were few large-scale software-intensive systems. Today they pervade the public and private sectors.(p. 165)

Thus, as "both disciplines address the same subject, the creation of complex software-intensive systems, albeit from different perspectives" (p. 165), they pose a unified software systems engineering. A similar rationality is argued by Boehm (2000): "A unified culture of systems and software engineering can tame the rapid changes in information technology" (p. 114). For Boehm, "organizations can change from slow, reactive, adversarial, separated software and systems engineering processes to unified, concurrent processes. These processes better suit rapid development of dynamically changing software-intensive systems involving COTS, agent, Web, multimedia, and Internet technology" (p. 114). A software-intensive system of systems (SISOS) concept and other core trends for the mutual interaction of SwE and SE disciplines are also proposed in a later study (Boehm, 2006). The rationale for improving the SwE acquisition process in the new scenario of SISOS is expanded in Boehm and Lane (2006). For Sommerville (1998), engineering complexity is manifested through the sociotechnical software-intensive systems. These systems can be described as:

Systems where some of the components are software-controlled computers and which are used by people to support some kind of business or operational process .... [Such] systems, therefore, always include computer hardware, software which may be specially designed or bought-in as off-the-shelf packages, policies and procedures

and people who may be end-users and producers/consumers of information used by the system .... Socio-technical systems normally operate in a "systems-rich" environment where different systems are used to support a range of different processes. (p. 115)

Somerville argues that SE foundations are needed in SwE programs because a computer science approach is reductionistic and isolates students from the organizational and human-based complexities in developing large-scale software development. The SE discipline contributes to improve the engineering rigor of SwE practices. This enhanced curriculum then revalues the SwE methods and tools and incorporates well-tested management engineering approaches.

In the domain of information systems, most studies have been focused on behavioral and managerial perspectives (Hevner, March, Park, & Ram, 2004), rather than engineering complexity issues. A seminal study (Nunamaker, Chen, & Purdin, 1991) introduced the system development process, from the engineering and software engineering disciplines, as a research method for IS to be used jointly with theoretical, observational, and experimental research. But the proposal is focused in the study of the final artifacts rather than in the study of the engineering and design methods to cope with CITOS. Consequently, behavioral-oriented research is ultimately stressed, and the development engineering process is offered as a mediator rather than a primary research goal.

Other authors (Hevner & March, 2003; Hevner et al., 2004; March & Smith, 1995), using the core foundations for a design science established by Herbert A. Simon (1969), have formulated a design research paradigm in information systems, one different from a routine design paradigm based in the application of the existent knowledge for building an artifact. A theoretical framework that justifies behavioral (called *natural*) and design dimensions to study IT in organizations is reported by March and Smith (1995). The behavior (*natural*) dimension accounts for the formulation and testing (justifying) of theories about how and why IT works or does not work in an organizational setting. In the design dimension, the building and evaluation of IT artifacts are conducted. These authors classify IT artifacts (e.g., the research products) as constructs, models, methods, and instantiations. Hevner and March (2003) and Hevner et al. (2004) extend the study proposing design principles and research validation methods for the IS domain. They assert that the design issue is a core topic in the engineering discipline, but it has been rarely explored in the IS domain (the design references used by these authors come from the computer science, software engineering, artificial intelligence, and political science domains). Complexity and wicked problem concepts (Rittel & Weber, 1973) are also described by these authors. From a systems approach, a similar construct: messes as a system of problems, has also been defined (Ackoff, 1973).

These few studies, then, contribute directly to the IS domain introducing the engineering complexity issue of CITOS. Paradoxically, despite some SE literature reports that the incremental deployment of IT is generating SoS (Carlock & Fenton, 2001; Keating et al., 2003), CITOS as SoS are still not studied in the IS domain.

The Management and Behavioral Complexity of CITOS

According to Sterman (2001), managerial complexity is of two types: combinatorial or dynamic. Relevant for the IS domain is that combinatorial managerial complexity is most perceived by organizational managers but dynamic managerial complexity affects them more. As Sterman (2001) indicates:

Most people think of complexity in terms of the number of components in a system or the number of possibilities one must consider in making a decision. The problem of optimally scheduling

an airline's flights and crews is highly complex, but the complexity lies in finding the best solution out of an astronomical number of possibilities. Such problems have high levels of combinatorial complexity. However, most cases of policy resistance arise from dynamic complexity—the often counterintuitive behavior of complex systems that arises from the interactions of the agents over time. Dynamic complexity can arise even in simple systems with low combinatorial complexity. (p. 11)

This resistance (often occurring as a delay by decision makers to make critical managerial choices regarding courses of action and to intervene in critical situations) happens because complex side effects are generated in messy organizational systems. These complex organizational systems, characterized by an underlying structure of mechanisms that is highly coupled, dynamic, adaptive, self-organizing, and with emergent counterintuitive behaviors (Sterman, 2001), demand at least a similar complex systemic solution as the controller system (Bar-Yam, 2003b).

In the domain of information systems, the management complexity is manifested in failed IT projects (CIO UK Web site, 2007; Standish Group, 2003). Failed IT projects are defined as projects where there are cost over-runs, large schedule delays, incomplete delivery of systems, system underutiliztion, or cancellations before completion or early system disposal (Ewusi-Mensah, 1997; Wallace & Keil, 2004). A common issue reported in such studies is the critical influence of management inadequacies during the implementation life cycle. Management complexity is important for CITOS deployment because an information system comprises technology, procedures, data, software, and people. Moreover, the technical, socio-economical, and political-cultural components of the CITOS environment are factors whose influences must be identified for reaching a successful system deployment (Gelman, Mora, Forgionne, & Cervantes, 2005;

Mora, Gelman, Cervantes, Mejia, & Weitzenfeld, 2003). This view is supported by a vast literature on IS implementation research (Kwon & Zmud, 1987). A holistic view of the phenomenon, then, requires the inclusion of managerial complexity and interactions with engineering complexity. Efficiency and efficacy engineering project success metrics (on time project completion, on budget, and with a high percentage of expected requirements delivered) must be complemented with system effectiveness metrics (associated to managerial complexity) to manage CITOS projects.

In the SE domain, managerial complexity is manifested when large-scale but simple systems become a SoS and when the usual technical, operational, economical, and political (TOEP) feasibility priority order shifts to a political, economical, operational, and technical (PEOT) order (Carlock & Fenton, 2001). Then, "seamless interoperability and acceptable performance to all users at an acceptable cost are the most important priorities" (Carlock & Fenton, 2001, p. 245). Managerial complexity can be addressed through enterprise systems engineering (a natural extension of SE) for an updated and adequate engineering management of SoS development or procurement. In a SoS, each system component is also conceptualized as a system comprised of hardware, software, facilities, procedures, and people. Such a whole SoS operating is linked to needed support systems. Facilities and support systems have usually been ignored in the IS literature. An exception is the "SERVQUAL" concept to measure IS service quality. Management complexity of SoS is then addressed through an extended management SE life cycle involving strategic, project management (midlevel), and implementation/operational levels. Recent evidence of managerial complexity issues manifested in large-scale software systems projects and solved through a SE enhancement is reported in Hole, Verma, Jain, Vitale, and Popick (2005). Critical deficiencies in the older SwE methodology are identified as follows:

In its existing state (pre-SE&A) this framework did not address requirements, architecture development, integration, and verification as part of a coherent Systems Engineering methodology. Existing descriptions of these SE&A practices were general and open to interpretations, and often "hidden" in broader activities and work products. (Hole, Verma, Jain, Vitale, & Popick, 2005, p. 80)

Enhancements, such as project mission awareness over the traditional project objectives, non-negotiable mission-critical requirements, better requirements traceability activities, project manager and lead systems engineer roles, disciplined change impact analysis, and tangible scored reviews (also generated independently in the SwE domain but rarely used in the IS domain), are reported as contributions from the SE domain. IT and IS architecture views of the full enterprise, such as Zachman's Framework (Sowa & Zachman, 1992; Zachman, 1987), have received little attention in the IS domain.

The complexity behavioral dimension in the management of large-scale projects has been reported also in the SwE domain (Curtis, Krasner, & Iscoe, 1988). An implicit holistic multilayer model (business, company, project, team, and individual milieus) is used to study the behavioral interactions in the system. Findings suggest that large-scale software development demands that the learning, negotiation, communication, and customer interactions activities (that are not usually considered in SwE management projects) be accommodated explicitly in the process.

The Research Complexity of CITOS

Because CITOS is concerned with engineering and managerial/behavioral complexity, a research complexity inherently appears when CITOS are investigated. Comprehensive IS research frameworks that recognize behavioral and engineering perspectives are recent (Hevner & March, 2003;

Hevner et al., 2004; March & Smith, 1995). As noted earlier, much research in the IS discipline dilutes IT artifacts and consequently their engineering characteristics or when focused on the computational view, this ignores the behavioral issues (Orlikowski & Iacono, 2001). CITOS' complexity demands mutual research interaction from both sides.

A comprehensive IS research framework (Mora, Gelman, Forgionne, Petkov, & Cano, 2007a) uses a critical realism stance (Bhaskar, 1975) and a multimethodology research worldview (rationalized by Mingers, 2000, 2001, 2002), to frame some core ideas of the Theory of Systems (Ackoff, 1971; Gelman & Garcia, 1989; Mora et al., 2003), as a proposal to accommodate the disparate and conflicting research stances (positivist, interpretative, and critical). Four postulates are articulated in Mora et al. (2007a) to frame such disparate philosophical stances. Postulate P4 posed as integrator and underpinned in critical realism says that:

The world is intelligible for human beings because of its stratified hierarchy of organized complexities—the widest container is the real domain that comprises a multi-strata of natural, man-made and social structures as well as of event-generative processes that are manifested in the actual domain that in turn contains to the empirical domain where the generated events can or cannot be detected. (p. 3)

For Bhaskar (1975), reality is independent of human beings: "a law-governed world independently of man" (p. 26), but the social structures and their generative mechanisms are conditioned to the existence of human beings at first and then these really exist and can be studied and intervened. Bhaskar also explains that "it is not the character of science that imposes a determinate pattern or order in the world; but the order of the world that, under certain determinate conditions, makes possible the cluster of activities we call science" (p. 30). Accordingly to Mingers (2002):

CR (critical realism) recognizes the existence of a variety of objects of knowledge—material, conceptual, social, psychological—each of which requires different research methods to come to understand them. And, CR emphasizes the holistic interaction of these different objects. Thus it is to be expected that understanding in any particular situation will require a variety of research methods both extensive and intensive. (p. 302)

Other IS frameworks and models based on the Theory of Systems as the most adequate models to cope with the complexities faced by IS practitioners and academicians have been reported (Alter, 2001, 2003, 2007; Bacon & Fitzergarld, 2001; Ives, Hamilton, & Davis, 1980; Nolan & Whetherbe, 1980). Such IS systemic research frameworks are usually ignored in IS research, which is still guided by a reductionistic view.

Thus, there is an extensive granularity manifested by a vast array of IS relevant topics, but the topics are disconnected as a whole. An IS body of knowledge from an accumulation research tradition is missing. A plausible reason, according to Mora, Gelman, Cano, Cervantes, and Forgionne (2006a) and Mora et al. (2007a), is that the holistic view of the IS discipline has been lost from its original conceptualization in the 1960s. Consequently, research topics appear disconnected from a general standardized research framework (as SE and SwE have through a BoK).

These large unconnected research topics, the infrequently-used underlying microtheories, the broad background of IS researchers, the lack of finding accumulation, and the engineering and managerial richness of the phenomena involved are also manifestations of research complexity in the IS domain.

Consequences and Initial Interdisciplinary Efforts for CITOS Complexity

Table 1 summarizes the new engineering, managerial, and research challenges faced by practitioners and researchers in CITOS (and from the software engineering and systems engineering disciplines oriented to CITOS).

Such challenges imply a need for human resources with the adequate competencies for the development (e.g., the engineering and research design view) and the management (e.g., the behavioral research view) of CITOS. Nevertheless, the IS curriculum literature has not addressed it enough. An OECD (2004) study, for instance, claims that most critical technical competencies in ICT must be learned directly from organizations:

The need for ICT skills can be satisfied in part through education and training. Full-time education does not appear to be the most important path to obtaining general and advanced skills. As schools become well equipped, however, students develop at least basic ICT skills, and ICT-related degrees can be obtained through formal education. For specialist skills, however, sector-specific training and certification schemes may be more effective, given the rapid changes in skills needs and the constant introduction of new technologies. (p. 12)

Given such complexity, our position is that an interdisciplinary (e.g., a systemic integration of several disciplines) IS graduate curriculum is a plausible course of action that will enable practitioners and researchers to acquire a holistic view of such phenomena.

Proposals for a mutually enhanced SwE and SE curriculum (Bate, 1998; Brown & Scherer, 2000; Denno & Feeney, 2002; Hecht, 1999; Johnson & Dindo, 1998; Rhodes, 2002; Sommerville, 1998) and a new unified software systems engineering discipline (Andriole & Freeman, 1993; Boehm,

Table 1.

Engineering challenges	• A myriad of mature and affordable ICT as building blocks are available for system designers. • A high variety of ICT capabilities are demanded for internal and external system users. • Organizational SoS and FoS are required to be engineered to integrate multiple autonomous large-scale systems from several providers. • Multiple international standards are available to system designers. • Multiple system engineering methodologies and particular vocabulary exists in engineering specialties.
Managerial challenges	• Dynamic complexity of CITOS is not as easily perceived as the combinatorial complexity. • Unexpected counterintuitive behavior are exhibited in CITOS. • TOEP project order of priorities is shifted to POET. • Multiple technologies to be evaluated, acquired, deployed, trained for, and managed are available. • Effectiveness (holistic) metrics are required besides traditional efficiency and efficacy ones. • A large variety of skilled ICT and operational human resources are demanded for CITOS. • Control and coordination scales are increased with CITOS as SoS and FoS.
Research challenges	• The understanding of CITOS demands design and behavioral research modes to be conducted. • The engineering and managerial/behavioral richness of phenomena demands a pluralist and multimethodology approach. • There are still an extensive use unique of multiples disconnected microtheories. • An IS BoK or general conceptual framework is still missing. • The big picture of the IS phenomena has been lost.

2000; Thayer, 1997, 2002) are initial efforts to cope with IT complex systems.

The SE discipline is also being fostered to extend its coverage from an enterprise level focused perspective (Farr & Buede, 2003) (taught in engineering management or industrial engineering disciplines) and to redefine its identity (Emes, Smith, & Cowper, 2001). Such expansion might enable SE to strengthen a systems view for managing the complete organization and the traditional technical processes for engineering a product or service (Arnold & Lawson, 2004; Bar-Yam, 2005; Emes et al., 2001). In particular, Bar-Yam (2005), using a trade-off design between the variety (number of different and highly-independent actions pursued by the components) and scale (number of elementary components performing the same core task) of a system, suggests an evolutionary SE to design and manage complex systems where simultaneous designs, competitive teams, and ongoing fielded and virtual tests are conducted. Other proposals argue for a new SE education focused on complex systems (Beckerman, 2000; Cleary, 2005; Franke, 2001) and the systems approach (Frank & Waks, 2001). For instance, Moti and Waks (2001) report: "technological systems grow larger, more complex, and interdisciplinary, electronics and high-technology industries face a growing demand for engineers with a capacity for systems thinking" (p. 361). These authors also suggest that the SE knowledge about domain specializations (software, computer systems, etc.) be about the (1) complexity of the system; (2) interconnections of lower and upper level systems; and (3) functional domains and constraints. In the IS domain, few similar direct or indirect arguments have been reported (Hevner et al., 2004; Mingers, 2001; Mora, Gelman, Macias, & Alvarez, 2007c). Hence, our **Proposition P1 (the interdisciplinary IS BoK proposition)** that argues that an interdisciplinary body of knowledge is called for in information systems because of the raising of systems engineering, management, and research complexity of CITOS can be supported.

A COMPARATIVE ANALYSIS OF INFORMATION SYSTEMS, SOFTWARE ENGINEERING, AND SYSTEMS ENGINEERING DISCIPLINES

To support **Proposition P2 (the systems approach foundation proposition)** that argues that **P1** is plausible through incorporating foundations, principles, methods, tools, and best practices from the systems approach by way of the systems engineering and software engineering disciplines, this article continues and enriches three recent reports (Mora et al., 2006a; Mora, Cervantes, Gelman, Forgionne, & Cano, 2006b; Mora, Gelman, O'Connor, Alvarez, & Macías, 2007b) and is developed under the following rationale: (1) the engineering, management, and research complexity of the issues involved with the emergent CITOS is beyond the scope of the traditional monodisciplinary and reductionistic view of IS (from proposition **P1**); (2) the IS discipline is so fragmented that it has become in disconnected islands in a knowledge sea; and (3) an interdisciplinary, systemic approach (Ackoff, 1960) provides the adequate philosophical paradigm and methodological research tool to cope with the phenomena of interest.

A historical review of the origins of the SE, IS, and SwE disciplines shows that SE is the oldest (from late 1930s) followed by IS (late 1950s), and then SwE (late 1960s). We consider SE the most mature discipline, as evidenced by the existence of large-scale projects using standardized theories, methods, and tools (Honour, 2004), followed by the IS discipline. SwE, by its separation from computer science (Denning, Comer, Gries, Mulder, Tucker, Turner, & Young, 1989) as an independent discipline, can be considered the newest and less mature area of study. Using several sources (Editorial policy statement, 2006; INCOSE, 2004; SEI, 2003) and the PQR concept (Checkland, 2000), it is possible to compare general definitions (shown in Table 2) for these disciplines. At first glance, the three disciplines study disparate systems: a well-defined physical system, a computer software system, and an IT-based organizational system.

Recent SE (Rhodes, 2002) and SwE (Boehm, 2000) literature has noted the increasing inclusion of software and IT components in current and emergent complex systems. Rhodes (2002), for instance, remarks that software is a critical component, like hardware and people, in the entire artificial organizational system developed by systems engineers. Also as mentioned earlier (Sommerville, 1998, p. 115), the SwE discipline has suggested that software systems must be considered as sociotechnical software-intensive systems. From an IS perspective, this definition of software systems corresponds to what is considered an information system (Mora et al., 2003). The SE discipline has also identified that the usual technical, operational, economical, and political (TOEP) order of priorities (Carlock & Fenton, 2001) has been changed to a political, economic, operational, and technical (PEOT)

order when complex and large systems are designed. It is usually accepted that SE (Hitchins, 2003) can be deployed in different hierarchical levels: (1) the Artifact SE; (2) the Project SE; (3) the Business SE; (4) the Industry SE; and (5) the Socio-economic (environment) SE. Therefore, the increasing inclusion of software components suggests a needed interaction between SE and SwE, and between these disciplines and IS to cope with the same object of study under different systemic scales.

Table 3 updates the analysis (Mora et al., 2007c) of the relations of these disciplines with their reference disciplines. A qualitative 5-point scale from 1 (very low support) to 5 (very high support) is used to report the current support level assessed by the authors and based on the different studies reviewed (Buede, 2000; Emes et al., 2005 for SE; Glass, 2003; Sage, 2000; SWEBOK (IEEE, 2001) for SwE; and Culnan & Swason, 1986; Glass et al., 2004; Vessey, Ramesh, & Glass, 2002; for IS). A grey shading is also used in the cells to show the recommendations. Six relevant implications for an interdisciplinary IS BoK can be reported.

Table 2. A systemic comparison of the conceptual definition of the SE, SwE, and IS disciplines

PQR-system Construct	Discipline		
	<S: Systems Engineering>	<S: Software Engineering>	<S: Information Systems>
<S> is a system to do <P> ...	... is an interdisciplinary approach and means to <P: enable the realization of successful systems>	... is the technological and managerial discipline concerned with <P: systematic production and maintenance of software products>	... [is the discipline] <P: concerning [to IT-based systems]>
through <Q> ...	... <Q: [the integration of] all the disciplines and specialty groups into a team effort forming a structured development process that proceeds from concept to production to operation [and] considers both the business and the technical needs of all customers>	... that are <Q: developed and modified>	... <Q: [the scientific study and] the development of IT-based services, the management of IT resources, and the economics and use of IT>
in order to contribute to achieving <R>	... with the <R: goal of providing a quality product that meets the user needs>	... on <R: time and within cost estimates>	... <R: [positive] managerial and organizational implications>

Table 3. Reference disciplines for SE, SwE, and IS disciplines

Disciplines of Reference	SE	SwE	IS
Industrial & Manufacturing Engineering	● ● ● ● ●	●	●
Management Sciences & Operations Research (MS&OR)	● ● ● ● ●	●	● ● ●
Business/Organizational Sciences (B&OS) (Economy, Accounting, Marketing, Finance)	● ● ●	●	● ● ● ● ●
Social/Behavioral Sciences (S&BS) (Psychology, Sociology, Political Sciences, Law)	● ● ●	●	● ● ● ● ●
Mathematics and Statistics	● ● ●	●	● ● ●
Other Engineering and Physical Sciences (Electronic Engineering, Electrical Engineering, Mechanical Engineering, Quality Engineering)	● ● ●	●	●
Systems Sciences (Systems Thinking, Systems Dynamic, Soft Systems, Critical Systems)	● ● ●	●	●
Computer Sciences (CSc)	●	● ● ● ● ●	● ● ●
Management and Business Process Engineering	● ● ●	●	●
Services Science (Management & Engineering)	?	?	?
Software Engineering	● ● ●	● ● ● ● ●	●
Systems Engineering	● ● ●	● ● ●	●
Information Systems	?	● ● ●	● ● ● ● ●

First, the SE and IS disciplines have been shaped by at least two basic disciplines (Industrial & Manufacturing Engineering Management Science & Operations Research for SE; Behavioral & Social Science and Business & Organizational Science for IS) while SwE has been formed from just one discipline (Computer Science). SwE has been largely considered as a research stream and body of knowledge for Computer Science (Denning et al., 1989). In the last decade, the usefulness of other reference disciplines for SE and IS (Fuggetta, 2000; Kellner, Curtis, deMarco, Kishida, Schulemberg, & Tully, 1991), has been recognized in SwE. A recommendation for the IS discipline is to lessen the variety of such interactions from a high (manifested by excessive MBA courses in the curriculum of graduate MIS programs) to a

normal level of support. High-level support from multiple disciplines generates a loss of identity for the IS discipline when the IT artifacts are diluted (Orlikowski & Iacono, 2001). Given the managerial and engineering complexity shown in CITOS and their technical and sociopolitical inter-relationships with their upper and lower level systems, a holistic research approach that combines behavioral and design research approaches supported for the SE and SwE disciplines is encouraged.

Second, while systems science was a core discipline of reference for IS (Ives et al., 1980; Nolan & Wetherbe, 1980) and SE (Sage, 2000), now systems science is scarcely used in IS research. To cope with CITOS, recent proposals to re-incorporate this original foundation (Alter, 2001,

2003, 2007; Gelman et al., 2005; Lee, 2000; Mora et al., 2003, 2007a) have been reported. In the SwE discipline, there has been little evidence of such incorporation, but the recent unification efforts with SE could implicitly link SwE with systems science. Such support should be encouraged in SE and extended in IS and SwE. Third, management engineering and business process engineering, which has been typically incorporated in the SE and partly in SwE curriculum (through CMMI and ISO 15504 initiatives), has been largely ignored in IS with the exceptions of BRP, ERP process modeling, and emergent ITIL initiatives (Mora et al., 2007b). According to Farhoomand and Drury (1999, p. 16), the highest average percentage (25%) of the themes published in eight premier IS journals and the ICIS proceedings during the 1985-1996 period were from "reference disciplines" while the "information system" themes reached 14%. These authors also report that "there seems to be a shift from technical themes towards non-technical themes" (p. 17). Similarly, Orlikowski and Iacono (2001) found in 177 research articles (published in the ISR journal during the period 1990-1999) that the highest percentage (25%) of papers corresponds to a "nominal" view of IT (e.g., IT is absent). Although we accept the relevance of such domains for the IS discipline to understand the suprasystems served by CITOS, we suggest a better balance between organizational, behavioral, and social sciences and management engineering. This balance provides the IS discipline with new conceptual tools for CITOS management and engineering. Then, the server system (CITOS) and the served system (business process) are system components that are studied and designed or redesigned simultaneously.

Fourth, while the SE discipline is strengthened through the interaction of other engineering and management disciplines and permits a normal self-reference, the IS and SwE disciplines have been extensively self-referenced (Glass et al., 2004). We consider that it has had more negative than positive consequences when an imperialist

and nonpluralist view of models, frameworks, and theories are encouraged. Self-reference in a domain is positive when there is an open interaction with related domains and when there is an evolutionary accumulation of knowledge rather an iterative decreasing increment of knowledge. The acceptance of qualitative and interpretative research methods was largely rejected in the IS discipline through the self-reference of repetitive specific quantitative methods. Another example is the service science engineering and management initiative (Chesbrough & Spohrer, 2006), rarely addressed in IS but incorporated in the SE and SwE domains in a seamless way.

Fifth, although IS and SwE have been outside the engineering specialties, such as electronic, electrical, and mechanical, CITOS demands at least knowledge on the foundations of such disciplines for IS researchers and practitioners to get an understanding of IT limitations. Consider, for example, developing automated vision recognition systems combined with RFID and secure mobile IT for airports, banks, or worldwide package delivering services. A CIO or research academic should be able to understand the scope, limitations, and costs of deploying such solutions and collaborating in their design (as systems engineers do). IT technical knowledge must not be diluted in IS research. Another example is the planning, design, and management of a data center where network-critical physical infrastructure (NCPI) issues include electrical power, environmental cooling, space, racks and physical infrastructure, cabling, grounding, fire protection, and other issues related to design security (Industry Report, 2005).

Sixth, while SE and SwE have mutually acknowledged the need for interaction, the IS discipline still ignores possibly beneficial interdisciplinary communications. New theoretical incorporations to the IS discipline from the design/engineering paradigm could be obtained by incorporating the systems science paradigm, including the complex systems intellectual movement.

Table 4 exhibits the BoK and general research themes derived for these disciplines. From Table 4, the first inference is that SE and SwE, by their engineering heritage, are most likely to interact in the next 25 years. The IS discipline, in contrast, seems to be unaware of the dramatic changes and challenges that world organizations are demanding due to complex sociotechnical information systems. In the cells with very low interaction (value of 1 point), more interaction is suggested to expand the body of knowledge. According to the systems approach, a system is understood only if it is studied: (1) from two perspectives (like a unitary whole and as a set of interdependent parts) and (2) within its wider system and comprising internal subsystems (Ackoff, 1971; Gelman & Garcia, 1989).

Details of the need for a systems approach in the IS discipline have already been reported (Alter, 2001, 2003, 2007; Bacon & Fitzgerald, 2001; Gelman et al., 2005; Lee, 2000; Mora et al., 2003, 2007a). A second finding from Table 4 is the lack of teaching and research of IS frameworks and standards/models of processes (e.g., CobIT, ITIL, ITSEC, and ISO 20000). The SE and SwE disciplines have developed their rigor through the development, deployment, and compliance with standards of process, but such a movement has largely been ignored in the IS discipline (Beachboard & Beard, 2005). A third finding from Table 4 is that SE has fewer missing interactions than the other two disciplines. A strong implication is that systems engineers are more holistically trained, studying and carrying out large-scale and complex systems (Frank & Waks, 2001), than software engineers and information systems practitioners (Mora et al., 2006b).

In Table 5, the research approaches used in the three disciplines are reported under the same scale. The main categories of research are adapted from Denning et al. (1999), Nunamaker et al. (1991), Hevner and March (2003), Vessey et al. (2002), and Glass et al. (2003, 2004). Theoretical, conceptual, and modeling approaches can be considered pieces

of conceptual research that study concepts, constructs, frameworks, methodologies, algorithms, and systems without using data directly from real artifacts. Engineering and behavioral approaches, in contrast, are empirical research methods that take data directly for artifacts, people, or organizations. The holistic systems approach is considered a pluralist and multimethod research paradigm that uses diverse research approaches relying on the complexity of the research situation, goals, and availability of resources.

Table 5 shows that the SE research is conducted mostly through modeling, but it also uses the other research approaches in a more balanced way than the other two disciplines. According to Glass et al. (2004), theoretical/conceptual studies are more frequent than engineering studies in SwE. For the case of IS, because of its strong historical dependence on business and organizational sciences, most studies are empirical (behavioral approach). Recent studies have argued the need for using modeling/simulation (Mora et al., 2007a) and engineering/design approaches (Hevner & March, 2003; Hevner et al., 2004; Nunamaker et al., 1991) in the IS discipline. In a similar way, other studies have suggested that SwE must conduct empirical research (Kitchenham et al., 2002) and expand the few modeling/simulation studies performed (Madachy, 1996).

Table 5 also shows a behavioral approach bias for the IS discipline, even though the table covers positive and interpretative stances. We believe that the understanding of, and the effective intervention in, single large scale and emergent CITOS demand an interdisciplinary and multimethodological research approach. Mingers (2001) has reported extensively the relevance and need of such an approach for the IS research stream through a critical realism philosophical stance (Mingers, 2002). Previous analysis for management science and operations research (highly linked to SE) has also been reported (Mingers, 2000, 2003). In this article, we support this proposal and believe that the systems approach can glue the disparate,

Table 4. Main BoK and research topics for SE, SwE, and IS disciplines

Main BoK and Research Themes for SE, SwE, and IS	SE	SwE	IS
Systems Engineering Foundations	• • • • •	• •	•
System of Systems and Complex Systems Engineering	• • •	•	?
Model and Simulation of Systems	• • •	•	?
Frameworks and Standards/Models of Processes for SE	• • •	• • •	•
Systems Engineering Quality and Management	• • •	•	?
Human Systems Engineering	• • •	• • •	•
Simulation of Systems	• • • •	•	•
Systems Thinking and Systems Foundations	• • •	?	•
Engineering Design	• • •	•	?
Business Process/Workflow Systems Management and Engineering	• • •	•	•
Risk and Project Management	• • •	• • •	• • •
Control Theory	• • •	?	?
Operations Research	• • •	• • •	• • •
Integration, Verification, and Validation of Systems	• • •	•	•
Software Engineering Foundations	•	• • • • •	•
Software Engineering Tools and Methods	•	• • • •	• •
Software Engineering and Management	•	• • •	•
Frameworks and Standards/Models of Processes for SwE	•	• • •	•
Software Engineering Economic	•	•	•
Information Systems Foundations	•	•	• • • • •
Information Systems Engineering	•	•	•
Information Systems Management	•	•	• • •
Frameworks and Standards/Models of Processes for IS	?	?	?
Information Technologies Tools	•	• • •	• • •
Behavioral Issues in Information Systems			• • • • •
Business Organizational Foundations			• • • • •
Knowledge Management Systems	•	•	•
Specific Domains and Careers of Applications	• • •	• • •	• • •
Service Engineering and Management	?	?	?

Table 5. Main research approaches for SE, SwE, and IS disciplines

Research Paradigms	SE	SwE	IS
Formal Theoretical Approach (Theorem Proving & Mathematical Analysis)	• • •	•	•
Conceptual Analysis (Description, Formulation, or Evaluation of Concepts, Frameworks, Models, Methods)	• • •	• • • • •	• • •
Modeling Approach (Conceptual Modeling, Mathematical Modeling, Simulation)	• • • • •	•	•
Engineering Approach (Design of Artifacts, Concept Implementation)	• • • • •	• • •	•
Behavioral Approach (Survey, Case Studies, Social Experiments)	•	•	• • • • •
Holistic Systems Approach (Multimethodological, Interdisciplinary, Critical Realism-based Research)	•	•	•

conflicting, and still monodisciplinary views of our discipline (Mora et al., 2007a). In concordance with some SE and SwE studies, we also believe the IS behavioral approach can enhance the SE and SwE disciplines. For the IS discipline, a more balanced use of the engineering, modeling, conceptual, and formal theoretical and behavioral approaches under the holistic paradigm of the systems approach is encouraged.

DISCUSSION AND IMPLICATIONS FOR AN INTERDISCIPLINARY BOK IN INFORMATION SYSTEMS

To complete this proposal, we identify the investment resources required, the potential benefits, and the effectiveness measures for developing the aforementioned interdisciplinary IS BoK to cope with the emergent CITOS. In the investment dimension, hard and soft issues can be considered. Hard issues are the financial resources required to redesign a graduate curriculum, prepare new human resources, and deploy integrated labs for

students. Under the assumption of an already existent common core engineering curriculum for SE and SwE, the challenge is the incorporation or adaptation of it to the IS domain. It is typical for academic institutions to share labs and library resources, so the new investments in better equipped and integrative labs would be a worthy investment. The virtualization and distribution of ICT resources actually can allow a large campus to share valuable ICT resources with small academic units. For instance, a small business data center lab (not a computer network lab) could be developed for training SE, SwE, and IS graduate students in different problem and solution domains. For the SE and IS disciplines, the ICT architectural planning of the data center as well as the managerial and financial operations of the ERP installed in the data center lab are adequate issues for several courses in the curriculum suggested in Table 4. For SwE students, the same lab can be useful to deploy and test service-oriented application software and middleware. An implicit benefit for SE, SwE, and IS graduate students is the interdisciplinary team interaction for solving problems under a systems approach. The investment needed will be determined by supply and demand. Regarding soft issues, the existence of such an interdisciplinary IS BoK—that should support the current IS 2006 curriculum (Gorgone, Gray, Stohr, Valacich, & Wigand, 2006)—with the SE and SwE foundations, principles, and methods also demands: (1) adjustments in the faculty power relationships; (2) the management risk of identity loss in the discipline; and (3) the compliance with national accreditation boards procedures. Soft issues, then, become the main barriers for this proposal.

The main benefits estimated from this interdisciplinary IS BoK are (1) the revaluing of the IS discipline as a core element in modern society and business organizations; (2) the update of the body of knowledge required by IS practitioners and academics to cope with the emergent CITOS; (3) the joint development under a holistic view of the rich and complex phenomenon of CITOS with the two highly related disciplines of SE and SwE; (4) the development of a shared mindset of concepts and worldviews with SE and SwE practitioners and academics; and (5) the increased supply of a new generation of IS professionals and researchers to meet growing organizational demands.

Figures of merit to evaluate the effectiveness of this proposal are the following: (1) trends of the critical failures reported in CITOS (similar to the measures reported in Standish Group (2003); (2) trends of the financial losses derived by failures in CITOS; (3) trends of the simplification and standardization of principles, foundations, and methods shared for the three disciplines used in organizations; (4) trends of the high-quality and relevant research conducted and published under the paradigm proposed in this article; and (5) trends in the enrollment of IS graduate programs under this new interdisciplinary IS BoK. Hence, we understand that this proposal is a challenge that will generate positive as well as negative reactions. However, our academic responsibility and final purpose is to strengthen the IS discipline to

face the challenges of business organiztions and society. The IT artifact as an artificial symbolic processor for processing, storage, and transport data, information, and knowledge (Simon, 1999) has transformed the world. Now, the challenge for the IS discipline (Ackoff, 1967) will be to transform such IT artifacts and align them with the emergent CITOS concept.

Proposition P2 (the systems approach foundation proposition), that poses that **P1** is plausible through incorporating foundations, principles, methods, tools, and best practices from the systems approach by way of systems engineering and software engineering disciplines, can also be supported.

CONCLUSION

We have developed this article with the aim to improve and reposition the IS discipline to accommodate the emergence of CITOS. In this proposal, we articulate the rationale for two propositions. **P1** argues that a plausible response to cope with CITOS is an interdisciplinary IS, SE, and SwE engineering and management body of knowledge, and **P2** argues that such realization is plausible through incorporating foundations, principles, methods, tools, and best practices from the systems approach by way of the systems engineering and software engineering disciplines. We believe the evidence articulated from the comparison of the IS discipline with SE and SwE (through the structured systemic definitions, the disciplines of reference, the knowledge for research and teaching, and the main research paradigms used) supports our claim. It has also been argued that:

The ability of science and technology to augment human performance depends on an understanding of systems, not just components. The convergence of technologies is an essential aspect of the effort to enable functioning systems that include human beings and technology; and serve the human beings to enhance their well-being directly and indirectly through what they do, and what they do for other human beings. The recognition today that human beings function in teams, rather than as individuals, implies that technological efforts that integrate human beings across scales of tools, communication, biological and cognitive function are essential. (Bar-Yam, 2003b, p. 1)

Thus, we believe that this position paper and the new *International Journal in Information Technology and the Systems Approach* are robust academic efforts toward this essential purpose. The road, however, will not be an easy academic endeavor.

ACKNOWLEDGMENT

We thank the several senior colleagues that helped us with the insightful review, expertise, and wisdom in systems engineering, information systems, and software engineering, to eliminate conceptual mistakes, improve the clarity of the article, and focus on the relevant issues for IS, SE, and SwE disciplines from a systems approach.

REFERENCES

Ackoff, R. (1960). Systems, organizations and interdisciplinary research. *General System Yearbook, 5,* 1-8.

Ackoff, R. (1967). Management misinformation systems. *Management Science, 14*(4), 147-156.

Ackoff, R. (1971). Towards a system of systems concepts. *Management Science, 17*(11), 661-671.

Ackoff, R. (1973). Science in the Systems Age: Beyond IE, OR, and MS. *Operations Research, May-Jun,* 661-671.

Alter, S. (2001). Are the fundamental concepts of information systems mostly about work systems? *CAIS, 5*(11), 1-67.

Alter, S. (2003). 18 reasons why IT-reliant work systems should replace "the IT artifact" as the core subject matter of the IS field. *CAIS, 12*(23), 366-395.

Alter, S. (2007). Could the work system method embrace systems concepts more fully? *Information Resources Management Journal, 20*(2), 33-43.

Andriole, S., & Freeman, P. (1993, May). Software systems engineering: The case for a new discipline. *Software Engineering Journal,* pp. 165-179.

Arnold, S., & Lawson, H.W. (2004). Viewing systems from a business management perspective: The ISO/IEC 15288 standard. *Systems Engineering, 7*(3), 229-242.

Avison, D., Wood-Harper, A., Vidgen, R., & Wood, J. (1998). A further exploration into information systems development: The evolution of multiview2. *Information Technology & People, 11*(2), 124-139.

Bacon, J., & Fitzgerald, B. (2001). A systemic framework for the field of information systems. *The DATA BASE for Advances in Information Systems, 32*(2), 46-67.

Bar-Yam, Y. (2003a). When systems engineering fails: Toward complex systems engineering. In *Proceedings of the International Conference on Systems, Man & Cybernetics* (pp. 2021-2028). Piscataway, NJ: IEEE Press.

Bar-Yam, Y. (2003b). Unifying principles in complex systems. In M.C. Roco & W.S. Bainbridge (Eds.), *Converging technology (NBIC) for improving human performance* (pp. 1-32). Kluwer.

Bar-Yam, Y. (2005). About engineering complex systems: multiscale analysis and evolutionary engineering. In S. Brueckner et al. (Eds.), *ESOA 2004*, Heidelberg, Germany (pp. 16-31). Springer-Verlag. Lecture Notes in Computer Science, 3464.

Bate, R. (1998, July-August). Do systems engineering? Who, me? *IEEE Software*, pp. 65-66.

Beachboard, J., & Beard, D. (2005). Innovation in information systems education-II: Enterprise IS management: A capstone course for undergraduate IS majors. *CAIS, 15*, 315-330.

Beckerman, L. (2000). Application of complex systems science to systems engineering. *Systems Engineering, 3*(2), 96-102.

Boehm, B. (2000, March). Unifying software engineering and systems engineering. *Computer*, pp. 114-116.

Boehm, B. (2006). Some future trends and implications for systems and software engineering processes. *Systems Engineering, 9*(1), 1-19.

Boehm, B., & Lane, J. (2006, May). 21st century processes for acquiring 21st century software-intensive systems of systems. *Crosstalk: The Journal of Defense Software Engineering*, pp. 1-9.

Brown, D., & Scherer, W. (2000). A comparison of systems engineering programs in the United States. *IEEE Transactions on Systems, Man and Cybernetics (Part C: Applications and Reviews), 30*(2), 204-212.

Buede, E. (2000). *The engineering design of systems* (Wiley Series in Systems Engineering). New York: Wiley.

Calvano, C., & John, P. (2004). Systems engineering in an age of complexity. *Systems Engineering Journal, 7*(1), 25-34.

Carlock, P., & Fenton, R. (2001). System of systems (SoS) enterprise system engineering for information-intensive organizations. *Systems Engineering, 4*(4), 242-261.

Carr, N. (2003). IT doesn't matter. *Harvard Business Review, 81*(5), 41-49.

Checkland, P. (2000). Soft systems: A 30-year retrospective. In P. Checkland (Ed.), *Systems thinking, systems practice* (pp. A1-A65). Chichester, UK: Wiley.

Checkland, P., & Holwell, S. (1995). Information systems: What's the big idea? *Systemist, 7*(1), 7-13.

Chesbrough, C., & Spohrer, J. (2006). A research manifesto for services science. *Communications of the ACM, 49*(7), 35-40.

CIO UK Web site. (2007). Late IT projects equals lower profits. Retrieved July 10, 2007, from *http://www.cio.co.uk/concern/resources/news/index.cfm?articleid=1563*

Cleary, D. (2005). Perspectives on complex-system engineering. *MITRE Systems Engineering Process Office Newsletter, 3*(2), 1-4.

Culnan, M., & Swason, B. (1986). Research in management information systems 1980-1984: Points of work and reference. *MIS Quarterly, 10*(3), 289-302.

Curtis, B., Krasner, H., & Iscoe, N. (1988). A field study of the software design process for large systems. *Communications of the ACM, 31*(11), 1268-1287.

Demirkan, H., & Goul, M. (2006). AMCIS 2006 panel: Towards the service oriented enterprise vision: Bridging industry and academics. *CAIS, 18*, 546-556.

Denning, P., Comer, D., Gries, D., Mulder, M., Tucker, A., Turner, J., & Young, P. (1989). Computing as discipline. *Communications of the ACM, 32*(1), 9-23.

Denno, P., & Feeney, A. (2002). Systems engineering foundations of software systems integration. In J.-M. Bruel & Z. Bellahsène (Eds.), *OOIS 2002 Workshops*, Heidelberg, Germany (pp. 245–259). Springer. Lecture Notes in Computer Science, 2426.

Dugmore, J. (2006, May). Benchmarking provision of IT services. *ISO Focus*, pp. 48-51.

Editorial policy statement. (2006). *MIS Quarterly, 23*(1), iii.

Emes, M., Smith, A., & Cowper, D. (2001). Confronting an identity crisis: How to "brand" systems engineering. *Systems Engineering, 8*(2), 164-186.

Ewusi-Mensah, K. (1997). Critical issues in abandoned information systems development projects. *Communications of the ACM, 40*(9), 74-80.

Farhoomand, A., & Drury, D. (1999). A historiographical examination of information systems. *CAIS, 1*(19), 1-27.

Farr, J., & Buede, D. (2003). Systems engineering and engineering management: Keys to the efficient development of products and services. *Engineering Management Journal, 15*(3), 3-9.

Feingenbaum, D. (1968). The engineering and management of an effective system. *Management Science, 14*(12), 721-730.

Frank, M., & Waks, S. (2001). Engineering systems thinking: A multifunctional definition. *Systemic Practice and Action Research, 14*(3), 361-379.

Franke, M. (2001). The engineering of complex systems for the future. *Engineering Management Journal, 13*(12), 25-32.

Fuggetta, A. (2000). Software process: A roadmap. In *Proceedings of the ICSE International Conference on Software Engineering*, Limerick, Ireland (pp. 25-34). ACM Digital Library.

Gelman, O., Mora, M., Forgionne, G., & Cervantes, F. (2005). Information systems and systems theory. In M. Khosrow-Pour (Ed.), *Encyclopedia of information science and technology* (vol. 3, pp. 1491-1496). Hershey, PA: Idea Group.

Gill, G. (1995). Early expert systems: Where are they now? *MIS Quarterly, 19*(1), 51-81.

Glass, R. (1998). *Software runaways*. London: Prentice.

Glass, R., Ramesh, V., & Vessey, I. (2004). An analysis of research in computing disciplines. *Communications of the ACM, 47*(6), 89-94.

Glass, R., Vessey, I., & Ramesh, V. (2002). Research in software engineering: An analysis of literature. *Information & Software Technology, 44*(8), 491-506.

Gorgone, J., Gray, P., Stohr, E., Valacich, J., & Wigand, R. (2006). MSIS 2006: Model curriculum and guidelines for graduate degree programs in information systems. *CAIS, 17*, 1-56.

Goul, M., Henderson, J., & Tonge, F. (1992). The emergence of artificial intelligence as a reference discipline for decision support systems. *Decision Sciences, 11*(2), 1273-1276.

Hecht, H. (1999, March). Systems engineering for software-intensive projects. In Proceedings of the *ASSET Conference*, Dallas, Texas (pp. 1-4).

Hevner, A., & March, S. (2003, November). The information systems research cycle. *Computer*, pp. 111-113.

Hevner, A., March, S., Park, J., & Ram, S. (2004). Design science in information systems research. *MIS Quarterly, 21*(8), 75-105.

Hitchins, D.K. (2003). *Advanced systems thinking, engineering and management*. London: Attach House.

Hole, E., Verma, D., Jain, R., Vitale, V., & Popick, P. (2005). Development of the ibm.com interactive solution marketplace (ISM): A systems engineering case study. *Systems Engineering, 8*(1), 78-92.

Honour, E. (2004). Understanding the value of systems engineering. In *Proceedings of the IN-COSE Conference* (pp. 1-16). INCOSE.

IEEE. (2001). *SWEBOK: Guide to the software engineering body of knowledge.* Los Alamitos, CA: IEEE Computer Society Press.

INCOSE. (2004). *Systems engineering handbook.* Author.

Industry Report. (2005). Facility considerations for the data center version 2.1 (White paper). APC and PANDUIT Companies. Retrieved July 10, 2007, from *http://www.apc.com*

Ives, B., Hamilton, S., & Davis, G. (1980). A framework for research in computer-based management information systems. *Management Science, 26*(9), 910-934.

Jackson, M. (1991). *Systems methodology for the management sciences.* New York: Plenum.

Johnson, K., & Dindo, J. (1998). Expanding the focus of software process improvement to include systems engineering. *Crosstalk: The Journal of Defense Software Engineering*, 1-13.

Keating, C., Rogers, R., Unal, R., Dryer, D., Sousa-Poza, A., Safford, R., Peterson, W., & Rabadi, G. (2003). System of systems engineering. *Engineering Management, 15*(3), 36-45.

Kellner, M., Curtis, B., deMarco, T., Kishida, K., Schulemberg, M., & Tully, C. (1991). Non-technological issues in software engineering. In *Proceedings of the 13th ICSE International Conference on Software Engineering* (pp. 149-150). ACM Digital Library.

Kitchenham, B., et al. (2002). Preliminary guidelines for empirical research in software engineering. *IEEE Transactions on Software Engineering, 28*(8), 721-734.

Kwon, T., & Zmud, R. (1987). Unifying the fragmented models of information systems implementation. In J. Boland & R. Hirshheim (Eds.), *Critical issues in information systems research* (pp. 227-251). New York: John Wiley.

Lee, A. (2000). Systems thinking, design science and paradigms: Heeding three lessons from the past to resolve three dilemmas in the present to direct a trajectory for future research in the information systems field. Retrieved July 10, 2007, from *http://www.people.vcu.edu/~aslee/ ICIM-keynote-2000/ICIM-keynote-2000.htm*

Madachy, R. (1996). Modeling software processes with system dynamics: Current developments. In *Proceedings of the System Dynamics Conference* (pp. 1-23). Cambridge, MA: Systems Dynamic Society.

Magee, C., & de Weck, O. (2004). Complex system classification. In *Proceedings of the 14th Annual INCOSE International Symposium of the International Council on Systems Engineering* (pp. 1-18). INCOSE.

March, S., & Smith, G. (1995). Design and natural science research on information technology. *Decision Support Systems, 15*, 251-266.

Mingers, J. (2000). The contributions of critical realism as an underpinning philosophy for OR/MS and systems. *Journal of the Operational Research Society, 51*, 1256-1270.

Mingers, J. (2001). Combining IS research methods: Towards a pluralist methodology. *Information Systems Research, 12*(3), 240-253.

Mingers, J. (2002). Realizing information systems: Critical realism as an underpinning philosophy for information systems. In *Proceedings of the 23rd ICIS International Conference in Information Systems* (pp. 295-303).

Mingers, J. (2003). A classification of the philosophical assumptions of management science methodologies. *Journal of the Operational Research Society, 54*(6), 559-570.

Mora, M., Cervantes, F., Gelman, O., Forgionne, G., & Cano, J. (2006b). The interaction of systems engineering, software engineering and information systems disciplines: A system-oriented view. In *Proceedings of the 18th International Conference on Systems Research, Informatics and Cybernetics*, Baden-Baden, Germany (pp. 1-5).

Mora, M., Gelman, O., Cano, J., Cervantes, F. & Forgionne, G. (2006a). Theory of systems and information systems research frameworks. Proceedings from the *50th Annual Meeting of the International Society for the Systems Sciences*, paper 2006-282, (1-7), Somona State University, CA.

Mora, M., Gelman, Cervantes, F., Mejia, M., & Weitzenfeld, A. (2003). A systemic approach for the formalization of the information system concept: Why information systems are systems. In J. Cano (Ed.), *Critical reflections of information systems: A systemic approach* (pp. 1-29). Hershey, PA: Idea Group.

Mora, M., Gelman, O., Forgionne, G., Petkov, D., & Cano, J. (2007a). Integrating the fragmented pieces of IS research paradigms and frameworks: A systems approach. *Information Resources Management Journal, 20*(2), 1-22.

Mora, M., Gelman, O., Macias, J. & Alvarez, F. (2007c). The management and engineering of IT-intensive systems: a systemic oriented view. Proceedings from the *2007 IRMA International Conference*, (1448-1453), Vancouver, BC, Canada: Idea Group.

Mora, M., Gelman, O., O'Connor, R., Alvarez, F., & Macías, J. (2007b). On models and standards of processes in SE, SwE and IT&S disciplines: Toward a comparative framework using the systems approach. In K. Dhanda & R. Hackney (Eds.), *Proceedings of the ISOneWorld 2007 Conference: Engaging Academia and Enterprise Agendas.* Information Institute. Track in System Thinking/Systems Practice, Las Vegas, USA, April 11-13, 2007, paper-49, 1-18.

Mulder, M., et al. (1999). ISCC'99: An information systems-centric curriculum '99: program guidelines for educating the next generation of information systems specialists, in collaborating with industry. Retrieved July 10, 2007, from *http://www.iscc.unomaha.edu*

Nolan, R., & Wetherbe, J. (1980, June 1). Toward a comprehensive framework for MIS research. *MIS Quarterly,* pp. 1-20.

Nunamaker, J., Chen, T., & Purdin, T. (1991). Systems development in information systems research. *Journal of Management Information Systems, 7*(3), 89-106.

OECD. (2004). Highlights of the OECD information technology outlook 2004. Retrieved July 10, 2007, from *http://www.oecd.org*

Orlikowski, W., & Iacono, S. (2001). Desperately seeking the IT in IT research. *Information Systems Research, 7*(4), 400-408.

Prewitt, E., & Cosgrove, L. (2006). *The state of the CIO 2006* (CIO Report pp. 1-8). Retrieved July 10, 2007, from *http://www.cio.com/state*

Rittel, H., & Weber, M. (1973). Dilemmas in a general theory of planning. *Policy Sciences, 4*, 155-169.

Sage, A. (2000). Systems engineering education. *IEEE Transactions on Systems, Man and Cybernetics (Part C: Applications and Reviews), 30*(2), 164-174.

Sage, A., & Cuppan, D. (2001). On the systems engineering and management of systems of systems and federations of systems. *Information, Knowledge, Systems Management, 2*, 325-345.

SEI. (2003). *SEI annual report 2003*. Pittsburgh: Software Engineering Institute.

Shenhar, A., & Bonen, Z. (1997). The new taxonomy of systems: Toward an adaptive systems engineering framework. *IEEE Transactions on Systems, Man and Cybernetics (Part A: Systems and Humans), 27*(2), 137-145.

Simon, H. A. (1969). *The sciences of the artificial.* Cambridge, MA: MIT Press.

Sommerville, I. (1998). Systems engineering for software engineers. *Annals of Software Engineering, 6*, 111-129.

Sowa, J., & Zachman, J. (1992). Extending and formalizing the framework for information systems architecture. *IBM Systems Journal, 31*(3), 590-616.

Standish Group International. (2003). *The extreme CHAOS report.* Author.

Sterman, J. (2001). Systems dynamic modeling: Tools for learning in a complex world. *California Management Review, 43*(4), 8-25.

Thayer, R. (1997). Software systems engineering: An engineering process. In R. Thayer & M. Dorfan (Eds.), *Software requirements engineering* (pp. 84-109). Los Alamitos, CA: IEEE Computer Society Press.

Thayer, R. (2002, April). Software systems engineering: A tutorial. *IEEE Computer*, pp. 68-73.

van Bon, J., Pieper, M., & van deer Veen, A. (2006). *Foundations of IT service management, based in ITIL*. San Antonio, TX: Van Haren.

Vessey, I., Ramesh, V., & Glass, R. (2002). Research in information systems: An empirical study of diversity in the discipline and its journals. *Journal of Management Information Systems, 19*(2), 129-174.

von Bertalanffy, L. (1972, December). The history and status of general systems theory. *Academy of Management Journal*, pp. 407-426.

Wallace, L., & Keil, M. (2004). Software project risks and their effect on outcomes. *Communications of the ACM, 47*(4), 68-73.

Zachman, J. (1987). A framework for information systems architecture. *IBM Systems Journal, 26*(3), 276-292.

ENDNOTE

[1]   The title and aim of this article was inspired in Feigembaum's (1968) paper.

This work was previously published in International Journal of Information Technologies and Systems Approach, Vol. 1, Issue 1, edited by J.D. Paradice; M. Mora, pp. 1-24, copyright 2008 by IGI Publishing (an imprint of IGI Global).

Chapter 2
A Question for Research:
Do We mean Information Systems or Systems of Information?

Frank Stowell
University of Portsmouth, UK

ABSTRACT

In this chapter I raise questions about the nature of Information systems, the way that they are designed and developed and suggest areas that IS researchers may wish to investigate. A concern is raised about the way we think about the domain of information systems and a suggestion made that rather than think of it in terms of the mnemonic IS, with is association with IT, it should be thought of in terms of as systems of information. This suggestion is made as a means of highlighting considerations developers have to take into account which go beyond those of technology alone. As a means of instigating this proposition four questions are raised in the chapter which are intended to stimulate further information systems research. These questions are about the nature of IS, design Methods, the underpinning philosophy and finally, IS failure.

GENERAL INTRODUCTION

Information Systems, as a domain on knowledge, is rarely satisfactorily explored in the literature. There are papers which discuss IS within the context of a particular area of application e.g. Management Information Systems but few deal with the nature of Information Systems. Although IS researchers and practitioners refer to IS theory rarely do they define what they mean. The dearth of discussion about the constituents of the subject itself implies that there is universality of understanding about the nature and composition of IS. It is true that the range of knowledge and the variety of skills that IS embraces makes its definition, in terms familiar to the more traditional areas of expertise, difficult to achieve. The lack of a common and acceptable description has vexed the IS community for some years and a sound theory of IS is still elusive (see also Gregor, 2006, p612).

We can argue that the mnemonic IS, which is the common way of referring to the area, has added

DOI: 10.4018/978-1-60566-976-2.ch002

to this difficulty which is compounded by many examples in the literature where IS and IT are used interchangeably. This apparent confusion between IS and IT may be seen, by some academics and practitioners, as an indication that Information Systems is a transitory domain of knowledge created by the accessibility of Information Technology which will disappear as the technology itself becomes part of the cultural infrastructure e.g. much like users of mobile telephones have become expert in their usage with little practical guidance.

So what can we say about Information Systems? We can say is that Information Systems is a general term that defines our branch of learning, our discipline. It is an area of knowledge which is concerned with the way of using technology which is determined by purposeful (willed activity becomes action) human activities. We can argue that the practice is concerned with gaining understanding about *systems of information*. The separation of the two terms serves to emphasise our branch of learning i.e. IS relates to the intellectual underpinning and associated learning about the domain, and systems of information refers to our area of interest i.e. any situation where we take action and from that action learn from it which in turn contributes to the formation and reformation of knowledge about the domain itself.

Rather than attempt to define Information Systems I have chosen instead to raise 4 issues in the form of questions which I believe reflect important areas of research and practice which will contribute to the body of knowledge. The first question is: What is an Information System – how do we recognise one?; the second question relates to Methods - How do we set about designing Information Systems and what approaches do we use that are distinctly IS?; the third question relates to what philosophical ideas underpin IS as a subject domain and, finally; What constitutes a failure? The latter being important as there are many IT/IS failures reported where there is no actual failure of the IT but the "system" as a whole is deemed

by the users as having failed. Is it failure which acts to differentiate between IT and IS?

WHAT IS AN INFORMATION SYSTEM?

Whilst we can accept that computers are at the heart of most businesses it should not lead us to assume that Systems of Information (our territory) is solely about computing any more than it is about marketing or stock control. It is about all of the components that together make up a system of information for the collective clients. It is worth reminding ourselves that there is a difference between data processing and information; people are interested in identifying and understanding what a data object means but computers only need to identify data objects. Traditional Data Processing (DP) is not concerned with information because it produces data which are used to guide routine activities without being explicitly interpreted to the activity and the human actors informed by the data. It seems axiomatic that such straightforward considerations are incomplete when looking at the "system" as a whole.

The knowledge base of Information Systems (as a discipline) is concerned with information technology (IT) but it also requires an equal knowledge of other areas including social and management science and of business practices. The development of IT systems require skills that focuses upon the technology and the way that it might be used to assist the client (end user) undertake some tasks, whereas IS is concerned with knowledge and skills required first, to gain understanding and then to be used as a means of improving the clients system of information as a whole.

The failure of commentators (and some academics), to differentiate between IT (data) and IS (information) has resulted in a profusion of reports in the literature (both academic and practitioner) that are referring to data processing

Figure 1. A notional information system

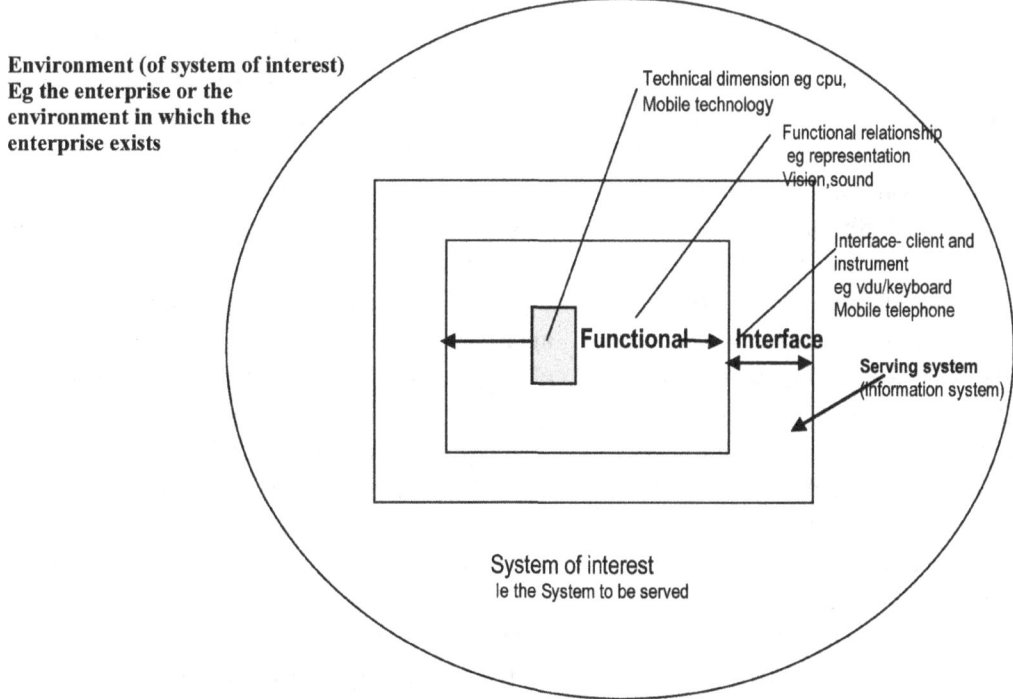

but are often, describing them as information systems. If we think of the "territory" in which we work in terms of a *system of information* and consider the whole in a mereolgical sense; that is to say that the relations of part to whole and the relations of parts within a whole. (Varzi, 2004) then it amplifies what our task is as Information Systems professionals i.e. to consider all parts, human and technical and the relationships between them (see figure 1).

So where does this leave us? We can say that systems of information are formed from interrelated information units and the relations between them and as a consequence our primary task is to understand the purposeful actions of those that make up the system of interest and describe the systems of information that enable it. In other words our first task is to consider what system it is that is to be served (Winter et al, 1995, Checkland and Scholes, 1999) and then model the serving systems (systems of information) – i.e. first "at-

tempt to identify the whole relevant system and its components" (Checkland and Scholes, 1999).

The importance of understanding the relationship between information and associated connections within an enterprise cannot be over-stressed. It is this distinction that makes IS practice different from that of IT. IT practitioners are more concerned with what they perceive as the technological needs of the situation. If this is the case then our first challenge is to ask ourselves if there is an over reliance upon methods of design and development that come from computing and software engineering and if so are such technically oriented approaches suited to the development of *systems of information*? One potential area for further research and publication lies in greater analysis of development methods shaped by the practice, in particular where the lessons have been learnt from the development of public information systems.

RE-THINKING IS METHODS OF DESIGN

Many methods used in Information Systems development have their origins in the early years of computing and even those developed in recent years owe more to satisfying data processing needs than defining a system of information. One major obstacle with technically oriented methods of design is that it is difficult for the technical expert to avoid their position as expert being used to impose personal values or ideologies. Too often the design methods used effectively promote the technical viewpoint instead supporting the users' desires and values. It could be that it is the technical domination and unintentional sidelining of the end-user which is the underlying cause of IS failure. This may well be the case for those large public information systems which are reported as failing (see Cross, 2005)

The impact and number of IS failures (e.g. see Fortune and Peters, 2005) has precipitated much research into ways of involving the end-user in the design process (e.g. Delone and McLean, 1992, 2003; Cavaye, 1995; Bødker, 1996; Lynch and Gregor, 2004) because this is seen as one way of improving satisfaction. Examples include high level architectural specification techniques such as Jackson Development (Jackson, 1983) which endeavours to define how the human, the hardware and the software interact. Joint Application Design, (McConnell, 1996), includes a brainstorming element to enable some users to communicate ideas, to reduce resistance to change and help the end-users to understand the development process but the method is not completely client driven and defaults to technical requirements. Structured Systems Analysis and Design Method (SSADM, 1990), is a modified version of traditional information systems development life cycle but it is still a data-driven approach based on data flow diagrams and entity life histories.

Efforts were made in the 1980's and 90's to add softer methods to enrich the IS definition (Boland, 1985, 1991; Stowell, 1985; Avison and Wood-Harper, 1986; HMSO, 1993; West, 1991; Stowell and West, 1994; Dobbin and Bustard, 1994, Savage, and Mingers,1996; Bell and Wood-Harper, 1998,) but this had little impact upon the success of the IT based systems of information. Object Oriented Analysis seemed to offer a better way of defining an information system (West et al, 1996; Liang et al, 1997; Graham, 1998) but this too is technically oriented. Although the OO approach consists of five major activities: defining class and objects; identifying structures; identifying subjects; defining attributes and defining services, it is strongly algorithmic with complicated calculations that are not straightforwardly broken down into objects and it is expensive to develop. Like many methods used for information system development OO is a means to a programming end.

In more recent times the best known of the lightweight methodologies XP (extreme programming) focuses on building a person-to person mutual understanding of the problem environment through what they describe as minimal formal documentation and maximum on-site interaction (Highsmith, 2000). There are four values involved; communication, simplicity, feedback and courage. Communication is between customers and programmers and design is kept simple, and also embraces constant feedback from the customers. Using the approach the programmers claim that they are able to respond, 'courageously', to changing requirements and technology. Bødker, (1996) description of the XP development cycle shows that it begins promisingly be getting end users to provide "stories" to estimate development time and provide a plan for the release of applications. This is because it makes it easier to get feedback from users as the whole application is developed, but the criteria for success are measured in technological terms. For example, it is measured in terms of early and continuous delivery of valuable software. (Agile, 2004).

The above summary is not intended to provide an exhaustive analysis of attempts to embrace user requirements nor of more recent thinking about software development life cycles but to serve as an illustration of the way that the end-user is becoming more involved in the development of the software. The involvement of the end user (or client) in the technical side of the situation underlines the importance attached by designers in obtaining a successful result. Nevertheless, in many instances success is measured by the working software and it is questionable if this is a good indicator for measuring its usefulness to a system of information. It could be argued that such methods of software development is what computer hackers have done for years; namely, shape the programme as they go along. As Bustard and Keenan (2005) point out it is not clear in this kind of approach who the customers are, how they are selected and how the suitability of the resultant software is assessed, moreover is it as they point out it is not clear from the methods used if the design process is requirements or product driven.

From a development perspective, one way of involving the end-user in the process might be to undertake it with them and, together, envisage how the technology can be assimilated into their system of information. There is no universal formula to do this but we can say that the approach used should be flexible and capable of authentication (see Champion and Stowell, 2001). The pioneering work of Churchman, (1971); Ackoff, (1978); Vickers (1981) and Checkland (1981) all provide important insights into methods of inquiry, analysis and design. But selecting the right method is not a trivial undertaking and care must be taken to ensure that tools used do not get in the way of the learning about the system of interest (Churchman, 1971; Checkland, 1981/99; West and Stowell, 2000).

The suggestion being made here is that we need to research and test ideas which provide the end user, or client, with greater control and which

will also produce a specification for the information system as a whole, including the technical and non-technical activities. There are problems with this idea as many researchers have found it difficult to translate aspects of the information requirements into technical support (see, Rashid, et al 2006 as one example). The translation of the definition of requirements, in natural language, into a technical specification is sometimes referred to as "bridging the gap" (Champion and Stowell, 2001; Champion et al, 2004; Stowell and Cooray, 2006) and often it is this aspect where the richness of the client definition is lost in favour of the technical requirements.

To this end we could re-examine the way we view information systems development and the lessons learnt from past attempts at embracing both the technical and non technical dimensions. For example, Langefors (1995) early work attempted to relate information requirements through to a technical specification using what he called an Infological approach. Infology was suggested as a means of thinking through the possibilities of translating natural language into technical specification. Mumford and Henshall's, (1979) and Mumford's, (1995), work on socio-technical systems is a source of lessons learnt about participation. Avison and Wood-Harper's (1986) ideas, expressed as Multiview, provides an example of the way that the despite best efforts technical requirements still tend to dominate the end result. Similarly Stowell's, (1985, 1997, 2006) attempt at linking Soft Systems Methodology (SSM) to dataflow diagrams and latterly OO modelling provides lessons about the difficulties of maintaining the continuity of "soft systems" thinking through to technical specification. Lyytinen's (1988) discussion of SSM and Information System failure provides further insights into ways of using non reductionist ideas in design and development. The linking of SSM to SSADM (HMSO, 1993) is also an important milestone in the recognition of the failure of purely technical methods as the means of Information system development (see Lewis, 1995). Revisiting Hirschheim et al, (1995) we can consider some philosophical foundations of data modelling which will help in thinking about the different perspectives that underlie approaches that have emerged over the past twenty years. Stamper's (1997) ideas on Organisational Semiotics also contribute to the difficulty of solving "soft" problems that require practical solutions. Walsham's (1991) ideas of Organisation and Alter's paper (1999) similarly provide useful discussions on the thinking about Information Systems from a theoretical basis. By revisiting these ideas and their applications we may find lessons that will contribute to the development of methods more suited to understanding and designing systems of information capable of satisfying clients needs.

SOME THOUGHTS ABOUT AN UNDERPINNING PHILOSOPHY

Human situations are (usually) complex so it is reasonable to suggest that a clear-cut application of the empirical methods of natural science into organizational intervention is problematical. The constituent parts of human social networks are difficult to distinguish and even when we are able to do so they do not easily provide an understanding of the whole. Social networks are complex because we humans are not entirely subjected to physical laws as are other living creatures. For example, we may choose a course of action that might be judged as being irrational by others. Even so, we are able to reflect upon our actions and upon our surroundings and through that understanding chose a course of action which will allow us to maintain our relationship with our environment (see Vickers, 1983, 1991, for a discussion on Appreciative Systems and Relationship Maintenance, both ideas important to IS). Recognising this is critical to Information Systems development in that there should be an 'Appreciation' of the *systems of information* that exist as a precursor to

action so that whatever is developed or changed it will ensure that the relationships of the system of interest and its environment is maintained.

Understanding an enterprise and how it operates cannot be replicated within a laboratory and hence the method of inquiry employed must be flexible and capable of addressing "real world" complexity. To gain an "Appreciation" of the system of information, how it operates and how it can be improved means being a part of that system rather than a dispassionate observer. By thinking in terms of systems of information and the need to gain an "Appreciation" (ibid) of what that means to the clients might help to avoid the premature imposition of purely technical solutions.

IS practitioners should have knowledge of non-reductionist methods as well as the reductionist techniques associated with computing and software engineering. In Churchman's (1971), classic text "Design of Inquiring Systems" he reflected upon the deficiencies of the methods of science when applied to human situations, he said that "As a system, science cannot discuss social change (implementation) in any but a very restricted sense......science has no adequate way of studying the elusive, since it always aims for precision, and hence in some real sense science is alienated from nature" (Churchman 1971). Churchman's point has resonance for Information Systems since success seems to be as much about understanding the "elusive" as it does the supporting technology.

IS professionals should be skilled in using ideas capable of taking on the complexity of human interactions and their information needs. This is a tall order but Systems thinking may offer such an intellectual device. "Systems" is an epistemology, a theory of knowledge which we might adopt to help us understand the world. The noun System comes from the Greek, "Syn", to declare; addition, i.e. the action of putting something(s) together and "Histemi", which means "to stand", so Systima (the combining of Syn and Histemi), means to put something together that "stands" (i.e. has a purpose), which those who subscribe to the notion of wholeness refer to by the noun System.

The notion of Systems as a way of understanding the world as Bullock and Trombley, (2000) put it is an "an approach to the study of physical and social systems which enables complex and dynamic situations to be understood in broad outline". By adopting the Systems epistemology as our underpinning idea we might find that it provides a means of addressing some of the issues that exercises our domain.

Yet the way that the term System is used can also create difficulties. For example, the adjectives systemic and systematic are often confused. Whilst both terms are adjectives from the same noun (i.e. 'system') they are frequently used badly even by those who subscribe to the notion of "system". When we refer to taking a systems approach or adopting an holistic approach to problem solving we would use the word 'systemic' (e.g. we might claim that we have undertaken a systemic analysis of a business problem) rather than undertaking a systematic analysis. We use the term 'systematic' when adopting an ordered, step-by-step, methodical approach. In some respects the way that the term is used can reveal the kind of thinking of the individual that uses them. For example, systematic can be associated with a functionalist way of approaching analysis whereas systemic can be associated with a more holistic approach.

But using Systems ideas is not without difficulties and questions have been raised about the problem of assessing the quality the work (e.g. Klein and Myers 1999). There are also problems with combining methods from different intellectual origins such as the notion behind multi methodology (Mingers and Gill, 1997) as each will refer to its own paradigm which means that inevitably there is some sort intellectual compromise (see Midgley, 2000 for further discussions on paradigm incommensurability). Whilst criticisms about quality are important and are not to be dismissed lightly neither should an apparent intellectual impasse be the cause of the abandoning

attempts to use them. Ideas such as Checkland's notion of recoverability, Naughton's constitutive and strategic rules (Checkland 1999, Checkland and Holwell, 1998) and Champion and Stowell's notion of authentication (Champion and Stowell, 2001) contribute to the rigour of using an interpretivist systems approach to IS development.

By revisiting the ideas within the context of reported IS failure and lessons learnt from earlier research (e.g. the brief references in design methods above) new ideas about IS development and the relationship between client needs and technical specification may emerge. For example, the work of Stowell (1985-2007) who in 1985 proposed using Soft System Methodology as a means of assisting in IS design, Avison and Wood-Harper (1986) had similar ideas and the work of Bell and Wood-Harper (2003) and Cornford and Smithson (2006) also refer to the use of soft systems with the development process. In 1988 Lyytinen proposed an holistic, systems oriented approach as a preferred way of researching IS failure. Since that time attempts to use systems ideas have met with varying degrees of success but the lessons learnt from this early work provide the basis for moving forward and an underpinning philosophy for the discipline.

WHAT IS FAILURE?

IT projects have a poor record throughout the world. Cross reports that 70% of projects in the USA fail to meet their timetable or budget or come up to specification (Cross, 2005). A survey carried out by Dunleavy and Margetts in 2004 and across seven countries show that the UK has the highest record of failure with Japan and the Netherlands with the best records. OECD reports only 28% of all IT projects in the US were successful in 2000 (Kristensen and Buhler, 2001). The reports give several reasons for lack of success including differences between public and private sector projects, project size, project isolation, government interface with IT industries, market and technical dominance of IT companies, modernity of equipment, and the extent to which government retains IT expertise.

The above reports indicate that failure results from a mixture of things but provide little evidence that the failure is of a technical origin so one conclusion that can be formed is that reported failure is the result of human error or discontent and not failure of the technology. One example where mismatch between technical provision and user needs is highlighted comes from a quotation reported in the National Press, by the chairman of a public accounts committee who stated "One of the major problems was the 'horrible interface' between civil servants, who understand all there is to know about, for example, the national insurance system but know little of how a computer works, and technicians who just know the reverse", he went on … "they don't spend enough time at the start of the project explaining where they are both coming from" (Morris and Travis, 2001).

The likelihood of avoiding future failures is doubtful and there is little to suggest much has been learnt from the recent past. But what is an IS Failure? Previous attempts to define IS failure tend to focus on a failure to the take up the technology and its associated software, but success and failure of systems of information are not as easy as this to measure. In a recent text Fortune and Peters (2005) characterise IS failure as a "product of outputs which are considered to be undesirable by those involved". They cite Vickers (1981) who perhaps proves a richer and more experiential definition as "A human system fails if it does not succeed in doing whatever it was designed to do; or if it succeeds but leaves everyone wishing it had never tried".

One of the largest public IT programmes in the world (the UK National Health IT project called the National Project for IT) approved in 2002 by the then UK Prime Minister Tony Blair, is late, over budget and facing difficulties of implementation. It has been reported that the future of the

NHS's £12.7bn computer programme is in doubt after its managers acknowledged further delays in introducing a system for the electronic storage and transmission of patients' records. Connecting for Health, the NHS agency responsible for the world's biggest civil IT project, said "it was no longer possible to give a date when hospitals in England will start using the sophisticated software that is required to keep track of patients' medical files." Moreover the Treasury is reported to be earmarking health service IT as a candidate for cuts to compensate for the billions spent on the bailout of the banks.(Carvel, 2008).

A senior figure in the programme is reported as criticising staff for resisting the need for change. Amid growing dissent over the way the Government's reforms are being pushed through he went on to say that there is "an unwillingness to "embrace" the Government's £12 billion IT upgrade for the health service" (Daily Telegraph, 2007). In London, attempts to install the system at the Royal Free hospital and Barts caused weeks of confusion and disruption. Other trusts that were next in line were so alarmed that they pleaded for postponement (Carvel, 2008) It could be that a prime reason for failure is that IS development is often undertaken as a technological project and because of this there is a failure to distinguish between the systems of information and the technology that supports it.

Importantly failure is often a failure to engage the end-user, or client, in the development process and as a consequence there is a failure to take ownership. Returning to the national project for IT a survey carried out by MEDIX in 2004 showed that 40% of doctors stated that they had no information about the project with one practitioner stating "as a practicing clinician, I am concerned that this IT programme has all the hallmarks of previous governmental IT failures, for example, failure to consult with end-users about how it will integrate with their daily work and make their work easier. If it is perceived as management or government driven additional tasks (which it is currently, by the few who have heard of it), then it will fail." There has been more recent reported concern about a lack of clarity and poor communications. Some of those interviewed said they felt disempowered and frustrated because decisions were being made by Connecting for Health and local IT service providers without consulting key NHS staff. Collins, T., (2007).

It would seem that despite observations made decades before little has moved on as Lyytinen observed two decades ago "An IS fails when it cannot fulfil the expectations and this incapability calls for stakeholder action" (Lyytinen, 1988). But if the solution is to engage the clients more how can the user, stakeholder, client, engage in something as complex as IS development which necessitates specialist knowledge of problem solving, technology, working practices and some degree of clairvoyance? It would seem this is an area for more research.

Clearly a failure can be caused for a variety of quantifiable reasons such as being over budget, technology breakdown and so on but it is the impact upon the users and intended consumers that causes the greatest problems. Whilst the specific cause of failure differs with each information system what they have in common is economic and social calamity. The failure of a large public system of information effects not just those who operate or use it as a management tool it also frustrates the sponsors, the designers, those who managed the project and, increasingly, the general public.

There are numerous reports of failure (see Fortune and Peters, 2005; Cross,2005; Myers 1994) but the puzzle is why do they continue? Lyytinen and Robey suggest that many ISD organisations appear unable or unwilling to adjust their practices even when they fail to produce beneficial results (Lyytinen and Robey, 1999). It would seem that lessons about failure are not being learnt and research into design methods not taken up, a situation which one commentator describes as being "like a computer virus, endlessly replicate the mistakes of the past" (Caulkin, 2004).

CONCLUSION

In this chapter we have considered elements of the IS domain with a suggestion that it might be useful to consider IS in terms "Systems of information" (SI) as a means of thinking about the wider implications for definition and design. Questions have been raised about the relationship between the cause of information system failure and the methods used for SI development suggesting that the methods themselves may impose restrictions because of the in-built bias towards satisfying the technology. The Suggestion is made that designing an information system requires consideration of the system as a whole and its relationship to its environment and the argument goes on to question whether the predominant approaches to design are appropriate to the task. Clearly this is a difficult proposition to satisfy as it brings with it the need for knowledge of many areas including organisational analysis, information theory, group interaction, anthropology, inquiring systems, and technical knowledge.

The logical outcome of the assertions in this chapter is to suggest that the design methods should be able to take into account the end-user and consumers views and specify the IT support. Moreover, it should be possible for the end user/client to authenticate the outcome i.e. the design should be represented in a form that a non technical person can readily understand.

It is suggested that using Systems ideas might provide an intellectual framework for SI development. But the difficulties of doing so are raised by drawing attention to the problem combining two apparently incompatible concepts into one seamless method i.e. the need to satisfy the client and the resultant technical specification. But it is important not to let such difficulties be the cause of the abandoning attempts to create methods suited to IS.

Systems Thinking and Practice provides ideas and methods of making sense of the complexities of systems of information. It is hoped that by

Table 1. Suggested topics for research

Topic	Associated question
Information Systems	• Is Information Systems a discipline? IS papers can be found in a variety of journals dealing with the technical aspects of information system provision through to the IS management. But what is it that binds these papers together under the general heading of Information Systems? • Is Information Technology and Information Systems the same thing? • Is IS a part of Management Science or Computing? • We welcome papers dealing the educational requirements of the IS professional • Papers dealing with the social and cultural context of Information Systems are particularly welcome, e.g. the effects upon social cohesion
Systems of Information	• How are Systems of Information understood? • To paraphrase Stafford Beer "Given information systems what are organisations now?" The relevance of organisational models to systems of information? • Research dealing with Information, Systems and Democracy and freedom • Systems of Information for specialist areas e.g. the Law, medicine, security • Research dealing with the impact of IS upon social cohesion
IS Design and Development	• There is a significant literature based on interpretivist ideas but when these ideas are translated into methods they often fall short. We would like to see more research in this area. • What is the relationship between an information system and the supporting technology? • We welcome papers reporting Action Research projects • Research into methods of Inquiry Problem identification and Analysis and their suitability to large and small scale projects.
Philosophy	• Has IS an underpinning Philosophy? IS draws upon a wide rage of theory does it contribute to the confusion between IT and IS? • To what school of thought does IS belong? We welcome a critical review of its intellectual home
Failure	• Why do Information Systems fail? • By what criteria should we judge success? • We welcome papers dealing with the impact and causes of IS failure upon the wider community e.g. welfare and medical systems • The social and economic impact of failure

raising questions about the nature of Information Systems it may serve to reassess the way in which we set about defining what it is and designing how these systems might operate. The intention is that by encouraging researchers to re-visit the way that we set about development it will provide opportunities for researchers and practitioners to contribute ideas about the development and distinctiveness of the discipline. To this end we should also reflect upon IS curricular we teach in our Universities and ask if it reflects the distinctiveness of IS or is it here that the confusion between IS and IT begins? (see Work, 1997 for an interesting discussion on IS curricular).

The following table provides some suggestions for IS research, including research concerned with the theory and research from the practice. The suggestions are intended to encourage submissions to journals from a variety of areas of interest where the implications of Information Systems design and development are explored.

REFERENCES

Ackoff, R. L. (1978). *The art of problem solving*. New York: John Wiley and Sons.

Agile Technology. (2004) http://www.agilealliienceeurope.org.

Alter, S. (1999). A general, yet useful theory of information systems. *Communications of the Association for Information Systems, 1*, 13.

Avison, D.E., & Wood-Harper, A.T. (1986). Multiview: An exploration in information systems development. *The Australian Computer Journal, 18*(4).

Avison, D. E., & Wood-Harper, A. T. (1990). *Multiview: An Exploration in Information Systems Development.* Maidenhead: McGraw-Hill.

Bell, S., & Wood-Harper, A. T. (1998). *Rapid information systems development.* Maidenhead: McGraw Hill.

Bell, S., & Wood-Harper, A. T. (2005). *How to set up information systems: A non-specialist guide to the multiview approach.* London: Earthscan publications.

Bødker, S. (1996). Creating conditions for participation: conflicts and resources in systems development. *Human-Computer Interaction, 11*, 215–236. doi:10.1207/s15327051hci1103_2

Boland, R. (1985). Phenomenology: A preferred approach to research in information systems. In E. Mumford, R.A. Hirschheim, G. Fitzgerald, & A.T. Wood-Harper (Eds.), *Research methods in information systems.* North-Holland: Amsterdam.

Boland, R. (1991). Information systems use as a hermeneutic process. In H. Nissen, H. Klein & R. Hirschheim (Eds.), *Information systems research: Contemporary approaches and emergent traditions.* North-Holland: Amsterdam.

Bullock, A., & Tromby, S. (2000). *New Fontana dictionary of modern thought* (p. 443). London: Harper Collins.

Bustard, D. W., & Keenan, F. (2005). Strategies for systems analysis; Groundwork for process tailoring. *IEEEE, ECBC.*

Carvel, J. (2008). Bank bailout puts £12.7bn NHS computer project in jeopardy. *Guardian newspaper.* Wednesday 29 October 2008 (see also http://www.guardian.co.uk/society/2008/oct/29/nhs-health)

Caulkin, S. (2004, May 2). Why IT just doesn't compute: Public sector projects even more likely to fail than private. *The Observer* (p. 9).

Cavaye, A. L. M. (1995). User participation in system development revisited. *Information & Management, 28*(5), 311–323. doi:10.1016/0378-7206(94)00053-L

Champion, D., & Stowell, F. A. (2001). Pearl: A systems approach to demonstrating authenticity in information system design. *Journal of Information Technology, 16*, 3–12. doi:10.1080/02683960010028438

Champion, D., Stowell, F. A., & O'Callaghan, A. (2004). Client led information system creation (CLIC): Navigating the gap. *Information Systems Journal, 15*(3), 213–231. doi:10.1111/j.1365-2575.2005.00191.x

Checkland, P. B. (1981). *Systems thinking, systems practice.* Chichester: Wiley and Sons.

Checkland, P. B. (1999), *Systems thinking, systems practice, includes 30 year retrospective.* Chichester: Wiley and Sons.

Checkland, P. B., & Holwell, S. E. (1998). *Information, systems and information systems.* Chichester: Wiley and Sons.

Checkland, P. B., & Scholes, J. (1999). *Soft systems methodology in action: 30 Year retrospective.* Chichester: Wiley and Sons.

Churchman, C. W. (1971). *The design of inquiring systems: Basic concepts of systems and organisation* (p. 18). New York: Basic Books.

Collins, T. (2007). Academic study finds that NHS IT programme is "hampered by financial deficits, poor communication and serious delays." *Computer Weekly*. Retrieved from http://www.computerweekly.com/blogs/tony_collins/2007/05/academic-study-finds-that-nhs-1.html

Cornford, T., & Smithson, S. (2006). *Project research in information systems*. Basingstoke: Palgrave Macmillan.

Cross, M. (2005, October). Public sector IT failures. *Prospect*, 48-52.

Daily Telegraph. (2007, April 25). NHS staff block reforms, says ex-minister. Retrieved from http://www.telegraph.co.uk/news/main.jhtml?xml=/news/2007/04/24/nreforms24.xml

Delone, W. H., & Mclean, E. R. (1992). Information systems success: The quest for the dependent variable. *Information Systems Research*, *3*(1), 60–95. doi:10.1287/isre.3.1.60

Delone, W. H., & Mclean, E. R. (2003). The Delone and Mclean Model of Information System Success: A Ten Year Update. *Journal of Management Information Systems*, *19*(4), 9–30.

Dobbin, T. J., & Bustard, D. W. (1994, August). Combining soft systems methodology and object-oriented analysis: the search for a good fit. In *Proceedings of the 2nd Information Systems Methodologies Conference*.

Dunleavy, P., & Margettes, H. (2004, September 1-5). *Government IT performance and the power of the IT industry: A cross national analysis*. Paper presented the Annual meeting of American Political Science Association, Chicago.

Fortune, J., & Peters, G. (2005). *Information systems, achieve success by avoiding failure*. Chichester: John Wiley & Sons.

Graham, I. (1998). *Requirements engineering and rapid development*. London: Addison Wesley.

Gregor, S. (2006). The nature and theory of information systems. *MIS Quarterly*, *30*(3), 611–642.

Highsmith, J. A. (2000). *Adaptive software development: A collaborative, approach to managing complex systems*. Dorset House.

Hirschheim, R. (1985), Information systems epistemology: A historical perspective. In E. Mumford, R. Hirschheim, G. Fitzgerald, & A. Wood-Harper (Eds.), *Research methods in information systems* (pp. 1199-1214). Amsterdam: Elsevier.

Hirschheim, R., Klein, H. K., & Lyytinen, K. (1995). *Information systems development and data modelling: Conceptual and philosophical foundations*. Cambridge: The University Press.

Hirschheim, R.A., & Klien, H.K. (1989). Four paradigms of information system development. *Social aspects of computing, 32*(10), 1199-1214

HMSO. (1993). Applying soft systems methodology to an SSADM feasibility study. *Information systems engineering library* (series). London: HMSO.

Jackson, M. A. (1983). *System development*. London: Prentice Hall

Klein, H. K., & Myers, M. D. (1999). A set of principles for conducting and evaluating interpretive field studies in information systems. *MIS Quarterly*, *23*(1), 67–93. doi:10.2307/249410

Kristensen, J.K., & Buhler, B. (2001, March). The hidden threat to e-government. Avoiding large government IT failures. *OECD Public Management Policy Brief, Policy Brief no. 8*.

Langefors, B. (1995). *Essays of infology* (B. Dahlbom, ed). Lund: Studentlitteratur.

Lewis, P. J. (1995). Applying soft systems methodology to an SSADM feasibility study: CCTA. *Systems Practice, 8*(3), 337–340. doi:10.1007/BF02250482

Liang, Y., West, D., & Stowell, F. A. (1997). An interpretivist approach to IS definition using object modelling. *Information Systems Journal, 8*, 63–180.

Lynch, T., & Gregor, S. (2004). User participation in decision support systems development: Influencing system outcomes. *European Journal of Information Systems, 13*, 286–301. doi:10.1057/palgrave.ejis.3000512

Lyytinen, K. (1988). Stakeholders, information system failures and soft systems methodology: An assessment. *Journal of Applied Systems Analysis, 15*, 61–81.

Lyytinen, K., & Robey, D. (1999). Leaving failure in information systems development. *Information Systems Journal, 9*, 85–101. doi:10.1046/j.1365-2575.1999.00051.x

MConnell, S. (1996). *Rapid development*. Microsoft Press Books.

MEDIX. (2004). *Medix Survey for the BBC of Doctor's Views about the National Programme for IT*. Retrieved from www.medix-uk.com

Midgley, G. (2000). *Systemic intervention*. New York: Kluwer Academic/ Plenum Publishers.

Mingers, J., & Gill, A. (Eds.). (1997). Multi methodology. Chichester: Wiley.

Morris & Travis. (2001, February 16). Straw Calls Halt to £80m IT System. *The Times*.

Mumford, E. (1995). *Effective requirements analysis and systems design: The ETHICS method*. Basingstoke: Macmillan.

Mumford, E., & Henshall, D. (1979). *A participative approach to computer system design*. London: Associated Business Press.

Myers, M.D. (1994). A disaster for everyone to see: An interpretive analysis of a failed project. *Accounting, management and information technology, 4*(4), 185-210.

Rashid, A., Meder, D., Wiesenberger, J., & Behm, A. (2006). Visual requirement specification in end-user participation. *MERE workshop 2006*. IEEE. Retrieved from http://www.collabawue.de/cms/dokumente/Rashid_et_al_visual_requirement_specification_%20in_end-user_participation.pdf

Savage, A., & Mingers, J. (1996). A framework for linking soft systems methodology (SSM) and Jackson system development (JSD). *Information Systems Journal, 6*, 109–129. doi:10.1111/j.1365-2575.1996.tb00008.x

SSADM (1990, July). *SSADM Version 4, Reference Manuals*. Oxford: NCC, Blackwell.

Stamper, R. (1997). Organisational semiotics. In J.M. Mingers & F.A.Stowell (Eds.), *Information systems: An emerging discipline?* (pp. 267-283). Maidenhead: McGraw Hill.

Stowell, F. A. (1985). Experiences with SSM and data analysis. *Information technology training* (pp. 48-50).

Stowell, F. A. (1991). Towards client-led development of information systems. *Journal of Information Systems, 1*, 173–189. doi:10.1111/j.1365-2575.1991.tb00035.x

Stowell, F. A., & Cooray, S. (2006) Client led information systems creation (CLIC) reality or fantasy? *15th International Conference on Information System Development Methods and Tools and Theory and Practice*, Budapest Hungary. August 30-September 1, 2006.

Stowell, F. A., & Cooray, S. (2006, August 9). Addressing IS failure - Could a "systems" approach be the answer? *Special Symposium on Information Systems Research and Systems Approach*. Baden Baden, Germany.

Stowell, F. A., & West, D. (1994). *Client-led design: A systemic approach to information systems definition*. Maidenhead: McGraw-Hill.

Varzi, A. (2004). Mereology. In E. N. Zalta (Ed.), *The Stanford Encyclopedia of Philosophy (Fall 2004 Edition)*. Retrieved from http://plato.stanford.edu/archives/fall2004/entries/mereology/

Vickers, G. (1981). The poverty of problem solving. *Journal of Applied Systems Analysis, 8*, 15–22.

Vickers, G. (1983). *The art of judgement*. London: Harper Rowe.

Vickers, G. (Ed.). (1991). *Rethinking the Future*. New Brunswich: Transaction Publishers.

Walsham, G. (1991). Organisational metaphors and information systems research. *European Journal of Information Systems, 1*, 83. doi:10.1057/ejis.1991.16

West, D. (1991). *Towards a subjective knowledge elicitation methodology for the development of expert systems*. Unpublished PhD thesis, University of Portsmouth.

West, D., Liang, Y., & Stowell, F. A. (1996). Identifying, selecting and specifying objects in object-oriented analysis: An interpretivist approach. In J.M. Ward (Ed.), *Proceedings of the 1st UKAIS Conference*. Cranfield University.

West, D., & Stowell, F. A. (2000). Models and diagrams and their importance to information systems analysis and design. In D.W. Bustard, P. Kawalek, & M.T. Norris (Eds.), *Systems modelling for business process improvement* (pp. 295-31). London: Artech.

Winter, M. C., Brown, D. H., & Checkland, P. B. (1995). A role for soft systems methodology in information systems development. *European Journal of Information Systems, 4*, 130–142. doi:10.1057/ejis.1995.17

Wood-Harper, A. T. (1985). Research methods in information systems: Using action research. In E. Mumford, R., Hirschheim, G. Fitzgerald & A.T. Wood-Harper (Eds.), *Research methods in information systems*. Amsterdam: Elsevier.

Work, B. (1997). Some reflections on information systems curricular. In J.M. Mingers & F.A. Stowell (Eds), *Information systems: An emerging discipline?* (pp. 329-360). Maidenhead: McGraw Hill.

Chapter 3
On the Study of Complexity in Information Systems

James Courtney
University of Central Florida, USA

Yasmin Merali
Warwick Business School, UK

David Paradice
Florida State University, USA

Eleanor Wynn
Intel Corporation Information Technology, USA

ABSTRACT

This article addresses complexity in information systems. It defines how complexity can be used to inform information systems research, and how some individuals and organizations are using notions of complexity. Some organizations are dealing with technical and physical infrastructure complexity, as well as the application of complexity in specific areas such as supply chain management and network management. Their approaches can be used to address more general organizational issues. The concepts and ideas in this article are relevant to the integration of complexity into information systems research. However, the ideas and concepts in this article are not a litmus test for complexity. We hope only to provide a starting point for information systems researchers to push the boundaries of our understanding of complexity. The article also contains a number of suggested research questions that could be pursued in this area.

INTRODUCTION

This article reflects some thoughts of the editorial review board for the complexity area of this new journal. We are pleased to see a journal introduced whose mission is to truly emphasize a systems approach in the study of information systems and information technology. Within this area of the journal, we will focus on the issue of complexity. We think it is befitting of the area that this article was a group effort. Complexity has many aspects, and we are eager to receive submissions that are truly informed by a systems approach in general and a complexity perspective in particular.

In the sections that follow, we will outline some thoughts on what complexity is, what it can mean when used to inform information systems research, and how some individuals and organizations are using notions of complexity. We provide some comments on how organizations are dealing with technical and physical infrastructure complexity, as well as the application of complexity in specific areas such as supply chain management and network management to more general organizational issues. We offer these pages as a beginning of a dialog on the topic, not as an exhaustive or restrictive set of criteria. We believe the concepts and ideas in this article are relevant to the integration of complexity into information systems research and that, in most cases, some aspect of these topics will be apparent in future submissions. However, the ideas and concepts in this article are not a litmus test for complexity. We expect, and hope, that information systems researchers will push the boundaries of our understanding of complexity through their efforts, which they report in this journal.

COMPLEXITY CONSIDERED

Human life is frequently described as becoming more and more complex, and rightly so. It seems that the terms "complex" or "complexity" appear everywhere. In some part, this is because life really is complex! But this conclusion is also driven by the fact that over the last few decades, we have learned more about the nature of complexity and the role that complexity plays in our lives. Complexity is a feature of all living and natural systems. The approach we speak of has permeated the natural sciences as a way of understanding natural order. However, its application to human systems is to date fragmented.

A recent issue of the journal *Complexity* (Complexity at large, 2007) provides a glimpse of this phenomenon. The first seven pages provide an index into complexity studies from a wide range of disciplines. Here we find news about studies in biodiversity, weather prediction, stem cells, learning, gene therapy, battlefield operations, algorithm development, morality, neural activity in primates, topographical issues in anthropology, organ development, consciousness, robotic reasoning, human moods, and, appropriately, complexity measures. Presumably, the common thread in all of the articles referenced is some notion of complexity.

The focus of this area in the *International Journal of Information Technology and the Systems Approach* (IJITSA) cannot, unfortunately, be so broad. We must limit our scope to topics in information technology. That, however, will not be a serious constraint. The application of complexity theory to information system design, implementation, testing, installation, and maintenance is well within the scope of this IJITSA area. Fundamental issues related to definition, measurement, and application of complexity concepts are valid areas of inquiry. In looking at complexity in information technology, however, we cannot overlook the organizational structures that technology supports, in the image of which information technology is designed.

Information technology underlies and supports a huge part of the operations of modern organizations. By extrapolation, therefore, the role of information systems as they support complex organizational processes is well within our scope. Simon (1996) argued that complexity is a necessary feature of organizations and Huber (2004), in a review of management research, underscores the importance of recognizing that organizational decision making in the future will occur in an environment of growing and increasing complexity.

Indeed, information technology underlies a large part of life itself for young people today. Their lives are entwined in online social networks. They may have a "relationship" with hundreds of other people who they have never met. Their identity may be connected to online activities in ways that no other prior generation has ever experienced. Concepts such as "network" and "relationship" are fundamental to complexity. Investigations of information technology supported communities through a complexity theory lens are certainly within the scope of this area of IJITSA. But complexity and interdependency underlie "normal" social science as well. Granovetter's seminal work (1973, 1983) on "weak ties" in social networks remains a model today in social network theory (Watts, 2003). As well, Lansing's study of Balinese farming reflects a complex systems approach to traditional society (Lansing, 2006).

COMPLEXITY EXPLORED AND DESCRIBED

But let us not get ahead of ourselves, for our understanding of complexity is still evolving. A good starting point for this area is to define, to the extent that we can, what our terms mean. A distinction has to be made between a system having many different parts—complexity of detail and a system of dynamic complexity. In the case of complexity of detail, the system may be treated by categorization, classification, ordering, and systemic-algorithmic approach. A system has dynamic complexity when its parts have multiple possible modes of operation, and each part may

be connected, according to need, to a different part. Dynamic complexity exists when a certain operation results in a series of local consequences and a totally different series of results in other parts of the system (O'Connor & McDermott, 1997). So we see that even constructing a definition is no small task when dealing with the topic of complexity. In fact, we will not be surprised to publish papers in the future that clarify or expand the definitions we offer today.

Complexity is a characteristic that emerges from the relationship(s) of parts that are combined. The idea that the "whole is greater than the sum of the parts" is fundamental to considerations of complexity. Complex describes situations where the condition of complexity emerges from that being considered. Complexity cannot be foreseen from an examination of the constituent parts of a thing. It is a characteristic that emerges only after the parts are entwined in a way that subsequent separation of the parts would destroy the whole. We can see hints of this characteristic even in descriptions of situations that are not focused specifically on complexity. For example, Buckland (1991) writes of information systems that support libraries: "By complexity, we do not simply mean the amount of technical engineering detail, but rather the diversity of elements and relationships involved" (p. 27). He further observes that systems that are provided on a noncommercial basis are necessarily more complex than commercial systems due to the political dimension of their provision. Clearly, this notion of complexity goes well beyond the hardware and software and considers a much broader system in use.

One widely accepted definition of a complex adaptive system comes from Holland (1995), as cited in Clippinger (1999). A complex adaptive system is said to be comprised of aggregation, nonlinearity, flows, diversity, tagging, internal models, and building blocks. What these mean in the context of information systems is the subject of an entire paper. The basic principle is that complex systems contain many interaction variables that interact together to create emergent outcomes. Initial conditions may be local and small in scale, but may gain nonlinearity due to aggregation, and so forth.

Thinking in terms of complexity and some of the concepts and metaphors that are emerging in the study of complexity is a departure from some traditional scientific thinking. Many approaches to understanding that are "scientific" have involved decomposing some thing into its parts so that the parts may be better understood. This reductionism in understanding often sacrifices as much as it gains by losing the richness of context in which the object studied exists. Such an approach provides great knowledge about parts, but little about the whole. It assumes that each part has its own trajectory unaffected by other parts. Moreover, this approach is limited by relying entirely on countable "units" as opposed to analog conditions.

The dynamics of interaction between elements gives rise to a number of features that are difficult to reconcile with some of the tenets of the "classical" IS paradigm and its methods for dealing with complexity (see Merali, 2004, for more detail). Schneider and Somers (2006) identify three "building blocks" of complexity theory: nonlinear dynamics, chaos theory, and adaptation and evolution. By nonlinear dynamics, they refer to dissipative structures that exhibit an inherent instability. These structures may be easily affected by a small change in the environment. They do not tend toward equilibrium. Rather, they go through transitions, typically moving into conditions of greater complexity both quantitatively and qualitatively. This is fundamentally different from the General Systems Theory inclination toward equilibrium.

Chaos is a deterministic process that is progressively unpredictable over time. Chaos theory provides a basis for the study of patterns that initially seem random, but upon closer inspection turn out to be nonrandom. Schneider and Somers

observe that under chaos, a basis of attraction is formed that brings about the nonrandomness. A "strange attractor" accounts for the system's bounded preferences.

Chaos is critical to the process of adaptation and evolution. Schneider and Somers (2006) observe that complex adaptive systems (CAS) reflect an ability to adapt through the emergent characteristic of self-organization. Karakatsios (1990) has developed a simple illustration of how order can emerge from chaos or randomness in such systems. First, a matrix is randomly populated with a binary variable, say zeroes and ones. Let a zero value represent the notion of "off" and a one value represent the notion of "on". Next, the following algorithm is iteratively applied to the matrix:

For each cell in the matrix
If 3 or fewer neighboring cells and this cell are on, set this cell to off.
If 6 or more neighboring cells and this cell are on, set this cell to on.
If 4 neighboring cells are on, turn this cell on.
But if 5 neighboring cells are on, turn this cell off.
Repeat until no changes occur.

Some of us have tried it and found that the matrix typically stabilizes in as few as five or six iterations. However, not all systems have the capacity to adapt. Some systems find small changes in the environment too disruptive to ever evolve to another state. Catastrophe theory studies systems that may transition into one of two states, one stable and the other highly chaotic. Whether a system enters a chaotic state or remains stable may be highly sensitive to initial conditions, so sensitive in fact that it may not be possible to know inputs precisely enough to predict which state the system will enter. This may appear to be troublesome to those attempting to manage organizational systems, but work in the area of complex adaptive systems tells us that systems can adapt and learn and information can be fed back to the control mechanism (management) to keep the organization on a relatively stable path. On the other hand, other systems are too stable and do not react to the environment in any meaningful way. These systems are essentially inert. They continue in their current behavior oblivious to the environment around them. Somewhere between these two extremes are systems that are able to react to the environment in a meaningful way. Kauffman (1995) suggests it is the systems "poised" at the edge of chaos, the ones that are not too stable and not too instable, that have the flexibility to evolve. He theorizes a set of variables that affect the degree of chaos/nonchaos in a system, and hence its ability to evolve. The application of chaos theory to information systems design, implementation, testing, installation, and maintenance is well within the scope of IJITSA.

With the impressive growth of the field of complex systems, the lack of a clear and generally accepted definition of a system's complexity has become a difficulty for many. While complexity is an inherent feature of systems (Frank, 2001), a system may be complex for one observer while not for another. This is not due to subjective observation, but due to the observers' scales of observation. A system that is highly complex on one scale may have low complexity on another scale. For example, the planet Earth is a simple dot—a planet moving along its orbit—as observed on one scale, but its complexity is substantial when viewed in terms of another scale, such as its ecosystem. Thus, complexity cannot be thought of as a single quantity or quality describing a system. It is a property of a system that varies with the scale of observation. Complexity, then, can be defined as the amount of information required to describe a system. In this case, it is a function of scale, and thus a system is to be characterized by a complexity profile (see Bar-Yam, 1997, 2002a, 2002b, 2004).

COMPLEXITY AS A LENS FOR INVESTIGATION

Complexity concepts have been deployed to study complex systems and their dynamics in two ways. The first is through the direct use of complexity concepts and language as sense-making and explanatory devices for complex phenomena in diverse application domains. To capture the "unfolding" of the *emergent* dynamics, we need to have methods that can provide a view of the *dynamics* of the *changing* state in continuous time. The complex systems approach to doing this is by describing state cycles using mathematical models or by running simulations.

The second is through agent-based computational modeling to study the dynamics of complex systems interactions and to reveal emergent structures and patterns of behavior. Agent-based computational modeling has characteristics that are particularly useful for studying socially embedded systems. Typically agent-based models deploy a diversity of agents to represent the constituents of the focal system. The modeler defines the environmental parameters that are of interest as the starting conditions for the particular study. Repeated runs of the model reveal collective states or patterns of behavior as they emerge from the interactions of entities over time. Agent-based models are very well-suited for revealing the dynamics of far-from equilibrium complex systems and have been widely used to study the dynamics of a diversity of social and economic systems.

With the escalation of available computational power, it will be possible to build bigger models. The mathematicians and the natural scientists have a powerful battery of technologies for studying dynamical systems. However, for social systems, the specification of the components for the construction of agent based models is a challenging prospect. The challenge of creating entire mini-economies in silicon is not one of processing power, but one of learning how to build sufficiently realistic agents.

The science of complexity allows us to consider the dynamic properties of systems. It allows us to explore how systems emerge and adapt. When viewed as a complex adaptive system, it provides us a mechanism for dealing with both the technical and the social aspects of systems. We have new metaphors for articulating how IS are used and how they evolve. We move from concepts embedded in an assumption of stable hierarchies to ideas embedded in an assumption of networks of dynamic relationships. With this, we move closer to a unified view of IS and management.

Simon (1996) writes: "Roughly, by a complex system I mean one made up of a large number of parts that have many interactions" (p. 183). This simple definition can be readily applied to organizations and their information systems. Thus, an organization is a complex system if it has many units (departments, for example) and there are many interactions among units. A complex information system is one that has many elements (programs, modules, objects, relationships, attributes, databases, etc.) that interact in many ways.

At the most fundamental level, technological developments have the potential to increase connectivity (between people, applications, and devices), capacity for distributed storage and processing of data, and reach and range of information transmission and rate (speed and volume) of information transmission. The realization of these affordances has given rise to the emergence of new network forms of organization embodying complex, distributed network structures, with processes, information, and expertise shared across organizational and national boundaries. The network form of organizing is thus a signature of the Internet-enabled transformation of economics and society. Merali (2004, 2005) suggests conceptualizing the networked world as a kind of global distributed information system.

Yet, this only begins to get at the complexity of complex systems. Systems have boundaries that separate what is in the system from what is outside—in the environment. Environments themselves may be complex, and the system, the organization, or the information system may interact with the environment in many ways. Moreover the interactions themselves may be complex.

An information system that exists with a particular organization (ignoring inter-organizational systems, for the moment) has the organization as its environment. If the organization and its information requirements are stable, then the information system itself has relatively little need to change, other than to keep up with changes in relevant hardware and software technologies (which may be no mean feat in and of itself).

However, it seems to be the norm today for organizations and their environments to be in a state of constant change. Organizations must adapt to environmental changes in order to survive, not to mention thrive. The same can be said for information systems in organizations. Organizations may even rely upon their information systems in order to understand, analyze, and adapt to such changes. Thus, we say that organizations and information systems are one form of complex adaptive systems, a topic of great interest today among those interested in systems theory.

Simon (1996) describes three time periods in which there were bursts of interest in studying complex systems. The first followed World War I and resulted in the definition of "holism" and an interest in Gestalts, and a rejection of reductionism. The second followed World War II and involved the development of general systems theory, cybernetics, and the study of feedback control mechanisms. In one perspective, in the second era, the information system of an organization is viewed as a feedback mechanism that helps managers guide the enterprise towards its goals.

We are now in a third era. The foundation had been laid for the development of the concept of complex adaptive systems, elements of which include emergence, catastrophe theory, chaos theory, genetic algorithms, and cellular automata. Complex adaptive systems receive sensory information, energy, and other inputs from the environment, process it (perhaps using a schema in the form of an updatable rule-base), output actions that affect the environment, and feedback control information to manage system behavior as learning occurs (update the schema).

Complex adaptive systems are reminiscent of the concepts of organizational learning and knowledge management, which have been viewed from the perspectives of Churchman's (1973) inquiring systems which create knowledge or learn and feed that knowledge back into an organizational knowledge base (Courtney, 2001; Courtney, Croasdell, & Paradice, 1998; Hall & Paradice, 2005; Hall, Paradice, & Courtney, 2003). Mason and Mitroff (1973), who studied under Churchman as he was developing the idea of applying general systems theory to the philosophy of inquiry, introduced this work into the IS literature early on, and it has ultimately had great influence on systems thinking in IS research.

Complexity in this context is in the form of "wicked" problems (Churchman, 1967; Rittel & Weber, 1973). In sharp contrast to the well-formulated but erratically behaving deterministic models found in chaos theory, in a wicked situation, "formulating the problem *is* the problem," as Rittel and Weber put it (1973, p. 157, emphasis theirs). The question that arises here is whether problems in management domains that involve human behavior are of such a different character that elements of complexity theory and chaos may not apply. This is clearly an open question and one that can only be addressed through additional research.

WHAT DOES THIS MEAN FOR IS?

There is no question that information systems in organizations, as they have been defined, are complex. The very basis of information systems, the underlying technologies, programs, machine language, and so forth, are inherently ways of dealing with complexities of calculation and the complexity of the use contexts, in this case, the organization. What has not been included in the description of information systems as "systems" are several key notions from complex adaptive systems and current compute models that directly or indirectly reflect complex systems modeling. These include machine learning, Bayes nets, inferencing algorithms, complex calculations for science applications, visualization, virtualization schemes, network traffic modeling, social networking software, and diverse other areas.

Organizational analysis as we know it, even in its evolution to be inclusive of multiple paradigms of research, has failed to acknowledge that organizations are inherently complex. Organizations defy simplification, and the only way to deal with this fact is to embrace and manage complexity. Structuration theory and actor network theories applied to organizations both begin to cope with this reality that the whole is greater than the sum of the parts and that outcomes are emergent.

While visionary management authors like Wheatley (1992, 2006), Weick and Sutcliffe (2001), Axelrod and Cohen (2000), and others have written directly on the topic, the application of their thinking is not evident in the ordinary management situation. There is some adoption on the edges in areas where complexity is defined by the behavior of objects, like supply chain management, RFID tagging and tracking, and network traffic. However, these applications often occur without recognition of the greater framework they represent. Further, attempts to generalize from these technically specific domains to the overall behavior of the organization have not been accepted easily.

Figure 1.

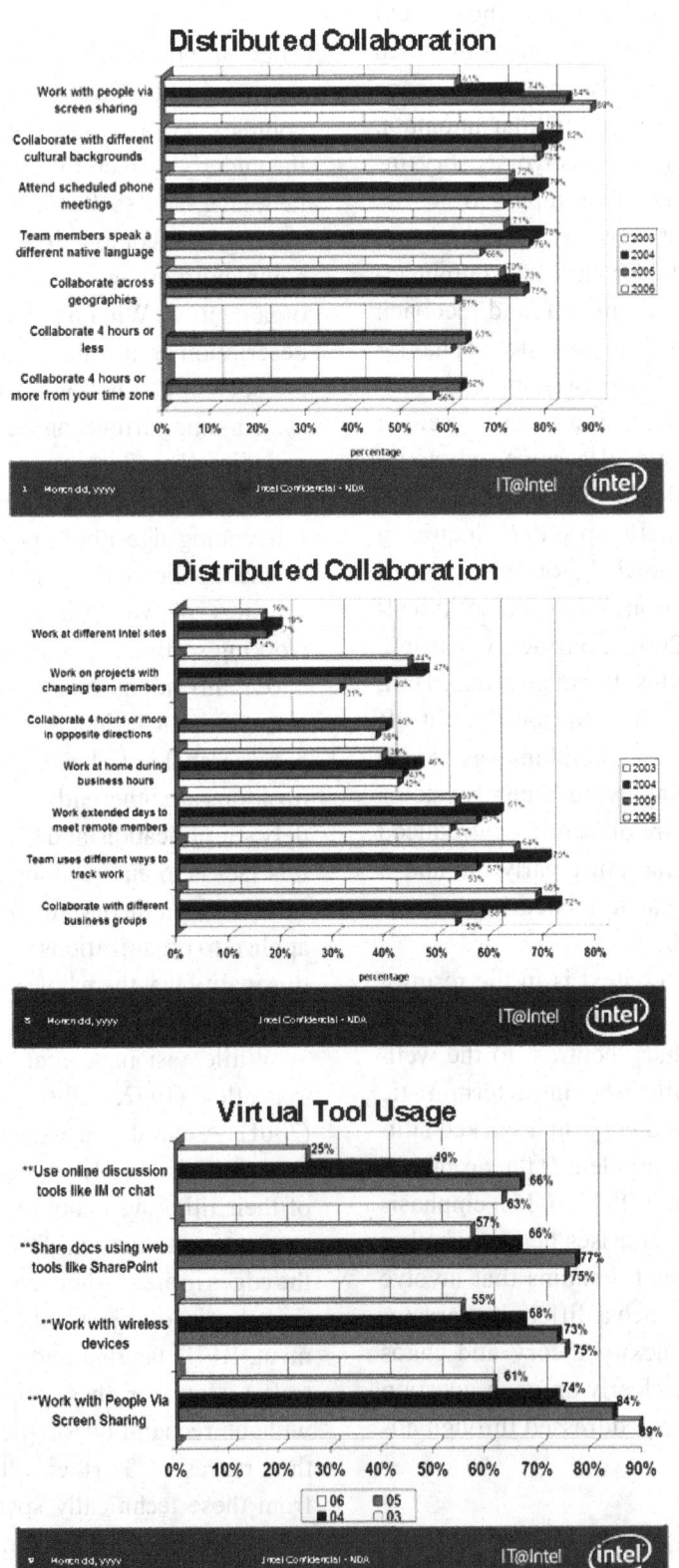

What is missing from the computational paradigms that do use complexity in their mode of operation is simply the recognition that this is so. It is as if connectionists have entered into the world of dealing with complexity as a "natural environment", like air or water, which ceases to be noticed.

At this point in history, the organization and its information systems are inextricable. There is no turning back, as there may have been as late as the 1970s when paper systems were still an option. Aside from back-ups for legal purposes, all large organizations are fully committed to their information systems environments as infrastructure. Indeed, technical infrastructure encroaches on physical infrastructure with outsourcing, telecommuting, globalization of work, and other major trends. As information systems facilitate more and more networked operations and distributed work, as enterprise applications emerge that serve one and all, the very functioning of the organization, especially a large one, becomes impossible without an information system. Studies of Intel's workforce find that on six dimensions of time, space, organizational affiliation, software tools, culture, and number of projects, the workforce can be said to be operating approximately 2/3 in "virtual mode"—across time, space, culture, multiple teams, and so forth (Wynn & Graves, 2007).

This means that the workforce coordinates itself mostly on the network. If the medium of coordination, action, and work production is primarily a network, it more and more resembles a rapidly changing complex system that has the possibility of being self-organizing in a very positive way. Indeed, that is the case. But without the recognition that virtuality equates with greater capacity for self-organization (and that self-organization is adaptive), then this enormous potential will be not only underutilized, but at times interfered with, sub-optimized, and cut off from its latent functionality.

The interesting thing is that the science is there; the systems are there; the computational capacity is there. All that is lacking is the consciousness to apply them. Some notable exceptions exist, however. The Department of Defense Command Control Research Project has a number of publications that apply a self-organizing system concept to hierarchical command and control systems. Boeing PhantomWorks (Wiebe, Compton, & Garvey, 2006) has drawn out the Command and Control Research Program (CCRP) scheme into a large system dynamic model. In short, there is no lack of research and conceptual material.

But getting this across to people responsible for the stock price and cost containment of a very large organization is no simple matter. It seems risky, even though it is likely much less risky than acting as if the world were a stable place and a linear progress model will provide a safe approach to operations. As a defense strategy organization, CCRP recognizes acutely that they are dealing with volatile, rapidly changing, network-based asymmetrical conflicts that also have great potential for reaching critical mass and nonlinear effects so large they could overwhelm conventional forces, or at least those using conventional methods.

The large organization lives in very much the same world as the military organization, only effects are slower to take hold and direct loss of life is not normally a risk. However, there are environmental instabilities in global politics, competition and licensing, labor forces, currency and liquidity, stock market fluctuations, energy costs, supply chains that reach across the globe, transportation, changing demand, and of course, competitors. All of these elements together, and others not noted, comprise a highly complex and turbulent environment. That is the external environment. The internal environment of the organization and its information system can create the adequate response to the external environment. For that to happen, both workforce and information systems need to be seen as comprising an adap-

Table 1. Some research questions related to complexity and information systems

Does chaos theory really apply to the IS domain? IS seems to have characteristics more resembling those of wicked problems where formulating the problem is the problem. Chaos consists of well-specified models whose behavior gets less predictable over time because of nonlinearities. The two do not seem to be isomorphic.
How does one go about modeling agents in IS problems? Modeling computer systems may not be so difficult, but modeling human actors seems to be problematic. How, for example, do you model changing schemata as learning occurs? Human agents have almost limitless attributes. What characteristics of human agents are important to model? How do you model the exchange of knowledge among agents?
As organizations and IS become more intertwined, it becomes increasingly important that the IS be reliable. Does chaos theory make sense here, in that the organization's behavior may be unpredictable if the IS fails?
How does management's attitude about importing innovation from the environment affect the IS function? Does sharing or importing innovations help the organization fit the environment?
How do we define and measure organizational and information systems complexity? How do we test complex models of organizations and information systems? We need to measure to be able to test.
Is it possible to organize and manage so that the organization and its information systems co-evolve and emerge together? Can prototyping help support co-evolution?
From Rouse (2007, pp. 16-17): What architectures underlie the physical, behavioral, and social phenomena of interest? How are architectures a means to achieve desired system characteristics? How can architectures enable resilient, adaptive, agile, and evolvable systems? How can and should one analytically and empirically evaluate and assess architectures prior to and subsequent to development and deployment? What is the nature of fundamental limits of information, knowledge, model formulation, observability, controllability, scalability, and so on? How can decision support mechanisms be developed for multistakeholder, multi-objective decisions?

tive resource. This is where explicit recognition of complexity can make the difference.

A recent special issue of the journal *Information Technology & People* (Jacucci, Hanseth, & Lyytinen, 2006) took a first step in applying this approach to what we know about information systems research (Benbya & McElvey, 2006; Kim & Kaplan, 2006; Moser & Law, 2006). However, a journal that is regularly dedicated to this theme is needed both to publish available research and to foster further research on this important topic.

We offer a set of possible research questions in Table 1. This list is by no means exhaustive, and we welcome work on these and others that our audience may conceive.

CONCLUSION

Few would argue that complexity is not inherent in living today. As networked information environments become more integrated into both our social and our working lives, the number of relationships

with others may grow, and the relationships we have with them may become more complex. We exist, along with our relationships, in an environment of equal or greater complexity.

We strive to understand what complexity means and what it implies for us. We believe that a better understanding of complexity will give us a better ability to function more effectively and achieve our goals, both personal and professional. We welcome research that will broaden our understanding of complexity, help us understand how to embrace a notion such as emergence in complexity, show us how to use complexity to inform our social and our work lives, leverage the self-organizing capabilities of complex adaptive systems to achieve personal and organizational goals, and apply metaphors from chaos and other complexity-oriented theories to better describe and understand our world. We look forward to publishing the best work in these areas and in others that will surely emerge.

REFERENCES

Alberts, D., & Hayes, R. (2005). *Power to the edge: Command and control in the information age* (CCRP Publication Series). CCRP.

Atkinson, S., & Moffat, J. (2005). *The agile organization: From informal networks to complex effects and agility* (CCRP Information Age Transformation Series). CCRP.

Axelrod, R., & Cohen, M. (2000). *Harnessing complexity: Organizational implications of a scientific frontier*. New York: Basic Books.

Bar-Yam, Y. (1997). *Dynamics of complex systems*. Reading, MA: Perseus Press.

Bar-Yam, Y. (2002a). Complexity rising: From human beings to human civilization, a complexity profile. In *Encyclopedia of Life Support Systems (EOLSS)*. Oxford, UK: UNESCO, EOLSS Publishers.

Bar-Yam, Y. (2002b). General features of complex systems. In *Encyclopedia of Life Support Systems (EOLSS)*. Oxford, UK: UNESCO, EOLSS Publishers.

Bar-Yam, Y. (2004). Multiscale variety in complex systems. *Complexity, 9*, 37-45.

Benbya, H., & McKelvey, H. (2006). Toward a complexity theory of information systems development. *Information Technology & People, 19*(1), 12-34.

Buckland, M. (1991). *Information and information systems*. New York: Praeger Publishers.

Churchman, C.W. (1967). Wicked problems. *Management Science, 4*(14), B141-B142.

Churchman, C.W. (1971). *The design of inquiring systems: Basic concepts of systems and organization*. New York: Basic Books.

Clippinger, J.H. (1999). *The biology of business: Decoding the natural laws of enterprise*. Jossey-Bass.

Complexity at large. (2007). *Complexity, 12*(3), 3-9.

Courtney, J.F. (2001). Decision making and knowledge management in inquiring organizations: A new decision-making paradigm for DSS [Special issue]. *Decision Support Systems, 31*(1), 17-38.

Courtney, J.F., Croasdell, D.T., & Paradice, D.B. (1998). Inquiring organizations. *Australian Journal of Information Systems, 6*(1), 3-15. Retrieved July 10, 2007, from http://www.bus.ucf.edu/jcourtney/FIS/fis.htm

Frank, M. (2001). Engineering systems thinking: A multifunctional definition. *Systemic Practice and Action Research, 14*(3), 361-379.

Granovetter, M. (1973). The strength of weak ties. *American Journal of Sociology, 78*, 6.

Granovetter, M. (1983). The strength of weak ties: A network theory revisited. *Sociological Theory, 1*.

Hall, D.J., & Paradice, D.B. (2005). Philosophical foundations for a learning-oriented knowledge management system for decision support. *Decision Support Systems, 39*(3), 445-461.

Hall, D.J., Paradice, D.B., & Courtney, J.F. (2003). Building a theoretical foundation for a learning-oriented knowledge management system. *Journal of Information Technology Theory and Applications, 5*(2), 63-89.

Holland, J.H. (1995). *Hidden order: How adaptation builds complexity*. Helix Books.

Huber, G. (2004). *The necessary nature of future firms*. Thousand Oaks, CA: Sage Publications.

Jacucci, E., Hanseth, O., & Lyytinen, K. (2006). Introduction: Taking complexity seriously in IS research. *Information Technology & People, 19*(1), 5-11.

Karakatsios, K.Z. (1990). *Casim's user's guide.* Nicosia, CA: Algorithmic Arts.

Kauffman, S. (1995). *At home in the universe: The search for laws of self-organization and complexity.* Oxford University Press.

Kim, R., & Kaplan, S. (2006). Interpreting socio-technical co-evolution: Applying complex adaptive systems to IS engagement. *Information Technology & People, 19*(1), 35-54.

Lansing, S. (2006). *Perfect order: Recognizing complexity in Bali.* Princeton University Press.

Lissack, M.R. (1999). Complexity: The science, its vocabulary, and its relation to organizations. *Emergence, 1*(1), 110-126.

Lissack, M.R., & Roos, J. (2000). *The next common sense: The e-managers guide to mastering complexity.* London: Nicholas Brealey Publishing.

Merali, Y. (2004). Complexity and information systems. In J. Mingers, & L. Willcocks (Eds.), *Social theory and philosophy of information systems* (pp. 407-446). London: Wiley.

Merali, Y. (2005, July). Complexity science and conceptualisation in the Internet enabled world. Paper presented at the *21st Colloquium of the European Group for Organisational Studies,* Berlin, Germany.

Moser, I., & Law, J. (2006). Fluids or flows? Information and qualculation in medical practice. *Information Technology & People, 19*(1), 55-73.

O'Connor, J., & McDermott, I. (1997). *The art of systems thinking.* San Francisco: Thorsons.

Rittel, H.W.J., & Webber, M.M. (1973). Dilemmas in a general theory of planning. *Policy Sciences, 4,* 155-169.

Rouse, W.B. (2007). *Complex engineered, organizational and natural systems: Issues underlying the complexity of systems and fundamental research needed to address these issues* (Report submitted to the Engineering Directorate, National Science Foundation, Washington, DC).

Schneider, M., & Somers, M. (2006). Organizations as complex adaptive systems: Implications of complexity theory for leadership research. *The Leadership Quarterly, 17,* 351-365.

Simon, H.A. (1996). *The sciences of the artificial.* Boston: The MIT Press.

Watts, D. (2003). Six degrees: The science of a connected age. New York: W.W. Norton & Co.

Webster. (1986). *Webster's ninth new collegiate dictionary.* Springfield, MA: Merriam-Webster, Inc.

Weick, K., & Sutcliffe, K. (2001). *Managing the unexpected: Assuring high performance in an age of complexity.* Jossey-Bass Publishers.

Wheatley, M. (1992, 2006). *Leadership and the new science: Discovering order in an age of chaos.* Berrett-Koehler Publishers.

Wiebe, R., Compton, D., & Garvey, D. (2006, June). A system dynamics treatment of the essential tension between C2 and self-synchronization. In *Proceedings of the International Conference on Complex Systems,* New England Complex Systems Institute, Boston, Massachusetts.

Wynn, E., & Graves, S. (2007). *Tracking the virtual organization* (Working paper). Intel Corporation.

This work was previously published in International Journal of Information Technologies and Systems Approach, Vol. 1, Issue 1, edited by D. Paradice; M. Mora, pp. 37-48, copyright 2008 by IGI Publishing (an imprint of IGI Global).

Chapter 4
Importance of Systems Engineering in the Development of Information Systems

Miroljub Kljajić
University of Maribor, Slovenia

John V. Farr
Stevens Institute of Technology, USA

ABSTRACT

The interrelationship between Information Systems (IS), Systems Engineering (SE), and Information Systems Development (ISD) is discussed from past, present, and future perspectives. While SE is relatively a well-established discipline based upon an interdisciplinary approach to enable the realization of successful systems, ISD has evolved to a variant of SE applied mainly for the development of IS. Given the growth in complexity, need for enterprise wide solutions, and the cost and schedule overruns that have be common place for modern software centric systems, well-established tools, techniques, and processes are needed for the development of good IS. Similarities and differences of methodology as well as their evolution and perspectives are also presented herein. We found a positive trend in the evolution of research methodology and published material in SE and its use in IS.

INTRODUCTION

Our task for this chapter is to analyze the significance of IS for SE and its methods as well as the relevance of systems approach (SA) and SE for the development of IS. It is impossible to utilize SE and SA methods without proper IS, and conversely you cannot develop cost effective and efficient IS without disciplined SE or SA. All of them are mutually dependent. However, there has historically

DOI: 10.4018/978-1-60566-976-2.ch004

been a difference in the SE methods used for the IS development and other socio-technical problems. These differences in tools, techniques, and processes were caused by user experience, traditional domain stovepipes. Fortunately, established domain centric stovepipe practices will ultimately converge because of economic forces. Our aim in this article is to generalize and highlight these different methodologies and their relevance for research of the vast variety of processes where IS and system's methodologies are essential conditions. Evidently, IS are an essential part of any real process: biological, organizational and technical that enable data collection, processing, storing and retrieving, while SE and SA are general methodology for system "construction" and deployment. Although IS has its' own uniqueness and logic of functioning it is always part of the systems; to enable communications and are inseparable from systems itself. In general, IS has properties and function developed trough evolution as a part of the organizational systems as the results of the social relationship, technology development, and methodology. Presently, the advances in net information technology and software science have been tremendously impact on the organizations structure and functioning as well as implementation of IS. Various kinds of IS, such as Enterprises Resource Planning systems, Decision Support Systems, Group Decision Support, and Knowledge Management Systems, have become recognized as indispensable in enabling organizations to survive. In order to cope with huge variety of IS produced by the Internet, the www ontology of IS was developed. A formal concepts of the sets of objects of interest and the relationships that hold among them. A conceptualization is an abstract, simplified view of the world that we wish to represent for some purpose (Gruber, 1995). This definition is in turn similar to yet more specific to the context of domain of interest. This concept is similar to the definition of abstract general systems (Mesarović and Takahara, 1989) and has origin in it and philosophy yet more pragmatic. In the article we will try to highlight the relation between these two important phenomena: IS and their methodologies.

The world is rapidly changing with outsourcing, globalization, network centricity, and complexity being the mantra for 21st century engineering. Services dominant the economies of most countries (see Figure 1) with IS being the key business enabler. Also, as shown in Figure 2, the operating environment is also changing. Gone are the days of an engineer working at the component level in a stovepipe organization. In fact, a more relevant definition of all engineering

should be "in today's global business environment, engineers integrate hardware, software, people, and interfaces and to produce economically viable and innovative applications while ensuring that all pieces of the system are working together." (Farr, 2007)

As shown by Figures 1 and 2, one can conclude how important are IS and methodologies to handle IS in order to cope with complex society. The world crisis just started could be caused by structure change in world economy as shown in Figure 1 and our limited knowledge to cop with new reality. In the rest of the paper we will try to highlight systematically the role of methodology (SE and SA) for development of IS and vice versa.

The word "system" can broadly be defined as an integrated set of elements that accomplish a defined objective (INCOSE 2004). Simply put, a system is a whole consisting of parts and being more than sum of its parts. That was an axiom of ancient philosophers, which accurately anticipated the contemporary definition of systems. Only order, structure, and behavior were added to the meaning of systems in cybernetics and general systems theory. Complex systems are usually understood intuitively, as a phenomena consisting of a large number of elements organized in a multi-level hierarchical structure where elements themselves could represent systems (Mesarović and Takahara, 1989). The word "complex" is used only to indicate that the problem treated here cannot be expressed only in hard (quantitative) relations and those many relevant characteristics are qualitative. With a conception of complex systems, we think about a system within which a main role is played by a complexity of control and information processes. We also now understand that for a system to operate at optimum efficiency that the components or sub-systems must operate sub-optimally. Undoubtedly, existing SE methodology is applied to small, medium, large scale and complex process but with complex systems, SE moves to a SA methodology. Fortunately, these

Figure 1. Growth of services for the U.S. economy (Michigan Tech, 2008)

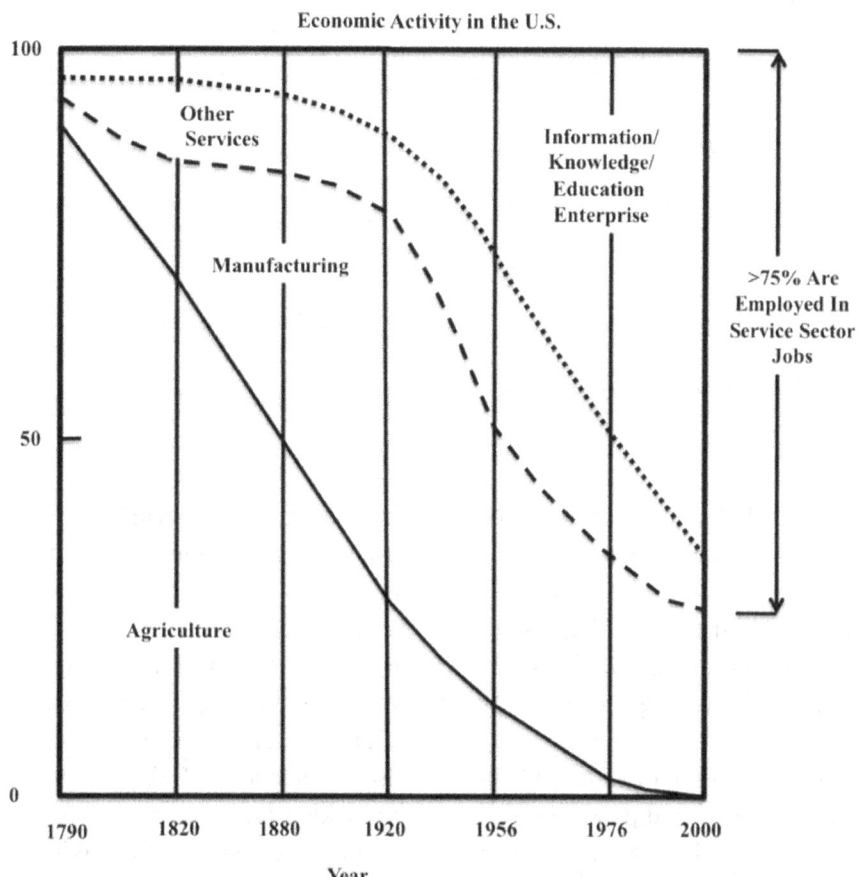

same SE techniques that have been successfully applied to complex systems are also being applied to systems of systems (SoS) and large enterprises with limited success.

A paradigm of SE has played important role in the dealing with different aspect of human activity. In the beginning, it was based on empirical knowledge and heuristics in the building of human-made objects like pyramids, fortifications, tools, etc. Industrial production and scientific organization began with Henry Ford and F. W. Taylor, who contributed to work specialization, planning, and control. The result was mass production, standardization, and higher productivity at defined quality levels. This period of world-view in science and production was known as

the Machine Age and was marked by its use of classical analysis for problem solving (Akoff, 1999). Systems engineering was subsequently born in the telecommunications industry of the 1940s and nurtured by the challenges of World War II, when project managers and chief engineers with the assistance of key subsystem leads, oversaw the development of aircraft, ships, etc. The post-war creation of more complex systems mainly in defense and communication systems led to the formalization of SE as an engineering discipline. Its relevance became indispensable after WWII, when technical solutions and organizational interactions become highly complex. A landmark for systems philosophy was founded in General Systems Theory (Bertalanffy, 1968) and

Figure 2. The current environment for developing new products and services (modified from Stevens Institute of Technology, 2007)

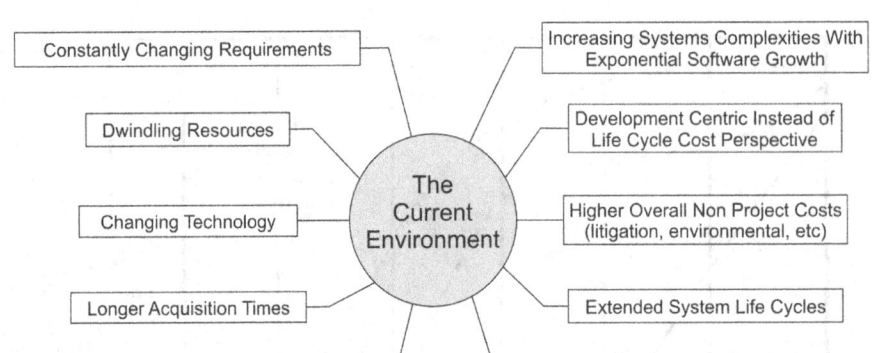

Cybernetics (Wiener, 1948) and continued to be adapted to the different contexts and tools, taking new meaning and significance with successive ages as indicated by Akoff (1999). The history of civilization development and growth is closely related to the history of working methodology and organization.

Although modern definitions of SE are of later date, there are several books and papers on this topic, which discuss SE and the systems engineer in great detail; these include (Thome, 1993; Martin, 1996; Sage, 2000). Some examples of modern SE definitions are shown in Table 1. These definitions are derived from the root of two words: engine and systems. An engine is a device consisting of different parts. Engineers are those who construct engines according to their knowledge of natural sciences and economic law. "Systems" means a whole consisting of parts. Systems engineering is a methodology of how to "construct" purposeful systems in their surroundings. It is obvious that the meaning of SE has changed with the complexity of the man-made systems and social changes in its surroundings. As a comprehensive definition of SE as methodology, we took Thome's (1993): "Systems engineering consists of applying a System Approach to the engineering of systems." Figure 3 shows a graphical representation of what might constitute a SA to the engineering of a system. Its

domain is the engineering of solutions to systems problems independent of employing a certain technology for realizing systems functions and properties. In this definition SE was understood as composition of SA and engineering of solution for systems problems independent of type of process. However, a SA could be considered also as enhanced SE for complex problem solving, taking into accounts not only stakeholders' requirements but also the environments requirements. That means considering a complex system from all relevant points of view in its environment during developing, maintaining and functioning. The similarity and difference of methodology titles were discussed in (Lazanski and Kljajić, 2006) where the triadic principle of Peirce (1998) was used to explain the meaning of methodology in a context of problem solving. According to Pierce (1998), meaning is a triadic relation between a sign, an object, and an interpretant. A general meaning can always be found in genuine triadic relations, but can never be found in degenerate triadic relations. Only a subject gives real value and meaning to the model and methodology in a frame of a context of the problem. Basic principles and requirements for SE and its translation to practice as well as for SE education are described in Martin (1996) and Sage (2000).

Table 1. Standard definitions of SE

International Council on Systems Engineering (INCOSE, 2004)	Systems engineering is an interdisciplinary approach and means to enable the realization of successful systems.
Military Standard on Engineering Management 499A (USAF, 1974)	The application of scientific and engineering efforts to: (1) transform an operational need into a description of system performance parameters and a system configuration through the use of an iterative process of definition, synthesis, analysis, design, test, and evaluation; (2) integrate related technical parameters and ensure compatibility of all related, functional, and program interfaces in a manner that optimizes the total system definition and design; (3) integrate reliability, maintainability, safety, survivability, human, and other such factors into the total technical engineering effort to meet cost, schedule, and technical performance objectives.
Department of Defense (DoD, 2004)	Systems engineering is an interdisciplinary approach or a structured, disciplined, and documented technical effort to simultaneously design and develop systems products and processes to satisfy the needs of the customer. Systems engineering transforms needed operational capabilities into an integrated system design through concurrent consideration of *all* Lifecycle needs
NASA (NASA, 1995)	Systems engineering is a robust approach to the design, creation, and operation of systems.

Figure 3. Relationship between the traditional SE functions (center column), cost (left column), and supportability and logistics (right column) (Stevens Institute of Technology, 2007)

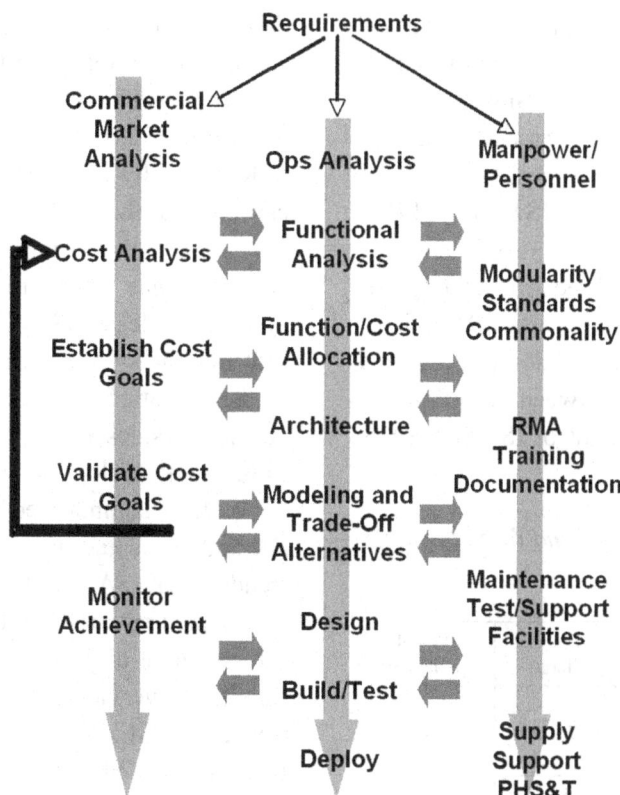

The relevance and growth of SE and its variation for complex problem solving and its management could be clarified by a number of articles published in the last decade. Our research results from the Engineering Village (Engineering Village, 2007) and the Web of Science (WoS Expanded, 2007) are shown in Table 2. The Engineering Village database is large and represents all articles from conference proceedings, journals from Journal Citation Records (JCR), secondary databases and doctoral dissertations. Different results were observed from the Web of Science (WOS) database since it covers only journals from JCR.

Although SE and SA represent just methodologies and IS real systems, we compare their portions in order to see how frequently IS has been used besides established methodology. As shown in Table 2, the sum of relative frequency of SA and SE is 77% and IS =23% from the Engineering Village database. Since our primary interest is IS and its associations with research methodologies like ISD, SE, SA, Simulation and its combination, within the database n1= 442,767 articles were found:

ISD = ISD&IS=29%, IS&SE= 47%, IS&SA = 18%, IS&SE & SA=6%, and for IS&SE&Simulation only three articles. Regarding ISD or ISD&IS, the answer was 29% in both cases. It means that ISD methodology was used exclusively with IS and is between IS&SE=47%, and IS&SA=18%. The result of IS&SE&SA is

Table 2. Engineering Village and Web of Science subject searches

	Engineering Village	Web of Science
Articles in Database (1969 to 2007) Keywords IS, SE, and SA	1,926,146	22,615
% SE	58	5
% IS	23	80
% SA	19	15

5.8% while IS&SE&Simulation had only three articles.

Because our primary interest is the use of research methodology like: ISD, SE, SA and its combination with IS from the WoS database for n = 18,055 articles, the following result has been obtained: IS&SA = 0.4%, IS&SE = 0.4%, ISD = IS&ISD = 2%, IS&Simulation = 3.5% and for IS&SE&SA and IS&SE&Simulation only two and five articles were find respectively. From these findings, only 6% of articles use established methodology with regard to IS all other 94% articles do not use labels like ISD, SE, and SA or similar. This finding does not mean that methodology is not used at all but perhaps not explicitly.

This research clearly shows two different trends in publications. In the broader Engineering Village database, the majority of publication contents keywords is by SE at 58% and then IS at 23% followed by SA at 19%, while from WOS database; from the Journal references listed in the Journal of Citation Reports (JCR) most dominant Keywords is IS at 80% followed with SA at 15% and SE at 5%. Keeping in mind that articles in JCR is usually represents finished research, we can analyze the WOS database where IS=80% (n1=18,055) as meaning that these articles considered topics where information systems play dominant role like: Production IS, management information systems, Medical IS, Educational IS etc. SA at 15% could mean that articles also consider besides SA methodology another process or just methodology with SA dominant.

We have summarized key concepts for SE and SA and presented an overview of the research trends in SE, SA, and IS as reflected by the two established research databases. The remainder of this chapter is divided into three sections: Section 2 contains overview of some theories and methods relevant for SE; Section 3 deals with the anticipative concept of SE and a simulation approach to SE; and finally, Section 4 provides some concluding remarks and ideas for further research.

OVERVIEW OF SOME THEORIES AND METHODS RELEVANT FOR SE

Even though methodologies do not belong directly to systems theory, they are its products in searching for the means of complex problem solving. The diversity of systems phenomena created a variety of concepts and theories to describe them. The description of a system depends on the describer's point of view, interests, culture and time (Koizumi, 1993). Experience, learning, knowledge and motives, influence an individual's consciousness and consequently society's awareness, which results in a certain choice of action. A social reality, which is a consequence of a compromise, is an organization and is measurable by its goals and means for achieving these goals. This is an objective matter, although there is a problem of measurability, scale ordering and subjective understanding of an individual and his value. The objective exists in time and place and not separated from them, even though it is only partly described. Awareness that partial description is not wholeness and the fact that we can more or less get close to this wholeness requires a SA (Kljajić, 1994). In this way Miller's Living Systems (Miller, 1978) represents a comparative analogy among the structures, functioning and processing of energy and information of different living phenomena. A comparative scheme is just an analogy without the power for deeper understanding of the phenomena. Even though we can find some useful similarities among an organism, an organ, and an organization, we can say that these are actually different systems with regards to their behavior. This approach can be partly useful as an analogy with organizational science. Organizational systems are complex goal-oriented systems (Ackoff, 1999) designed to achieve certain purposes. As such organization is a function of the past; present and future state and represents an anticipatory system. Therefore, the basic principles of systems development are essentially anticipatory as consequences of decision-making

based on anticipated and feedback information. To estimate the consequence of decision-making, the decision maker needs a model of the system and the environment. System Dynamics (Forester, 1961), and System Thinking (Senge, 1994) are equivalent and can be unified within the systems concept (Kljajić, 1994). Some relevant paradigms for analysis were described in (Rosenhead, 1989; Flood and Carson, 1988), including: soft systems analysis, hard systems analysis, critical thinking, strategic options development and analysis. It is not surprising that a number of works have been dedicated to these topics. There are almost no differences among them; different names are a result of the complex context and the author's point of view. As Forrester (1994) states, all these titles have one and only one aim "to emphasize that this is the wish following an integral research of complex phenomena through its feedback connections." It is the eternal wish of a human being for the complete yet never-ending description of his surroundings. The cybernetics and general system theory expose these wishes even more. We can accept a SA or systems point of view as being proper. Even more: thinking and rethinking is the mental process of a human being. It can be true or false in relation to a matter of thinking. If taken terminologically, it is a metaphor, with which we would like to expose a working method for mathematics (mathematical thinking) or philosophical method (philosophical thinking). This is the reason that the basic concept of General System Theory (GTS) was the interdisciplinary work for complex problem solving. It is obvious that we cannot find an actual solution with just formal methods. Abstract matters need concrete ones and vice versa. The philosophy of SA is typical for complex problem solving and can be expressed with two words: interdisciplinary methodology + context problem solving = systems approach. Its openness and transparency satisfy Popper's requirements for provability (Popper, 1973: p. 131): "Within a methodology we do not define only a problem and search for a solution, but also set conditions for verification of solutions and validation of alternatives." All complex phenomena are systems in their essence, whose methodology derives from Cybernetics and GST.

In order to illustrate the interconnection of the above theory, ISD methodology relevant for IS and its evolution the articles (Xu, 2000; Zhu, 2000; Jan and Tsai, 2002) will be analyzed. As a good example of Miller's living systems analogy and Ackoff's (1999) lucid systems classification in the paper (Jan and Tsai, 2002) a three-stage ISD has analyzed: methodology for the IS as machine, methodology for the IS as part of organization as well as IS as the part of social system. The study investigates the changing roles and missions of IS for the three stages and explores the evolution of ISD strategies. In early phase of IS development for the organization as machine, IS was developed by information specialists. The role of the IS is to support transaction processing systems and operational control. In the organic stage, the IS role is to support transaction processing systems at all organization and all levels of management. In the social stage, the IS role is to support organization as a social system and its mission should take account of organization as part of larger systems even ecological ones. Similarly in (Xu, 2000) author reviewing the contribution of systems science to information systems research stressed how concepts and findings in systems science have to be applied, extended and refined in IS research.

Zhu (2000) presents in an article titled "WSR: A System Approach for ISD a Systems-Based Approach", which is derived from traditional Oriental thinking and contemporary practice in that social-cultural setting. As a philosophical framework, wuli-shili-renli (WSR), contends four principles; seeing ISD as a differentiable whole, treating ISD methods as complementary opposites, conducting ISD as a spiral bubble-management process, and searching for ISD methodologies in a form not independent from that of general management approaches. In this way, ISD researchers should develop methodologies in a form familiar to users and, at the same time, incorporate the best aspects of various methods, which nicely coincide with Ackkof's (1999) classifications. Samaras and Horst's (2005) SE perspectives on the human-centered design of health IS are described. With human-centered design, authors require that SE method to take into account human ergonomics (although cognitive aspects would be better), which in other words mean systems approach. An example of an IS for community nursing is presented in (Šušteršič et al. 2002). The goal of IS, in this case, is to reduce the workload with modern information and communication technologies and to improve the quality of nurses' work. It relies on an integrated and structured information picture, with special emphasis on transparency and interpretability. In (Mouratidis et al. 2003) the security of information systems is considered as an integral part of the whole system development process. The above-mentioned articles, dedicated mostly to IS and its developing methodologies, clearly show the evolution of ISD to a SA methodology.

ANTICIPATIVE NATURE OF SA AND SE METHODOLOGIES

Many problem-solving methodologies are roughly similar regardless of the type of the process or purpose of the article. Three types of articles could be found on SE topics: SE Methodology, SE Application and SE Education. This triad of methodology, application, and education are inter-related; each methodology is context dependent as well as on education curricula. Therefore, common bases of all articles are also at high levels. Within this rough classification there are large variation and combinations of methods. The best example of that is the fact that almost 94% of articles devoted to IS do not use in the keywords ISD, SE or SA. With all respect to the keywords, in these

94% of articles some implicit methodology has to be used.

Most common terms used with respect to SE can be divided into three stages: the initiation stage, the growth stage, and the maturity stage, which correlate to Jenkins' four phases cited in (Flood and Carson, 1988): Systems Analysis, Systems Design, Implementation and Operation. Each of these phases could have several sub phases for detail analyses. Thus, in (Martin, 1996) SE is defined with three parts:

A. SE Management Plans (organizes, controls and directs the technical development of a system or its product);
B. Requirements and Architecture Definition (defines the technical requirements based on the stakeholders requirements); and
C. System Integration and Verification (integrates the components at each levels of the architecture and verifies that the requirements for those component are met).

SE is normally presented in terms of a flowchart describing a process. Figure 4 presents one such process and is loosely refereed to as the "Vee"

process. For this representation, the SE process begins at the upper left with the definition of user requirements and of system concepts that meet those requirements. It continues down through system design and fabrication, then up through testing, integration, verification, and delivery of a product. Since SE encompasses the entire system life cycle, many SE diagrams continue to the right with segments representing system upgrades, maintenance, repair, and finally, disposal.

The relevance of concurrent engineering in order to reduce design changes in the developing phase is stressed in (Martin, 1996). In (Sage, 2000) in a similar way, the problem of SE education was dealt with from the point of view of SE as method, process and management. In the context of education of SE, Sage (2000) elaborates three groups of knowledge: a natural science basis, an organizational and social science basis, and an information science and knowledge basis. This curriculum was further elaborated (Shenhar, 1994) into courses: mathematics and statistics, technical and engineering disciplines, economic and financial disciplines, management and organization theory and SE procedures. Asbjornsen and Hamann (2000) approach unified SE education

Figure 4. The Vee model of SE (modified from Forsberg and Mooz, 1992)

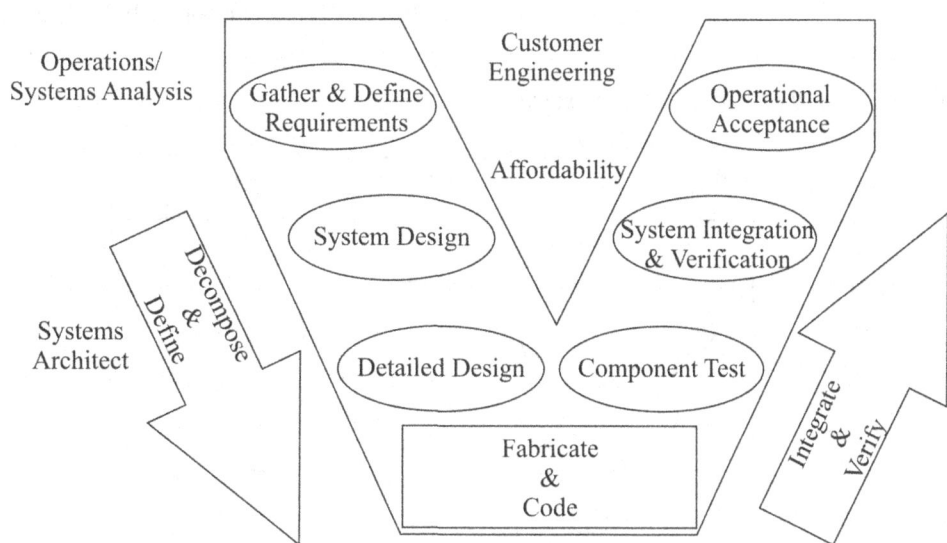

from a systems theory concept appealing to the SA concept to define the ratio between theory and applicative knowledge in curricula.

Perkins (2002) presents an educational program in an industrial process SE was analyzed. His analyses show what is common in SE education curricula as well as differences caused by specific of chemical process. Finally (Brown and Scherer, 2000) compare SE programs in the United States. They note that there are relatively small numbers of students in SE as well as that none of the associations covering SE have a successfully defined core body of SE knowledge embraced by academic institutions. The INCOSE (2007) has developed a standard reference for graduate programs. However, SE programs can range from operations research, control theory, information systems engineering, to industry standards of SE. Yet in the future the first concerns of the "integration of information technology subject into the SE curricula" must be addressed. Similarities in these diverse processes suggest that there is a general process that might be closely related to human thinking (Bahill and Gissing, 1998). Bahill and Gissing defined these procedure with acronym SIMILAR which means: State the problem, Investigate alternatives, Model the system, Integrate, Launch the system, Assess performance, and Re-evaluate.

From a decision point of view, SA to SE as described by SIMILAR could be unified as in Figure 5. As portrayed, the process represent the progress of problem solving at anticipated systems performance in environment X (systems requirements in Figure 3), decision means action U according chosen methodology while feedback means interactive control and adaptation of realized task Y in phases of design, development and deployment. Figure 3 clearly shows the interdependence of problems to be solved (process) users as decision makers and methodologies used. Yet the simulation method for dynamic testing of alternatives for the anticipated system performance could be a very useful tool within SE.

SIMULATION METHODS AS A PART OF SA METHODOLOGY

The role of the simulation methodology in the understanding x systems is constantly evolving and increasing. Today in moderns organizations two words are dominant: change and learning from which are derived change management and learning management. Human knowledge, the simulation model and decision methodology combined in an integral information system offers a new standard of quality in management problem solving (Simon, 1967). The simulation

Figure 5. General model of a goal oriented system

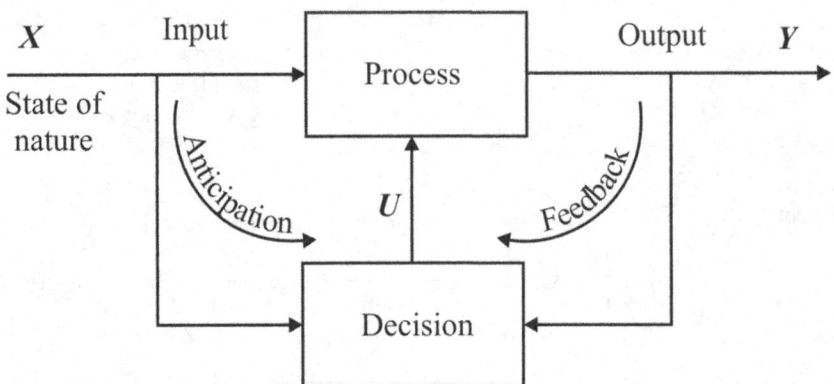

model is used as an explanatory tool for a better understanding of the decision process and/or for defining and understanding learning processes. An extensive study on using the simulation method in enterprises can be found in (Gopinath and Sawyer, 1999). Information systems and decision support is an important area in Management Information Systems (MIS) as the part of complex SE. That could be clarified by number of articles associated with IS and Simulation: 3.5%. The majority of them discuss IS related to decision assessment. For example, in (Mora et al, 2003; Mora et al, 2005) a new framework of identifying and classifying the support capabilities provided by the full range of decision-making support systems is posed with special regards to the information and knowledge representation and processing capabilities. However, only a few papers were used for simulation methods and SE in IS research. In (Gao

and Li, 2006), business process re-engineering (BPR) is regarded as a revolution of enterprise management. Advances in modeling languages such as the Unified Modeling Language (UML) is an industry standard that is used in modeling business concepts when building software systems in an object-oriented manner has also become mainstream for most SE organizations. Recently, XML has gained ground in becoming a key enabler of these systems in terms of transport of information and commands. All of this demonstrates the growth in tools and languages needed to describe and model a system.

Using a hybrid simulation model, decision assessment of BPR was analyzed Kljajić et al (2000). The decision assessment has been organized at two hierarchical levels. The model at the upper level is used for the assessment of an enterprise's strategy (continuous simulation). At

Figure 6. The principle diagram of simulation approach for decision support in enterprises

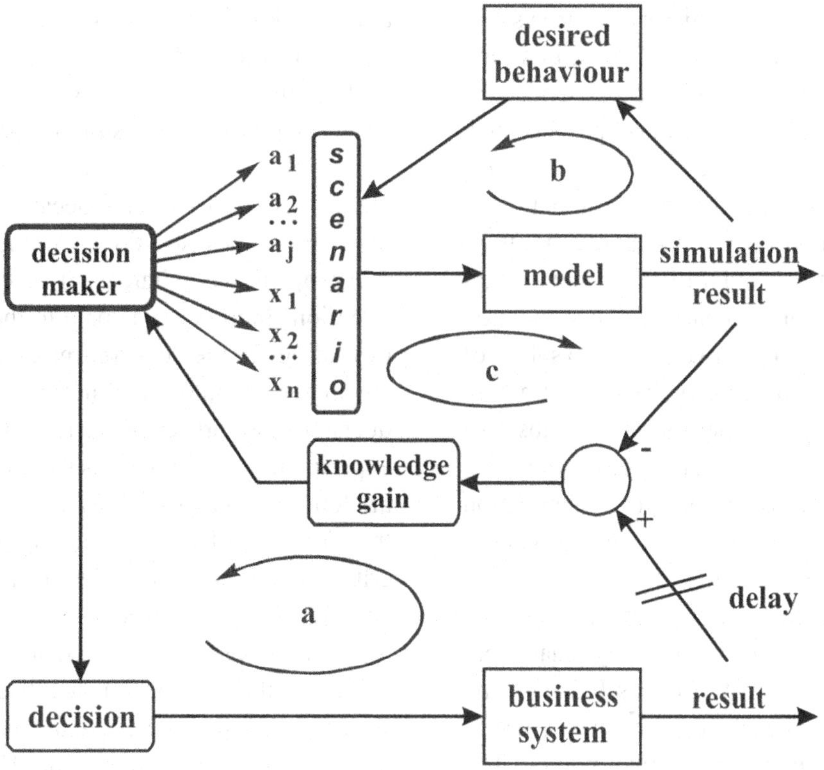

the lower level, the model is used for discrete event simulation, necessary for operations planning and testing production performance. The simulation approach seems to be an appropriate methodology for obtaining anticipative information for decision-making as shown in Figure 5. Roughly speaking, this involves the concepts of state, goal, criteria, alternative and the state of nature combined in a dynamic model interacting with decision-making groups. In both of these, simulations interacted with human experience create a new quality. The representation of the proposed approach is shown in Figure 6 adapted according (Kljaić, 1994).

Figure 6 shows the interaction between the user, simulation model and scenario in the process of seeking a solution to a managerial problem as decision support in a business system. The following three basic loops are emphasized:

a) The causal or the feed-back loop, representing the result as a consequence of former decision-making, and being a part of management experience and history of the system. From the learning point of view, this loop could be named "learning by experience".

b) The anticipative or intellectual feedback loop, which provides the feed forward information relevant to the formulation of the system strategy. This loop consists of the simulation model of the system, criteria function and scenarios. The simulation scenarios consist of two subsets: a subset of input xi that anticipates the state of nature (or exogenous scenarios) and a subset of alternatives aj (or endogenous scenarios). The generation of scenarios of the simulation system that responds to the what-if is based on different scenarios anticipating future impacts of the environment and desired performance of the system. They usually represent the extrapolation of past behavior and an expert evaluation of development targets employing the brainstorming method. The most delicate part of this circle is above all (principally) the methodology of the system simulation that facilitates "experimenting" on the system model and the model of the process itself.

c) The a posteriori information concerning model applicability and former decision-making. This loop represents the pragmatic validation of the model. The comparison of the prior information concerning the impact of the selected strategy on the system behavior with the achieved results allows us to gain knowledge and evaluate the value of the model and improve it. In this way, learning is facilitated on the basis of a priori assumptions on the model and not just on the basis of empirical experiences, which is usually delayed.

Loops a) and c) are the basic ones for the knowledge generation and experience for learning and quality decision-making. Loop b) represents the knowledge validation. In literature, major attention has been paid to the methodology of design, testing and evaluation of the model. As such a simulation could be very useful in testing of alternatives in the similar methodology or merely SA. The user is, however, the key element of the three circles because he/she is the one who makes decisions. As most of simulation projects necessitate teamwork, considerable attention should also be paid to the presentation of findings in the decision-making process. The advantage of simulation method application for decision support is proved by the laboratory experiment on the business simulator (Škraba et al, 2003; Škraba et al, 2007). The authors tested the efficiency of decision making (DM) (value of criteria function and its variation) at three different conditions: a) DM-based on problem understanding, b) DM-based on problem understanding and using simulation model as feedback and c) DM, which in addition to individual feedback also use information of group decisions. The best results were achieved for condition c) than b) and the worst at the group a).

Advantage of simulation model as a part of SA lays in fact that problem defined in natural language could be easily transformed in directed graph convenient for qualitative analysis and then transformed in computer program. In this case user always can check correctness of stated problem within certain theory and further its translation to computer programming.

This is important especially in case of complex problem where feedback loop and stochastic relation are present, no mother if the process is continuous or discrete event. Rich graphic presentations and 3D animation of simulated process make this technique unique for testing systems performance in phase of system design and deployment.

Modeling and simulation (M&S) have become ever more central to the development of modern systems. Unprecedented advances in digital processing have made high fidelity representation of systems and subsystems in computer models possible from the simplest of our systems to the most complex. This has made it possible to examine the projected performance of systems over wide excursions of design and environmental assumptions very early in the development process when key resources are committed. Today's M&S tools make it possible to perform extensive SoS and enterprise wide simulations and evaluate alternate architectures at affordable cost and early enough to make a difference.

SUMMARY AND CONCLUSIONS

In this paper the relationship among Information systems, SE, SA, and ISD was discussed from past, present and future perspectives. While SE is a well-established methodology for developing of different kind of man-made objects from components to enterprises, ISD is a variant of SE applied for IS development. Based on a library study, it is possible to see a positive trend in the evolution of research methodology in IS and the use of IS for SE implementation toward the SA methodology. All classical methods initially developed for specific problems and processes converge with development of information technology (IT) and society into one holistic methodology colored with specific problems (context) and user preferences. A common name for SE or ISD could be SA or more precisely SA to SE and SE with SA as a holistic methodology for complex problem solving. A core substance for management of all these methods is IS. Conversely, when IS itself should to be developed and deployment in such complex situation as internet, intranet, e-everything, geographic information systems, e-market, and management information systems on a global enterprise level, then ISD have to move to SA combining with principles of SE but combined with different methods and tools for dynamic testing of IS in all phase of SE by means of system simulation. One cannot imagine how to test reliability, stability, functioning and behavior of global (complex) systems, where IS represents backbone and the central nervous systems, after disaster impact without SA methodology combined with systems simulation.

No matter how we describe a methodology for complex systems managing: systems thinking, SA, soft systems methodology, critical systems methodology or dialectic systems, the essence of such methodology has to be intention to cope with wholeness of the process in its environment. Or more precisely, environment is a relevant part of the tangible problem solving local or global. Advances in basic and applied science along with a host of computer aided design tools offer solutions in developing information technology for more complex IS of the enterprise. Because of complexity, we must adopt a holistic perspective and remain focused on end user requirements. It means that it has to contain all existing particular methodologies (context oriented) in one working methodology where the user has to be in the first plan - anthropocentric orientation.

In our opinion SA, as world-view on systems dynamic originated from general systems theory and SE with its tradition in production process, is a proper candidate for that. It can unify all methods dealing with complex systems such like ISD, SE, Software Engineering, and Operations Research. Systems approach in natural way provides synthesis of structure, behavior and utility via goal, state, criteria and feedback control at anticipated property of the complex systems. But, only at the condition that user oriented solutions are the main goal from cognitive, interpretive, ergonomic and functional point of view.

REFERENCES

Ackoff, R. L. (1999). *Ackoff's best - His classical writings on management*. John Wiley and Sons.

Asbjornsen, O. A., & Hamann, R. J. (2000, May). Toward a unified systems engineering education. *IEEE TSMC Part C-Applications and Reviews*, *30*(2), 175–182.

Bahill, A. T., & Gissing, B. (1998, November). Re-evaluating systems engineering concepts using systems thinking. *IEEE TSMC Part C-Applications and Reviews*, *28*(4), 516–527.

Bertalanffy, L. V. (1968). *General system theory*. New York: George Braziller.

Brown, D. E., & Scherer, W. T. (2000, May). A comparison of systems engineering programs in the United States. *IEEE TSMC C-Applications and Reviews*, *30*(2), 204–212.

Department of Defense. (2007) *Defense acquisition guidebook*. Retrieved June 12, 2007 from https://akss.dau.mil/dag/TOC_GuideBook. asp?sNode=RandExp=Y,2004

Engineering Village. (2007). Retrieved March 26, 2007 from http://www.engineeringvillage2.org/controller/servlet/Controller?EISESSION=1_ebf0681117f03f83db5dses2andCID=quickSearch anddatabase=1

Farr, J. V. (2008, October). *System life cycle costing: economic analysis, estimation, management*. Draft textbook.

Flood, R. L., & Carson, E. R. (1988). *Dealing with complexity: An introduction to the theory and application of systems science*. New York and London: Plenum Press.

Forrester, J. W. (1994, Summer-Fall). System dynamics, systems thinking, and soft OR. *System Dynamics Review*, *10*(2-3), 245–256. doi:10.1002/sdr.4260100211

Forsberg, K., & Mooz, H. (1992). The relationship of systems engineering to the project life cycle. *Engineering Management Journal*, *4*(3), 36–43.

Gao, X., & Li, Z. (2006, December). Business process modeling and analysis using UML and polychromatic sets. *Production Planning and Control*, *17*(8), 780–791. doi:10.1080/09537280600875273

Gopinath, C., & Sawyer, J. E. (1999). Exploring the learning from an enterprise simulation. *Journal of Management Development*, *18*(5), 477–489. doi:10.1108/02621719910273596

Gruber, T. (1995). Novemer). Toward principles for the design of ontologies used for knowledge sharing. *International Journal of Human-Computer Studies*, *43*(5-6), 907–928. doi:10.1006/ijhc.1995.1081

INCOSE (2004, June). *INCOSE Systems engineering handbook*. (Version 2A).

INCOSE (2007). Proposing a framework for a reference curriculum for a graduate program in systems engineering. *INCOSE Academic Council Report*. Version 2007-04-30.

Jan, T. S., & Tsai, F. L. (2002, January - February). A systems view of the evolution in information systems development. *Systems Research and Behavioral Science, 19*(1), 61–75. doi:10.1002/sres.441

Kljajić, M. (1994). *Theory of system*. Kranj, Slovenia: Moderna organizacija.

Kljajić, M., Bernik, I., & Škraba, A. (2000). Simulation approach to decision assessment in enterprises. *Simulation, 75*, 199–210. doi:10.1177/003754970007500402

Koizumi, T. (1993). *Interdependence and change in the global system*. Lanham, New York, London: University Press of America.

Lazanski, T. J., & Kljajić, M. (2006). Systems approach to complex systems modeling with special regards to tourism. *Kybernetes, 35*(7-8), 1048–1058. doi:10.1108/03684920610684779

Li, Z. B., & Xu, L. D. (2003, May). Polychromatic sets and its application in simulating complex objects and systems. *Computers & Operations Research, 30*(6), 851–860. doi:10.1016/S0305-0548(02)00038-2

Martin, J. N. (1996). *Systems engineering guidebook: A process for developing systems and products*. Boca Ration, FL: CRC Press.

Mesarović, M. D., & Takahara, Y. (1989). *Abstract systems theory*. Springer-Verlag.

Michigan Tech (n.d.). Retrieved December 11, 2008 from http://www.sse.mtu.edu/sse.html

Miler, J. G. (1978). *Living systems*. McGraw-Hill Book Company.

Mora, M., Forgionne, G., Cervantes, F., Garrido, L., Gupta, J., & Gelman, O. (2005). Toward a comprehensive framework for the design and evaluation of intelligent decision-making support systems (i-DMSS). *Journal of Decision Systems, 14*, 321–344. doi:10.3166/jds.14.321-344

Mora, M., Forgionne, G., Gupta, J., Cervantes, F., & Gelman, O. (2003). A framework to assess intelligent decision-making support systems, knowledge-based intelligent information and engineering systems (LNAI 2774 (PT 2), pp. 59-65).

Mouratidis, H., Giorgini, P., & Manson, G. (2003). Integrating security and systems engineering: Towards the modeling of secure information systems. *Advanced Information Systems Engineering* (LNCS 2681, pp. 63-78). Retrieved from http://www.sse.mtu.edu/sse.html

National Aeronautics and Space Administration (NASA). (1995, June). *Systems Engineering Handbook*. SP-610S.

Peirce, C. S. (1931). The essential Peirce: Selected philosophical writings. *The Peirce Edition Project 1998*. Indiana University Press.

Perkins, J. (2002). Education in process systems engineering: Past, present and future. *Computers & Chemical Engineering, 26*, 283–293. doi:10.1016/S0098-1354(01)00746-3

Poper, K. (1973). *The logic of scientific discovery*. (Nolit, Belgrade, Trans.) (Original work published 1968, London, Hutchinson, 3rd Edition).

Rosenhead, J. (1989). *Rational analysis for a problematic world*. West Sussex: John Wiley.

Sage, A. (2000, May). System engineering education. *IEEE TSMC Part C-Applications and Reviews, 30*(2), 164–174.

Samaras, G. M., & Horst, R. L. (2005). A systems engineering perspective on the human-centered design of health information systems. *Journal of Biomedical Informatics, 38*, 61–74. doi:10.1016/j.jbi.2004.11.013

Senge, P. (1994). *The fifth discipline: The art and practice of the learning organization*. Doubleday.

Shenhar, A. (1994, February). Systems engineering management: A framework for the development of multidisciplinary discipline. *IEEE TSMC*, *24*(2), 327–332.

Simon, H. (1967). *Model of man*. (5th printing) John Wiley and Sons.

Škraba, A., Kljajić, M., & Borštnar Kljajić, M. (2007, January). The role of information feedback in the management group decision-making process applying system dynamics models. *Group Decision and Negotiation*, *16*(1), 77–95. doi:10.1007/s10726-006-9035-9

Škraba, A., Kljajić, M., & Leskovar, R. (2003). Group exploration of system dynamics models - Is there a place for a feedback loop in the decision process? *System Dynamics Review*, *19*(3), 243–263. doi:10.1002/sdr.274

Stevens Institute of Technology (2007). *SDOE 625 - Class notes for systems design and operational effectiveness*.

Šušteršič, O., Rajkovič, V., Leskovar, R., Bitenc, I., Bernik, M., & Rajkovič, U. (2002, May-June). An information system for community nursing. *Public Health Nursing (Boston, Mass.)*, *19*(3), 184–190. doi:10.1046/j.0737-1209.2002.19306.x

Thomé, B. (1993). *Systems engineering: Principles and practice of computer-based systems engineering*. Chichester: John Wiley and Sons.

United States Air Force. (1974, May 1) *Military standard - engineering management*. MIL-STD-499A.

Wiener, N. (1948). *Cybernetics*. John Wiley and Sons.

WOS EXPANDED (2007). Retrieved Match 26, 2007 from http://wos.izum.si/CIW.cgi

Xu, L. D. (2000, March-April). The contribution of systems science to information systems research. *Systems Research and Behavioral Science*, *17*(2), 105–116. doi:10.1002/(SICI)1099-1743(200003/04)17:2<105::AID-SRES287>3.0.CO;2-M

Zhu, Z. C. (2000, March-April). WSR: A systems approach for information systems development. *Systems Research and Behavioral Science*, *17*(2), 183–203. doi:10.1002/(SICI)1099-1743(200003/04)17:2<183::AID-SRES293>3.0.CO;2-B

Chapter 5
Towards a Wider Application of the Systems Approach in Information Systems and Software Engineering

Doncho Petkov
Eastern Connecticut State University, USA

Denis Edgar-Nevill
Canterbury Christ Church University, UK

Raymond Madachy
Naval Postgraduate School, USA

Rory O'Connor
Dublin City University, Ireland

ABSTRACT

The chapter provides possible directions for the wider application of the systems approach to information systems development. Potential improvement of software development practices is linked by some leading experts to the application of more systemic ideas. However, the current state of the practice in software engineering and information systems development shows the urgent need for improvement through greater application of systems thinking.

INTRODUCTION

Information Technology (IT) articles often include statements along these lines: "systems development continues to be challenging. Problems regarding the cost, timeliness, and quality of software products still exist." (Iivari and Huisman, 2007, p.35).

Such a statement justifies the continuous search for improvement of Information Systems Development (ISD). Boehm, one of the founding fathers of Software Engineering, stressed in a recent interview the importance of the systems approach to achieve improvements in software development (see Lane, Petkov and Mora, 2008). These are some of the origins for the motivation for this paper.

DOI: 10.4018/978-1-60566-976-2.ch005

Glass, Ramesh and Vessey (2004) provide an analysis of the topics covered by the three computing disciplines - Information Systems (IS), Software Engineering (SE) and Computer Science (CS) - and show overlaps between them all in the area of systems/software concepts. They also demonstrate that CS has only minor regard of the issues and concerns of systems/software management. Sommerville (2007) states that CS is concerned with the theories and methods that underlie computers and software systems rather than the engineering and management activities associated with producing software. Whilst acknowledging that CS, SE and IS do have a considerable overlap, the practices of both IS and SE have to deal with common matters such as the management of huge development projects, human factors (both software developers and software end users), organizational issues and economic aspects of software systems development and deployment (Van Vilet, 2000).

For the reasons stated above we will concentrate here only on SE and IS and their links to systems thinking. We will consider as a starting point the reality that the whole computing field has evolved historically as several 'stovepipes of knowledge'; CS, SE and IS (Glass et al., 2004). Whether the separation or integration of computing disciplines will prevail is a complex issue. Integration has yet to be achieved as a consequence of the sets of values central to each area. We believe, along with others, that a systems approach may lead to improvement of the development and management of software systems and to a greater integration of computing. One might expect that the use of the word "system" in various contexts today leads to more "systems thinking", but is this true?

A reflective history of the IS field is presented in Hirschheim and Klein (2003, p.244-249). According to them, because of its roots in multiple disciplines, "such as computer science, management, and systems theory, it is hardly surprising that the field of IS cast a wide net when defining its boundaries, sweeping in many themes and boundaries" (Hirschheim and Klein, 2003, p.245). In that light, it is somehow striking to note the conclusion about a lack of a systems approach in IS research according to Lee (2004, p.16). Alter (2004, p.757) is even more specific claiming that "the information systems discipline is ostensibly about systems, but many of our fundamental ideas and viewpoints are about tools, not systems".

The systems approach has been acknowledged, in the SE literature, as providing an insight into the factors that influence the success or failure of computer technologies (Mathieu, 2002, p.138). It is symbolic that the 2006 special issue of the IEEE Computer magazine on the 60th anniversary of the IEEE Computer Society was dedicated to the past and future of SE. A brief examination of the papers in that issue shows that four of them are dealing with some systems features and the other three give examples of tool thinking. None of the seven papers issue had a reference to any source from the field of systems thinking and only one paper (Baresi, Di Nitto and Ghezzi, 2006) had references to several classic SE sources dealing with fundamental systems ideas. This does not advance the ideas suggested by Boehm (2006a) and Sommerville (2007) that there is need to integrate SE with Systems Engineering; a branch of systems thinking (see Jackson, 2003).

The contribution of this research is in the identification of areas where a systems approach would lead to improvements in ISD within a point of view that favors implicitly the integration of the IS and SE disciplines. The paper will proceed with an analysis of how links between software development and systems thinking were perceived in the fields of IS and SE. This is done predominantly with the intention of exploring the application of systems ideas to software development separately in the two fields, outlining the success stories and the open problems. At the end we will propose possible directions for future research in software development within SE and IS associated with the systems approach.

ON INFORMATION SYSTEMS DEVELOPMENT AND SYSTEMS THINKING

A review of the history of various IS development methods is presented in Avison and Fitzgerald (2003). Iivari and Huisman (2007) point out, however, that the research literature on IS development is scarce. This is most evident for the period after 1990. Prior to that point the origins of IS research were associated more strongly with issues on building information systems. However, one sub-area of IS development grew significantly in the UK and elsewhere over the last twenty years; incorporation of Soft Systems Thinking (SST) into IS (see Checkland, 2001).

Soft Systems Thinking and Social Science and their Influence on IS

Stowell and West (1996) argued in the mid 1990's that practices of IS design had not appeared to have progressed since 1979; despite attempts in several proposals to embrace the social aspects of an information system, most seem to be based upon a functionalist view. Stowell and West (1996) explored the shift towards anti-positivism in the mid 1980's which resulted in a number of suggested methodologies that focussed upon the social implications of computer systems design. As examples they point out Soft Systems Methodology (SSM) (see Checkland, 1999), the MULTIVIEW approach (Avison, 2000), participative systems design and others (see also Avison and Fitgerald, 2003).

SSM evolved originally from experience within interventions in various management problems in public administration and industrial companies. However, subsequently it evolved more towards the field of IS (see Checkland and Holwell, 1998). Stowell (1995) presents a collection of papers analyzing various aspects of the contribution of SSM to IS. SSM seems to be the most well researched interpretive systems approach used in the field of IS (see Holwell (2000) for a detailed account of the literature on SSM and Checkland and Poulter (2006) for a contemporary presentation of SSM ideas).

The relevance of SSM to the field of IS has been explored in two directions. One way is to apply SSM on its own in some IT related aspect; e.g. extend the standard SSM method to specify the information requirements of the system (see Wilson, 1990). The use of SSM in data modelling is explored by Lewis (1995). A further application of SSM for improvement of software quality is presented in Sweeney and Bustard (1997). A second direction of using SSM in Information Systems is through the linking of SSM to existing design methods. An overview and detailed analysis of using SSM with structured analysis and design is provided by Mingers (1995). Several authors have covered aspects of combining the Unified Modelling Language (UML) with SSM. A recent paper by Sewchuran and Petkov (2007) analyses the related theoretical issues and shows a practical implementation of a combination of UML and SSM within a Critical Systems Thinking (CST) (see Jackson, 2003) framework justified by Multimethodology (see Mingers, 2001).

On Critical Systems Thinking, Multimethodology and IS

Multimethodology is a meta theory for mixing methods from different methodologies and paradigms in the same intervention (Mingers 2001). It seems to be an attractive vehicle for further research in systems thinking and IS research. Further refinement of the ideas on pluralist interventions can be found in a recent paper on Creative Holism (Jackson, 2006). Details on three cases illustrating how Multimethodology and CST were practiced in separate systemic interventions in the Information and Communications Technologies sector can be found in Petkov, Petkova, Andrew and Nepal (2007).

In his paper on the links between CST and IS research, Jackson (1992) demonstrates the power of an integrated critical approach in the IS field. However, there have been relatively few subsequent publications on the practical application of CST in IS. Some of them are surveyed in Ngwenyama and Lee (1997), a paper demonstrating the significant relevance of CST to IS. Another interesting example, exploring how Triple Loop Learning (Flood and Romm, 1996) can be applied to the complexities during systems development is given in Finnegan, Galliers and Powell's (2002) work. Further papers on systems thinking and IS can be found in proceedings of several meetings on the philosophical assumptions of IS research that took place after 1997; including the UK annual systems conference, the European Conference on Information Systems, the Australasian Conference on IS and Americas Conference on Information Systems (AMCIS).

CST provides both theoretical sophistication and practical directions for future research that are applicable to IS. Jackson (2003) cautions that whatever argument is made in favour of pluralism, it is bound to run up against objections from those who believe in the incommensurability of paradigms. The latter notion is linked to the assumption that if paradigms have distinct and opposing philosophical foundations, applying them together is impossible. This issue has been addressed by several authors in the past (see Jackson, 2003). Zhu (2006), however, questioned recently the relevance of concerns about paradigm incommensurability from a practical point of view; another issue for possible further research. His view on paradigm incommensurability is similar to that of the pragmatic pluralism approach. This is based on the assumption that we are witnessing the end of a particular reading of theory and that there is no single truth and no single rationality (White and Taket, 1996, p.54).

Both pragmatism and functionalism are often criticised in systems thinking (see Jackson, 2003). However, an interesting and relevant new systems approach in IS, the work system method (Alter, 2007), has emerged recently that may be linked to the pragmatic school of thought.

The Work System Method and IS

Alter (2006) stresses that past dominance of single ideas like Total Quality Management and Business Process Reengineering are not sufficient to influence the IS field profoundly. The work system method provides a rigorous but non-technical approach to any manager or business professional to visualize and analyze systems related problems and opportunities (Alter, 2006). This method is more broadly applicable than techniques "designed to specify detailed software requirements and is designed to be more prescriptive and more powerful than domain-independent systems analysis methods such as soft system methodology" (Alter, 2002).

We may note that making comparisons between the work system method and soft systems methodology requires a broader investigation of their philosophical assumptions and scope. A possible starting point for comparing their areas of applicability could be the classification of strategies for doing systems analysis provided by Bustard and Keenan (2005). SSM has been attributed by them to the situation when the focus is on development of a long term vision of the environment in which a computer system is to be used with identification of appropriate organizational changes (see Bustard and Keenan, 2005). Where does Alter's approach stand in the Bustard and Keenan (2005) classification is an open question for research requiring both theoretical work and field experimentation. Petkov and Petkova (2008) consider the systemic nature of the work system method and its applicability to understanding business and IS problems to be its most distinctive and important characteristics. Possible research directions for incorporating the work system method in systems analysis and design are presented in Petkov, Misra and Petkova (2008). Though the work system method has a relatively short history and a small group of followers for now, the multifaceted scale of Alter's work, bringing together systems ideas with methods for deeper understanding of work systems and ISD, has strong appeal.

On sticking to a single research tradition in IS. Bennetts, Wood-Harper and Mills (2000) provide an in-depth review of combinations of SSM with other IS development methods supporting multiple perspectives along the ideas of Linstone (1984). Thus they brought together two distinct traditions in IS research: the former practiced in UK/Europe/Australia where SSM has found significant acceptance, and the latter being pursued predominantly in the USA. Linstone's ideas are strongly related to the influence of Churchman whose analysis of Inquiring Systems was a starting point for some significant IS research that followed (see as an example Vo, Paradice & Courtney, 2001).

It is interesting to note that Bennetts et al. (2000) have examined sources not only from IS but also from the CS and SE literature. This raises a question that is hard to answer in a simple way. We observe that often authors of SE articles belong to CS or IS departments, rather than engineering schools (Dietrich, Floyd & Klichewski, 2002) (Aurum & Wohlin, 2005). On the other hand, it seems that publications on IS development written by US scholars often use references only from IS or from SE disciplines; depending on the field of the authors (a refreshing exception is a series of articles written over many years by R. Glass and I. Vessey with several collaborators (Glass et al., 2004)). The reason could be the lack of communication between CS, SE and IS (see Glass, 2005). Another possible reason is the growing concern within the separate computing fields for promoting and protecting their own paradigms (Bajaj et al, 2005).

Maybe similar paradigmatic concerns have led Allen Lee to formulate his first idea from an advice to IS researchers: "practice paradigm, systems thinking and design science" (Lee, 2000). These are seen as a recipe to address the three dilemmas

that are as relevant today as they were in 2000: the rigor versus relevance debate in IS research; the "reference discipline" versus "independent discipline" dilemma; the technology versus behaviour as a focus for IS research dilemma.

So far we have considered the second of Lee's ideas and its relevance to IS development over the last 15 years and to a lesser degree some issues related to scientific paradigms in terms of Kuhn (1970). Further details on earlier contributions of Systems Science in the 1970's and 1980's can be found in comprehensive reviews related to the fields of IS research (see Xu, 2000); Decision Support Systems (see Eom, 2000) and Information Resources Management (see McLeod, 1995). Mora, Gelman, Forgionne, Petkov and Cano (2007) presented a critique and integration of the main IS research paradigms and frameworks reported in the IS literature using a systems approach. We briefly comment below on design science, a more recent trend in IS research.

On Design Science as One of the Directions to Resolve the Three Dilemmas in IS

According to Hevner, March, Park and Ram (2004), IS related knowledge is acquired through work in behavioral science and design science paradigms. They point out that "behavioral science addresses research through the development and justification of theories that explain phenomena related to the identified business need, while design science addresses research through the building and evaluation of artifacts designed to meet the particular need". Another relevant detail is the differentiation that Hevner et al. (2004) make between routine design and system building from design science. The former is associated with application of existing knowledge to organizational problems, while the latter is associated with unique (often wicked or unresolved) problems that are associated with the generation of new knowledge. The latter idea is similar to the main thesis in Hughes and Wood-Harper (1999).

Hevner et al. (2004) laid the foundation for a significant boost in IS research on issues related to IS development, including systems analysis and design science. The journal Communications of AIS started a series of articles in 2005 on this topic; the first of which was Bajaj et al. (2005). We may note that in spite of progress in applying action research in IS in theory (see Baskerville and Wood-Harper (1998) and in practice (see the IbisSoft position statement on environment that promotes IS research) the dominant IS research trend has been of a positivist behavioral science type which is another challenge for the proponents of a systems approach.

A substantial attempt to provide suggestions towards resolving the three dilemmas in IS research mentioned by Lee (2000) is discussed in Hirschheim and Klein (2003). They identify a number of disconnects between various aspects of IS research and outline a new body of knowledge in IS development (Iivari, Hirschheim and Klein, 2004). They suggest there are five knowledge areas in ISD: technical knowledge, application domain (i.e. business function) knowledge, organizational knowledge, application knowledge and ISD process knowledge. Further, according to Hirschheim and Klein (2003) "ISD process knowledge is broken down into four distinctive competencies that IS experts are suggested to possess: (1) aligning IT artefacts (IS applications and other software products) with the organizational and social context in which the artefacts are to be used, and with the needs of the people who are to use the system as identified through the process of (2) user requirements construction... (3) organizational implementation from which (4) the evaluation/assessment of these artefacts and related changes is factored out ... These competencies are ... at best weakly taken into account in the ten knowledge areas of SWEBOK" (see for comparison SWEBOK, 2004). Hirschheim and Klein (2003) present comprehensive proposals

for strengthening the IS field. Their work was partly motivated by a widely discussed paper by Benbasat and Zmud (2003) on the identity crisis in the IS discipline. Both papers provide important background details about the IS research environment in which one may pursue the main ideas of this paper.

ON SOFTWARE ENGINEERING AND SYSTEMS THINKING

Software Engineering has a primary focus on the production of a high quality technological product, rather than on achieving an organisational effect, however increasing emphasis in SE is being given to managerial and organisational issues associated with software development projects. Cornford and Smithson (1996) observe that SE "can never encompass the whole range of issues that need to be addressed when information systems are studies in the full richness of their operational and organisational setting".

Weinberg (1992) writes about systems thinking applied to SE. It is an excellent introduction to systems thinking and quality software management dealing with feedback control. It has a close kinship with the concepts of systems thinking and system dynamics in Madachy (2008), even though it is almost exclusively qualitative and heuristic. Weinberg's main ideas focus around management thinking about developing complex software systems; having the right "system model" about the project and its personnel.

Systems thinking in the context of SE as described in Madachy (2008) is a conceptual framework with a body of knowledge and tools to identify wide-perspective interactions, feedback and recurring structures. Instead of focusing on open-loop event-level explanations and assuming cause and effect are closely related in space and time, it recognizes the world really consists of multiple closed-loop feedbacks, delays, and non-linear effects.

Lee and Miller (2004) advocate a systems thinking approach in their work on multi-project software engineering pointing that "in general, we are able to make better, more robust, and wiser decisions with systems thinking, since we are considering the problem by understanding the full consequences of each feasible solution".

Other details on systems thinking with links to other books and articles can be found through practitioner's web sites such Weinberg (2007), Developer (2007) or Yourdon (2007). The interest of software practitioners in systems ideas is a significant fact; in light of the previously mentioned debate about relevance in the IS literature. However, systems thinking is not mentioned by Reifer (2003) in his taxonomies of the SE theory state-of-the-art and SE state of practice. In relation to that, we will discuss below whether systems ideas are promoted in SE education.

Software Engineering Education and Systems Thinking

The coverage of systems concepts in leading SE textbooks is possibly another indicator about the way the systems approach is perceived within the SE community. We considered books by several well established authors: Sommerville (2007), Pressman (2009) and Pfleeger (2008) amongst many. Table 1 shows a summary of findings related to the treatment of several typical systems notions in those books.

Table 1 shows that the systems concepts covered in the three widely used textbooks are mostly related to introductory notions from systems thinking. There is nothing about open and closed systems, about the law of requisite variety or any other aspect of cybernetics, very little about socio-technical systems and nothing about soft systems methodology or CST. In our

Table 1. Systems features covered in popular software engineering textbooks

Notions covered	Author		
	Sommerville	Pressman	Pfleeger
System definition	Yes	Yes	Yes
Boundary	Implied	Yes	Yes
Open vs Closed systems	No	No	No
Relationships	Implied	Implied	Yes
Interrelated systems	Implied	Implied	Yes
Emergent property	Yes	No	No
Decomposition	Yes	Yes	Yes
Coupling	No	Yes	Yes
Cohesion	No	No	Yes
Hierarchy	Yes	Yes	Yes
System behaviour	Yes	Yes	Yes
Law of requisite variety	No	No	No
Socio-technical systems	Yes	No	No
Systems engineering	Yes	To some extent	To some extent

opinion these are unexploited notions that have some potential to introduce fresh ideas in SE after further research.

Crnkovic, Land and Sjogren (2003) question whether the current SE training is enough for software engineers. They call for making system thinking more explicit in SE courses. They claim that "the focus on modifiability (and on other non-functional properties) requires more of a holistic and system perspective" (Crnkovic et al., 2003). Similar thoughts are shared more recently by others in engineering like Laware, Davis and Perusich (2006).

The narrow interpretation of computing disciplines is seen as a contributory factor to the drop in student enrolments in the last five years. Denning (2005) hopes that students will be attracted by a new educational approach promoted by the ACM Education Board that relies on four core practices: programming, systems thinking, modelling and innovating. It has now been four years since those ideas were stressed by ACM but there is little evidence that systems thinking has become a core practice emphasized in teaching in any of the three computing disciplines.

In the UK the Quality Assurance Agency (which monitors and quality assures all UK university programmes) recently published the updated version of the Computing benchmark statement (encompassing IS, SE and CS) on the content and form of undergraduate courses (QAA, 2007). Although not intended to be an exhaustive list but "… provided as a set of knowledge areas indicative of the technical areas within computing" it fails to make explicit reference to systems thinking or systems approaches and makes only one reference to "systems theory" under a more general heading of "systems analysis and design". Perhaps the answer is to explore how to introduce these concepts earlier in pre-university education or to continue to try to convince the broader academic community of the importance of systems thinking.

One promising systems approach used for education of software engineers is the Model-Based System Architecting and Software Engineering (MBASE) framework being used at USC, and also adapted by some of their industrial affiliates. According to Boehm (2006c), MBASE integrates the systems engineering and SE disciplines, and considers stakeholder value in the system development. The MBASE framework embodies elements of agile processes, and teaches students to "learn how to learn" as software development will continue to change. Valerdi and Madachy (2008) further describe the impact of MBASE in education.

On Software Engineering and Systems Engineering

Systems Engineering is concerned with all aspects of the development and evolution of complex systems where software plays a major role. Systems engineering is therefore concerned with hardware development, policy and process design and system deployment, as well as software engineering. System engineers are involved in specifying a system, defining its overall architecture and then integrating the different parts to create the finished system. Systems engineering as a discipline is older than SE; people have been involved in specifying and assembling complex industrial systems such as aircraft and chemical plants for more than a hundred years (Sommerville, 2007).

A thought provoking comparison of SE culture versus systems engineering culture is presented by Gonzales (2005). This work points out to where we should strive to change the perceptions of the SE student entering the IT profession. We agree with Gonzales (2005, p.1) that we "must continue the dialogue and ensure that we are aware of strides to formalize standard systems engineering approaches and generalize software engineering approaches to capturing, specifying and managing requirements". We would also suggest that this dialogue should be supported by more work on the application of a systems approach to SE; stimulated by journals such as the International Journal on Information Technologies and the Systems Approach (IJITSA).

Boehm (2006b) concludes that "The push to integrate application-domain models and software-domain models in Model Driven Development reflects the trend in the 2000's toward integration of software and systems engineering". Another reason he identifies is that other surveys have shown that the majority of software project failures stem from systems engineering shortfalls. A similar thought is expressed by Boehm and Turner (2005), who state that there is a need to move towards a common set of life-cycle definitions and processes that incorporate both disciplines' needs and capitalize on their strengths.

Boehm (2006a) points out that "recent process guidelines and standards such as the Capability Maturity Model Integration (CMMI), ISO/IEC 12207 for software engineering, and ISO/IEC 15288 for systems engineering emphasize the need to integrate systems and software engineering processes". He further proposes a new process framework for integrating software and systems engineering for 21st century systems, and improving the contractual acquisition processes.

A very recent development illustrates the increasingly recognized importance of applying systems thinking to large and complex acquisition processes for software-intensive systems. The United States Department of Defense (DoD) just created a long-term Systems Engineering Research Center (SERC) as a consortium of universities. The SERC leverages developments in systems architecting, complex systems theory, systems thinking, systems science, knowledge management, and software engineering to advance the design and development of complex systems across all DoD domains (Stevens Institute of Technology, 2008).

As the first research centre focused on systems engineering, it is specifically concerned with integrating systems and SE. Some research areas include software-unique extensions and modern software development techniques and how they relate to systems engineering; flexible systems engineering environments to support complex software systems and commercial-off-the-shelf hardware and software integration; and other aspects involving SE and IS (Stevens Institute of Technology, 2008).

Part of the SERC acquisition research is to further develop the Incremental Commitment Model (ICM) (Boehm & Lane, 2007) for better integrating system acquisition, systems engineering, and SE. The ICM is a risk-driven process generator for incremental development of complex systems that uses the principles of MBASE with both plan-driven and agile process components.

An issue is how to capitalize on these new and upcoming developments in SE as will be discussed in the next section.

The Evolution of Plan-Driven and Agile Methods in SE and System Thinking

The traditional software development world, characterised by SE, advocates use plan-driven methods which rely heavily on explicit documented knowledge. Plan-driven methods use project planning documentation to provide broad-spectrum communications and rely on documented process plans and product plans to coordinate everyone (Boehm & Turner, 2004). The late 1990s saw something of a backlash against what was seen as the over-rigidity contained within plan-driven models and culminated in the arrival of agile methodologies, which rely heavily on communication through tacit, interpersonal knowledge for their success.

Boehm and Turner (2004, p.23) quote Philippe Kruchten (formerly with IBM Canada and now a professor at UBC in Vancouver) who has likened the Capability Maturity Model (CMM) – a plan-drive approach - to a dictionary; "that is, one uses the words one needs to make the desired point- there is no need to use all the words available". They conclude that processes should have the right weight for the specific project, team and environ- ment. Boehm and Turner (2004) have produced the first multifaceted comparison of agile and plan-driven methods for software development. Their conclusions show that neither provides a 'silver bullet' (Brooks, 1987). Some balanced methods are emerging. We need both agility and discipline in software development (Boehm & Turner, 2004, p.148).

Boehm (2006b) presents a deep analysis of the history of SE and of the trends that have emerged recently. These include the agile development methods; commercial off-the-shelf software and model driven development. The same author points out that the challenges are in capturing the evolving IT infrastructure and the domain restruc- turing that is going on in industry. In our opinion it is necessary to investigate further if systems thinking may play a role in integrating agile and plan-driven methods (see Madachy, Boehm & Lane (2007) as an application of systems thinking to this problem). It has also been speculated that systems thinking could be relevant to Extreme Programming (XP) as it supports building relevant mental models (see Wendorff, 2002).

Kroes, Franssen, van de Poel and Ottens (2006) deal with important issues in systems engineering; such as how to separate a system from its environment or context. They conclude that the idea that a socio-technical system can be designed, made and controlled from some cen- tral view of the function of the system, has to be given up as many actors within the socio-technical system are continuously changing (redesigning) the system. This is an important issue deserving further investigation in light of software systems and the methods implied by agile development frameworks.

Systems Dynamics and SE

A widely publicized idea is modelling software development processes through systems dynamics (see Abdel-Hamid & Madnick (1991), Madachy (2008) and others). The differences and relation- ships between systems dynamics and systems thinking are detailed in Richmond (1994) and others. Systems dynamics is a tool that can as- sist managers to deal with systemic and dynamic properties of the project environment, and can be used to investigate virtually any aspect of the software process at a macro or micro level. It is useful for modeling socio-technical factors and their feedback on software projects. The systems dynamics paradigm is based on continuous sys- tems modeling, which has a strong cybernetic thread. Cybernetic principles are relevant to many types of systems including software development systems, as detailed in Madachy (2008).

The primary purposes of using systems dynam- ics or other process modeling methods in SE as summarized from Madachy (2008) are strategic management, planning, control and operational management, process improvement and technol- ogy adoption and training and learning. Example recent work by Madachy (2006) focuses on the use of systems dynamics to model the interaction between business value and the parameters of a software process for the purpose of its optimiza- tion. Another application of systems dynamics to assess a hybrid plan-driven and agile process that aims to cope with the requirements of a rapidly changing software environment while assuring high dependability in Software-Intensive- Systems-of-Systems (SISOS) is presented in Madachy, Boehm and Lane (2007).

On Other Methods of Systems Thinking Applicable to SE

The development of understanding of a particular software project for making better judgments about the cost factors involved in cost and effort estimation is supported also by the work of Petkova and Roode (1999). They implemented a pluralist systemic framework for the evaluation of the factors affecting software development productivity within a particular organizational environment. It combines techniques from several paradigms; stakeholder identification and analysis (from SAST, see Mason and Mitroff, 1981), from SSM (Checkland, 1999), Critical Systems Heuristics (Ulrich, 1998) and the Analytic Hierarchy Process (Saaty, 1990).

While we could not find any specific earlier accounts of the use of SSM in the mainstream SE literature, it is significant that Boehm has recognised its potential as he quotes its originator in a recent paper: "… software people were recognizing that their sequential, reductionist processes were not conducive to producing user-satisfactory software, and were developing alternative SE processes (evolutionary, spiral, agile) involving more and more systems engineering activities. Concurrently, systems engineering people were coming to similar conclusions about their sequential, reductionist processes, and developing alternative 'soft systems engineering' processes (e.g., Checkland, 1999), emphasizing the continuous learning aspects of developing successful user-intensive systems" (Boehm, 2006a).

One does not need always to have a systems philosophy in mind to generate an idea that has a systemic nature or attempts to change the current thinking in SE. Thus, Kruchten (2005) presents, under the banner of postmodernist software design, an intriguing framework for software design borrowed from architecture. One may investigate how such an approach is different from a systemic methodology and what are their common features. Starting from a language-action philosophy point of view, Denning and Dunham (2006) develop a framework of innovation based on seven practices that are interrelated in their innovation model – every element is in a relationship with all others, thus fulfilling the criterion for "systemicity" by Mitroff and Linstone (1993). We need more

analogical examples of systemic reasoning or even just of alternative thinking related to every aspect of the work of a software engineer and IS developer demonstrating the power of innovative interconnected thinking. The analysis so far allows us now to formulate some recommendations in the following section.

CONCLUDING RECOMMENDATIONS ON THE NEED FOR MORE RESEARCH LINKING SOFTWARE ENGINEERING, INFORMATION SYSTEMS DEVELOPMENT AND SYSTEMS THINKING

We may derive a number of possible directions for future work from the analysis of research and practice in ISD and systems thinking within the fields of IS and SE. Alter (2004) has produced a set of recommendations for greater use of systems thinking in the IS discipline which incorporate various aspects of the work system method. We believe that Alter's proposals are viable and deserve the attention of IS and SE researchers.

Boehm and Turner's (2005) suggestions to address management challenges in integrating agile and plan-driven methods in software development will be used by us as an organizing framework for formulating directions for research on integrating IS, SE and the systems approach. The five main points below are as defined originally by Boehm and Turner (2005) for their purpose, while we have provided for each of them suggestions promoting such integration along the aims of this paper:

1. **Understand how communication occurs within development teams:** There is a need to continue the work on *integrating systemic methods promoting organizational learning* (see Argyris and Schon, 1978) like systems dynamics, stakeholder analysis, soft systems methodology, critical systems thinking and others to identify the advantages of using

specific methods and their limitations when dealing with uncovering the micro climate within a software development team. Most of the previously mentioned applications of systems methods for this purpose have had limited use and little experimental evaluation. *More case studies need to be conducted in different software development organizations to validate the claims for the applicability of such methods and to distil from the accumulated knowledge best practices and critical success factors* relevant to flexible, high quality software development teams. We may *expand further the boundary of investigations with respect to what is happening at the level of systems-of-systems* (see Sage, 2005). An example of related relevant ideas on cost estimation for large and complex software projects can be found in Lane and Boehm (2007). Another direction is to *explore information systems development as a research act*, as suggested by Hughes and Wood-Harper (1999) and Hevner et al. (2004, as well as the philosophy of integrating practice with research in the field of software and management, promoted by IbisSoft.

2. **Educate stakeholders:** This is probably the most difficult task of all. It needs to be addressed at several levels:

    ◦ **Implement changes in educational curricula:** it is essential to introduce the systems idea in relatively simple forms at undergraduate level and in more sophisticated detail at masters' level. There is *a need to create the intellectual infrastructure for more doctoral dissertation projects in IS or SE involving systems thinking*. Teaching could be supported by *creating an accessible repository for successful utilization of systems ideas in IT education*. Among the many examples we may mention here the use of SSM in

project-based education at a Japanese university (Chujo & Kijima, 2006), on integrating systems thinking into IS education (see Vo, Chae & Olson, 2006), or the use of MBASE in student projects (see Boehm (2006c) and Valerdi & Madachy, 2007).

◦ **Broaden the systems knowledge of IS and software engineering educators:** the current situation in some of the computing disciplines can be compared to a similar one in Operations Research (OR) in the 1960s, which had evoked a sharp critique by Ackoff (1999) in his famous paper "The future of operational research is past", published originally in 1979. Ackoff (1999, p.316) points that survival, stability and respectability took precedence over development and innovativeness in OR in the mid 1960's and its decline began. The *challenge however is not just to bring systems thinking to IS and SE education beyond several elementary concepts of general systems theory but to keep up to date with the latest body of knowledge in the systems field*. For a comprehensive overview see Jackson (2003) and for recent developments in systems science, see Barton, Emery, Flood, Selsky and Wolstenholm (2004)

◦ **Empower IT developers to practice systemic thinking:** a significant role here needs to be played by research on the most *suitable forms for continuing professional education on IT and the systems approach, supported by professional meetings and journals for mixed audiences* like this one, that are oriented to academia and industry practice. Ackoff (2006, p.707) underlines that one of the reasons why systems ideas are adopted by few organizations is that "very little of the systems literature and lectures are addressed to potential users". Further he stresses the need to analyse management failures systemically, pointing out that there are two types of failures: errors of commission and errors of omission. In spite of publications analysing software failures like Glass (2001), *there is still room for systemic analysis of IT failures and there are very few accounts of errors of omission* in software projects.

  ○ **Change the attitudes of clients in managerial and operational user roles:** viable research and practical activities in this direction could *use the work system method* (Alter, 2006) and other relevant methods to develop better understanding of organizational problems and to improve their communication with software developers.

3. **Translate agile and software issues into management and customer language:** We may suggest several possible directions here:

  ○ **Investigate in a systemic way the existing agile and plan-driven models** for software development and continue with the work started in Boehm (2006a) on creating new process models integrating not just SE and systems engineering ideas but other applicable systems concepts as well.

  ○ **Explore the applicability of "Sysperanto"** (see Alter, 2007) to foster a common language for all stakeholders in software development.

  ○ **Build methods and tools to facilitate the communication process between software developers, and**

customers and supporting multiple perspective representations of problem situations as proposed by Linstone (1984).

4. **Emphasize value for every stakeholder:** Design science research and agile methods place high emphasis on this idea. There is a *need for more research on systemic identification of stakeholder values*. Further there is a need for research on methods to model and help the effective analysis and *better systemic understanding of all aspects of software development*, related to the technical product attributes, the project organizational attributes, the developers attributes and the client features in a particular project or system-of-projects.

5. **Pick good people, reward the results and reorient the reward system to recognize both individual and team contribution:** These suggestions can be categorized as human resource management issues and hence are also suitable for *investigation through suitable systemic approaches and problem structuring methods, including multi-criteria decision analysis, promoting evaluation and decision making*.

One of the limitations of the scope of our proposals is that we have provided suggestions reflecting only on the above five ideas by Boehm and Turner (2005). A systemic investigation of all aspects of ISD could lead to a much broader set of considerations integrating SE, IS and systems thinking. We believe, however that the examples we have provided here can lead to easier adaptation and development of other relevant ideas serving a similar purpose. Another possible limitation is that we have produced our suggestions for future research on integrating SE, IS and the systems approach by assuming that the current state of the art and practice in SE and IS are known and we have focused rather only on identifying examples of the use of a systems approach in IS or SE. As we have pointed earlier, we have relied on the comprehensive analysis of the state-of-the-art of the IS discipline provided by Hirschheim and Klein (2003). We have also reflected on trends in SE (see Reifer (2003), Boehm (2006a, b) and Boehm and Turner (2004)) and on the comparative analysis of research in the three computing disciplines by Glass et al. (2004). It would be interesting to conduct a further investigation of IS implementation as a whole that goes beyond the existing disciplinary boundaries and takes a systems approach as an organizing viewpoint.

Most of our recommendations on integrating IS, SE and systems thinking relate to issues of organizational learning where contemporary systems methods have a significant history of achieving improvement. The relevance of this paper is supported by Boehm's recent interview, mentioned earlier (see Lane et al., 2008). The challenge for IS and SE practitioners, researchers and educators is not just to investigate the issues we discussed in this paper but also to practice the systems approach for improved ISD.

ACKNOWLEDGMENT

The authors are very grateful to D. Bustard, I. Bider, M. Mora and D. Paradice for their insightful comments that helped improve an earlier version of the paper that was published in IJITSA in 2008.

REFERENCES

Abdel-Hamid, T. K., & Madnick, S. E. (1991). *Software project dynamics.* Englewood Cliffs, NJ: Prentice-Hall.

Ackoff, R. (1999). *Ackoff's best: his classic writings on management.* New York: Wiley.

Ackoff, R. (2006). Why few organizations adopt systems thinking. *Systems Research and Behavioral Science, 23*(5), 705–708. doi:10.1002/sres.791

Alter, S. (2002). The work system method for understanding information systems and information systems research. *Communications of the AIS, 9*(6), 90–104.

Alter, S. (2004). Desperately seeking systems thinking in the information systems discipline. *Proceedings of Twenty-Fifth International Conference on Information Systems,* 757-769.

Alter, S. (2006). *The work system method: connecting people, processes, and IT for business results.* Lakspur, CA: Work System Press.

Alter, S. (2007). Could the work system method embrace system concepts more fully? *Information Resources Management Journal, 20*(2), 33–43.

Argyris, C., & Schon, D. A. (1978). *Organizational learning. A theory of action perspective.* Addison- Wesley.

Aurum, A., & Wohlin, C. (Eds.). (2005). *Engineering and managing software requirements.* Heidelberg: Springer.

Avison, D. (2000). *Multiview: An exploration in information systems development* (2nd ed). Alfred Waller Ltd.

Avison, D. E., & Fitzgerald, G. (2003). Where now for development methodologies? *Communications of the ACM, 46*(1), 79–82. doi:10.1145/602421.602423

Bajaj, A., Batra, D., Hevner, A., Parsons, J., & Siau, K. (2005). Systems analysis and design: Should we be researching what we teach? *Communications of the AIS, 15,* 478–493.

Baresi, L, Di Nitto, & Ghezzi, C. (2006). Toward open-world software: issues and challenges. *IEEE Computer, 39*(10), 36-43.

Barton, J., Emery, M., Flood, R. L., Selsky, J., & Wolstenholm, E. (2004). A maturing of systems thinking? Evidence from three perspectives. *Systemic Practice and Action Research, 17*(1), 3–36. doi:10.1023/B:SPAA.0000013419.99623.f0

Baskerville, R., & Wood-Harper, A. T. (1998). Diversity in information systems action research methods. *European Journal of Information Systems, 7*(2), 90–107. doi:10.1057/palgrave.ejis.3000298

Benbasat, I., & Zmud, R. (2003). The Identity Crisis within the IS Discipline: Defining and Communicating the discipline's Core Properties. *MIS Quarterly, 27*(2), 183–194.

Bennetts, P. D. C., Wood-Harper, T., & Mills, S. (2000). A holistic approach to the management of information systems development. *Systemic Practice and Action Research, 13*(2), 189–206. doi:10.1023/A:1009594604515

Boehm, B. (2006a). Some future trends and implications for systems and software engineering processes. *Systems Engineering, 9*(1), 1–19. doi:10.1002/sys.20044

Boehm, B. (2006b). A view of 20th and 21st century software engineering. In *Proceedings of the 28th international conference on Software Engineering* (pp. 12-29).

Boehm, B. (2006c). Educating students in value-based design and development, keynote address. In *Proceedings of the 19th Conference on Software Engineering Education and Training (CSEET)*.

Boehm, B., & Lane J. (2007). Using the incremental commitment model to integrate system acquisition, systems engineering, and software engineering. *USC-CSSE-TR-2207* (Short Version in Cross Talk, October 2007, pp, 4-9)

Boehm, B., & Turner, R. (2004). *Balancing agility and discipline - a guide for the perplexed*. Boston: Addison-Wesley.

Boehm, B., & Turner, R. (2005). Management challenges for implementing agile processes in traditional development organizations. *IEEE Software, 5*, 30–39. doi:10.1109/MS.2005.129

Brooks, F. P. (1987). No silver bullet: Essence and accidents of software engineering. In *Proceedings of the IFIP Tenth World Computing Conference* (pp. 1069-1076).

Bustard, D. W., & Keenan, F. M. (2005). Strategies for systems analysis: groundwork for process tailoring. In *Proceedings of 12th Annual IEEE International Conference and Workshop on the Engineering of Computer Based Systems (ECBS 2005)* (pp. 357-362). Greenbelt, MD, USA, 3-8 April.

Checkland, P. (1999). *Systems thinking, systems practice*. Chichester: Wiley.

Checkland, P., & Holwell, S. (1998). *Information, systems and information systems*. Chichester: Wiley.

Checkland, P., & Poulter, J. (2006). *Learning for action: A short definitive account of soft systems methodology and its use by practitioner, teachers and students*. Chichester: Wiley.

Checkland, P. B. (2001). Soft systems methodology. In J. Rosenhead & J. Mingers (eds), *Rational analysis for a problematic world revisited*. Chichester: Wiley

Chujo, H., & Kijima, K. (2006). Soft systems approach to project-based education and its practice in a Japanese university. *Systems Research and Behavioral Science, 23*(1), 89–106. doi:10.1002/sres.709

Cornford, T., & Smithson, S. (1996). *Project research in information systems*. MacMillan.

Crnkovic, I., Land, R., & Sjögren, A. (2003). Is software engineering training enough for software engineers? In *Proceedings 16th International Conference on Software Engineering Education and Training*. Madrid, March 2003. IEEE.

Denning, P. J. (2005). Recentering computer science. *Communications of the ACM, 48*(11), 15–19. doi:10.1145/1096000.1096018

Denning, P. J., & Dunham, R. (2006). Innovation as language action. *Communications of the ACM, 49*(5), 47–52. doi:10.1145/1125944.1125974

Developer (2007). *Developer*, an online magazine for software developers at http://www.developerdotstar.com/mag/categories/systems_software_series.html.

Dietrich, Y., Floyd, C., & Klichewski, R. (2002). *Social thinking-software practice*. Boston: MIT Press.

Eom, S. (2000). The contribution of systems science to the development of the decision support systems subspecialties: an empirical investigation. *Systems Research and Behavioral Science, 17*, 117–134. doi:10.1002/(SICI)1099-1743(200003/04)17:2<117::AID-SRES288>3.0.CO;2-E

Finnegan, P., Galliers, R. D., & Powell, P. (2002). Planning electronic trading systems: Re-thinking IS practices via triple loop learning. In S. Wrycz (Ed.), *Proceedings of the Tenth European Conference on Information Systems* (pp. 252-261). Retrieved from http://www.csrc.lse.ac.uk/asp/aspecis/20020114.pdf

Flood, R. L., & Romm, N. R. A. (1996). *Diversity management: Triple loop learning*. Chichester: Wiley.

Glass, R. (2001). *Computing failure.com*. Upper Saddle River, NJ: Prentice Hall.

Glass, R. (2005). Never the CS and IS Twain shall meet? *IEEE Software*, 120–119.

Glass, R., Ramesh, V., & Vessey, I. (2004). An analysis of research in computing disciplines. *Communications of the ACM, 47*(6), 89–94. doi:10.1145/990680.990686

Gonzales, R. (2005). Developing the requirements discipline: Software vs. systems. *IEEE Software*, (March/April): 59–61. doi:10.1109/MS.2005.37

Hevner, A. R., March, S. T., Park, J., & Ram, S. (2004). Design science in information systems research. *MIS Quarterly, 28*(1), 75–105.

Hirschheim, R., & Klein, H. K. (2003). Crisis in the IS field? A critical reflection on the state of the discipline. *Journal of the Association of Information Systems, 4*(5), 237–293.

Holwell, S. (2000). Soft systems methodology: Other voices. *Systemic Practice and Action Research, 13*(6), 773–797. doi:10.1023/A:1026479529130

Hughes, J., & Wood-Harper, T. (1999). Systems development as a research act. *Journal of Information Technology, 14*, 83–94. doi:10.1080/026839699344764

IbisSoft. (2007) *Environment that promotes IS research*. Retrieved from http://www.ibissoft.se/english/index.htm?frameset=research_frame.htm&itemframe=/english/about_isenvironment.htm.

Iivari, J., Hirschheim, R., & Klein, H. (2004). Towards a distinctive body of knowledge for information systems experts: Coding ISD process knowledge in two IS journals. *Information Systems Journal, 14*, 313–342. doi:10.1111/j.1365-2575.2004.00177.x

Iivari, J., & Huisman, M. (2007). The relationship between organizational culture and the deployment of systems development methodologies. *MIS Quarterly, 31*(1), 35–58.

Jackson, M. C. (1992). An integrated programme for critical thinking in information systems research. *Journal of Information Systems*, *2*, 83–94. doi:10.1111/j.1365-2575.1992.tb00069.x

Jackson, M. C. (2003). *Systems thinking. Creative holism for managers.* Chichester: Wiley.

Jackson, M. C. (2006). Creative holism: A critical systems approach to complex problem situations. *Systems Research and Behavioral Science*, *23*(5), 647–657. doi:10.1002/sres.799

Kroes, P., Franssen, M., van de Poel, I., & Ottens, M. (2006). Treating socio-technical systems as engineering systems: Some conceptual problems. *Systems Research and Behavioral Science*, *23*(6), 803–814. doi:10.1002/sres.703

Kruchten, P. (2005). Casting software design in the Function-Behavior-Structure framework. *IEEE Software*, (March-April): 52–58. doi:10.1109/MS.2005.33

Kuhn, T. S. (1970). *The structure of scientific revolutions* (2nd ed.). University of Chicago Press: Chicago.

Lane, J. A., & Boehm, B. (2007). System-of-systems cost estimation: Analysis of lead system integrator engineering activities. *Information Resources Management Journal*, *20*(2), 23–32.

Lane, J.-A., Petkov, D., & Mora, M. (2008). Software engineering and the systems approach: A conversation with Barry Boehm. *International Journal of Information Technologies and Systems Approach*, *1*(2), 99–103.

Laware, J., Davis, B., & Peruisch, K. (2006). Systems thinking: A paradigm for professional development. *The International Journal of Modern Engineering*, *6*(2).

Lee, A. (2000). Systems thinking, design science, and paradigms: Heeding three lessons from the past to resolve three dilemmas in the present to direct a trajectory for future research in the information systems field. *Keynote Address at 11th International Conference on Information Management, Taiwan, May 2000.* Retrieved from http://www.people.vcu.edu/aslee/ICIM-keynote-2000

Lee, A. (2004). Thinking about social theory and philosophy for information systems. In J. Mingers & L. Willcocks (Eds.), *Social theory and philosophy for information systems* (pp. 1-26). Chichester: Wiley, Chichester.

Lee, B., & Miller, J. (2004). Multi-project software engineering analysis using systems thinking. *Software Process Improvement and Practice*, *9*(3). doi:10.1002/spip.204

Lewis, P. (1995). New challenges and directions for data analysis and modeling. In F. Stowell (Ed.), *Information systems provision: The contribution of soft systems methodology* (pp. 186-205). London: McGraw-Hill.

Linstone, H. A. (1984). *Multiple perspectives for decision making. Bridging the gap between analysis and action.* New York: North Holland.

Madachy, R. J. (2006). Integrated modeling of business value and software processes. In *Unifying the software process spectrum* (LNCS 3840, pp. 389-402).

Madachy, R. J. (2008). *Software process dynamics.* Wiley. IEEE Press.

Madachy, R. J., Boehm, B., & Lane, J. A. (2007). Software lifecycle increment modeling for new hybrid processes. In *Software Process Improvement and Practice.* Wiley. Retreived from http://dx.doi.org/10.1002/spip.332

Mason, R., & Mitroff, I. (1981). *Challenging strategic planning assumptions.* New York: Wiley.

Mathieu, R. G. (2002). Top-down approach to computing. *IEEE Computer, 35*(1), 138–139.

McLeod, R. (1995). Systems theory and information resources management: integrating key concepts. *Information Resources Management Journal, 8*(2), 5–14.

Mingers, J. (1995). Using soft systems methodology in the design of information systems. In F. Stowell (Ed.), *Information systems provision: The contribution of soft systems methodology* (pp. 18-50). London: McGraw-Hill.

Mingers, J. (2001). Multimethodology- mixing and matching methods. In J. Rosenhead & J. Mingers (Eds), *Rational analysis for a problematic world revisited*. Chichester: Wiley.

Mitroff, I., & Linstone, H. (1993). *The unbounded mind*. New York: Oxford University Press.

Mora, M., Gelman, O., Forgionne, G., Petkov, D., & Cano, J. (2007). Integrating the fragmented pieces of IS research paradigms and frameworks: A systems approach. *Information Resources Management Journal, 20*(2), 1–22.

Ngwenyama, O. K., & Lee, A. S. (1997). Communication richness in electronic mail: critical social theory and the contextuality of meaning. *MIS Quarterly, 21*(2), 145–167. doi:10.2307/249417

Petkov, D., Misra, R., & Petkova, O. (2008). Some suggestions for further diffusion of work system method ideas in systems analysis and design. *CONISAR/ISECON Proceedings*, Phoenix, November.

Petkov, D., & Petkova, O. (2008). The work system model as a tool for understanding the problem in an introductory IS Project. *Information Systems Education Journal, 6*(21).

Petkov, D., Petkova, O., Andrew, T., & Nepal, T. (2007). Mixing multiple criteria decision making with soft systems thinking techniques for decision support in complex situations. *Decision Support Systems, 43*, 1615–1629. doi:10.1016/j.dss.2006.03.006

Petkova, O., & Roode, J. D. (1999). An application of a framework for evaluation of factors affecting software development productivity in the context of a particular organizational environment. *South African Computing Journal, 24*, 26–32.

Pfleeger, S. L. (2008). *Software engineering theory and practice*. Upper Saddle River, NJ: Prentice Hall.

Pressman, R. (2009). *Software Engineering, A practitioner's approach* (7th ed.). New York: McGraw Hill.

QAA. (2007). *Subject benchmark statements: Computing*. Quality Assurance Agency, UK. Retrieved from http://www.qaa.ac.uk/academicinfrastructure/benchmark/statements/

Reifer, D. (2003). Is the software engineering state of the practice getting closer to the state of the art? *IEEE Software, 20*(6), 78–83. doi:10.1109/MS.2003.1241370

Richmond, B. (1994, July). System dynamics/systems thinking: Let's just get on with it. In *Proceedings of the 1994 International System Dynamics Conference*, Sterling, Scotland. Retrieved from http://www.hps-inc.com/st/paper.html

Saaty, T. (1990). *Multicriteria decision making - The analytic hierarchy process* (2nd ed.). Pittsburgh: RWS Publications.

Sage, A. P. (2005). *Systems of systems: Architecture based systems design and integration*. Keynote address. International Conference on Systems, Man and Cybernetics, Hawaii, USA.

Sewchurran, K., & Petkov, D. (2007). A systemic framework for business process modeling combining soft systems methodology and UML. *Information Resources Management Journal, 20*(3), 46–62.

Sommerville, I. (2007). *Software Engineering* (8th ed.). Harlow: Pearson.

Stevens Institute of Technology. (2008). *Systems Engineering Research Center*. Retrieved from http://www.stevens.edu/sercuarc/

Stowell, F. (Ed.). (1995). *Information systems provision: The contribution of soft systems methodology*. London: McGraw-Hill.

Stowell, F., & West, D. (1996). Systems thinking, information systems practice. In *Proc. 40th Conference of the Int. Society For Systems Sciences*.

SWEBOK. (2004). *Software engineering body of knowledge*. Defined by the IEEE CS and ACM, http://www.swebok.org/ironman/pdf/SWEBOK_Guide_2004.pdf

Sweeney, A., & Bustard, D. W. (1997). Software process improvement: making it happen in practice. *Software Quality Journal, 6*, 265–273. doi:10.1023/A:1018572321182

Ulrich, W. (1998). *Systems thinking as if people mattered: Critical systems thinking for citizens and managers*. Working paper No.23, Lincoln School of Management.

Valerdi, R., & Madachy, R. (2007). Impact and contributions of MBASE on software engineering graduate courses. *Journal of Systems and Software, 80*(8), 1185–1190. doi:10.1016/j.jss.2006.09.051

Vilet, V. (2000). *Software engineering: Principles and practices*. Wiley.

Vo, H. V., Chae, B., & Olson, D. L. (2006). Integrating systems thinking into IS education. *Systems Research and Behavioral Science, 23*(1), 107–122. doi:10.1002/sres.720

Vo, H. V., Paradice, D., & Courtney, J. (2001). Problem formulation in inquiring organizations: A multiple perspectives approach. In *Proceedings of the Seventh Americas Conference on Information Systems*, Boston, MA.

Weinberg, G. (1992). *Quality Software Management, Volume 1: Systems Thinking*. Dorset House Publishing, New York

Weinberg, G. (2007). a site for books, articles and courses at http://www.geraldmweinberg.com/

Wendorff, P. (2002). Systems Thinking In Extreme Programming. In *Proceedings of the Tenth European Conference on Information Systems*, 203-207, available at http://www.csrc.lse.ac.uk/asp/aspecis/20020124.pdf

White, L., & Tacket, A. (1996). The End of Theory? *Omega, 24*(1), 47–56. doi:10.1016/0305-0483(95)00048-8

Wilson, B. (1990). *Systems: Concepts, Methodologies and Applications*. Second ed., Wiley, Chichester.

Xu, L. D. (2000). The contribution of systems science to information systems research. *Systems Research and Behavioral Science, 17*(2), 105–116. doi:10.1002/(SICI)1099-1743(200003/04)17:2<105::AID-SRES287>3.0.CO;2-M

Yourdon, E. (2007). a site for books, articles and blogs at http://www.yourdon.com

Zhu, Z. (2006). Complementarism versus pluralism: are they different and does it matter? *Systems Research and Behavioral Science, 23*(6), 757–770. doi:10.1002/sres.706

Chapter 6
Pluralism, Realism, and Truth:
The Keys to Knowledge in Information Systems Research

John Mingers
University of Kent, UK

ABSTRACT

The aim of this article is to outline some of the key themes that I believe are important, first, in apply-ing the systems approach to produce high quality IS research in general and, second, to consider more specifically some of the questions and debates that are of interest within the philosophy of IS and of the systems approach. Four themes are identified: being systemic, being critical and realist, being pluralist in approach, and having a concern for truth and knowledge.

INTRODUCTION

It gives me great pleasure to welcome a new jour-nal devoted to the systems approach, in this case applied in the domain of information systems and information technology. I have been appointed as Senior Associate Editor for the area of Information Systems Philosophy and the Systems Approach, and I have been asked to contribute a short posi-tion article for this, the inaugural issue of the *International Journal of Information Technologies and the Systems Approach* (IJITSA).

The aim of this article is to outline some of the key themes that I believe are important, first, in applying the systems approach to produce high quality IS research in general and, second, to consider more specifically some of the ques-tions and debates that are of interest within the philosophy of IS and of the systems approach. This article explores my own personal position and particularly themes such as pluralism, critical realism, and multimethodology, so I need to state that these are not necessarily those of the journal as a whole. The official Call for Papers makes it clear that IJITSA is interested in a wide range of research areas and methodological approaches, so potential authors should not feel circumscribed by what I say, but nevertheless I believe these themes are very much in accord with the thinking of the founders of the journal and do not get aired as much as they should in other IS journals. Papers along these lines would therefore be very much welcomed.

PHILOSOPHY AND SYSTEMS

I would like to begin by giving a brief review of the role of philosophy with respect to systems research with the help of Figure 1, which draws in part on the ideas of both Bhaskar (1978) and Checkland (1999).

We can begin in the bottom of Figure 1 with the ongoing flux of events and ideas. Following Bhaskar, we can term this the domain of the Actual, the actual occurrences and non-occurrences of the everyday world. We can then see that these events are the manifestations of underlying mechanisms or systems, often unobservable, which through the interactions of their properties and powers, generate the events. We should note two things: that the events are part of the causal dance in that they can be triggers of the underlying mechanisms and that human beings are also part of this picture as powerful generative mechanisms.

Moving up Figure 1, we can see that science emerges as a domain of reflective action in which people try to understand and explain the workings of the everyday world. This involves observation and interrogation of the Actual and the Real, as well as attempts to test and validate theories and, in the case of action research, explicitly to bring about change. In Bhaskar's terms, this is the domain of the Empirical in which a small subset of all the actual events that occur is captured for scientific activity.

We can now move to another metalevel and consider the emergence of philosophy, more specifically the philosophy of science, as a domain of reflective action that considers the nature of science and research and, in particular, tries to offer guidance about how science can and/or should be carried out. The main philosophical

Figure 1. The domains of science and philosophy with respect to the Real, the Actual and the Empirical

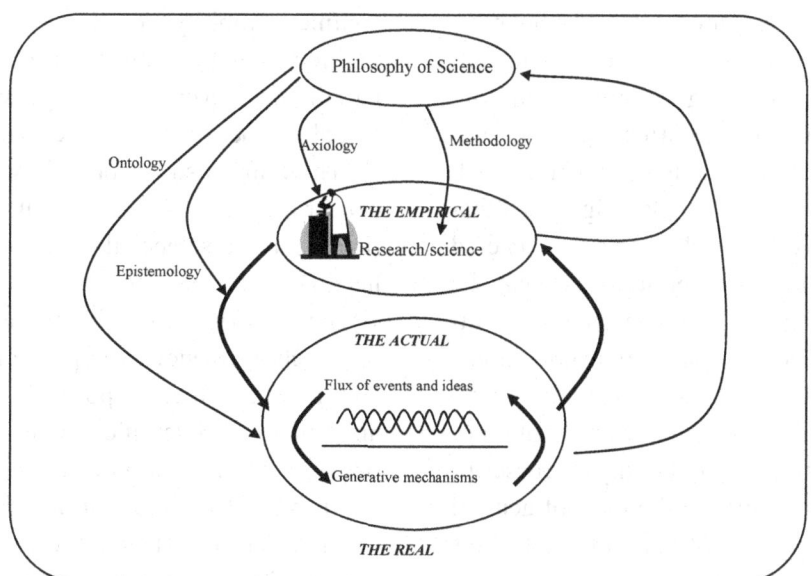

Note that in fact the Empirical is a subset of the Actual and Philosophy of Science is a subset of the Empirical but they have been drawn apart for clarity. The whole constitutes the Real.

questions that arise can be classified in terms of (Mingers, 2003a):

- **Ontology**: what kinds of objects or entities may be taken to exist, and what are their types of properties or forms of being?
- **Epistemology**: what is our relationship, as human beings, to the objects of our knowledge (including ourselves), and what distinguishes valid (i.e., true) knowledge from belief or opinion?
- **Methodology:** Given the first two, what methods should we use to acquire valid knowledge?
- **Axiology:** what are the purposes or values of science? What are the ethical or moral limits of science (if any)?

We can see from Figure 1 how these four elements relate to various aspects of the world and of science. The whole of Figure 1 constitutes what Bhaskar would call the domain of the Real.

A particular set of assumptions about these four questions can be called a paradigm (Burrell & Morgan, 1979; Kuhn, 1970). Kuhn originally used the term to describe the historical development of one set of theories (e.g., the theory of relativity) from another (e.g., Newtonian physics) but now, especially in the social sciences, paradigms are often seen to be co-existing. For much of the 20th century, the prevailing scientific paradigm was empiricism, more specifically positivism, which sees science as limited to explaining events that can be empirically observed. Events are expected to display regularities or patterns that can be explained as being particular instances of universal laws of the form "given certain conditions, whenever event X occurs then event Y will occur". Science is seen as the systematic observation of event regularities (Humean causality), the description of these regularities in the form of general laws, and the prediction of particular outcomes from the laws.

This view of science was extensively critiqued. The idea of pure, objective perception and observation was exploded by psychologists, sociologists, and philosophers; others showed that observational terms were not an atomistic picturing of reality but part of a pre-given linguistic structure—in short, that all observation was theory-dependent; and Popper (1959, 1969), based on Hume, rejected the possibility of verification and induction, replacing it with falsification and deduction. In response, there developed the standard "deductive-nomological (D-N)" or "hypothetico-deductive" method.

This view leads to a much greater recognition of the social and psychological nature of scientific activity. The idea of paradigms replacing each other over time has developed, particularly within social science, to the idea of there being competing paradigms existent at the same time (e.g., positivist, interpretive, and critical). This is often combined with the claim that paradigms are incommensurable. Clearly, the Kuhnian view has major relativistic implications for empiricism. It highlights the constructed, conventional nature of scientific theorising, and truth is that which is accepted by a scientific community rather than correspondence to some external reality. The incommensurability thesis is even more undermining since it makes it impossible to judge between paradigms or even assert that a later paradigm is actually superior to an earlier one.

There are also major philosophical debates concerning the nature of social science in relation to natural science that can only be sketched here. Broadly, there are three possible positions: (1) the *naturalist* view that there is one general approach to science that applies to all domains. Within this category, positivists hold that for anything to be scientific, it must follow the canons of positivism/empiricism and thus be based on universal generalisations from empirical observations. (2) The antithesis is the view that the social world is intrinsically different to the natural world, being constituted through language and meaning, and thus involves entirely different hermeneutic, phenomenological, or social constructivist approaches. The argument here would be the idealist one that ontologically social objects do not exist in the way physical ones do (i.e., as subject independent) and that epistemologically there is no possibility of facts or observations that are independent of actors, cultures, or social practices. (3) The most radical position denies the possibility of objective or scientific knowledge at all, in either domain. Arguments here come from the strong sociology of knowledge program: poststructuralists such as Foucault (1980) and, more generally, postmodernists (Best & Kellner, 1991) who attempt to undermine even the most basic categories of modernist rationality.

In the rest of the article, I want to put forward four themes which I value in terms of generating significant and reliable knowledge: systems thinking, pluralism, critical realism, and truth. Whilst research published in IJITSA clearly does not have to conform to all of these, I would expect that it does accord with at least some.

THE SYSTEMS APPROACH

The systems approach at its most general involves deciding explicitly to employ systems concepts in theorising and research. Some of the basic concepts are a system as a collection of elements that are linked together such that the behaviour or characteristics of the whole depend on the relationships between the parts rather than the nature of the parts themselves; a boundary separating the system from its environment; subsystems and wider systems that form a nested hierarchy; multiple-cause relations and feedback loops; communication and control systems; the inevitable entanglement of the observer with the observed; and a commitment to holism rather than reductionism. Beyond these basic concepts, there are many particular systems approaches that have been developed, for instance, systems

engineering, cybernetics, system dynamics, soft systems, general systems theory (GST), living systems theory, or complexity theory.

How does systems thinking relate to the philosophical questions and paradigms outlined above? I would argue that it could be conceived of as a paradigm in the sense that it does have implications for each of the levels, but in a sense, it contains within it the particular paradigms described above. There are empiricist, interpretive, and critical versions of systems thinking just as there are its opposite—reductionism or individualism.

To what extent is the systems approach evident in the IS/IT literature? At one level, one could say that systems thinking is at least implicit in much of the IS research. The discipline is called Information *Systems* after all, and I suspect that few academics would say they were reductionists, or deny that they assumed a systems approach. However, the number of papers that formally or explicitly claim to use systems seems very small. Although I have by no means conducted a thorough survey, looking through some of the main IS journals over the last five years revealed precious few papers. Perhaps one of the most relevant to this article and IJITSA is by Mora, Gelman, Forgionne, Petkov, and Cano (2007) which aims to use the systems approach to provide a unifying framework for the disparate research paradigms within IS. Porra, Hirschheim, and Parks (2005) examined the history of Texaco's IT function using GST, interestingly using several different "lenses," thus showing that GST can be used in an interpretive manner. Champion, Stowell, and O'Callaghan (2005) developed a framework for supporting the client during the process of systems design based on Checkland's (1999) Soft Systems methodology and Vickers' appreciative systems concepts. McBride (2005) used the ideas of chaos and complexity theory to examine the implementation of an IS strategy in the U.K. Probation Service. Johnstone and Tate (2004) developed a systems model of human information processing behaviour that incorporated both qualitative and quantitative views of the same problem. Finally, we can sympathise with Alter (2004), whose ICIS paper was titled "Desperately Seeking Systems Thinking in the IS Discipline", who argued for a systems thinking approach as opposed to a "tool thinking" approach within IS.

Hopefully, IJITSA can to some extent redress this by ensuring that all the papers it publishes are based explicitly in systems thinking.

CRITICAL REALISM

Critical realism (CR) is a theory within the philosophy of science that has been developed primarily by Roy Bhaskar over the last 30 years (Bhaskar, 1978, 1979, 1993). In essence, it maintains a position that is opposed to both positivism (empiricism) and interpretivism (idealism) while accepting elements of both. CR strongly espouses ontological realism—that is, the existence of a reality independent of our knowledge of it, yet also accepts epistemological relativism—that is, that our knowledge of that reality will always be locally situated and provisional. Critical realism is becoming increasingly influential within information systems (Dobson, 2001a, 2001b; Klein, 2004; Longshore Smith, 2006; Mingers, 2004b; Monod, 2004; Mora et al., 2007; Mutch, 1999; Pather & Remenyi, 2004; Wikgren, 2005) and management, more generally (Ackroyd & Fleetwood, 2000; Contu & Willmott, 2005; Fleetwood & Ackroyd, 2004; Hunt, 2005; Mutch, 2005; Reed, 2001, 2005; Willmott, 2005).

In fact, we can distinguish between Critical Realism and "critical" "realism". What I mean is that CR is one specific philosophical approach but there are other ways of being both *critical*, for example, critical hermeneutics (Myers, 2004) or critical theory (Klein & Huynh, 2004) or *realist*, for example, scientific realism (Hunt, 2005) or critical scientific realism (Niiniluoto, 2002).

The idea of "being critical" or of critique has a long history rooted in the work of both Kant and Marx. At its most general, we can see it as a deep form of questioning, of refusing to take things for granted, or accepting the status quo (Mingers, 2000). We can distinguish two main senses, one linked to epistemology and one to axiology in Figure 1. The first, Kantian idealist sense, is concerned with knowledge and the limits to or conditions of knowledge. Here we are questioning particular forms or types of knowledge; the explicit, or often implicit, assumptions made by particular theories; or indeed the transcendental limits of knowledge itself. The second, Marxist materialist sense, involves questioning the oppressive nature of society and its institutions and organisations. The two senses can be seen to come together in the work of Habermas (1978, 1984, 1987) and later Foucault (1988a, b) who had epistemological concerns about the nature of knowledge and portrayed how knowledge was constituted and distorted by the operation of power within society.

Within IS, there has been a long, if somewhat thin, line of critically-inspired research based mainly on the work of Habermas going back to Mingers' (1980) comparison of Habermas and soft systems methodology; Mumford, Hirschheim, Fitzgerald, and Wood-Harper's (1985) exploration of alternative research approaches; and Lyytinen and Hirschheim's (1988) application of communicative action theory. Klein and Huynh (2004) provide a comprehensive coverage of Habermas's theory and its relevance for IS while Sheffield (2004) and Heng and de Moor (2003) are recent empirical studies. Foucault has been less well used although Zuboff's (1988) seminal book is inspired by his ideas. Willcocks (2004) provides an exceptionally clear introduction to Foucault's complex works and covers applications within IS and more broadly within management studies. There are other philosophers whose work falls within the critical realm but who have made less impact on information systems so far, including

Adorno (Probert, 2004) and Callon and Latour (actor-network theory) (Mitev & Wilson, 2004; Walsham, 1997). Howcroft and Trauth (2005) have produced a handbook on critical research in information systems which covers theories as well as practical applications.

I am particularly concerned that the research published in IJITSA should be (self-) critical at least in the epistemological sense. All research is inevitably embedded within the messy real world of power, politics, and interest, yet it is often described in papers as though it were pure and perfect. This is a delusion that we all collectively maintain. As authors, we do it because we fear our papers will be rejected, and as referees/editors, we do it because we fear the journal's reputation for rigorous research will be tarnished. In reality, however, as has been continually shown by the critical research discussed above, it is a fallacy to suppose that rigour comes from sanitised results; that facts can be divorced from values; or that research can be isolated from power and interest (Backhouse, Hsu, & Silva, 2006; Cordoba & Robson, 2003; Doolin, 2004). It is much better to be open and explicit about the real context of the research so that readers can judge for themselves any potential effects or biases. A framework of questions for assisting this process can be found in Mingers (1997).

PLURALISM

We have seen that different paradigms developed within IS and other social science disciplines—positivism, interpretivism, critical, and perhaps postmodernism—and for a period these became like silos, each separate and isolated from the others (Burrell & Morgan, 1979). Each paradigm developed its own research methods and research had to be conducted from within one paradigm. There were divisive debates between the paradigms (Mingers, 2004a).

However, this polarisation could not continue and it came to be recognised within social science generally that the world does not divide itself up neatly into hard and soft, quantitative and qualitative, and that to understand its full complexity we need to combine methods from a variety of perspectives. Tashakkori and Teddlie (1998) published one of the first methodology books entirely on mixed methods in social science based on many studies that had already been carried out, and Mingers and Gill (1997) published a book on "multimethodology", that is, combining management science methodologies. This movement has grown considerably with a new handbook of mixed methods (Tashakkori & Teddlie, 2003); a new journal—*Journal of Mixed Methods Research*; and discussion of mixed methods in general research methodology textbooks, for example, Bryman and Bell (2003).

There are several different positions with respect to mixed methods or multimethodology, and debate in this area is certainly welcome in IJITSA. My own view is a strong one in that I argue for plurality not just of methods but of all the four philosophical dimensions in Figure 1. Moreover, I suggest that this stance actually follows from both a systems perspective, seeing the world in a holistic way, and a critical realist perspective which accepts a plurality of objects of knowledge. To employ Habermas's (1984) model (discussed further below), we can say that communications and actions relate to three different "worlds"— the objective or material world that consists of all actual or possible states of affairs; the social or normative world that consists of accepted and legitimate norms of behavior; and the subjective or personal world that consists of individuals' emotions, feelings, and ideas. As human beings, we have different epistemological access to each of these ontological domains: we *observe the* material world, we *participate in our* social world, and we each *experience "my"* own personal world. This in turn generates the need for distinctive methodologies which then need to be combined together to synthesise our understanding of the whole. Finally, we have distinct axiological relations to the worlds (Habermas, 1993), pragmatic with respect to the material world, moral for the social world, and ethical for the personal world. Finally, this leads into the view that there are actually different forms or types of knowledge and that these are distinguished by different forms of truth (Mingers, 2003b).

TRUTH

It is paradoxical that "truth" is one of the great unspoken concepts of the IS literature. If we assume that the purpose of scholarly research is the generation of valid and reliable knowledge, and that truth is a central characteristic of valid knowledge, then we might expect that there should be discussion and debate about the nature of truth and procedures for discovering the truth. In fact, the subject is never mentioned in the IS literature, even in the literature on knowledge management where one might expect that a concern with the validity of knowledge was central. I searched the ISI Web of Knowledge using the very general key words "information systems" and "truth" going back to 1970. Amazingly, only two proper IS papers were returned (Sheffield, 2004; Zhou, Burgoon, & Twitchell, 2004), neither of which had any sustained discussion of truth.

This situation could be explained if the concept of truth was essentially simple and uncontested, yet this is actually far from the case. Differing positions in terms of ontology and epistemology inevitably lead to different views on the nature of truth, and even as to whether truth is a necessary or attainable characteristic of knowledge. The most common view, in Western philosophy, is that knowledge is *true, justified, belief (TJB)*. This stems from Plato's *Theaetetus* where Socrates argues that knowledge implies not just that a belief be true but also that there be a rational explanation for it. These three conditions have been taken to

be both necessary and sufficient for a proposition to count as knowledge. In other words, to validly assert "I know that p ..." implies:

1. You must sincerely believe that p is the case.
2. You must have justifiable grounds, evidence, or explanation for p.
3. p must, indeed, be true.

Although this sounds straightforward, there are in fact many problems with each condition as well as their conjunction. For instance, there is much debate about what would constitute proper justification for such a belief—empirical evidence, rational argument, personal experience, perception, or what? How in any case can we determine if something is actually true? More fundamentally, however, there are several different and competing theories of truth (see Mingers, 2003b, for a fuller discussion and references).

Correspondence theories (Popper, 1959; Russell, 1912; Tarski, 1944; Wittgenstein, 1974) are the main and most obvious view of truth. They hold that truth (and falsity) is applied to propositions depending on whether the proposition corresponds to the way the world actually is. *Coherence theories* (Bradley, 1914; Putnam, 1981; Quine, 1992) stress the extent to which a proposition is consistent with other beliefs, theories, and evidence that we have. The more that it fits in with other well-attested ideas, the more we should accept it as true. This approach avoids the need for a direct comparison with "reality". However, it is more concerned with the justification of beliefs rather than their absolute truth. *Pragmatic theories* (James, 1976; Peirce, 1878; Rorty, 1982) hold that truth is best seen in terms of how useful or practical a theory is—that which best solves a problem is the best theory. A version of this is instrumentalism, which holds that a theory is simply an instrument for making predictions and has no necessary connection to truth at all. This also leads into consensus theories. An obvious argument against this view is that a *true* theory is likely to be most useful and powerful[1] and therefore should be an important component of a *useful* theory.

Consensus or discursive theories (Gadamer, 1975; Habermas, 1978) hold that truth is that which results from a process of enquiry resulting in a consensus amongst those most fully informed—in the case of science, scientists. At one level, we can see that this must be the case if we accept with critical realism the impossibility of proving correspondence truth. But, today's accepted truth is usually tomorrow's discarded theory and so this does not guarantee truth. Finally, there are *redundancy, deflationary,* and *performative theories* (Frege, 1952; Horwich, 1991; Ramsey, 1927; Strawson, 1950) which argue, in different ways, that the whole concept of truth is actually redundant. If we say "it is true that snow is white", we are saying no more than that "snow is white"; the two propositions will always have the same truth values and are therefore equivalent.

Turning now to critical realism, what view of truth does it espouse? The first thing to say is that the whole approach is fallibilist. That is, since CR accepts epistemic relativity, the view that all knowledge is ultimately historically and locally situated, it has to accept that theories can never be proved or known certainly to be true. Thus, if provable truth were to be made a necessary criteria for knowledge, there could be no knowledge within critical realism.

Bhaskar does discuss the notion of truth and comes up with a multivalent view involving four components or dimensions (Bhaskar, 1994, p. 62) that could apply to a judgment about the truth or falsity of something. The first level is sincerity or trustworthiness: truth as being that which is believed from a trustworthy source—"trust me, I believe it, act on it."; second is warranted or justified: based on evidence and justification rather than mere belief—"there's sound evidence for this."; third is weak correspondence: corresponding to or at least being adequate to some intransitive object of knowledge. Whereas the first two dimensions

are clearly in the transitive domain and strongly tied to language, this aspect moves beyond to posit some sort of relation between language and a referent. Finally, ontological and alethic: this level is the most controversial (Groff, 2000) as it moves truth entirely into the intransitive domain. It is the truth of things in themselves, and their generative causes, rather than the truth of propositions. It is no longer tied to language, although it may be expressed in language.

It is also interesting to consider Habermas's theories of knowledge and truth as his work has been applied within information systems. His first framework was known as the theory of knowledge-constitutive interests (KCI) (Habermas, 1978). This suggested that humans, as a species, had needs for, or interest in, three particular forms of knowledge. The *technical* interest in molding nature led to the empirical and physical sciences. For Habermas, these were underpinned by a pragmatist philosophy of science (inspired by Peirce) and a consensus theory of truth. The *practical* interest in communication and mutual understanding led to the historical and interpretive sciences underpinned by a hermeneutic criterion of understanding. And the *emancipatory* interest in self-development and authenticity led to critical science which identified repressions and distortions in knowledge and in society. Its criterion of success was the development of insight and self-expression free from constraint.

This theory of transcendental interests was the subject of much criticism (see Mingers, 1997, for a review), and Habermas later transmuted it into the theory of communicative action (TCA) (Habermas, 1984, 1987). Utterances and, I would argue, actions as well, raise certain validity claims that must, if challenged, be justified. These claims are *comprehensibility, truth, rightness,* and *truthfulness (sincerity)*. This is premised on the argument that utterances stand in relation to the three different "worlds" discussed above. When such a claim is challenged, the process of justification must always be discursive or dialogical.

That is, there should ideally be a process of open debate unfettered by issues of power, resources, access, and so on until agreement is reached by the "unforced force of the better argument" (Habermas, 1974, p. 240), what Habermas calls the "ideal speech situation". Thus, Habermas held a consensus or discursive view of truth both in the moral or normative domain of what ought we to do, as well as in the material domain of external reality. To say of a proposition, "it is true" is the same as saying of an action "it is right", namely *ideal, warranted assertability.*

However, more recently, Habermas (2003) has returned to the issue of truth and now rejects his discursive theory for propositions about the material world in favor of one with an irreducible ontological (i.e., realist) component. In essence, Habermas now maintains that there is a substantive difference between the moral domain of normative validity which can only ever be established through discussion and debate within an ideal speech situation, and the domain of propositional truth where properly arrived at and justified agreement may still be proven wrong by later events. Our experience of living in and coping with the world shows us that even the most strongly held and well-justified views may turn out to be false. "The experience of 'coping' accounts for two determinations of 'objectivity': the fact that the way the world is is not up to us; and the fact that it is the same for all of us" (Habermas, 2003, p. 254). Thus, the whole question of truth is linked intimately with that of knowledge and thus with research and the whole scholarly and scientific enterprise.

CONCLUSION

This article has addressed two related issues: what makes for rigorous research using the systems approach within information systems, and what are the areas for debate and development within the philosophy of IS and systems. I have identified four themes—being systemic, being critical and realist, being pluralist in perspective and methodology, and being concerned with the nature of knowledge and truth—which I believe are the hallmarks for producing research that is both rigourous and relevant to the complexities and seriousness of the problems we face in the organisational world. Each of these themes is an area of research and debate in its own right and hopefully will generate interesting and scholarly papers for the journal.

REFERENCES

Ackroyd, S., & Fleetwood, S. (2000). *Realist perspectives on management and organisations.* London: Routledge.

Alter, S. (2004). *Desperately seeking systems thinking in the information systems discipline.* Paper presented at ICIS 25, Washington, DC.

Backhouse, J., Hsu, C., & Silva, L. (2006). Circuits of power in creating *de jure* standards: Shaping an international information systems security standard. *MIS Quarterly, 30,* 413-438.

Best, S., & Kellner, D. (1991). *Postmodern theory: Critical interrogations.* New York: Guilford Press.

Bhaskar, R. (1978). *A realist theory of science.* Hemel Hempstead: Harvester.

Bhaskar, R. (1979). *The possibility of naturalism.* Sussex: Harvester Press.

Bhaskar, R. (1993). *Dialectic: The pulse of freedom.* London: Verso.

Bhaskar, R. (1994). *Plato etc.* London: Verso.

Bradley, F. (1914). *Essays on truth and reality.* Oxford: Oxford University Press.

Bryman, A., & Bell, E. (2003). *Business research methods.* Oxford: OUP.

Burrell, G., & Morgan, G. (1979). *Sociological paradigms and organisational analysis.* London: Heinemann.

Champion, D., Stowell, F., & O'Callaghan, A. (2005). Client-led information system creation (clic): Navigating the gap. *Information Systems Journal, 15*(3), 213-231.

Checkland, P. (1999). *Systems thinking, systems practice: Includes a 30-year retrospective* (2nd ed.). Chichester: Wiley.

Contu, A., & Willmott, H. (2005). You spin me round: The realist turn in organization and management studies. *Journal of Management Studies, 42*(8), 1645-1662.

Cordoba, J., & Robson, W. (2003). Making the evaluation of information systems insightful: Understanding the role of power-ethics strategies. *Electronic Journal of Information Systems Evaluation, 6*(2), 55-64.

Dobson, P. (2001a). Longitudinal case research: A critical realist perspective. *Systemic Practice and Action Research, 14*(3), 283-296.

Dobson, P. (2001b). The philosophy of critical realism: An opportunity for information systems research. *Information Systems Frontiers, 3*(2), 199-210.

Doolin, B. (2004). Power and resistance in the implementation of a medical management information system. *Information Systems Journal, 14*, 343-362.

Fleetwood, S., & Ackroyd, S. (Eds.). (2004). *Critical realist applications in organisation and management studies.* London: Routledge.

Foucault, M. (1980). *Power/knowledge: Selected interviews and other writings 1972-1977.* Brighton: Harvester Press.

Foucault, M. (1988a). Truth, power, self: An interview with Michel Foucault. In L. Martin, H. Gutman & P. Hutton (Eds.), *Technologies of the self: An interview with Michel Foucault* (pp. 9-15). Amherst: University of Massachusetts Press.

Foucault, M. (1988b). What is enlightenment? In P. Rabinow (Ed.), *The Foucault reader* (pp. 32-50). London: Penguin.

Frege, G. (1952). *Translations from the philosophical writings of Gottlob Frege* (P. Geach & M. Black, Trans.). Oxford: Blackwell.

Gadamer, H. (1975). *Truth and method.* New York: Seabury Press.

Groff, R. (2000). The truth of the matter: Roy Bhaskar's critical realism and the concept of alethic truth. *Philosophy of the Social Sciences, 30*(3), 407-435.

Habermas, J. (1974). *Theory and practice.* London: Heinemann.

Habermas, J. (1978). *Knowledge and human interests* (2nd ed.). London: Heinemann.

Habermas, J. (1984). *The theory of communicative action: Vol. 1: Reason and the rationalization of society.* London: Heinemann.

Habermas, J. (1987). *The theory of communicative action: Vol. 2: Lifeworld and system: A critique of functionalist reason.* Oxford: Polity Press.

Habermas, J. (1993). On the pragmatic, the ethical, and the moral employments of practical reason. In J. Habermas (Ed.), *Justification and application* (pp. 1-17). Cambridge: Polity Press.

Habermas, J. (2003). *Truth and justification.* Cambridge: Polity Press.

Heng, M., & de Moor, A. (2003). From Habermas's communicative theory to practice on the Internet. *Information Systems Journal, 13*, 331-352.

Horwich, P. (1991). *Truth.* Oxford: Blackwell.

Howcroft, D., & Trauth, E. (Eds.). (2005). *Handbook of critical information systems research: Theory and application*. London: Edward Elgar.

Hunt, S. (2005). For truth and realism in management research. *Journal of Management Inquiry, 14*(2), 127-138.

James, W. (1976). *The meaning of truth*. Cambridge, MA: Harvard University Press.

Johnstone, D., & Tate, M. (2004). Bringing human information behaviour into information systems research: An application of systems modelling. *Information Research, 9*(4), 1-31.

Klein, H. (2004). Seeking the new and the critical in critical realism: Deja vu? *Information and Organization, 14*(2), 123-144.

Klein, H., & Huynh, M. (2004). The critical social theory of Jurgen Habermas and its implications for IS research. In J. Mingers & L. Willcocks (Eds.), *Social theory and philosophy for information systems* (pp. 157-237). Chichester: Wiley.

Kuhn, T. (1970). *The structure of scientific revolutions*. Chicago: Chicage University Press.

Longshore Smith, M. (2006). Overcoming theory-practice inconsistencies: Critical realism and information systems research *Information and Organization, 16*(3), 191-211.

Lyytinen, K., & Hirschheim, R. (1988). Information systems as rational discourse: An application of Habermas's theory of communicative rationality. *Scandinavian Journal of Management Studies, 4*(1), 19-30.

McBride, N. (2005). Chaos theory as a model for interpreting information systems in organizations. *Information Systems Journal, 15*(3), 233-254.

Mingers, J. (1980, April). Towards an appropriate social theory for applied systems thinking: Critical theory and soft systems methodology. *Journal of Applied Systems Analysis, 7*, 41-50.

Mingers, J. (1997). Towards critical pluralism. In J. Mingers & A. Gill (Eds.), *Multimethodology: Theory and practice of combining management science methodologies* (pp. 407-440). Chichester: Wiley.

Mingers, J. (2000). What is it to be critical? Teaching a critical approach to management undergraduates. *Management Learning, 31*(2), 219-237.

Mingers, J. (2003a). A classification of the philosophical assumptions of management science methods. *Journal of the Operational Research Society, 54*(6), 559-570.

Mingers, J. (2003b). *Information, knowledge and truth: A polyvalent view* (Working Paper No. 77). Canterbury: Kent Business School.

Mingers, J. (2004a). Paradigm wars: Ceasefire announced, who will set up the new administration? *Journal of Information Technology, 19*, 165-171.

Mingers, J. (2004b). Re-establishing the real: Critical realism and information systems research. In J. Mingers & L. Willcocks (Eds.), *Social theory and philosophical for information systems* (pp. 372-406). London: Wiley.

Mingers, J., & Gill, A. (Eds.). (1997). *Multimethodology: Theory and practice of combining management science methodologies*. Chichester: Wiley.

Mitev, N., & Wilson, M. (2004). What we may learn from the social shaping of technology approach. In J. Mingers & L. Willcocks (Eds.), *Social theory and philosophy for information systems* (pp. 329-371). Chichester: Wiley.

Monod, E. (2004). Einstein, Heisenberg, Kant: Methodological distinctions and conditions of possibility. *Information and Organization., 14*(2), 105-121.

Mora, M., Gelman, O., Forgionne, G., Petkov, D., & Cano, J. (2007). Integrating the fragmented pieces of is research paradigms and frameworks: A systems approach. *Information Resources Management Journal, 20*(2), 1-22.

Mumford, E., Hirschheim, R., Fitzgerald, G., & Wood-Harper, T. (Eds.). (1985). *Research methods in information systems.* Amsterdam: North Holland.

Mutch, A. (1999). Information: A critical realist approach. In T. Wilson & D. Allen (Eds.), *Proceedings of the 2nd Information Seeking in Context Conference* (pp. 535-551). London: Taylor Graham.

Mutch, A. (2005). Critical realism, agency and discourse: Moving the debate forward. *Organization, 12*(5), 781-786.

Myers, M. (2004). The nature of hermeneutucs. In J. Mingers & L. Willcocks (Eds.), *Social theory and philosophy for information systems* (pp. 103-128). Chichester: Wiley.

Niiniluoto, I. (2002). *Critical scientific realism.* Oxford: Oxford University Press.

Pather, S., & Remenyi, D. (2004). *Some of the philosophical issues underpinning research on information systems: From positivism to critical realism.* Paper presented at the SAICSIT 2004, Prague.

Peirce, C. (1878, January). How to make our ideas clear. *Popular Science Monthly.*

Popper, K. (1959). *The logic of scientific discovery.* London: Hutchinson.

Popper, K. (1969). *Conjectures and refutations.* London: Routledge and Kegan Paul.

Porra, J., Hirschheim, R., & Parks, M. (2005). The history of Texaco's corporate information function: A general systems theoretical interpretation. *MIS Quarterly, 29*(4), 721-746.

Probert, S. (2004). Adorno: A critical theory for IS. In J. Mingers & L. Willcocks (Eds.), *Social theory and philosophy for information systems* (pp. 129-156). Chichester: Wiley.

Putnam, H. (1981). *Reason, truth, and history.* Cambridge: Cambridge University Press.

Quine, W. (1992). *Pursuit of truth.* Boston: Harvard University Press.

Ramsey, F. (1927). Facts and propositions. *Proceedings of the Aristotelian Society, 7.*

Reed, M. (2001). Organization, trust and control: A realist analysis. *Organization Studies, 22*(2), 201-223.

Reed, M. (2005). Reflections on the "realist turn" in organization and management studies. *Journal of Management Studies, 42*(8), 1621-1644.

Rorty, R. (1982). *Consequences of pragmatism.* Minnesota University Press.

Russell, B. (1912). *The problems of philosophy.* Oxford: Oxford University Press.

Sheffield, J. (2004). The design of gss-enabled interventions: A Habermasian perspective. *Group Decision and Negotiation, 13*(5), 415-435.

Strawson, P. (1950). Truth. *Proceedings of the Aristotelian Society, 24.*

Tarski, A. (1944). The semantic conception of truth. *Philosophy and Phenomenological Research, 4*, 341-375.

Tashakkori, A., & Teddlie, C. (1998). *Mixed methodology: Combining qualitative and quantitative approaches.* London: Sage Publications.

Tashakkori, A., & Teddlie, C. (2003). *Handbook of mixed methods in social and behavioural research.* Thousand Oaks, CA: Sage.

Walsham, G. (1997). Actor-network theory and IS research: Current status and future prospects. In A. Lee, J. Liebenau, & J. DeGross (Eds.), *Information systems and qualitative research* (pp. 466-480). London: Chapman Hall.

Wikgren, M. (2005). Critical realism as a philosophy and social theory in information science? *Journal of Documentation, 61*(1), 11-22.

Willcocks, L. (2004). Foucault, power/knowledge and information systems: Reconstructing the present. In J. Mingers & L. Willcocks (Eds.), *Social theory and philosophy for information systems*. Chichester: Wiley.

Willmott, H. (2005). Theorising contemporary control: Some post-structuralist responses to some critical realist questions. *Organization, 12*(5), 747-780.

Wittgenstein, L. (1974). *Tractatus logico-philosophicus* (Rev. ed.). London: Routledge and Kegan Paul.

Zhou, L., Burgoon, J., & Twitchell, D. (2004). A comparison of classification methods for predicting deception in computer-mediated communication. *Journal of Management Information Systems, 20*(4), 139-165.

Zuboff, S. (1988). *In the age of the smart machine: The future of work and power*. New York: Basic Books.

ENDNOTE

[1] Although postmodernists argue that it is the theory that is deemed most powerful that is accepted as true.

This work was previously published in International Journal of Information Technologies and Systems Approach, Vol. 1, Issue 1, edited by D. Paradice; M. Mora, pp. 79-90, copyright 2008 by IGI Publishing (an imprint of IGI Global).

Chapter 7
Information–As–System in Information Systems:
A Systems Thinking Perspective

Tuan M. Nguyen
HCMC University of Technology, Vietnam

Huy V. Vo
HCMC University of Technology, Vietnam

ABSTRACT

This article investigates the complex nature of information in information systems (IS). Based on the systems thinking framework, this study argues that information in IS is a system in its own right. A conceptual model of information-as-system is built on the systems thinking perspective adopted from Gharajedaghi's holistic thinking rooted from Ackoff systems approach, which is developed through Peirce's semiotics with the validity support of Metcalfe and Powell's perspective of information perception, Mingers and Brocklesby's schema of situational actions, Toulmin's theory of argumentation and Ulrich's theory of systems boundary. The proposed model of information-as-systems is described in terms of triads–on the structure, function, and process, all interdependent–in a context of information-as-system in IS.

INTRODUCTION

Information is the central object in several fields of information-related studies (Tuomi, 1999), including information systems (IS). Drucker (1999) argued that one of the challenges of managements in the next century is just information, which is not relating to technologies, but focusing on how to satisfy information requirements for knowledge workers and business managers in various societies and organizations.

Lee (2004) stated that information itself is the very rich phenomenon of subject matter distinct from organizations' or information technology's one. Therefore, IS researchers should pay more attention to the nature of information, which so far is less studied than the problems of technologies, organizations, or systems. Meanwhile, Ulrich (2001) recommended that we need to consider not only the ways of using ISs but also, more basically, the ways information is defined and becomes very important socially. Similarly, Lauer (2001) proposed that information-oriented perspective is essential to IS; Metcalfe and Powell (1995) argued that the area that IS may claim as its very own one is just the information.

Unfortunately, little research has been paid to how to conceptualize the information itself (Lauer, 2001). It is warned that understanding the nature of information is even more important than the process of IS design (Metcalfe & Powell, 1995) because IS may not exist without information (Mingers, 1996). Likewise, to understand what the information is and how to use it shall support us to solve the problems of requirements engineering of ISs (Goguen, 1996). To these researchers, knowledge of the nature of information is non-trivial, but there is still no agreement on what information is.

Recently, several studies have attempted to construct conceptual models of information. For example, Callaos and Callaos (2002) proposed a systemic notation of information, which is based on a dialectic process, consisting of two components—data and information—as well as of two respective relationships—perceptions or sensations and actions. The data component represents the objective side of information and the information component represents the subjective side of data. Meanwhile, Buckland (1991) suggested that information in IS is just information as thing. In this sense, information may be a tangible form to represent objects and events and it is in nature situational, consensus-led evidence. From a different view, Mingers (2006) proposed a comprehensive theory of semantic and pragmatic information, in which information, an object of IS, is associated closely with meaning, an object of human cognition.

This article is an exploratory study of the nature of information in IS. The key thesis is that the information in IS is a system, or specifically, a meaning system (Mingers, 1995), a human activity

system (Checkland, 2000), an inquiring system (Churchman, 1971), or an open system (Emery, 2000) in its own right. Based on Gharajedaghi's (2005) holistic thinking developed from Ackoff's systems approach (1974), a conceptual model of information-as-system is developed through the pragmatism semiotics (Peirce, 1931) with the validity support of the perceiver-concerns model of information (Metcalfe & Powell, 1995), the schema of situational actions (Mingers & Brocklesby, 1997), the argumentation theory (Toulmin, 1964) and the theory of systems boundary judgment (Ulrich, 2003). The proposed model is described in terms of triads on the structure, function, and process and context of information in IS. The main research questions are both whether or not information could be seen as system and what would be the system's model of information?

The article is organized in the following manner. First, we discuss the complex phenomenon of information in information related studies, including IS. Second, a brief discussion of Peirce's semiotics is presented. Third, we propose a systems model of information-as-system based on Gharajedaghi's systems thinking framework. Next we make some comparisons of our model with others, especially with Mingers' comprehensive theory of semantic and pragmatic information to discuss the implications of our model. Finally, the article's findings as well as several theoretical and practical contributions are presented in the conclusion.

INFORMATION IN INFORMATION SYSTEMS

While IS researchers mostly investigate organizational contexts to develop computer-based systems relying on functions or roles of individuals in various organizations (Ellis, Allen, & Wilson, 1999), studies of information in IS focus on the pragmatic aspects and information related applications, particularly how information is used in organizations, interacted with organizational factors, and exchanged among stakeholders. More, the field of IS requires a multidisciplinary approach to studying the range of socio-technical phenomena that determine their development, use, and effects in organizations and societies (Ellis, Allen, & Wilson, 1999). Given the teleological nature of IS (Churchman, 1971), the concept of information would be around social aspects that are shown as resources and constructive forces in society (Braman, 1989).

The whole picture of the field of information-related studies are briefly as follows: (i) information theory focuses on the forms of information, the signal flow from source to destination, (ii) information science aims at the information content, chiefly on textual information to deliver information services such as libraries, reference database via communications channels like journals, books, conferences, and so on, and (iii) IS focuses on the pragmatic aspect of information to develop computer-based systems for various organizational contexts (Ellis, Allen ,& Wilson, 1999).

Therefore, the disciplines of information theory, information science and IS have the same object of study, which is information. However, the focal aspects of the so-called information investigated vary from discipline to discipline. At this point, foreshadowing the semiotic triangle as a triadic relation of three most fundamental categories of all of human experience (*Figure 1*), we can designate that information has a triadic nature, in which its syntactic dimension, with information theory, plays the role of firstness or signifier, its semantic dimension, with information science, the role of secondness or signified, and its pragmatic dimension, with IS, the role of thirdness or signification.

Peirce's Pragmatism Semiotics

First, according to Peirce, semiotics is the formal doctrine of signs in all their aspects and, in a different view; it is also the theory of logic in general (CP 2.93; 2.227)[1]. We may learn that Peirce's semiotics is a direct foundation of his theories, which will be employed in our next triads, such as theory of perception (CP 1.336), of inference (CP 2.435-444), and of inquiry (CP 5.374-377), as well as of cognition and knowledge (CP 1.537; 2.60-66).

Second, Everaert (2006) argued that Peirce's semiotics has three foundational categories, which are necessary and sufficient for all of human experience and designated as firstness, secondness and thirdness. In other words, representamen, object, and interpretant are in order firstness, secondness and thirdness (CP 2.242), which are the three most universal categories of elements of all experience (CP 1.417-418). From this, it is also obvious that the Peircean model of sign is

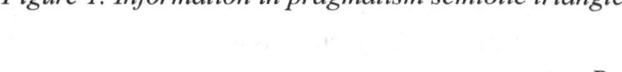

Figure 1. Information in pragmatism semiotic triangle

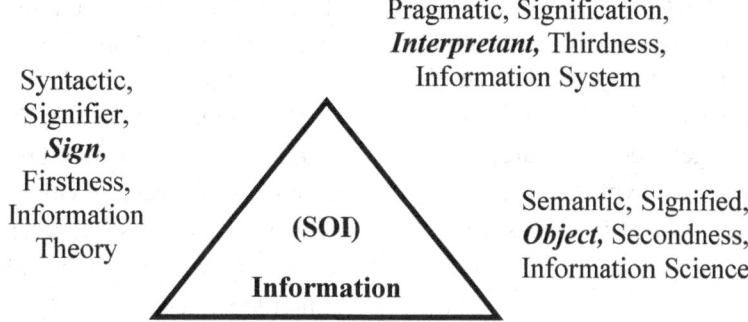

Syntactic, Signifier, **Sign,** Firstness, Information Theory

Pragmatic, Signification, **Interpretant,** Thirdness, Information System

Semantic, Signified, **Object,** Secondness, Information Science

(SOI)

Information

beyond Buhler's (1982) version in which sign, which is also admitted a kind of triadic pattern, turns out to be equal linguistic utterance, not a general expression of human experience. Next, in reference to Bhaskar's (1993) semiotic triangle of signifier, referent and signified, we may find that the Peircean version would be more generic and theoretically powerful, in the perspective of phenomenology in general and of IS foundations in specific. In fact, Bhaskar's conception of sign is put in the framework of critical realism, an underpinning philosophy for IS (Mingers, 2004), that Nellhaus (1998, p.21) remarked that:

Critical realism can fill many gaps in its semiotic theory by adopting a Peircean understanding of signs and semiosis, which grasps both the intransitive dimension of reality, and the semiotic nature of our mental access to it in the transitive dimension.

More, Nellhaus (p.3) argued that the Peircean theory of signs not only *"clearly satisfies critical realism"* but also *"possesses other critical realist features, including ones not envisioned by Bhaskar."*

Third, one more very interesting point in Peirce's work is just the *"love for triads,"* as Burks commented:

All three philosophers (Peirce followed Kant and Hegel) found triadic categorical schemes to be useful ways of structuring their theories of reality. (Burks, 1996, p.329, parenthesis added)

Specifically, Kant organized his table of judgments and categories in four groups of three and argued that, *In every group, the number of categories is always the same, namely, three. This is remarkable...* (Sowa, 2000, p.58). Hegel used his triadic logical schema of thesis-antithesis-synthesis iteratively to explain many aspects of reality (Burks, 1996), and Peirce himself confirmed that, *"every relation*

... of any greater number of correlates is nothing but a compound of triadic relations" (CP 1.347). Thus, to Peirce, there are only three, *"the number Three in philosophy"* (CP 1.355), categories in his schema, which beyond *"there is nothing else to be found in the phenomenon"* (CP 1.347). This is very distinctive, in comparison with Kant's table of categories where *"the third category is merely derivative"* (Sowa, 2000, p.58) and to Hegel's schema in which *"the three categories have not their several independent and irrefutable standings in thought"* (CP 5.91). Likewise, in reference to Morris' (1938) and then Nauta's (1972) five term model of sign (Mingers, 2006), the Peircean version of three-element sign would better meet the condition of *"the minimum number of entities that must be involved"* (Sowa, 2000, p. 61). As a summary, we would like to designate the Peircean version *pragmatism semiotics* to emphasize its nature as general, triadic, and pragmatic at once (Everaert, 2006).

SYSTEMS MODEL OF INFORMATION

Systems Descriptions of Information

First, for systems epistemology, system is a human construct (Metcalfe, 2004) of basic characteristics of general systems theory devised by von Bertalanffy (1968). Next, Metcalfe (2004) mentioned that the very first concept of systems theory is the boundary, which is the core idea of critical systems thinking such as Ulrich's (2003) and Midgley's (2003) ones and is also considered a separator between scientific thinking and systems thinking (Metcalfe, 2004).

The article prefers Gharajedaghi's schema of systems thinking growth. Generation 1, operations research, is to struggle with interdependences of mechanical systems. Generation 2, cybernetics and open systems, is to fight with double challenges of interdependence and self-organization in living or biological systems. Generation 3, design, is to cope with triple challenges of interdependence, self-organization and choice in socio-cultural systems that are bonded with information. With his schema, Gharajedaghi argued that systems thinking now in generation 3 is featured with interactive design, which Gharajedaghi intensively developed from Ackoff's (1993) design thinking. The design thinking is also emphasized in Churchman's (1971), Banathy's (1996), and Nelson's (1994) systems thinking and is essential for other disciplines opposed to natural sciences and living sciences (Simon, 1996). In such domains, we may recognize, in reference to traditional classification of systems thinking, that Gharajedaghi's design thinking is closer to critical systems thinking than soft or hard systems thinking. Moreover, Gharajedaghi's interactive design approach is supported with a comprehensive methodology that is applied broadly by Gharajedaghi himself.

Gharajedaghi emphasized on descriptions of social systems via plurality of structure, function, process, and their relationships in a specified context. This conception obviously reflects Midgley's (2003) systemic intervention in which three main concerns are boundary critique, theoretical and methodological pluralism, and action for improvement. Next, Gharajedaghi's descriptions are also quite similar to Banathy's (1992) three-viewpoint approach, which comprises viewpoint of system–environment, of function-structure and of process. Besides, the viewpoint is more or less familiar with Flood's (1999) four-window approach that applies ethical and critical thinking into systems thinking to show four windows simultaneously. Four windows are process, structure, meaning, and knowledge or power.

With the descriptions of process and context, Gharajedaghi's systems approach is able to accommodate Gunaratne's (2003) new systems thinking that is based on theories of complexity and non-linear dynamics and, hence, recognized the very important role of the factor of time in investigation of social systems. Moreover, Gharajedaghi's

systems approach is also closely associated with the continuity assumption, a continuity of the whole existing spatially as well as temporally. The continuity assumption, a core assumption of systems thinking (Barton, Emery, Flood, Selsky, & Wolstenholme, 2004), is an essential component in Peirce's principle of continuity (CP 1.171) and theory of evolution (CP 8.317-318), hence, some researchers suggest that the Peircean pragmatism is fundamentally a better choice of foundation for general systems thinking (Barton, 1999).

Structural Aspect of Information-as-System

Based on the semiotics model *(Figure 1)*, we view the structure of information-as-system as a triad of SOI including *sign* (S), *object* (O) and *interpretant* (I) or alternatively, including *signifier, signified* and *signification*. The respective relationship manifested by information is also a triadic one of syntactic, semantic, and pragmatic dimension of any communicative expression (Morris, 1938). The communication bears meaning of action of informing (verb) presented via the signifier and meaning of content of informing (noun) presented via the signified and meaning of effect of informing presented via the signification that is specified by the interpretant, and moreover, is an activity of interpretation beyond informing.

Note that a modified version of the triad can be applied when the object component (O) in the triad is replaced with the value component (V). This modification is adopted due to the insight provided by Goguen (1997) that the signified is specified with social values (V). In our triad model, the structure of information always includes the human component, because to Peirce, a sign is *"something which stands to somebody for something in some respect or capacity"* (CP 2.228). This idea is largely supported in the literature. For instance, relying on the need to include people in any information perspective, and human "concern" as a key to deal with information, Metcalfe and Powell (1995) proposed a *"perceiver-concerns perspective"* in which nothing rather than perceivers can generate information. Likewise, Mason (1978, p. 220) confirmed:

Every sign, to be 'informative' must signal, influence, persuade, indicate or otherwise affect the user. So the user becomes the sine qua non of any sign.

In the context of IS, the emphasis on human element in the structural dimension of information equals an emphasis on pragmatic aspect of information itself, namely, the *"rules governing the relationship between signs and their users"* (Mason, 1978, p. 220). However, the interpretant here should be extended into a *"community of competent inquirers"* to simultaneously interpret and justify the same phenomena (Ulrich, 2001). Rather similar to this is Goguen's (1996) viewpoint that information is generated by social groups (G). Thus, the interpretant component (I) could and should be extended into the component of community of inquirers (C) or social groups (G). This thought is in line with Peirce's pragmatic theory of reality and knowledge as expressed with Peirce's own words:

... the very origin of the conception of reality shows that this conception essentially involves the notion of a COMMUNITY, without definite limits,

and capable of a definite increase of knowledge. (CP 5.311)

After all, the original structural triad of SOI should be transformed into that of SOC (or SVG) to emphasize the importance of community rather than of individual (*Figure 2*), and our structural triad of information-as-system including individual or community would be a Mingers' (1995) human meaning system.

Functional Aspect of Information-as-System

The functional aspect of information-as-system is determined relying on Peirce's pragmatic theory of cognition or the triple model of human experience. According to Peircean epistemology, there are three worlds of experience and they are under the philosophical principle of continuity[2] (CP 7.438; 1.171). For the Outer, or the world of things, and the Inner, or the world of ideas, human experience is called "action" and "perception" respectively. Peirce explained that experience is *"the course of life"* (CP 1.426), is a consciousness of two varieties, which are both *"action, where our modification of other things is more prominent than their reaction on us, and perception, where their effect on us is overwhelmingly greater than our effect on them"* (CP 1.324). Meanwhile, specific to the third world, or the intermediate or logical world

Figure 2. Structural aspect (SOC) of information-as-system

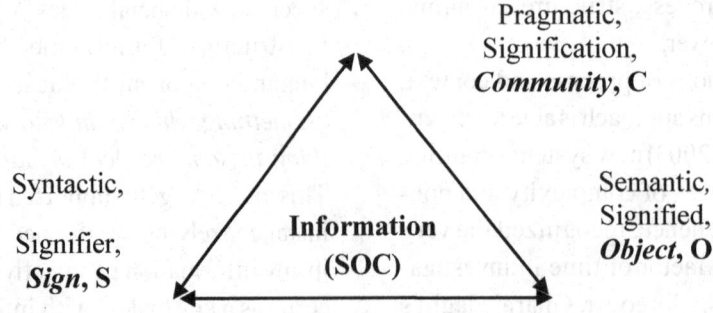

(CP 8.299), human experience turns into "cognition" in Peirce's term, which is a consciousness of synthesis, of mediation or of a process and *"this in the form of the sense of learning, of acquiring, of mental growth is eminently characteristic of cognition"* (CP 1.381-82).

Secondly, as warrants for our suggestion, we take, for instance, Boland, Tenkasi, and Te'eni's (1994) model of distributed cognition and Mingers and Brocklesby's (1997) schema of situational actions. In fact, distributed cognition could be considered the process through which individuals construct and reconstruct system of their roles through self-reflection (personal), dialogue (social), and action (material). Next, note that Mingers and Brocklesby (1997) added human means of accessibility into problem situations and hence made clear kinds of human actions for three real worlds of Habermas *(Table 1)*. From this, taking Ulrich's (2001) argument that Habermas's analysis starting from the pragmatic level in place of Peirces's one starting from the empirical level and that both shared the same *"discursive kernel of knowledge,"* which actually originated from Peirce, we upwardly map the triple of Peirce's worlds of human experience into the triple of Habermas' real worlds. As a result, the functional aspect of information-as-system can now be represented with a triad of social actions, here after referred to as the triad of DCC, that are design (D), creativity (C) and culture (C) *(Figure 3)*. Human design oriented to the material world, human creativity to the personal world, and human culture to the social world, respectively.

For the material world, action aims at building purposeful systems (Gharajedaghi, 2005) for social development. Such a development requires human to have activities of construction, or, in nature, activities of design that assist to understand system (Churchman, 1971), to change the existing (Ulrich, 2001), and to develop society and create future (Ackoff, 1974). Philosophically, as Churchman (1971) emphasized, systems design is a tool for improvements of human conditions.

Table 1. Situational actions in Habermas' three worlds

Analytical world	Epistemic relationships	Knowledge interests	Human means of accessibility
Material	Objective	Technical	Action
Social	Intersubjective	Practical	Languaging
Personal	Subjective	Emancipation	Emotion
[Adapted from Habermas (1984), Mingers and Brocklesby (1997)]			

Figure 3. Functional aspect (DCC) of information-as-system

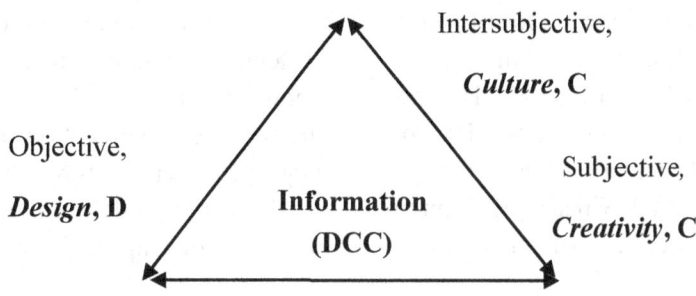

The improvement, to Churchman, implies learning and this turns out to be fit very well with the Peircean conception of "cognition" as a "sense of learning" just mentioned above.

For the personal world, action is a reflection or cognition accompanied with emotion (Gharajedaghi, 2005). It may be worth mentioning Dewey's (1997) approach of four types of thinking ranging from random thinking to reflective thinking where there is a continuous learning loop between thoughts and reflections on thoughts and thus knowledge creation becomes possible. Meanwhile, taking Ackoff's (1971) types of systems behaviors, Gharajedaghi (2005) concluded that action is type of behavior where outside changes are neither necessary nor sufficient conditions. The respective systems, purposeful systems, can select means as well as ends of changes and are able to generate the same outcome in different ways in the same environment and can produce different outcomes in the same and different environment. In sum, they are free will and *"such systems not only adapt and learn; they can also create"* (Gharajedaghi, 2005, p. 36). Therefore, action is featured as self-determined behavior of human beings, a typical purposeful system, and it is an autonomous activity resulting from creative capability including reasoning and emotion on the one hand, adaptation and learning on the other hand.

For the social world, communication needs to be accompanied with discourse to generate interactions (Ulrich, 2001) and hence to build culture (Gharajedaghi, 2005). Here, discourse is a way of construction of social reality (Habermas, 1984) or of organizational decision-making (Richardson, Courtney, & Haynes, 2006). By this, communications and discourses are linked into social interactions— between individuals, intra and inter-group, as well as organizations (Ulrich, 2001). Based on social interactions, shared image of a person could be formed, which is also human culture in society in order for members of social systems via their experience, belief, attitude, and ideals to get connected altogether (Gharajedaghi, 2005). For people, *"culture is the ultimate product and reflection of their history and the manifestation of their identity"* (Gharajedaghi, 2005, p. 121). Broadly, Gharajedaghi argued that social systems need to have learning capability, also called culture, which in turn consists of two aspects: (i) cognitive that includes attributes belonging to languages, meaning, thinking, and reasoning; (ii) normative that includes attributes belonging to values, belief, and social contracts. Therefore, the viewpoint of culture is seen by Gharajedaghi as an important dimension to build social systems by which culture creates human as well as human creates culture.

In summary, the functional aspect of information are reflected in social actions, and strongly expressed in terms of mutual relationships between individuals and their worlds of experience. In this sense, it is easily recognized that the functional points of design (D), of creativity (C) and of culture (C) are beyond "action," "emotion," or "self-reflection" and "languaging" or "dialogue" in Mingers and Brocklesby (1997) or Boland, Tenkasi and Te'eni (1994) models respectively in the context of sociocultural systems, which are information-bonded systems (Gharajedaghi, 2005). Because our functional triad is composed of social actions, in this respect, our model of information-as-system would be a Checkland's (2000) human activity system.

Process Aspect of Information-as-System

The process dimension of information-as-system, drawing on Peirce's pragmatic theory of knowledge, is divided into three (sub)processes: the process of perception, of abduction and of inquiry. Perception through observations of the outside world results in empirical data. Abductive operation of empirical data creates new ideas, hypotheses, or knowledge. Note that abduction is in nature both inference and insight (CP 1.332) and hence shows individual mental efforts (ego) to cope with the outside world (non-ego). Inquiry means to arrange, test, and confirm or refuse new ideas, hypotheses, or knowledge so as to draw some conclusions, or equally information that needs to be accepted by some community. Remind that, because inquiry is essentially inference or reasoning, inquiring methods are very important and identified by Peirce totally, such as method of tenacity, of authority, of a priori, and of science (CP 5.377-87). However, the very method of science is the only one acceptable by community of inquirers for scientific inquiry.

At the end of inquiry, information itself just formulated is ready for the next cycle of information formulation with influencing the following human perceptions and abductions. And the process of information-as-system or equally information formulation as just described is evolutionary theoretically indefinitely. In the light of semiotics, getting back to the structural aspect of information-as-system, information-as-system is just a typical semiotic triangle, hence, also just a semiosis, a Peirce's term, an infinite process of sign production and transmission (Mingers, 2006).

Support for our point is evident most in Toulmin's (1964) theory of argumentation. According to Toulmin's theory, information may be seen as an argumentative inquiry process. There are three key states in this process, which are data, warrant, and claim. Data are evidences or facts used as basis for argumentation and also called the start of reasoning. Warrant is a component of logic linking between data and claim and also plays the role of reasoning process through which speaker convinces audience of how to reach some conclusion from existing data. Finally, claim is result, conclusion, or goal of argumentation and is also what speaker would like to support. Therefore, as a process, information-as-system would have three states, hereafter referred to, for the sake of simplicity, as DWC, that are data (facts, given or D for short); warrant (backing, knowledge, or W for short); and claim (conclusion, information, or C for short) (*Figure 4*).

Courtney's (2001) and Richardson, Courtney, and Haynes' (2006) works are close to our descriptions about the states of the information process. For example, Courtney used Bock's (1998) and Nonaka's (1994) schemas of knowledge creation, through which information (explicit forms, in a shared way) is converted back and forth into knowledge (tacit, in heads of people). Hence, both schemas are essentially cyclical processes, which are similar to ours, but the relevant states of the processes seemed not to be specified in full. Similarly, Richardson, Courtney, and Haynes' theoretical principles for the Singerian inquiring systems development introduce the kinds of process of

Figure 4. Aspect of process (DWC) of information-as-system

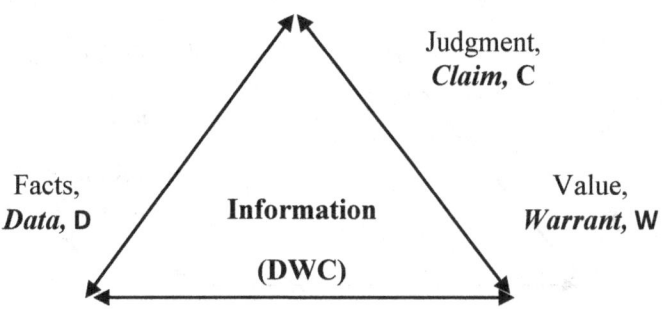

knowledge creation, which are put under the process of discussive action of Habermas. However, the process under discussion is limited into only generation of exoretic knowledge, the knowledge very essential for the Singerian inquiry, namely, for broad social and managerial problems and for more loosely coupled or community of practice. Thus, the transformation among the respective processes for various types of knowledge is not fully modeled. The special feature of our model of information as an argumentative inquiring process is that information may evolve over time and in space through the loops that exchange the roles of three different states: data (D), of warrant (W) and of claim (C). We will discuss time issue in the next paragraph, and space issue in the next section.

For the time perspective, Fenzl (2005) argued that there could be logical relationships between the past, present and future in discussion of information. Given that time only flows from past to future and, hence, on the one hand, human cannot get back to the past (Prigogine, 1997) or information is irreversible (Fenzl, 2005) and, on the other hand, social systems are open systems, far-from-equilibrium and historical (Gunaratne, 2003). Therefore, epistemically, information is the final state and the outcome of applying warrant onto given data. In other words, the following time relations are found (*Figure 5*): given data corresponding to objectivity or past tense; interpretive

knowledge to subjectivity or present tense; and resultant information or claim to intersubjectivity or future tense. Next, the resultant information or claim of previous iteration of argumentation process would be data of next iteration of the cyclical process of representation and interpretation of socio-cultural systems that are inherently bonded with information. Thus, briefly, we argue that the operation process of information as an argumentation process evolves over time and is irreversible.

For evolution in space, information-as-system could be seen in two points as follows. First, information, as reference system, may depend on observations (facts) or judgment (values) as indicated with Ulrich's (2003) boundary critique or, alternatively, interpretant (I) or community (C) may depend on sign (S) or object (O) in structure of information-as-system. As a result, information-as-system would evolve in space at boundary shift upon human or community interpretations. Second, there are two opposite hierarchies in the literature: data–information–knowledge (Ackoff, 1989), as well as knowledge–information–data (Tuomi, 1999) and Tuomi (1999) recognized that both hierarchies could be necessary. Thus, the roles among data, knowledge, and information can be transformed back and forth along the process of information formulation. And information evolution in space could be described a little bit more in the next section of contextual

Figure 5. Time aspect (PPF) of information-as-system

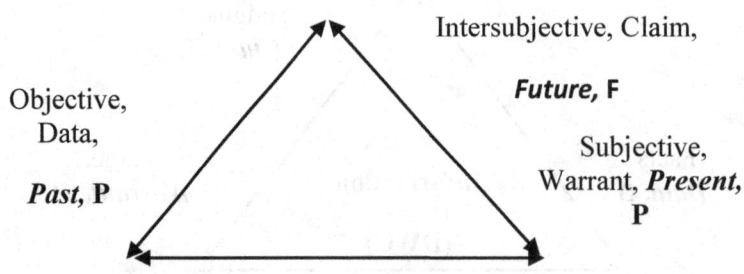

aspect of information-as-system. Further, in a meta-theoretical sense, the problem of evolution in space and over time seems, surprisingly, to be non-sense. That is because, to Peirce, *"space and time are relations connecting objects and events, and so ... cannot exist until objects and events exist"* (Burks, 1996, pp. 344-45).

Because our process triad of information-as-system is able to reach some convincing claim (or produce knowledge) for community, in this sense, it would be a Churchman' (1971) inquiring system.

Contextual Aspect of Information-as-System

The contextual aspect of information-as-system *(Figure 6)* is defined in a context relating to an IS (Lee, 2004) and a human (Metcalfe & Powell, 1995). Specifically, the context is an evolutionary triadic relation in which human is firstness, information secondness, and IS thirdness and they three co-evolve. In fact, to Lee, ISs consist of three (sub)systems, which are social system, technical system, and knowledge system. To be clear, the third system definitely comprises information as

its essential part because *"information cannot be neatly categorized under either the 'social system' heading or 'the technical system' heading"* (Lee, 2004, p. 13). On the other hand, human being is a familiar example of a purposeful system (Ackoff, 1971). From this and with our structural descriptions of information-as-system, we claim that the direct supersystem and the subsystem of information-as-system would be the IS and the interpretant or community of inquirers respectively. This relationship implies that, while any IS consists of at least a person (Mason & Mitroff, 1973), the human aspects are always embedded in the context of information. In fact, in reference to Mason and Mitroff's definition of an IS as five-element system, the human factor in IS is only manifested through the factor of information, which is in turn mentioned very clearly in the other elements of IS, which are psychological types, problems types, types of evidence generators and guarantors, organizational context, and modes of presentation.

Next, while the relationship between information-as-system and interpretants is already put in a semiotic triad as addressed in our structural, functional and process descriptions of informa-

Figure 6. Aspect of context (SIP) of information-as-system

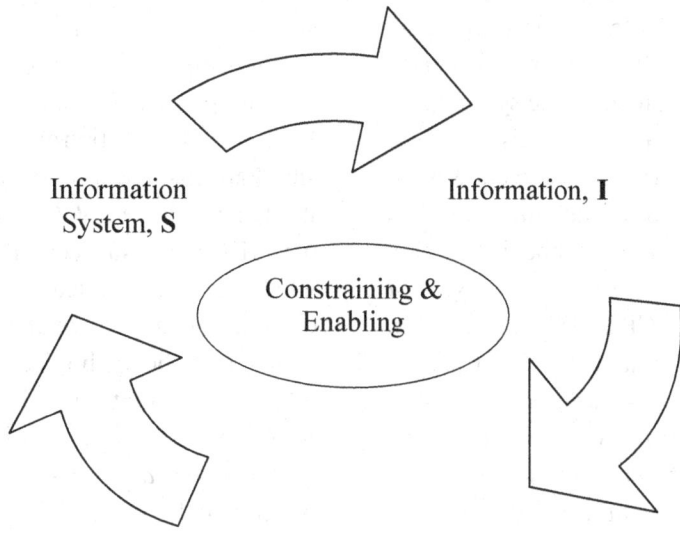

tion-as-system, the relationship between the IS and information needs to be elaborated more.

We make two following points. First, the IS itself, the IS social system, and the IS technical form a systemic triangulation, an Ulrich's (2003) concept, in which the IS plays the role of boundary judgment, the social system the role of value or evaluations, and the technical system the role of facts or observations. In Lee's (2004, pp. 11-2) words,

an information system is the result of an information technology enabling an organization, as much as an information system is the result of an organization enabling an information technology.

Lee's argument lends a full support to our boundary critique on IS and thus, our systemic triangulation of information-as-system, the IS social system and the IS technical system. Second, there may exist a recursive relationship between the IS and information-as-system, which is supported further with the arguments that both the IS and information-as-system consist of individual(s) and that in the extreme case of disappearing both the social and the technical system, IS would turn to mere information.

And last but not least, note that there may be a meta-theoretical analogy between IS as a systemic integration (i.e., system) of social system and technical system and information as a product (e.g., concerning a symbol) of connotation (depth) and denotation (breadth) (e.g., of the symbol). To be clearer, the former statement reads that IS is *"simply an instance of a sociotechnical system in general"* (Lee, 2004, p. 12) and the latter says that information concerning a symbol is *"the sum of synthetical propositions in which the symbol is subject or predicate"* (CP 2.418). Furthermore, while Lee (2004, p. 13) stated that IS *"would be the emergent result of the mutually and iteratively transformational interactions among the social system, the technical system and the knowledge system,"* Peirce considered information as a triadic

relation of concepts as concepts grow. More exactly, at this point, the analogy manifests an evolution from a level of system (i.e., information-as-system) to a higher level of system (i.e., IS).

At this point, we claim that, with our context triad referring to systems boundary, information-as-system would be an Emery's (2000) open system because it is able to meet the central theses for Emery's open systems thinking as follows: (i) there are systems and social environments as well as relationships between them; (ii) all systems are open and all boundaries are permeable; and (iii) people are at the heart of the system and create social environments and co-evolve with them (Barton, Emery, Flood, Selsky, & Wolstenholme, 2004, pp. 13-14).

DISCUSSIONS AND IMPLICATIONS

In this section, we shall make a couple of comparisons of our model to Buckland's (1991) concept of information-as-thing and Mingers' (2006; 1996) comprehensive theory of semantic and pragmatic information. Both versions are theoretically sound enough because they underlie some kinds of typologies, which are usually requested for formulation of scientific knowledge in general (Handfield & Melnyk, 1998). Besides, Mingers' works are at least much more typical and significant in IS because they have already covered evaluations of selected theories of information in the field.

First, our model of information is able to meet all four possible dimensions of information as Buckland's conception of information as thing in a simultaneous and all-in-one way. Particularly, with its structural triad SOC, information-as-system meets Buckland's aspects of *information-as-thing*, which may be included in its component S (sign or signifier) and of *information-as-knowledge*, which may be embodied in its component O (object or referent); with its process triad DWC, information-as-system satisfies Buckland's aspects of *information-as-process* (being informed), which may be the transformative flow from data to claim (i.e., information), and of *information processing* (i.e., data processing), which may be the application of knowledge onto data to produce something more significant. Our model of information-as-system not only includes *"the only form of information with which information systems can deal directly"* (Buckland, 1991, p. 359), but also covers *"important use of information... to denote knowledge imparted," "to denote the process of informing,"* and *"data processing."* Further, in an epistemic sense, our model of information-as-system is also compatible upward with Buckland's description of *"information by consensus"* because of intersubjectivity component as discussed earlier.

Therefore, from the practical view, our model of information-as-system could define classes of information-related activities expected by Buckland, such as various kinds of "information storage and retrieval systems," "knowledge bases for expert systems," and "statistical analysis" of "patterns in populations of objects and/or event" that belong to the objects of studies of various fields such as IS, information science, cognitive psychology, interpersonal communication and persuasion, and so on. And for the theoretical concerns, information-as-system can cover completely Buckland's four-aspect topography of information. Additionally, our functional triad crossing over all three worlds of human is in line fully with schema of three sources of information (Wersig, 1979, as cited in Buckland, 1991), the broader view of information that goes beyond Buckland's version of information-as-thing.

We next compare our conception with Mingers' (2006) very recent one, which we believe *"provides a clear and consistent conceptualization of the basic concepts of information"* (p. 128) because it is based on the evaluation of *"a comprehensive range of information theories and concepts"* (Mingers, 1996, p. 206). When comparing the two models, we expect not only to validate our conception but also to seek out some complementary aspects between them. And to be complete, our compari-

son shall be at two levels, meta-theoretical and theoretical level of the conception. For the former level, we consider the aspects of epistemology and ontology and for the latter level, the natures of and the relationships between information and its relatives as well as the aspects of process and context between two conceptions are compared. Note that our discussion here mainly refers to Peirce's semiotics, which is obviously a frame of reference for both models.

In an epistemic sense, our model of information seems to be more generic and simpler and hence more effective than Mingers' one in terms of the capability of theoretical explanation. However, Mingers' detailed accounts of kinds of sign are good enough to ontologically support our positioning basic concepts of information into semiotic triangle. In fact, whilst information in our conception is intersubjective, that of Mingers has both forms, subjective for *signification* and objective for the nested information or *intent*, which in turn has three possibilities that are analytic (i.e., by definition), nomic (i.e., by natural laws), and situational (social conventions and practices) consequences.

Rearranging Mingers' both forms, we suggest that there should be three new possibilities of information as follows: (i) *signification*, (ii) *analytic intent*, and (iii) *nomic*, as well as *situational intent*. Note that information in our model is Peirce's *interpretant* and hence, having three forms due to its nature of thirdness, and imagine that three possibilities of information just reordered are respectively three kinds of Peirce's *interpretant*, that is, *emotional* or *immediate, energetic* or *dynamical, logical* or *final interpretant*, the mapping between our conception and Mingers' one becomes direct as follows: (i) *signification* to *emotional interpretant*; (ii) *analytic* to *energetic*, and (iii) *nomic* as well as *situational* to *logical*. Hence, our conception of information as thirdness is supported perfectly with Mingers' forms of information.

Next, our term *knowledge* refers to the state of subjectivity whereas Mingers' *meaning* has also both dimensions; subjectivity for specific meaning or *import* and intersubjectivity for general meaning or *connotation system*. Meanwhile, remind that *knowledge* in our model is Peirce's *object* or *referent* and, hence, having two forms due to its nature of secondness, and also image that Mingers' two possibilities of *meaning* are correspondingly *immediate* and *dynamical* objects, our conception of *knowledge* as secondness matches successfully with Mingers' two forms of *meaning*.

One more point, take it that *meaning* is more subjective than *information*, and in reference to the very popular hierarchy of data-information-knowledge (Ackoff, 1989) and the reverse hierarchy of knowledge-information-data (Tuomi, 1999), for basic concepts of information in IS as data, information and knowledge (Courtney, 2001), we may prefer *knowledge* to *meaning*. From this, briefly, *information, knowledge*, and *data* are in line respectively with interpretant, object, and sign ontologically and with intersubjectivity, subjectivity, and objectivity epistemically as well as with thirdness, secondness, and firstness in the general sense of phenomenology. Getting back to Mingers' version, we may see that both *sign* and *meaning* in his model are in order assigned implicitly to *signifier* and *referent* in Bhaskar's semiotic triangle but *information* seems not to be mentioned very clearly in relation to *signified* in Bhaskar's.

At the theoretical level, first, the process aspect of information is the same basically in both models, but the transformation from *information* to *meaning* in Mingers' model is essentially a subprocess in our triad of process of information-as-system. Our triad model of process includes the process of abduction, of cognition, and of perception, which are in order the transformation from data to knowledge, then to information and then back to data newer, whereas Mingers' main process described in full is only the transformation from *information* to *meaning*, although he (Mingers,

2006, p.129) also recognized "*the role of the body and nervous system*" or "*embodied cognition*" in transformation of information back and forth. In addition, the introduction of time factor into our conception of information-as-system advanced our model in comparison to Mingers' one. With this, our model of information-as-system is more significant for development of new technologies employed in IS such as data mining, data warehouse, online analytical processing, and so on, which require a time component. As a result, our triads of process and of time may generally introduce the infinite evolution of information, in which information is any sign and for any human actions including but not limiting to human communications. Likewise, Mingers (2006) also referred to the similar growth but only for *meaning*, or seemingly emphasized on linguistic signs and intersubjective communications only.

Second, traditionally Mingers put information into the context of IS, which, as he emphasized, is really only a part of human meaning system. Meaning system, in turn, is composed of signs and signals, which are continually produced and interpreted and of course are carriers of information. Hence, for Mingers' version, information, IS, and meaning system are the nested levels and they all comprise the whole for human language and communication. However, Mingers also argued that information itself and even IS are meaningless until they get connected into the wider meaning system consisting of human beings. It is obvious that such conceptions of information as meaningless and of IS as non-human system are respectively opposed to the common sense usage of the term "information" and the traditional definition of IS as system consisting of at least a person (Mason & Mitroff, 1973).

Meanwhile, as developed earlier, our model puts person, information, and IS into a triadic relation in which person is firstness, information secondness, and IS thirdness. This means that these three levels are interdependent and each is significant in itself. Furthermore, for the general contextual problems of information with extra dimensions of truth, truthfulness and rightness, Mingers (2006) proposed Habermas's theory of communicative action as the most promising approach; whilst, to our version, such pragmatic dimensions are satisfied highly with the Peircean theories of reality, truth and knowledge, which have the same foundation, the pragmatism semiotics. One more final interesting point is that Habermas' validity claims of truth, truthfulness, and rightness, in the bottom line, can also be traced to the Peircean philosophical principle of continuity of the three worlds of experience, in which validity claim of truth belongs to Outer or the physical world, of truthfulness belongs to Inner or the personal world, and of rightness belongs to Logical or the social world. This may be not surprising to us because Peirce explained "*Logic is rooted in the social principle*" (CP 2.654) and Habermas also developed his "universal pragmatics" as universalism for social theory (Meadwell, 1994).

Third, regarding the nature of information, Mingers (2006) believes that *information* is a propositional content of signs but, for our conception, information is beyond *proposition*, which is secondness of Peirce's trichotomy of symbol, to reach to the status of an *argument*, which is thirdness of the trichotomy (Peirce, 1931). Simply put, we may find out some inference only in *argument*, not within *proposition*. For meaning, Mingers stated that the semantic content (meaning), not the information, must have some effect or action on a receiver. To our model, such actions (i.e., by the semantic dimension) have effects for the personal world only, whilst our information-as-system shows effects on all three worlds of experience of a receiver.

Fourth, concerning the relationships among basic concepts of information, we may see that the relation between *information* and *meaning* in Mingers' conception is opposed to the one between *information* and *knowledge* in ours, regardless of terminology. As Mingers pointed out that human cannot process *information*, only *meaning*, the human conversion from objective *information* (*intent*) to *meaning* is out of question and only subjective *information* (*signification*) should be considered instead. But, note that "*the receiver's knowledge, intentions and context determine what counts as information*" (Mingers, 2006, p. 124), the way from *information* to *meaning* clearly stated by Mingers needs also to be put in parallel with some way from *meaning* to *information*. Just the latter link is missing somehow in Mingers' model. Whilst, our triad of process shows clearly that *information* (claim) is the product of *knowledge* (*warrant*) and *data*, or equally, that there exists the way from *knowledge* to *information*. Of course, the transformation from *information* back to *data* and then to *knowledge* is also evident, as discussed earlier in the aspect of process. Furthermore, the crucial point we make here is that the two-way transformation from information to meaning and vice versa is insufficient to explain the relationships among concepts of information. That is, we need to take account of one more and only one more information-related concept that is merely the conception of data for our model.

Fifth, so far we have only discussed *information* and *knowledge* or *meaning*, not *data*. That is because the semiotic levels of syntactics and below are less significant to IS than to information science or information theory. In fact, our model has not gone far into empirics and syntactic level and hence, variety of signs in Mingers' typology of signs is worthy to be a complement to our model.

Finally, briefly, our model of information is more generic and simpler philosophically and firmer theoretically than Mingers' one. In regard to semantic and pragmatic aspects of information, ours has only three elements or states of affairs (data, knowledge, information) ontologically and three states of mind (objective, subjective and intersubjective), but the full process of information formulation or transformation of information back and forth is also determined completely and, especially, neither overlappings nor confusing between elements or states is shown. Hence, the clear cut as well as consistency between information and its relatives as well as between the respective processes is more easily established for our conception.

FINDINGS AND CONCLUSION

Although the significance of information in IS is early recognized, IS researchers so far have mostly focused on the subject matters of organizations, technologies, or systems in dealing with IS problems. Thus many problems with IS development, implementation, and usage could be rooted in the information itself. Our intention is to make contributions to this area of IS study.

Unlike other studies of ISs focused on either the technical or the social system, we attempt to investigate the concept of the knowledge system (Lee, 2004), which centered around the notation of information-as-system. Our argument is that future studies of IS should pay adequate attention to modeling of information-as-system, in addition to traditional models of technical and social systems. We believe that our proposed model creates a foundation for determining information necessary for business managers and to assist them in reformulating the problems of information significance and objectives for various societies, organizations, and individuals (Drucker, 1999).

The article is to critically explore the nature of information in IS and has developed a conceptual model of information-as-system. First, information could be considered a system in its own right that has aspects of structure, function, process, and context according to Gharajedaghi's (2005) systems thinking developed from Ackoff's (1974) systems approach.

Each aspect in turn could be represented in a semiotic triangle of Peirce. In terms of structure, that is a triple of interpretation—SOC— sign for representation of reality, object for reality itself,

and community of inquirers for interpretation of reality. In terms of function, that is a triple of social actions—DCC— design for the natural world, creativity for the mental, and culture for the social. In terms of process, that is a triple of states of argumentation DWC: data as a start, warrant as a link, and claim as a conclusion. In terms of time, we develop a continuous interval or continuum of time— PPF— of three referent points of time, namely, past, present, and future. In terms of context, we posit a recursive, not nested, structure of three levels—SIP—information system, information itself, and person all constrains and enables each other in an interdependent way. Finally, the triads of process, of time, and of context altogether are able to model the evolution of information over time and in space. Hence, information is investigated as an entity (i.e., a typical semiotic triangle) or a process (i.e., an indefinite semiosis). We believe that our general model of information-as-system is able to cover all types of signs of all three of human worlds, thus worth to be investigated more with care in IS and other disciplines such as social theories and organizational behaviours as Mingers suggested for his version of information (1996, p. 204).

We have shown that our triad model has a number of more advanced features than other models in the literature. First, our model of information is developed on the three-dimension descriptions (triadic relations) of the three aspects (structure, function, process) of information-as-system, regardless of its context. Only this way, our model of information is able to cover all three human worlds of experience fully but at a minimum effort and to show interdependent and evolutionary relationships among information-related phenomena. For example, our model is more than Callaos and Callaos' (2002) two component notation of information (e.g., data and information), completely covers Buckland's (1991) four aspect schema of information (e.g., as thing, as knowledge, as process, and information processing) and simpler as well as more clear cut than Mingers' comprehensive typology of signs. Second, our systems model cannot only go beyond the traditional dichotomy of objectivity and subjectivity existing in most current models, but also provide a comprehensive model for each one concept involved and hence avoid ambiguous statements such as "meaning is subjective or perhaps intersubjective," or "information is objective" but "mean different things to different people." Third, we believe our systems model far advances other models of information in its time-related nature of information. In the process aspect of information-as-system, we have already identified all three states of the process and they all are time-related: D (data, given) of past orientation, W (warrant, knowledge) of present orientation, and C (claim, information) of future orientation. Besides, the process is irreversible in its nature. This is similar to the concept of irreversible order of stages in IS analysis and design, implied in the traditional waterfall model. Last, while maintaining the nature of triads of information, our model emphasizes the "interpretant" or the pragmatic side of information, which stresses more on the intersubjectivity of information as social phenomenon.

To put the systems model of information into practice, implementation guidelines need to be developed in more detail. Any place information is needed; it should be designed on the aspects of its structure, function, process, and context. For example, one of the main problems in approaches of IS analysis and design is to determine information requirements of various users. Many IS analysis and design textbooks simply advise systems analyst to collect technical and non-technical requirements of the system. As implied in our model, this guideline is necessary but insufficient. Our model indicates that IS requirement is a complex object of information, which should be treated as a system in its own right. It means that we need to analyze, justify, and transform IS requirement with respect to its structure, function, process, and context. As Goguen (1996) argued,

understanding what information is and how to use it shall support us to solve the problems of requirements engineering of ISs. Our findings would facilitate systems analysts and designers at least in dealing with the problem of information requirements of various kinds of organizational and group ISs.

Ultimately, the article claims that, along with IS as a human meaning systems (Mingers, 1995), human activity system (Checkland, 2000), inquiring system (Churchman, 1971), or open system (Emery, 2000), information itself also needs to be considered such a system or such all systems at once, but definitely not an independent variable of IS. Information-as-system is both input and outcome of IS as well as of human actors. The interdependent relationships between human, information, and IS make the problems of three levels of these entities always be wicked or messy and, therefore, it is really necessary to have many deeper and broader researches on information itself in particular and ISs in general.

ACKNOWLEDGMENT

The authors thank three anonymous reviewers for their helpful comments on the article.

REFERENCES

Ackoff, R.L. (1993). Idealized design: Creative corporate visioning. *Omega, International Journal of Management Science, 21*(4), 401-410.

Ackoff, R.L. (1971). Towards a system of systems concepts. *Management Science, Theory Series, 17*(11), 661-671.

Ackoff, R.L. (1974). *Redesigning the future: A systems approach to societal problems.* New York: John Wiley & Son.

Ackoff, R.L. (1989). From data to wisdom. *Journal of Applied Systems Analysis, 16,* 3-9.

Banathy, B.H. (1992). *Systems design of education: Concepts and principles systems for effective practice.* New York: Educational Technology Publications.

Banathy, B.H. (1996). *Designing social systems in a changing world.* New York: Plenum Press.

Barton, J., Emery, M., Flood, R.L., Selsky, J.W., & Wolstenholme, E. (2004). A maturing of systems thinking? Evidence from three perspectives. *Systems Practice and Action Research, 17*(1), 3-36.

Barton, J. (1999). Pragmatism, systems thinking and system dynamics. In *Proceedings of the 17th Conference of the System Dynamics Society.* Palermo: System Dynamics Society.

Bhaskar, R. (1993). *Dialectic: The pulse of freedom.* London: Verson.

Boland, R.J., Tenkasi, R.V., & Te'eni D. (1994). Design information technology to support distributed cognition. *Organization Science, 5*(3), 456-475.

Braman, S. (1989). Defining information: An approach for policymakers. *Telecommunications Policy, 13*(1), 233-242.

Buckland, M.K. (1991). Information as thing. *Journal of American Society for Information Science, 42*(5), 351-360.

Bühler, K. (1982). *Karl Bühler: Semiotic foundations of language theory.* New York: Plenum Press.

Burks, A.W. (1998). Peirce's evolutionary pragmatic idealism. *Synthese, 106,* 323-372.

Callaos, N., & Callaos, B. (2002). Towards a systemic notion of information: Practical consequences. *Informing Science, 5*(1), 1-11.

Checkland, P.B. (2000). Soft systems methodology: A thirty year retrospective. *Systems Research and Behavioural Science, 17*(1), S11-S58.

Churchman, C.W. (1971). *Design of inquiring systems.* New York: Basic Books.

Courtney, J.F. (2001). Decision making and knowledge management in inquiring organizations: Toward a new decision making paradigm for DSS. *Decision Support Systems, 31,* 17-38.

Drucker, P. (1999). *Management challenges for the 21st century.* New York: Harper Business.

Ellis, D., Allen, D., & Wilson, T. (1999). Information science and information systems: Conjunct subjects disjunct disciplines. *Journal of the American Society for Information Science, 50*(12), 1095-1107.

Emery, M. (2000). The current version of Emery's open systems theory. *Systems Practice and Action Research, 13*(5), 623-643.

Everaert, D.N. (2006). Peirce's semiotics. In L. Hebert (dir.), *Signo* [online]. Rimouski, Quebec. Retrieved November 21, 2006, from http://www.signosemio.com.

Fenzl, N. (2005). Information and self organization of complex systems. In M. Petitjean (Ed.), *Proceedings of the Third Conference on the Foundations of Information Science* (pp. 1-11). Basel, Switzerland: Molecular Diversity Preservation International.

Flood, R.L. (1999). *Rethinking the fifth discipline: Learning within the unknowable.* London: Routledge.

Gharajedaghi, J. (2005). *Systems thinking – managing chaos and complexity: A platform for designing business architecture* (2nd ed.). Boston: Butterworth-Heinemann.

Goguen, J. (1996). Formality and informality in requirements engineering. In *Proceedings of the Second IEEE International Conference on Requirements Engineering* (pp. 102-108). IEEE Computer Society Press.

Goguen, J. A. (1997). Toward a social, ethical theory of information. In G.C. Bowker, S.L. Star, W. Turner, & L. Gasser (Eds.), *Social science, technical systems, and cooperative work: Beyond the great divide* (pp. 27-56). Mahwah, NJ: Lawrence Erlbaum Associates.

Gunaratne, S. (2003). Thank you Newton, welcome Prigogine: Unthinking old paradigms and embracing new directions. Part 1: Theoretical distinctions. *Communications, 28*, 435-455.

Habermas, J. (1984). *The theory of communicative action, Vol. 1: Reason and rationalization of Society.* London: Hienemann Education.

Handfield, R.B., & Melnyk, S.A. (1998). The scientific theory-building process: A primer using the case of TQM. *Journal of Operations Management, 16*, 321-339.

Jackson, M. (2000). *Systems approaches to management.* Boston, MA: Kluwer Academic.

Lauer, T.W. (2001). Questions and information: Contrasting metaphors. *Information Systems Frontiers, 3(*1), 41–48.

Lee, A.S. (2004). Thinking about social theory and philosophy for information systems. In J. Mingers & L. Willcocks (Eds.), *Social theory and philosophy for information systems* (pp. 1-26). Chichester, UK: John Wiley & Sons.

Meadwell, H. (1994). The foundations of Habermas's universal pragmatics. *Theory and Society, 23*(5), 711-727.

Metcalfe, M., & Powell, P. (1995). Information: A perceiver-concerns perspective. *European Journal of Information Systems, 4*, 121-129.

Metcalfe, M. (2004, November). Generalisation: Learning across epistemologies]. *Forum: Qualitative Social Research, 6*(1), Art. 17. Retrieved March 10, 2007, from: http://www.qualitative-research.net/fqs-texte/1-05/05-1-17-e.htm.

Midgley, G. (2003). Science as systemic intervention: Some implications of systems thinking and complexity for the philosophy of science. *Systemic Practice and Action Research, 16*(2), 77-97.

Mingers, J., & Brocklesby, J. (1997). Multimethodology: Towards a framework for mixing methodologies. *Omega, International Journal of Management Science, 25*(5), 489-509.

Mingers, J. (1995). Information and meaning: Foundations for an intersubjective account. *Information Systems Journal, 5*, 285-306.

Mingers, J. (2004). Realizing information systems: Critical realism as an underpinning philosophy for information systems. *Information and Organization, 14*(2), 87-103.

Mingers, J. (2006). *Realising systems thinking: Knowledge and action in management science.* New York: Springer.

Mingers, J. (1996). An evaluation of theories of information with regard to the semantic and pragmatic aspects of information systems. *Systems Practice, 9*(3), 187–209.

Morris, C. (1938). Foundations of the theory of signs. In O. Neurath (Ed.), *International Encyclopedia of United Science,* 1(2). Chicago: University of Chicago Press.

Nauta, D. (1972). *The meaning of information.* Mouton: The Hague.

Nellhaus, T. (1998). Signs, social ontology, and critical realism. *Journal for the Theory of Social Behaviour, 28*(1), 1–24.

Nelson, H. (1994). The necessity of being "un-disciplined" and "out-of-control:" Design action and systems thinking. *Performance Improvement Quarterly, 7*(3), 22-29.

Peirce, C.S. (1931-1958). *The collected papers of C.S. Peirce.* C. Hartshorne & P. Weiss (Eds.). (1931-1935). Vol. I-VI. A.W. Burks (Ed.). (1958). Vol. VII-VIII. Cambridge, MA: Harvard University Press.

Prigogine, I. (1997). *The end of certainty: Time, chaos, and the new laws of nature.* New York: The Free Press.

Richardson, S.M., Courtney, J.F., & Haynes, J.D. (2006). Theoretical principles for knowledge management systems design: Application to pediatric bipolar disorder. *Decision Support Systems, 42,* 1321-1337.

Simon, H.A. (1996). *The sciences of the artificial* (3rd ed.). Cambridge, MA: MIT Press.

Sowa, J.F. (2000). *Knowledge representation – logical, philosophical and computational foundations.* Pacific Grove, CA: Brooks/Cole.

Toulmin, S. (1964). *The uses of argument.* Cambridge: Cambridge University Press.

Tuomi, I. (1999). Data is more than knowledge: Implications of the reversed knowledge hierarchy for knowledge management and knowledge memory. *Journal of Management Information Systems, 16*(3), 103-117.

Ulrich, W. (2001). A philosophical staircase for information systems definition, design and development: A discursive approach to reflective practice in ISD (Part 1). *The Journal of Information Technology Theory and Application, 3*(3), 55–84.

Ulrich, W. (2003). Beyond methodology choice: Critical systems thinking as critically systemic discourse. *Journal of the Operational Research Society, 54*(4), 325-342.

von Bertalanffy, L. (1968). *General systems theory: Foundations, development, applications* (rev. ed.). New York: George Braziller.

ENDNOTES

[1]   Citations of Peirce's work in this article follow the form: "CP" (the abbreviation for Collected Papers) followed by *volume* and *paragraph* numbers, with a period between.

[2]   Note that continuity, according to Peirce, is *"simply what generality becomes in the logic of relatives, and thus, like generality, and more than generality, is an affair of thought, and is the essence of thought"* (CP 5.436).

This work was previously published in International Journal of Information Technologies and Systems Approach, Vol. 1, Issue 2, edited by M. Mora; D. Paradice, pp. 1-19, copyright 2008 by IGI Publishing (an imprint of IGI Global).

Chapter 8
An Analysis of the Imbursement of Currency in a Debt–Based Money–Information System

G. A. Swanson
Tennessee Technological University, USA

INTRODUCTION

The imbursement of currency into modern debt-based money-information systems is a concrete phenomenon. Economic jargon traffics in abstract concepts. This analysis attempts to bridge the gap between the abstract and the concrete and to provide practical insights into certain social consequences of different modes of currency imbursement.

Information science, like economics, has developed as an analytic science. Many reasons might explain that development. Not least among them is the daunting complexity of the matter-energy systems they concern. Button and Dourish (1996) provide an interesting view of the role of systems design to provide an opaque barrier for decision-makers against that complexity while enabling engagement. That conceptual distance can give a false sense that the design of information systems should be limited only by imagination. Many definitions of information have been proposed and some have gained acceptance in certain circles. In some highly abstracted systems, Bateson's (1972,

xxv-xxvi) "difference that makes a difference" has appeal. Nevertheless, when we consider that technology emerges in self-organizing, evolving living systems that exist in physical space-time, that definition has little explanatory power. Alternatively, Shannon's H restatement of the measure of entropy (Shannon, 1948) that became known as a measure of information is significantly explicatory of an essential connection between information and matter-energy processes. Shannon's treatment of information as a reduction of uncertainty has found wide application.

Miller (1978) makes the connection between information and the processes in which it emerges. He straight-forwardly endorses Shannon's definition while also defining information as the formal patterning in space-time of the elements of matter-energy that comprise a concrete system. Only when we recognize that all continuing material organization is made possible by the unidirectional entropic processes occurring in nature can we begin to recognize the fundamental connection between information and the evolving processes of humankind.

DOI: 10.4018/978-1-60566-976-2.ch008

An eon before the analytic sciences of economics and information came into being, information systems were emerging and forming human civilization. It is important for information systems scientists, from time to time, to contemplate their rolls in that continuing grand scheme.

Human civilization may be approaching an evolutionary cusp. The agents that usher in the new era will likely not be grand new political thinkers, or even the clash of philosophers and ethicists. The legal, accounting, and information scientists and technicians are the likely actors. Their work requires exacting specialization. The grand view is seldom taken. If history teaches us that an uninterrupted advance in civilization always occurs, we might lean heavily on the unseen hand always guiding to the benefit of the whole. But, history teaches the opposite. The grand question is: Will the evolutionary cusp advance or retard human civilization?

The insight of the medical profession, "Do no harm," should be considered by information systems scientists, technicians, and academics. Before replacing an evolved information system that supports life, whether biological or social, that system should be studied and understood.

This paper at first glance may seem inappropriate for the systems technologies as they are narrowly defined. It, however, attempts to describe in some significant degree two intertwined information systems embedded in the intercourse of social life. They are material processes of *exchange* and *money-information*. The analytics of information science often concern those processes. Their characteristics, consequently, constrain information systems development.

Information technology is never far removed from money-information processes. That fact notwithstanding, money-information processes are seldom a major subject of discussion except when the technology is directly involved with money transfers and accounting systems. Information systems are human artifacts and sometimes become prostheses. A journal that considers information technology in tandem with system approach provides opportunity to consider special information systems deeply embedded in societies. The exchange system and the money-information system have become societal prostheses performing vital and fundamental information processes. It is difficult to conceive of information, which is processed by more narrowly defined information systems, that is not in some way constrained by those deeply embedded systems. This paper considers certain patterns of control that emerge in the money-information processes of modern exchange-based societies. Those cybernetics should instruct certain design processes of information systems.

CYBERNETICS AND EXCHANGE-BASED SOCIETIES

Societies may be characterized by many different attributes and from a variety of perspectives. However one views a society, eventually the characterizations concerns a collection of living individuals in interaction, in physical space-time. Modern exchange-based societies are self-organizing and maintain a homeostasis, whereby complex internal changes occur among their components while they maintain a relative overall constancy. That relative constancy of the whole can only occur if a change in one component is met with opposite, reactive changes in the other components. From the time Cannon (1939, p. 293) gave us the term *homeostasis*, he anticipated the discovery of general principles that would apply not only to biological organization but also to "industrial, domestic, and social" ones.

General principles may be approached from an abstract perspective or from a concrete one. In both approaches, the object of the principles is ultimately concrete. That is so because human curiosity arises, as far as we may examine by science, in a material existence.

When we approach such principles from the concrete perspective, our first concerns are measurement and the avoidance of observer bias. From the abstract vantage, one is first concerned with the development of an understood conceptual system, usually a mathematical model, and then with a fitting of that system or model to measurements taken on concrete objects. The principles expressed in this paper are developed from the concrete perspective. That perspective demands measurement, and the models used are constrained by the properties of the measurement system employed. Ashby (1962), from the abstract vantage, emphasized the development of models based on *possible* patterns of change. Time and time again, he makes it clear that such models do not depend on the physical characteristics of any machine or dynamic system. They transcend those characteristics and model the behavior (how the object can change). His goal is to identify models with transformation properties that have exact parallelism with the behavioral properties of objects. In such an approach, the homeostasis, the complex interactions of its parts that form the observable whole, is encased in a black box. The daunting complexity of homeostasis is left to be explicated only on its general patterns of behavior—patterns of change exhibited by the system as a whole. Many systems that we investigate are in fact impenetrable, particularly those that involve mind and meaning. As might be expected, Ashby's "cybernetics" has proven fruitful in the development of artificial intelligence and other pattern recognition information systems.

Such abstracted approaches may be extended to grand human systems such as societies. Those extensions, however, may blind against many aspects of the reality encountered daily. The observer is in fact inside those systems and knows a lot about them. As a consequence, the limiting of our inquiry to the behavior of the whole neglects unnecessarily a wealth of information about the processes out of which the structures of societies emerge.

When we speak of action and reaction, we acknowledge that the condition of homeostasis can only occur inter-temporally. Movement in physical space always involves physical time. Process simply cannot occur absent time. Furthermore, to acknowledge action is to acknowledge motion, transportation in physical space-time. Having such powerful ideas now available, it is reactionary (to borrow a popular term) to rely on the more limited sense-response, action-reaction paradigm. The view of entities existing in space-time as processes—inputs, throughputs, and outputs—provides a much more robust inquiry. Not only can we speak of actions of one entity as it interacts with another but the interior of the entities is opened for investigation. There, processes of self-organization and homeostasis may be identified.

Analysis of the internal properties of homeostasis of modern exchange-based societies leads inevitably to two rudimentary lines of enquire. They are: What is the irreducible unit, the holon, of modern economic systems? And what is money? Upon sufficient answers to those questions, homeostatic processes themselves may be identified and examined. To that end, the following first concerns the rudimentary question of the irreducible unit. That discussion is followed by a short description of the internal control affected by exchange on societal components. The question of money is then engaged and the paper rounds out with more in-depth discussion of variations and adjustment processes that may occur due to different ways money-information is imbursed.

THE IRREDUCIBLE UNIT OF MODERN ECONOMIC SYSTEM

In the introduction, I have used without definition the term *modern exchange-based societies*. Such societies exist in space-time and are modern in the sense that they are of relatively recent emergence. Markets go back to pre-history, but societies that are dominantly controlled by market type processes (that are exchange-based) only go back a few centuries. I use the term *exchange-based* instead of the term *market* in part to avoid the value-laden language with which the latter has come to be associated. Mostly, however, the choice is one of exactness. The exchange is a specific process that occurs in the interactions of humans. Exchange is often associated with market environments. Exchange, however, need not be confined to them.

The exactness of definition that I seek is not evident in the various meanings of exchange found in the literature. Modern economic value theory uses the term in connection with the changes occurring in a subjective hierarchy or ranking of preferences or significance. Mises (1912) extends its use to intermediary trades that are taking on the characteristics of money. Its common use often implies a set of monetary reciprocating transactions.

The exchange to which I refer is a concrete *economic* phenomenon. It happens in physical space-time and may include both goods-services and certain forms of money. Ecology imposes economy. The environmental quality that gives rise to economy is defined by the law of entropy (Georgescu-Roegen, 1971; Swanson, Bailey, and Miller, 1997; Swanson, 2006). That which is used cannot be unused and, thus, scarcity of useful matter-energy (goods-services) ensues. In order for it to function in economies as goods-services do, money must also have utility and scarcity. Those qualities are imposed by human invention. The patterns by which money-information is imbursed affect patterns of economic processes, that in turn adjust certain social behavior.

The *exchange*, not generalized but specific, is the holon—the irreducible unit—of modern economic systems. Notwithstanding their immense complexity, those systems are combinations of specific and observable exchanges. While analytical models often conceptually divide exchanges into their component transactions, no dynamic

exist empirically to activate a single transaction. The fundamental dynamic of mutual benefit activates exchanges.

Simon (1962, p. 467), when discussing hierarchical system, asserts that there is "some lowest level of elementary subsystem." He associates this idea with that of *elementary particle* put forward by physicists. While recognizing that such lowest levels may be exceedingly complex in their own composition, he points to the fact that science accepts such cut-offs when they are carefully employed. My assertion that the exchange is the irreducible unit of modern economic activity is not unlike Simon's assertion and similar ones made in scores of disciplinal sciences.

An exchange consists of five elements—two transactions, reciprocity, a coupling relationship, and time immediacy. In it, two reciprocal transactions occur at a moment (without intervening time). An action is any change of matter-energy over time. A transaction is an action across the boundaries of interacting concrete systems. An exchange always occurs between two, and only two, entities. Compound exchanges of many different substances can occur, but always between only two entities. Related exchanges, those conditioned in some way upon other exchanges, often occur—but the exchanges themselves are always between only two entities.

THE CYBERNETICS OF THE EXCHANGE

The concept of exchange, as I have stated it, is straightforward, observable, and thus objective. That objectivity does not have to mean that we cannot explore the manner in which subjective, decisional aspects are connected to the objective. The exchange is a culmination of covenant and a consummation of contract. The behavioral and the legal are thus brought together in the economic. The behavioral and the legal are difficult to quantify. The economic is measurable.

Figure 1.

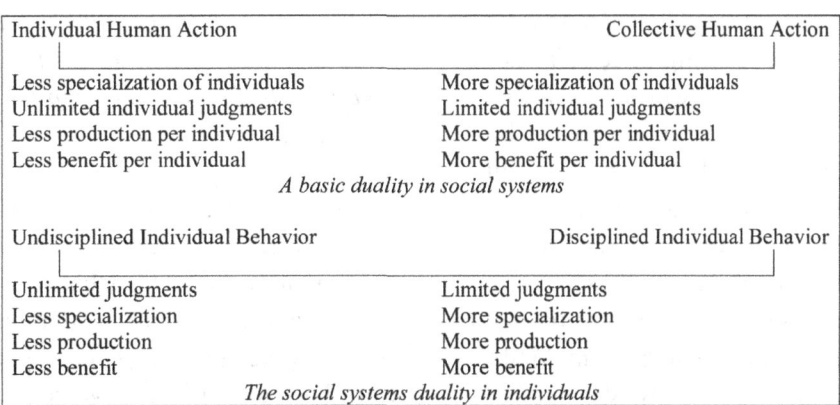

Individual Human Action	Collective Human Action
Less specialization of individuals	More specialization of individuals
Unlimited individual judgments	Limited individual judgments
Less production per individual	More production per individual
Less benefit per individual	More benefit per individual

A basic duality in social systems

Undisciplined Individual Behavior	Disciplined Individual Behavior
Unlimited judgments	Limited judgments
Less specialization	More specialization
Less production	More production
Less benefit	More benefit

The social systems duality in individuals

Individuals struggle between a duality of individual and collective human action. Some characteristics of the extremities of that duality are presented in Figure 1. In that tension, it is important to recognize that two very different actions are captured in the etymological roots of the terms *covenant* and *contract*. The meaning of covenant is "to come together," while that of contract is "to draw together." To come together, a near infinite array of possibilities present to each individual. To draw together implies limits, convergence. The covenanting process realistically may be studied no more concretely or exactly than that which stochastic, conjectural models may afford. Alternatively, the contractual process converges on an action. That action may be observed and measured in the case of the exchange.

The exchange forms an attractor, a vortex, drawing together in a decision structure the choices available to each of the exchanging entities. At the point that the determined value of that which each is offering equates, an exchange occurs. Individual entities simply will not give up more than they perceive that they are receiving. The values that, because of complexity and inaccessibility, may be indeterminable are in the exchange expressed. On a temporal map, the complex private decision processes subside as contract prevails. But the exchange is more than the exposition of internal

values. It is cybernetic. The possibilities that present to each entity in the covenant process are changed by new limits imposed by the economic objects surrendered and received.

The constraints exercised by exchange on societal components are important. Those constraints, however, are only part of the adjustment processes thus introduced. Patterns of exchanges impose adjustment processes on societal structure as well. With the advent of money, the patterns of exchanges can be designed to provide a primary means of societal control.

MONEY EMERGENCE

How does money emerge in economies? The answer to that question is colored by how one approaches economic science, whether by analytical or empirical means. Considerations of economic questions drift finally to questions of value. The analytic science of economics generally approaches the question of value from a subjective vantage—the value that individuals attach to economic actions. Such an approach is almost certain to ensure that the systems of such values are not isomorphic to any empirically discoverable system of objective values.

The choice to base the analytical science on subjective value is not based in ignorance of the objective values exhibiting in economies. It rather is an effort to explain the most fundamental aspects of economy. Those aspects always boil down to scarcity imposed by nature and utility arising in the minds and behaviors of humans. Auditors often "look through" the legal form of business organization to its "economic substance." Economists take a similar perspective but with reference to money instead of legal form. Models constructed with such a view may be expected to assign no economic value directly to the intermediary through which they gaze to the substance. The extension is that money flows themselves do not contain information. They rather provide a price index upon which certain judgments of value may be made.

While embracing completely the subjective value approach embodied in the marginal utility school of thought, Mises (1912) provides an objective explanation of how money emerges in economic activity. He views money simply as a medium of exchange. It arises according to Mises from an individual desiring a certain commodity for which he cannot directly trade a commodity in his possession. He must, therefore, make an indirect trade—that is, a trade for a third commodity that may in turn be traded for the commodity desired. Mises explains how certain commodities, being more "marketable," become the target of such indirect trade. He extends the use of the term exchange from subjective exchange to those indirect trades. That view of money persists to today in most economic theory. It is an easy step from that view to the treatment of money as exogenous to most economic models.

Mises (p. 51), furthermore, concisely states an objective theory of value as follows:

If . . . the possibility of an objective concept of commodity values is accepted, and exchange is regarded as the reciprocal surrender of equivalent goods, then the conclusion necessarily follows that exchange transactions must be preceded by measurement of the quantity of value contained in each of the objects that are to be exchanged. And, it is then an obvious step to regard money as the measure of value.

He, however, rejects such a theory because "scientifically" conceived value is subjective. Having further concluded that money has no utility (no subjective value) and thus must arise from objective exchange value, he imputes its current value back to its previous value as a commodity. This conclusion leads him to advocate a strong gold standard as the fundamental currency. Looking back, almost a century later, one has but to wonder what kind of a world would now exist if his advocacy had prevailed. His conclusion that money must arise from an objective exchange value, nevertheless, is interesting.

Mises relies on historic development in concluding a prominent place for gold in his hypothesis. Unfortunately, he neglected the reflexion of advancing human cognition in the developing money-related artifacts of the same period. When one includes that evidence, money takes on the character of information. With that quality, money is thrust to the heart of social organization and market activity very early in the development of human civilization.

Schmandt-Besserat (1992), from her study of the extant artifacts of a prehistoric accounting system, formulates a strong hypothesis that the evolution of human cognition moves from the concrete, objective to the abstract. She examines clay objects that began to appear with the advent of agriculture about 8000 B.C. and continued (with significant changes occurring) until about 3000 B.C. With the advent of monumental architecture about 3500 B.C., the accounting systems underwent major change. Schmandt-Besserat hypothesizes that both written language and abstract numbering emerge in that change. Those technologies are informational aids to human cognition. If her hypothesis holds, quantification and measurement clearly are not abstract additives of a much later emerging science. They are at the heart of human interaction and their fundamental expressions involve economy. And, to the point at hand, it is the informational content of certain artifacts that emerging into exchange processes distinguishes money from the economic goods-services it facilitates.

Swanson and Miller (1989) analyze the emergence of incipient forms of such money-information, termed *primitive money* by Grierson (1977), Einzig (1966) and others. It may be hypothesized that the inception of money-related artifacts occurred in the following order:

1. Bone tallies and pebbles used to count (about 10,000 B.C.)
2. Tokens used to account (about 8,000 B.C.)
3. Tokens used to count (about 3,500 B.C.)
4. Tokens used to account, count, and as a medium of transaction (about 3,500-3,000 B.C.)
5. Tablets used to separate the accounting artifact from the transaction artifact.
6. Precious metal artifacts used for long-distance exchanges.
7. Coins

From the coins, money-information markers continued to evolve as documented in the numismatic literature (Kagan, 1982). It is evident from that trace that the evolution of counting, and thus of measuring, are intertwined, with that of money (Table 1).

For the topic at hand, it is important to distinguish between two types of units of measurement, a counting unit and an accounting unit. A counting unit is one belonging to a quantified conceptual (abstract) system. Its value is de-

Table 1. An evolution of numbering

	Concrete System Elements	**Artifactual Abstractions (Models)**
Ordinal Numbering	Time ordering of creation rituals	Bone tallies, calendrical notations ONE-TO-ONE CORRESPONDENCE
Transition Numbering	Human counting systems— (fingers, toes, multiple persons)	Administrative token systems
	CONCRETE NUMBERING	ABSTRACT NUMBERING
	Trade of equal amounts of commodities	Abacus
	CONCRETE COUNTING	ABSTRACT COUNTING
Cardinal Numbering 1	Trade of differing amounts of commodities	Substitutions of differing amounts of administrative tokens representing different commodities
	CONCRETE COUNTING RATIOS	ABSTRACT COUNTING RATIOS
Cardinal Numbering 2	Trade on the basis of a common denominator	Administrative tokens representing common denominators
	ACCOUNTING UNITS	RUDIMENTARY MONEY-INFORMATION MARKERS 1
Cardinal Numbering 3	Accounting unit commodities used as medium of exchange, e.g., weighed metals	Tokens representing absolute (undifferentiated) quantities of transaction value
	PRIMITIVE MONEY RUDIMENTARY MONEY-INFORMATION MARKERS 2	RUDIMENTARY MONEY-INFORMATION MARKERS 3
Cardinal Numbering 4	COINS Undifferentiated quantities of exchange value calibrated on various monetary scales	

Swanson and Miller (1989, p. 42). Used by author permission.

termined entirely by the characteristics of that system. It has no empirical (concrete, material) content. Alternatively, an accounting unit is a unit of a convenient commodity used as a common denominator to establish relationships between the exchange values of other commodities. An accounting unit is always determined with reference to a concrete, material system element. It is an object among objects.

MONEY-INFORMATION

The information content of money should be further clarified. For that purpose, a clear distinction may be struck between the term *monetary information* and *money-information.* That distinction may be made on the basis of a subjective doubling of the terms *negentropy* and *information* introduced by Beauregard (1961). Shannon (1948) had called his H measure of entropy *information,* which term quickly became associated with the negative of entropy, with *negentropy.* If entropy is progress towards disorganization, negentropy is progress against disorganization. Negentropy came to connect the organization (the formal patterning) that characterizes physical systems with the more common understanding of the term information. Beauregard defined the relationship with two transitions (negentropy $\leftrightarrows$ information). The direct transition is *acquisition of knowledge* and monetary information is defined in that manner. Monetary information is a price index that is useful for all sorts of economic decision-making. Money-information, however, is something more. It corresponds with Beauregard's reciprocal transition *power to organize.* By its objective value in specific exchanges, money-information provides a motive force, a dynamic for actual economic activity.

The goods-services flows in economies are actual concrete (matter-energy) substances. Such substances accumulate in regions of space-time to form systems with interacting subsystems.

Information in such systems is defined as the arrangements of those matter-energy elements. Concrete information, consequently, is always borne in matter-energy markers. The term *marker* rather than that of *mark* is selected to indicate that information is a primal quality of such systems. Any matter-energy, by virtue of the changing arrangements of its elements, is a marker. In order to move in space or to endure in time concrete information must be borne on information markers. An economic power of organization is exercised by the introduction of money-information markers at the level of the exchange.

When money-information is recognized as to have power to organize, the quality of the objective value of accounting unit measurement becomes a significant ingredient of homeostasis. Furthermore, the homologies presenting in the money-information values of specific exchanges constitute a certain type of formal identity—a cybernetic system.

It is convenient to classify money-information markers (MIM) into two basic classes, time-lagged MIM and currency MIM. Time-lagged MIM (TLM) document executory contracts, which obligate the issuers to some future action. Currency MIM (CM) documents objective value free of any future obligation. For example, bonds obligate a transmission of a certain value at a certain future date and shares obligate managers to act in the financial interests of shareholders. That example points to the common subclasses of TLM. There is, however, an additional very important subclass of TLM. That subclass is *socialization documentation* such as tax and subsidy receipts and court orders that transfer wealth. This subclass should be disclosed in public financial statements, but it is not. Currently in the name of accountability, vast monetary values of socialization documentation are being introduced into the worldwide economic system with very little public disclosure. The stated purpose of such introductions is the injection of currency to get the credit markets functioning properly.

SOME ADJUSTMENT PROCESSES OF CERTAIN MIM IMBURSEMENTS

The particular patterns by which MIM are immited both have cybernetic effects on the decision processes of individuals and affect processes of homeostasis in societies. It is, consequently, informative to examine those patterns. Methods of mining such patterns in actual processes become daunting because of complexity. We may, therefore, resort to analytical simplification. The method of simplification is developed within a macro accounting conceptual framework built on four essential ideas. They are: 1. The irreducible unit of economic process is the exchange; 2. Complex modern economic processes are combinations of exchanges; 3. Within modern economic processes, money-information processes emerge from executory contracts documented by accounting instrumentation that temporally intervenes in the reciprocal transfers of trades; 4. Accounting documentation of debt is the basic form of modern money.

A trade is an exchange in which the reciprocating transactions are transfers of goods-services. The trade is primal. Out of it emerges all other types of exchanges. Contract consummation in a trade is determinate—it imposes a near infinite package of constraints from both an abstract vantage and a concrete one.

Economic activity is facilitated by relaxing, to some extent, those constraints by introducing MIM into trades. The resulting exchanges form chains of directed economic activity. Those interactions participate in the societal adjustment processes that maintain homeostasis. In the following paragraphs, various kinds of MIM introductions are presented. Each introduction relaxes certain constraints inherent in the trade which, in turn, increases uncertainty.

The notation used in the presentation can be demonstrated with a simple example from archeology. Weitemeyer (1962) describes a system of worker dockets that date in the Hammurabi and Samsuiluna reigns of the First Dynasty of Babylon. The dockets are pyramidal clay objects bearing impressions such as "The inscription mentions on the first side: 1 lu hun-ga 'one hired worker,' on the second side, the name of the man, and on the third side, month and day" (p. 12). A docket was given to a worker in exchange for his labor. The docket was then taken to the store where it was exchanged for probably grain (which became the accounting unit during that period). This is an example of an IOU temporarily intervening in the reciprocating transfers of a trade (the giving of labor for grain). It may be modeled as in Notation 1:

$$IOU^P_{-1}IOU^W_{+1}L^W_{-1}L^P_{+1} \quad IOU^W_{-1}IOU^S_{+1}G^S_{-1}G^W_{+1}$$

$$(1)$$

where: IOU is a type of promissory note, L is labor, G is grain, P is producer, W is worker, and S is stone. The notation always begins with an outflow (-) of a transaction from one entity followed by its inflow (+) to another entity. The reciprocating transaction follows in order (again beginning with outflow) to complete an exchange in four terms. Measurements are included by introducing quantitative subscripts. Very general terms are used to examine the cybernetic effects of money-information at a macro level—such as the general classes of MIM defined above. Additional terms are introduced as needed.

THE TRADE

The primal exchange is the *trade*. A trade is an exchange of goods-services for other goods-services. It is useful to define the transactions of a trade as *transfers*. The power of money-information to organize social processes originates in the temporal separation of the transfers of a trade. Notation 2 describes a trade.

$$GS_{-1}^{C1} \; GS_{+1}^{C2} \; GS_{-1}^{C2} GS_{+1}^{C1} \tag{2}$$

where C1 is one individual and C2 is another. The trade requires availability of certain goods-services, placing severe space and time restrictions on economic activity.

THE TRADE WITH DEBT INTERVENING

Such severe restrictions may be relaxed to a certain extent by the introduction of an intervening debt instrument. Such an instrument is introduced in Notation 1, but there it has more the characteristic of a receipt given by one component of an entity and received and destroyed by another. Even in that introduction, some release of the restrictions of direct trade are obtained. Debt instrumentation takes on a public characteristic when a means of civil enforcement of private contract is introduced. Its imission, then, may become the motive force (the dynamic) for more extended economic activity. For example, a merchant of medieval times passes through the countryside giving small amounts of gold to peasants in exchange for debt contracts, executed over their "marks," with their promises to deliver a certain volume of grain at harvest, and the promise is fulfilled (Notation 3).

$$GS_{-1}^{M} GS_{+1}^{P} TLM_{-1}^{P} TLM_{+1}^{M} \quad GS_{-1}^{P} GS_{+1}^{M} TLM_{-1}^{M} TLM_{+1}^{P} \tag{3}$$

where M is merchant and P is peasant. Time and space constraints are relaxed, but within certain limits. The new time constraints are defined by the executory contract underlying the debt instrument. That contract can also allow latitude in place of consummation. Uncertainty is introduced by the relaxed constraints. The risk of failure to repay is introduced, and it is possible for the merchant to transmit the TLM in return for other GS (giving the other person the right to the contracted harvest). However, the person accepting the TLM is subject to an increased risk of default (increased uncertainty) due to decreased personal knowledge concerning the payor. If the merchant does not exercise a choice of additional economic activity, the initiated chain of exchanges forms a circuit. The TLM introduced by the peasant is retrieved by him. The TLM immission is determinate. At the completion of only one trade, the circuit is closed and provides no further motive force for additional economic activity.

THE TRADE WITH DEBT AND CURRENCY INTERVENING

The incremental uncertainty of public versus private debt, perhaps more than anything else, explains why societies imburse currency. Caveat: Currency in macro accounting is not defined by the common usage of the term. *Currency* explicitly is MIM that carries no obligation for future action. Juxtaposed with debt, currency cannot be defaulted. All debt instruments risk the uncertainty of default. By accepting debt in exchange for currency, society bears the uncertainty of repayment.

Consider a situation in which individual initiative and trust emit the primary money-information as debt *and* society makes available currency to remove the risk of default (Notation 4).

$$TLM_{-1}^{C1} TLM_{+1}^{C2} GS_{-1}^{C2} GS_{+1}^{C1} \quad TLM_{-1}^{C2} TLM_{+1}^{S} CM_{-1}^{S} CM_{+1}^{C2} \tag{4}$$

$$CM_{-1}^{C2} CM_{+1}^{C1} GS_{-1}^{C1} GS_{+1}^{C2} \quad CM_{-1}^{C1} CM_{+1}^{S} TLM_{-1}^{S} TLM_{+1}^{C1}$$

Perhaps C1 gives a promissory note to C2 to obtain the labor of C2 and C2 obtains cash by giving the note to a social institution. (Society's money is actually processed by certain components). C2 then uses the cash to buy goods from C1, and C1 gives the cash to the social institution to satisfy the promissory note. In this particular chain, society releases the risk of default constraint incidental to debt but C1 does not take advantage of that freedom to motivate additional exchanges. Notice five aspects of this situation: 1. The immission of the original money (TLM) is from the entities engaged in exchange. Control of that initial process is at the societal component level. 2. The chain is determinate. Circuits are formed for both TLM and CM. Control may be imposed at both component and system levels. 3. The conversion of TLM to CM serves no obvious purpose, since only two entities are involved in a single trade. However, C2 shields itself against default of the TLM during the period that it holds CM. The availability of a shield may entice less trusting components into initiating economic activity. 4. The increased negotiability of the CM increases the probability that the motive force of the TLM will be extended to instigate additional trade during the life of the executory contract underlying the TLM. While in circulation, the currency provides immediacy of motive force. Notation 5 gives an example of a simple expansion instigated by available CM. 5. The motive force of the TLM has a discrete temporal limit. Debt contracts mature. Repayment is required at a moment certain. The temporal limit that is put on CM by introducing it in exchange for debt is a powerful control.

$$CM^C_{-1}CM^{Pb}_{+1}GS^{Pb}_{-1}GS^C_{+1} \quad CM^{Pb}_{-1}CM^{Pa}_{+1}GS^{Pa}_{-1}GS^{Pb}_{+1}$$

(5)

where a worker accepts a debt instrument from a producing entity (Pa) to supply labor. The worker then exchanges the debt for currency and uses the currency to purchase goods from a second producer (Pb). The second producer then purchases goods from the first producer, and the

first producer satisfies its debt that is now in the hands of society.

SOCIETY IMBURSED MIM

Thus far, only societal component initiated MIM has been considered. What happens when society decides to initiate MIM to increase production and commerce? Notation 6 maps a simple case of society initiated economic activity.

$$CM^S_{-1}CM^P_{+1}TLM^P_{-1}TLM^S_{+1} \quad CM^P_{-1}CM^C_{+1}GS^C_{-1}GS^P_{+1}$$

(6)

$$CM^C_{-1}CM^P_{+1}GS^P_{-1}GS^C_{+1} \quad CM^P_{-1}CM^S_{+1}TLM^S_{-1}TLM^P_{+1}$$

where society loans currency to producers, who in turn pay for labor and other factors of production. The laborer/consumers (C) then buy the product of the producers with the currency, and the producers repay society. Notice that in this circuit, the MIM initiated by society stimulates both production and commerce. Similar expansions to those possible in Notations 4 and 5 can occur here as well. In fact, since the motive force of all the represented goods-services activity is currency MIM, the array of possible expansion is increased. That characteristic notwithstanding, the prominent characteristic of debt-based currency is maintained. The currency is introduced for a discrete period. Its retrieval is predictable. It should also be noted that society is loaning currency to a societal component. Society is holding the debt. Society is not owing the debt.

CURRENCY DISCONNECTED FROM DEBT BY DEFAULT

Even though debt-connected currency provides predictability, that predictability is not determinate in the final sense. The culprit is risk of default. In the degree that debt may be defaulted, determinateness fails. A default, in essence, disconnects currency from the debt instrumentation that initiated it. From the side of the debt instrument, a default is tantamount to paying off the debt. That is so because the executory contract underlying the debt instrument terminates at its maturity date, whether by repayment or default. Since the currency has not been used to satisfy the debt, its flow is no longer predictable. The currency associated with defaulted debt remains in the economy to be used at the discretion of the holder. If the debt contract is between societal components, the lending component suffers the loss of currency. That currency left undirected is termed *local discretionary currency*. Such currency, it turns out, has no effect on the average ratio of currency to goods services (it is neither inflationary nor deflationary in the whole). Alternatively, defaulted debt contracts between components and society introduce *global discretionary currency* which, in fact, can affect changes in the ratio of total currency to total goods-services. Notation 7 maps that which happens upon default of debt to society.

$$CM_{-1}^C CM_{+1}^{Pb} GS_{-1}^{Pb} GS_{+1}^C \ /RSD/ \ CM_{+1}^{Pb}, CM_{-1}^S, TLM_{-1}^S, TLM_{+1}^{Pa}$$

$$\text{(7)}$$

$$CM_{-1}^S CM_{+1}^{Pa} TLM_{-1}^{Pa} TLM_{+1}^S \quad CM_{-1}^{Pa} CM_{+1}^C GS_{-1}^C GS_{+1}^{Pa}$$

where /RSD/ means residuals. The exchanges that could close the circuit are never made. The currency introduced in the initial exchange of the chain remains somewhere in the system. The debt is written off by society, and the limits imposed

by the temporal constraint of debt no longer exist. How the rogue currency affects further economic activity depends on many factors, not the least of which is economic structures designed to re-circulate such currency. Perhaps the most important point made by Notation (7) is that the money supply of CM in circulation is increased by defaults of debts to society.

HOLDING OF DEBT BY SOCIETY

That same effect (increased money supply) may be introduced by the action of society holding the debt that underlies the currency imbursement—by extending the maturity date of the debt instruments. Notation 7 holds but society does not write off the debt. Its increasing debt "burden" reflects the increasing currency supply in the economy.

Now that is a very interesting development. We all know that the national debt is *owed by* the society, not *owed to* the society. The common jargon associated with such processes conveniently clouds analysis of national debt and the money supply.

Some light may break through when we realize that governments (who owe the debt) are societal components that provide certain societal functions but only certain ones. Churches, synagogues, temples, and mosques provide other societal functions, and an exhausting list can be made of other types as well. In the realm of economic control through debt instrumentation, governments borrow from a pool of investors. The investors constitute a societal decider super-ordinate to that of the government. The national debt we commonly describe as *owed* is in fact held by that higher echelon of societal decider. What we inversely describe as the national debt is at least a surrogate for that described in Notation 7. It is the inversion that makes surrogating necessary, not the amount. The amount is actually an objective measurement established in exchange.

Now when we perceive in this manner, it becomes necessary to acknowledge a global integration of societal deciders super-ordinate to those of national governments. That global decider subsystem is rapidly becoming the final arbitrator of social order. The power to do so is exercised through debt instrumentation that subordinates the borrower to the lender. If the vast sums of debt instrumentation being used currently to imburse currency into national monetary systems are "sold" as usual to the class of global deciders, the locus of social order will likely be finalized at the supranational level. If, alternatively, semi-autonomous units such as the U.S. Federal Trade Commission hold the debt instruments exacted from societal components (leaving the currency in the economy), the headlong plunge to supra-nationalism may be abated.

REDUCING DEBT HELD BY SOCIETY

What, then, would happen if the national debt were paid off? The answer to that question is not simple, but we may examine some effects. In order for it to pay down the debt owed, the government taxes other societal components. Notation 8 presents this situation without regard to the beginning of the chain or its residuals.

$$TR_{-1}^{G} TR_{+1}^{C} CM_{-1}^{C} CM_{+1}^{G} \quad CM_{-1}^{G} CM_{+1}^{S} TLM_{-1}^{S} TLM_{+1}^{G}$$

$$(8)$$

were TR is tax receipt and G is government. Because a tax receipt is not a negotiable instrument, the CM taken from the consumer/laborer limits the choices of economic actions (remaining). It tends to slow economic activity. However, how it actually perturbs the system depends on whether the super-ordinate decider is endogenous or exogenous to the system. If endogenous, the CM comes right back into circulation (Notation 9,

again a partial chain).

$$TR_{-1}^{G} TR_{+1}^{C} CM_{-1}^{C} CM_{+1}^{G} \quad CM_{-1}^{G} CM_{+1}^{S/C} TLM_{-1}^{S/C} TLM_{+1}^{G}$$

$$(9)$$

where S/C indicates the dual role of the super-ordinate deciders as social lenders and consumer/laborers. Alternatively, if the super-ordinate decider is exogenous, the currency is removed from the system (Notation 10, partial chain).

$$TR_{-1}^{G} TR_{+1}^{C} CM_{-1}^{C} CM_{+1}^{G} \quad CM_{-1}^{G} CM_{+1}^{S/F} TLM_{-1}^{S/F} TLM_{+1}^{G}$$

$$(10)$$

where S/F indicates foreigner who will not re-introduce the CM in the economic system. In this case, the initial results of taxation (Notation 8) is not mitigated and the constraints imposed remain.

Notations 9 and 10 bring up the interesting question of whether or not truly exogenous government creditors exist. And that question begs an explanation of the function of national currency in the boundary of a society. Miller (1978; Swanson and Miller, 1989) define living systems at eight hierarchical levels of increasing complexity. The society is the seventh level. Systems at all of those levels exhibit twenty critical subsystems, without which they cannot endure in the environment of earth. The boundary is one of only two such subsystems that process both matter-energy and information. The others process one or the other. National currency (money-information) extends the boundary of a nation to all economic activity denominated in that currency. So, if such reasoning is correct, no truly exogenous government creditors exist. Nevertheless, one cannot neglect significant structural impediments to the free flow of national currency that is held beyond the delimited physical border of a nation.

CYBERNETIC DYNAMICS OF SUCH MOTIVE FORCES AS INTEREST, RENTS, AND PROFIT

Space does not allow a full discussion of the many effects of the introduction of these and other motive forces. Although the mutual benefit dynamic may suffice for an explanation of trades, seriously considering the value of time requires introductions of additional motive forces. Once a certain dynamic is accepted by economic participants, currency facilitation by government of economic processes must include the price of that dynamic. If it is not included, the price of the dynamic will inhibit the orderly retrieval of currency in the amount of the dynamic. Notation 11 describes one such situation.

$$CM^C_{-1.0}CM^P_{+1.0}GS^P_{-1.0}GS^C_{+1.0} \quad CM^P_{-1.0}CM^S_{+1.0}TLM^S_{-1.0}TLM^P_{+1.0}$$

$$(11)$$

where the TLM includes an obligation to pay interest to society. Society, however, fails to remit currency in the amount of the interest. If, after the interest part is defaulted, society insists on closing the circuit (removing the residual) several things might happen. For example, the producer may enter immediately into another debt contract with the government for another cycle of production and sales. Some of the cash received may then be remitted to the government to satisfy the interest owed. However, the producer will then be able to invest only 0.9 in the next round of production. The failure to introduce sufficient currency forces a slowing of economic activity. It is possible to manipulate the patterns of currency imbursement to retain the incentive/coercion of such lags without dampening economic activity.

Profit is a special case because it is introduced at the discretion of a participant in a chain of exchanges. A consequence of this is to relax the orderly processes that result when prices of specific dynamics may be anticipated. The relaxation of that constraint, however, has proven to provide a powerful motive force for innovative activity.

Similar patterns to those of profit emerge from wages, interest, rents, and other motive forces when participants in chains of exchanges are allowed the freedom to set their prices in specific exchanges. It is not difficult to see that the management of monetary and financial policy in such systems is forced to rely to a great extent on stochastic models. That fact notwithstanding, the objective values of each specific exchange in the trillions-plus exchanges per year that comprise a modern economic system are considered and observed—and, in most cases, are recorded. Each exchange works in tandem and in chains, and importantly, in circuits to produce a certain homeostasis of social interaction.

SUMMARY

The examples above provide some insights to certain cybernetic aspects of money-information imbursement. It is clear that the introduction of currency is not a necessary condition for economic activity to be facilitated beyond that allowed by trade. The negotiation of private debt can do that. Such facilitation is generally orderly. It is also clear that certain limits inherent in debt instrumentation can be relaxed by the imbursement of currency. As long as the currency introduced remains controlled by the temporal constraint of the executory contract underlying the debt instrument, the resulting economic processes are determinate within that limit. In the event that the debt instrument is defaulted, the temporal constraint is released, and the pattern of the motive force of the MIM is unpredictable.

The examples provided examine very simple expressions of some money-information marker imbursement and their contributions to societal homeostasis. Many more complex systems involving the dynamics of interest, taxes, rent, royalties, dividends, and profit have been modeled with the notation used (Swanson, 1993, pp. 97-149), as well as inter-societal exchanges (pp. 145-165).

IMPLICATIONS FOR INFORMATION SYSTEMS

Technology conceived broadly includes not only human artifacts but also processes and, yes, expanded perception (e.g., written language and mathematical systems). Information technology makes possible our modern modes of social order. The emerging global corporations are persons of law endowed with only the rights prescribed thereby. They are inventions of humans. They are technological informational entities. Those legal persons cohabit with natural citizens in modern societies, all interacting with each other and the environment.

The disciplines concerned with information science and technology have emerged only recently—actually in a single lifetime. As history goes, exchange-based societies emerged just a little earlier. The scientists, technicians, and academics (as with those in other disciplines) involved in information systems development are often concerned with narrow problems. Even though governments are some of the largest consumers of information technology, only recently have they begun to emphasize information systems that broadly and deeply integrate social processes. Those systems are being designed to gather and distribute information on already mature social systems. Such social systems could not exist without embedded information systems. I have attempted to draw attention to two such embedded information systems, exchange systems and money-information systems.

The concrete view of information in exchange-based societies has many implications for information systems design. Only five implications are presented here.

1. The money-information systems are particularly interesting because the constraints of those systems are pervasive and those constraints may be neglected in the design of particular information systems. The greater reason, however, that embedded societal information systems should be studied is that they often are the most efficient such systems and, by trial and error evolution, have incorporated satisfactory levels of redundancy. When such information systems exist, their strengths and weaknesses should be considered rigorously before substituting a designed information system.

2. The cybernetic effects of money information imbursement on both the individual entity's array of possibilities and on societal patterns of development may be important considerations in the design of information systems. A system approach suggests that, at the least, the possibility should be considered.

3. Information systems are often perceived to have changed the money-information markers to more energy efficient forms. They have done that, and that is important. It is also important that information systems have increased the rate of transmission of money-information. These advances, however, have much greater implications than simply obtaining energy-efficiency. In some situations, they have moved competition in the exchange system to a new technological level. Consider the super computers that engage arbitrage in the system of floating national currency values. The faster computer wins. Remember that debt instrumentation is the basic form of private money. In such information systems, the period of a debt instrument shrinks to nana-seconds. Hugh magnitudes of money are imbursed and transmitted. The higher technology systems work because they have incorporated the characteristics of the evolved exchange and money-information systems. The

magnitudes of money, however, if permitted into ordinary exchanges can have devastating effect.

The question of magnitudes of money useful to one section of a monetary system being destructive to another section has some relevance to the current "credit crisis." Are the trillions of dollars worth of currency imbursement the amounts needed to remove the drag of past unimbursed interest, profit, etc.? If so, in order to not simply compound the problem, semi-autonomous governmental units should hold and accountfor the debt instrumentation.

4. Democracy and exchanged-based societies have emerged more or less together. One reason is that in free exchange the societal decider subsystem is widely distributed to virtually all citizens. Emerging technology depends on the ability of individuals to account their ideas and skills to others. It is bottom-up. It is interesting, however, that the English language does not assign such a straight-forward meaning to the term accountability. It only allows "to hold accountable." Currently, evolving information systems seem overly concerned with top down control. It is true that "expert systems" and "open source" programs open up accountability from the bottom up, but even they can be used for top down control. Both bottom-up and top-down are necessary. Reaching the proper balance is difficult.

Because of the analytic science approach of information-related disciplines, it is easy to not consider sufficiently the cybernetics of the exchange processes themselves. Those systems are contributing to the homeostasis of societies. They may be overridden by information systems, the design of which overly connects critical elements, and that results in the unnecessary perturbation of societal processes. The current

sub-prime mortgage crisis provides an example of how this can happen. FASB157, par. 18, sec. a., moving towards a purely analytic concept of public financial reporting, stated: "Valuation techniques consistent with the market approach include matrix pricing. Matrix pricing is a mathematical technique used principally to value debt securities without relying exclusively on quoted prices for specific securities, but rather by relying on the securities' relationship to other benchmark quoted securities" (FASB, 2006).

Because FASB has the force of law, a fall in the credit rating of a company whose security you hold as an asset forces an immediate write-down of the value of that asset. No actual exchanges are required to bring about the falling value. However, the write-down can result in a lowering of certain financial ratios written into a debt covenant, which finally results in your creditor calling in the loan. Now you are forced on an untimely basis to provide cash to retrieve the note. The security that you hold has not been defaulted and yet despite your diligence in exchange, loss is imposed from above. Should the information systems be so tightly connected? At what point do the information systems that are used to set automobile insurance premiums based on fairness become unfair for their complexity beyond the control of policy holders? And, how do such systems undermine confidence of members of society at large?

5. An over-reliance of information science on its common analytic models and simulations may obscure some of the central characteristics of the concrete systems with which they are concerned. This paper has emphasized one such characteristic—the motive force of money-information in exchange. When models neglect the coupling relationship of the exchange, they neglect the dynamic that brings about the patterns being modeled. The models and simulations deriving from Leontief (1953) without doubt advanced the

art of governance. However, those models, as with most economic models, completely neglect the processes of money-information. There are at least as many types of money information as there are classes of goods-services. Information models that track those types are likely as necessary for good governance as those that model the good-services classes.

CONCLUSION

Money-information processes are important determinants of the homeostasis of modern exchange-based societies. While general principles of cybernetics may be approached from an abstract, holistic perspective, this discussion has approached certain aspects of social cybernetics from a concrete, internal perspective. From the chosen perspective, it is argued that the irreducible unit (the holon) of modern economic systems is the exchange. Highly complex economies are combinations of specific and observable exchanges. That complexity is facilitated by the introduction of money-information markers to temporally separate the reciprocating transfers of trades.

Ecology imposes economy by the entropy law. Human invention imposes scarcity on money-information markers, and they take on the motive force of commodities in exchange. A significant mechanism for imposing scarcity on money-information is the temporal limits of the executory contract underlying debt instrumentation. The temporal limits impose certain determinateness on the economic activity facilitated by the imbursement of money-information. Those limits can be released somewhat by the introduction of currency in a manner that ties it to debt instrumentation. The currency may, however, become discretionary (released from the temporal limits) upon default of the debt. In that case, the dynamics it may cause are indeterminate.

The system of notation developed in macro accounting provides a means of modeling different patterns of imbursing money-information. Some patterns result in chains of exchanges that form circuits—inserting and removing money-information to motivate specific economic activity. Other patterns result in chains that do not close—leaving residuals of discretionary currency and defaulted debt instruments. In each case, the models can provide significant cybernetic insight to the processes that concern policy makers at both the societal level and that of societal components.

Exchange and money-information systems, embedded deeply in social processes, constrain social activity. Information systems are designed to facilitate such activity. Consequently, their design should include consideration of the constraints imposed by exchange and money-information systems, and the patterns economic consequences imposed by various means of imbursing currency into social systems.

REFERENCES

Ashby, W. R. (1962). *An introduction to cybernetics.* London: Chapman.

Bateson, G. (1972). *Steps to an ecology of the mind.* New York: Ballantine.

Beauregard, O. C. (1961). *Sur l'equivalance entre information et entropie.* In J.G. Miller (Ed. And Trans.), *Living systems* (p. 42). New York: Mc-Graw-Hill.

Button, G., & Dourish, P. (1996). Technomethodology: Paradoxes and possibilities. In *Proceedings of the SIGCHI Conference on Human Factors in Computing Systems: Common Ground,* Vancouver, BC, Canada (pp. 19-26). New York: ACM. Retrieved from http://www.acm.org/sigchi/cyhi96/proceedings/papers/Button/jpd_txt.htm

Cannon, W. B. (1939). *Wisdom of the body.* New York: Norton

Einzig, P. (1966). *Primitive money* (2nd ed.). New York: Pergamon Press.

Georgescu-Roegen, N. (1971). *The entropy law and the economic process.* Cambridge, MA: Harvard University Press.

Grierson, P. (1977). *The origin of money.* London: The Athlone Press.

Kagan. (1982). The dates of the earliest coins. *American Journal of Archaeology, 86*(3), 343-360.

Leontief, W., et al. (1953). *Studies in the structure of the American economy.* New York: Oxford University Press.

Miller, J. G. (1978). *Living systems.* New York: McGraw-Hill.

Mises, L. V. (1912). *The theory of money and credit.* (H. E. Batson, Trans.). Indianapolis: Liberty Fund.

Schmandt-Besserat, D. (1992). *Before writing: From counting to cuneiform* (Vols. I-II). Austin, TX: University of Texas Press.

Shannon, C. E. (1948). A mathematical theory of communication. *The Bell System Technical Journal, 27,* 379–423, 623–656.

Simon, H. A. (1962). The architecture of complexity. *Proceedings of the American Philosophical Society, 106,* 467–469.

Swanson, G. A. (1993). *Macro accounting and modern money supplies.* Westport, CT: Quorum Books.

Swanson, G. A. (1998). Governmental justice and the dispersion of societal decider subsystems through exchange economics. *Systems Research and Behavioral Science, 15,* 413–420. doi:10.1002/(SICI)1099-1743(1998090)15:5<413::AID-SRES268>3.0.CO;2-F

Swanson, G. A. (2006). A systems view of the environment of environmental accounting. *Advances in Environmental Accounting and Management, 3,* 169–193. doi:10.1016/S1479-3598(06)03006-8

Swanson, G. A., Bailey, K. D., & Miller, J. G. (1997). Entropy, social entropy, and money: A living systems theory perspective. *Systems Research and Behavioral Science, 14,* 45–65. doi:10.1002/(SICI)1099-1743(199701/02)14:1<45::AID-SRES151>3.0.CO;2-Y

Swanson, G. A., & Miller, J. G. (1989). *Measurement and interpretation: A living systems theory approach.* New York: Quorum Books.

Weitemeyer, M. (1962). *Some aspects of hiring of workers in the Sippar region at the time of Hammurabi.* Copenhagen: Munksgaard.

Chapter 9
A Complex Adaptive Systems–Based Enterprise Knowledge Sharing Model

Cynthia T. Small
The MITRE Corporation, USA

Andrew P. Sage
George Mason University, USA

ABSTRACT

This paper describes a complex adaptive systems (CAS)-based enterprise knowledge-sharing (KnS) model. The CAS-based enterprise KnS model consists of a CAS-based KnS framework and a multi-agent simulation model. Enterprise knowledge sharing is modeled as the emergent behavior of knowledge workers interacting with the KnS environment and other knowledge workers. The CAS-based enterprise KnS model is developed to aid knowledge management (KM) leadership and other KnS researchers in gaining an enhanced understanding of KnS behavior and its influences. A premise of this research is that a better understanding of KnS influences can result in enhanced decision-making of KnS interventions that can result in improvements in KnS behavior.

CAS-BASED MODELING OF ENTERPRISE KNOWLEDGE SHARING

The enterprise KnS model developed here models enterprise knowledge sharing from a complex adaptive systems perspective. Hypothetical concepts that are fundamental to the development of this CAS-based model and to this research include:

1. Knowledge sharing is a human behavior performed by knowledge workers;
2. Knowledge workers are diverse and heterogeneous;
3. Knowledge workers may choose to share knowledge; and
4. The KnS decision is influenced by other knowledge workers and the KnS environment.

Enterprise knowledge sharing is the result of the decisions made by knowledge workers, individually and as members of teams, regarding knowledge sharing. As depicted in *Figure 1*, there are two major decisions (rectangles) that a knowledge worker makes: "Share Knowledge?" and "Type of Knowledge to Share?" This research models the KnS decisions as being influenced by the attributes of the individual knowledge worker, the KnS behavior of other knowledge workers, and the state of the KnS environment. Previous KnS studies and research identify factors that influence KnS behavior. However, few address the heterogeneity of knowledge workers and how the attributes of the individual knowledge worker, and knowledge worker teams, impact KnS behavior. The emergent enterprise KnS behavior, noted by the diamond shape in *Figure 1*, is the result of the interactions of the knowledge worker with the KnS environment and other knowledge workers. Relevant aspects of enterprise KnS behavior and the associated KnS influences are discussed in the sections that follow.

Enterprise KnS behavior takes on many forms. It can be a conversation around a water fountain, e-mail sent to a co-worker or a group forum, a presentation to a small group, an enterprise "best-practice" forum, or documents published to a corporate repository. Murray (2003) categorizes KnS activities into technology-assisted communication (videoconferencing, databanks/intranet, e-mail, and teleconferencing), meetings (face-to-face interaction, seminars and conferences, social events, and retreats), and training and development (mentoring, instructional lectures, video tapes, and simulation games). This research combines the two types of knowledge (tacit and explicit) and the ontological dimension (individual, group, and organization) of knowledge creation presented by Nonaka and Takeuchi (1995) to derive the types of KnS behavior for the model. The KnS behaviors investigated and incorporated in the enterprise KnS model are as follows:

1. **Individual tacit:** This behavior includes sharing tacit knowledge with an individual

Figure 1. Enterprise KnS influence diagram

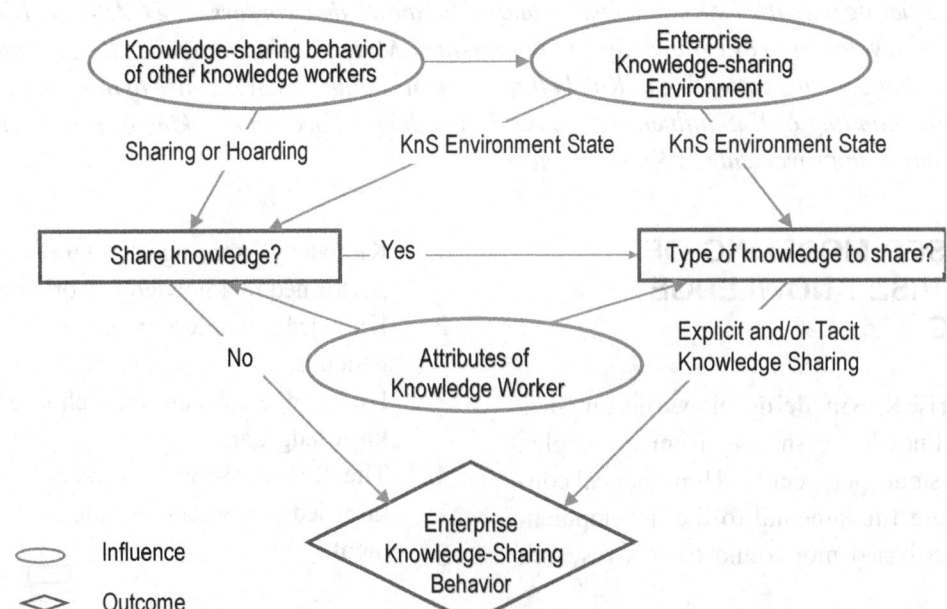

or individuals, such as face-to-face interactions in informal or formal meetings.

2. **Individual explicit:** This behavior includes sharing explicit knowledge with an individual or individuals, such as through sending e-mail or hard copy material to select individual(s).

3. **Group tacit:** This behavior includes sharing tacit knowledge with a group, such as face-to-face interactions with a community of interest, community of practice (CoP), or organizational unit.

4. **Group explicit:** This behavior includes sharing explicit knowledge with a group, such as posting or contributing to a community of interest, CoP, or organizational unit repository, Web site, or mailing list server.

5. **Enterprise tacit:** This behavior includes sharing tacit knowledge in an enterprise-wide forum, such as presenting at a technical exchange meeting or other forum that is open to the entire enterprise.

6. **Enterprise explicit:** This behavior includes sharing explicit knowledge in a manner that makes it available to anyone in the enterprise, such as publishing in a corporate-wide repository or enterprise-wide intranet.

While we investigate KnS behavior as being comprised of six different types, both tacit and explicit knowledge are often shared in a given situation. For example, in an enterprise KnS forum, tacit knowledge, such as unrehearsed oral presentations and responses to questions, and explicit knowledge, such as hard copy presentations, are generally both shared.

We investigate three major KnS influences on the associated sharing of knowledge:

1. The enterprise KnS environment,
2. KnS behavior of other knowledge workers, and
3. Attributes of the knowledge workers.

The KnS literature, such as reviewed in Small and Sage (2006), identifies many factors that influence KnS behavior. A discussion of each of the major influences is provided in the sections that follow.

The enterprise KnS environment is closely aligned to the Japanese concept of "ba" which translates into English as "place." Nonaka and Konno (1998) adapted this Japanese concept for their knowledge creation theory. "Ba," as described by Nonaka and Konno (1998), is the shared space for emerging relationships that can be physical, virtual, mental, or any combination of these. It is the place where knowledge is created, shared, and exploited. The "ba" is comprised of the knowledge resources and the people who own and create the knowledge. The KnS environment or "ba" is comprised of many factors that influence KnS behavior. There are at least six important influence factors in the KnS environment modeled and investigated here. A brief description of each of these factors is appropriate here:

1. **KnS technology:** KnS technologies are those technologies that allow knowledge workers to share tacit or explicit knowledge. Technologies and tools reported (APQC, 2000) as critical to knowledge sharing at best practice firms included: e-mail, intranets, document sharing systems, collaboration tools, and video conferences. Chu (2003) included e-mail, Internet, intranet, databases, and teleconferences in his listing of these. With the advent of Web 2.0, wikis, blogs, and social networking applications are being used to enable enterprise knowledge sharing (APQC, 2008)

2. **Leadership:** Leaders and managers in an organization impact KnS behavior by directing behavior, rewarding or recognizing behavior, and by setting KnS behavior examples. Many studies indicate that organizations with appropriate KnS leadership

behavior have more instances of appropriate KnS behavior than others.

3. **KnS culture:** Culture is an organization's values, norms, and unwritten rules. Most existing KM models and KnS investigations include culture as a critical enabler or influence on KnS behavior. Additionally, cultural issues are regularly cited as one of the concerns held by those implementing KM initiatives.

4. **Human networks:** This factor includes processes, technology, and resources that help to connect knowledge workers or support knowledge networks. Support for human networks, which includes informal and formal forums, is widely practiced among best practice organizations. They are often referred to as communities of practice or community of interests. Organizations can enable these networks with knowledge stewards, online collaboration tools, and tools to facilitate easy publishing.

5. **Rewards and recognition:** This factor includes the approaches organizations use to encourage or reinforce the discipline of knowledge sharing. Approaches include rewards, recognition, alignment with performance assessment and promotion, and conducting visible KnS events. When establishing rewards, organizations must consider the generic type of behavior they are trying to stimulate. Many organizations have instituted reward and award programs for knowledge sharing and/or have integrated incentives for knowledge sharing with performance appraisals and promotions.

6. **Alignment with strategy:** This refers to the alignment of knowledge sharing with business strategy. Best practice organizations do not share knowledge for the sake of knowledge. Rather, knowledge sharing is deemed critical to achieving business goals and is linked to the business strategy (APQC, 1999). The alignment of knowledge sharing

to business strategy can be either explicit or implicit. When organizations have explicit alignment, language regarding knowledge sharing can be found in documents such as strategic business plans, vision or mission statements, or performance measures. Organizations with implicit alignment are evidenced by knowledge sharing embedded in business practices. Fifty percent of the best-practice firms that participated in the APQC benchmarking study (APQC, 1999) on knowledge sharing were explicitly aligned, while the other half were implicitly aligned. Findings of two APQC benchmarking studies found that organizations where knowledge workers understood how knowledge sharing supported the business strategy had stronger KnS behavior.

The behavior of other knowledge workers within an organization affects the KnS decisions of a specific knowledge worker in many ways. Ford (2003) describes sharing knowledge as a risky behavior because the individual does not know how the shared knowledge will be used by the party who obtains it. Trust in, and some knowledge of, what the recipient of the shared knowledge will do with the shared knowledge are critical to knowledge sharing. From an enterprise perspective, knowledge workers must trust the organization not to cast them aside after the knowledge is harvested. From a peer interrelationship perspective, a knowledge worker must trust that the knowledge recipient will make ethical use of the shared knowledge (Bukowitz & Williams, 1999). If a knowledge worker shares and the knowledge recipient misuses the shared knowledge, from the perspectives of the intended purposes for sharing, then the knowledge worker may be reluctant to share knowledge in the future.

The KnS influence of individual knowledge workers attributes is very important because knowledge sharing is a human behavior in which the knowledge worker chooses to share. The decision to share is influenced by interactions. Leonard and Straus (1997), for example, assert that individuals have preferred habits of thought that influence how they make decisions and interact with others. Knowledge workers have many diverse attributes, some of which are fixed and others of which are variable. Some of the individual attributes or human factors identified in the KM and KnS literature include employees' means, ability, and motivation (Ives et al., 2000); job characteristics including workload and content (Chu, 2002); feelings of being valued and commitment to the project (Ipe, 2003); and conditions of respect, justice perception, and relationships with superiors (Liao et al., 2004).

Here, we model enterprise knowledge sharing as emergent behavior that is the result of decisions made by knowledge workers. The decisions, "Share Knowledge?" and "Type of Knowledge to Share?" depicted in *Figure 1* are based on dynamic interactions and are influenced by factors in the KnS environment, KnS behaviors of other knowledge workers, and the individual attributes

and perspectives of the knowledge worker. The CAS-based enterprise KnS model integrates the knowledge worker, KnS decisions, and the KnS influences into a CAS-based framework, which consists of two major components:

1. CAS-based enterprise KnS framework
2. Enterprise KnS simulation model (e-KnS-MOD).

A detailed discussion of each of the components is provided in the sections that follow.

CAS-BASED KNS FRAMEWORK

The CAS-based KnS framework is the most critical element of our CAS-based KnS model and distinguishes it from other KM models, such as those described in Small and Sage (2006). The CAS-based KnS framework describes enterprise knowledge sharing from a complex adaptive systems perspective. The properties of a CAS,

Figure 2. Major elements of the CAS-based KnS framework

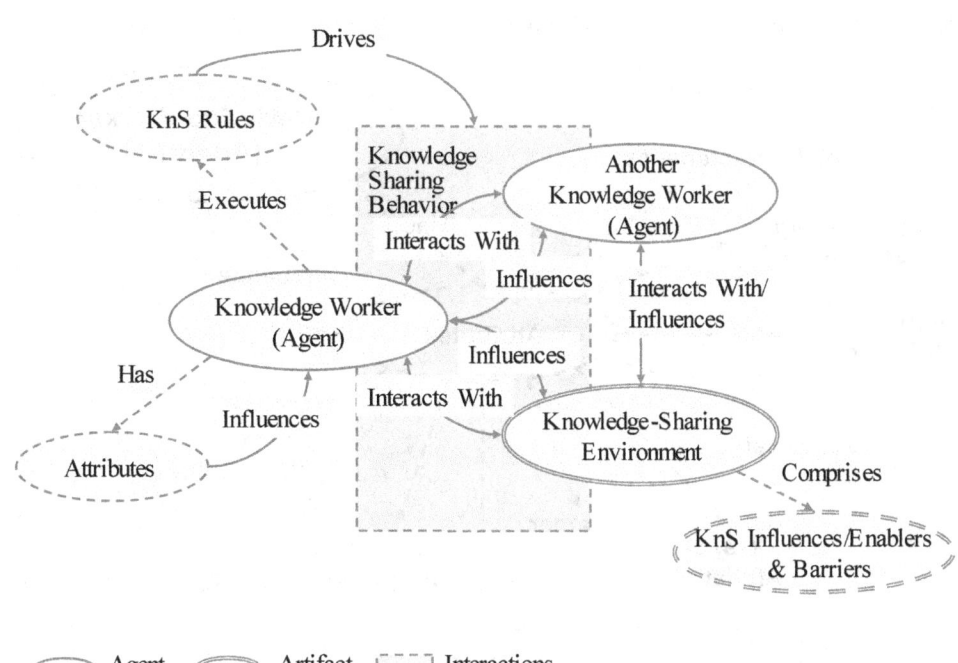

as described by Holland (1995), are aggregation, diversity, internal models, and non-linearity. Axelrod and Cohen (1999) identify variation, interaction, and selection as the hallmark of complex adaptive systems. Other important concepts of complex adaptive systems include the agent, strategy, population, type, and artifacts. For simplicity, the following constructs of a complex adaptive system have been addressed at the highest level of the enterprise KnS framework: agent, agent attributes, interactions, artifacts, and rules.

The CAS-based KnS framework, illustrated in *Figure 2*, is comprised of the following elements: knowledge worker(s); KnS environment (comprised of KnS influences/enablers and barriers); KnS behaviors; KnS rules; and attributes of the knowledge worker. The KnS behavior results from the interactions of the knowledge workers with each other and the KnS environment. The decision to share is influenced by individual attributes, KnS behavior of other knowledge workers, and the KnS environment. A mapping of the KnS influence diagram in *Figure 1* to the

CAS concepts used in the CAS-based framework of *Figure 2* is as follows:

- **KnS Influence Diagram Elements**
  - Knowledge workers
  - KnS Environment
  - KnS Decisions
  - Enterprise knowledge sharing
  - Knowledge worker attributes
- **CAS-Based KnS Framework Elements**
  - KnS Agents
  - KnS Environment (artifacts)
  - KnS Rules
  - KnS Behaviors (interactions)
  - KnS Agent attributes

The knowledge worker is the KnS agent within the CAS-based model. Critical to this concept is the diversity and heterogeneity of this KnS agent. The knowledge worker within an enterprise is diverse in many ways: personality, gender, role, and job level. *Figure 3* associates this segment of the KnS framework with the attributes of the knowledge worker. The KnS decisions (execution

Figure 3. Investigated attributes of knowledge worker

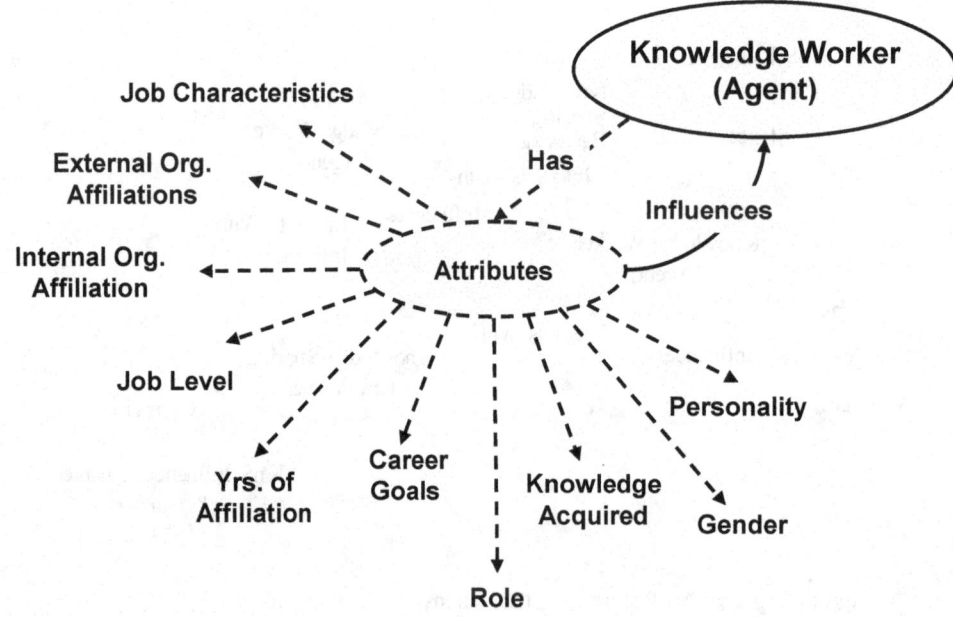

of rules) of a KnS agent depend on the agent's attributes and are influenced by the agents' interactions with other knowledge workers and the KnS environment.

The attributes of the knowledge worker investigated here include: personality, gender, level of knowledge acquired, years of affiliation, role, career goals, job level, internal organizational affiliation, external organizational affiliation, and job characteristics. These attributes are described as follows:

1. **Personality:** Such as introvert, extrovert, or a combination.
2. **Gender:** Male or female.
3. **Level of knowledge acquired:** The level of knowledge acquired over time (related to competency) by the knowledge worker.
4. **Years of affiliation:** The number of years a knowledge worker has been affiliated with the enterprise (i.e., number of years at the company).
5. **Role:** The role (s) the knowledge worker has within the enterprise, organization, or project. Examples include manager, technical leader, or technical contributor.
6. **Career goals:** The job or career-related goals possessed by the knowledge worker. Goals investigated as part of this research include: career growth (promotion), knowledge growth opportunities, satisfying customers, satisfying management, recognition, and reward.
7. **Job level:** The job level that is assigned by the company to a given knowledge worker, ranging from entry/junior level people to executive management.
8. **Internal organizational affiliation:** An enterprise usually consists of many organizations. This is the internal organization to which the knowledge worker is assigned.
9. **External organizational affiliations:** The number of external professional organiza-

tions with which the knowledge worker is affiliated.

10. **Job characteristics:** This includes number of tasks supported, workload, pace, and content of work.

KnS rules drive the decisions the knowledge worker makes. A knowledge worker has two fundamental KnS decisions: "Share Knowledge?" and "Type of Knowledge to Share?" The KnS rules are the same for all KnS agents. They are parameterized based on the attributes of the agents, behavior or other knowledge workers, and the state of the KnS environment.

An enterprise KnS environment consists of many factors that influence or enable KnS behavior. A KnS artifact is an entity in the enterprise (not a person) with which the knowledge worker interacts that either influences or enables their KnS behavior. An enterprise has many KnS artifacts, including information technology, performance and reward systems, knowledge repositories, and information help desk. The KnS influences or enablers examined here and illustrated in *Figure 4* include: KnS linked to corporate strategy, alignment of rewards and recognition, KnS embedded with work processes, KnS aligned with core values, enabling of human networks, and KnS technology (availability and ease of use). The artifacts that exist in an environment can have different enabling characteristics. A five-state characterization instrument was developed to characterize the KnS environment.

A knowledge worker (KW) gains or acquires knowledge by interacting with the environment and other knowledge workers. Knowledge sharing results in and from a KW interacting with another KW and/or with the KnS environment. Enterprise knowledge sharing is the result of knowledge workers interacting with other knowledge workers and the enterprise KnS environment. Included in the CAS-based framework are the following KnS behaviors: individual tacit, group tacit, enterprise tacit, individual explicit, group explicit, and enterprise explicit.

Figure 4. KnS influences/enablers investigated

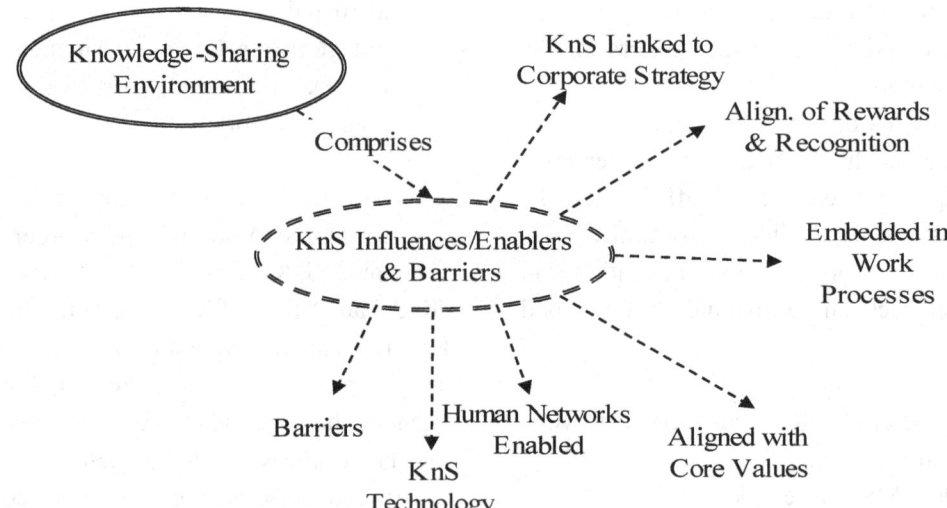

MULTI-AGENT ENTERPRISE KNS SIMULATION MODEL (E-KNSMOD)

The Enterprise KnS Model (e-KnSMOD) simulates enterprise knowledge sharing as the emergent behavior of knowledge workers, represented as agents, interacting with the KnS environment and other knowledge workers. The design of the e-KnSMOD is based on the CAS-based KnS framework described here. All of the constructs of the framework (KnS agent, agent attributes, KnS behavior, KnS environment, and rules) are implemented in the simulation model. For simplicity, the simulation model implements a subset of the attributes (level of knowledge, role, career goals, job level, and internal organizational affiliation) of the knowledge worker included in the CAS-based framework. The purpose of the model is to examine the effects of the KnS enterprise environment and behavior of other knowledge workers on the KnS behavior of a heterogeneous population of knowledge workers. Epstein and Axtell (1996) refer to agent-based models of social processes as artificial societies. The design and implementation of this model leverages the agent-based computer modeling of the artificial society known as The Sugarscape Model (Epstein

& Axtell, 1996) and the Sugarscape source code developed by Nelson and Minar (1997) using Swarm (Minar et al., 1996; Johnson & Lancaster, 2000; Swarm Development Group, 2004).

The e-KnSMOD model simulates a population of knowledge workers that work in an artificial enterprise. As with Sugarscape (Epstein & Axtell, 1996), the e-KnSMOD leverages the research results that have been obtained using cellular automata (CA) for agent-based modeling. KnS agents represent the knowledge workers, and the CA represents the artificial enterprise, KnS-scape. The KnS agents interact with each other and their environment as they move around the enterprise gaining valuable knowledge (a goal of many knowledge workers). Agents acquire knowledge by engaging in a knowledge creation opportunity or by receiving knowledge shared by other knowledge workers. In order to satisfy their goals, they must continue to generate new knowledge. As conceptually depicted in *Figure 5*, the e-KnSMOD consists of three major elements:

1. KnS agents ("knowledge workers")
2. The artificial enterprise or KnS-scape
3. Interactions (driven by rules).

Figure 5. Major elements of the e-KnSMOD

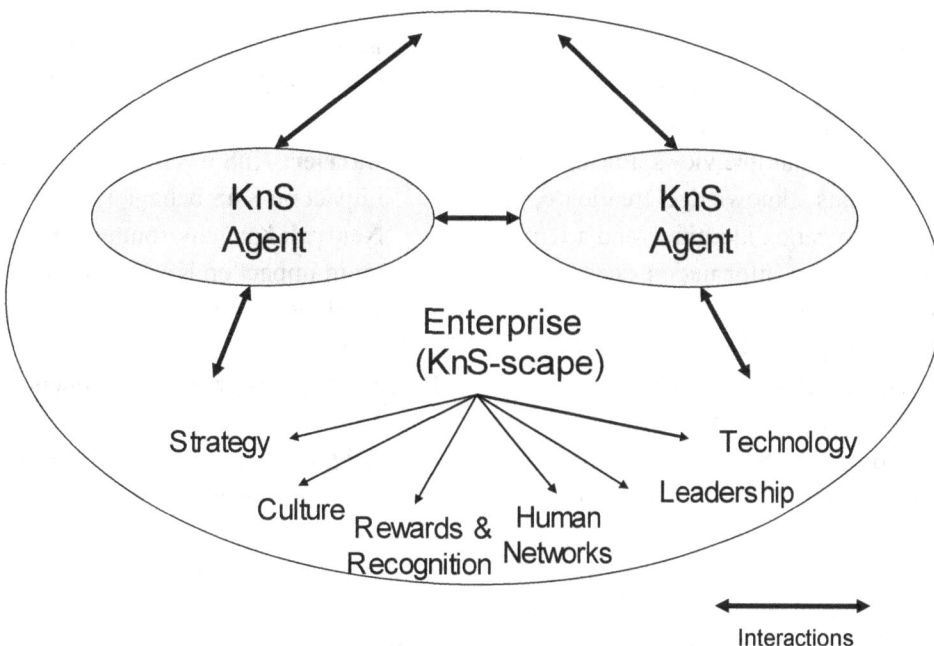

Each of these elements, as implemented in the e-KnSMOD, is described in the following subsections.

KnS Agent

A KnS agent represents a knowledge worker in the artificial enterprise. The KnS agents are heterogeneous. This implementation of e-KnSMOD models the following subset of attributes included in the CAS-based KnS framework: level of knowledge acquired, role, job level, and organization affiliation. Each KnS Agent is characterized by a set of fixed and variable states that vary among the agents. The fixed states include:

1. Level of knowledge acquired (competency)
2. Job level (vision is based on job level) in organization (e.g., Jr. Analyst, Sr. Analyst, Principal, Director)
3. Role in organization (manager, non-manager)
4. Organizational affiliation.

Each agent has the following variable states:

- New knowledge gained
- Location on the KnS-scape
- KnS indicator (indicates if the agent shared in the previous run cycle).

The KnS agent comes to the KnS-scape with a specified competency. Upon entry, the agent is assigned a vision and organizational affiliation. The job level is then based on vision. The KnS agent moves (changes location) around the enterprise in order to participate in knowledge-creation opportunities that allow the KnS agents to gain knowledge. The agent's vision restricts what knowledge creation events the agent can see. The agent decides to share or hoard the knowledge gained. If the agent decides to share, it can participate in one or more KnS behaviors: individual tacit, individual explicit, group tacit, group explicit, enterprise tacit, and enterprise explicit. The shared knowledge indicator is set when the agent shares knowledge.

KnS-scape: The Artificial "Ba"

The KnS-scape, which represents the "Ba," is represented by a two-dimensional (50 x 50) co-ordinate grid. The grid is built using the Swarm tool set. The grid has multiple views. Each point (x, y) on the grid has a knowledge-creation opportunity, an organization identifier, and a KnS environment state. The information needed by the model to create these views is read from data files, which can be specified at run time. A KnS agent is randomly placed on the KnS-scape. The organizational unit associated with the agent's initial location on the KnS-scape determines an agent's organizational affiliation. When a KnS agent engages in a knowledge-creation opportunity, it acquires the knowledge associated with the opportunity. An organization view of the KnS-scape would indicate that there are four different organizations within the enterprise. The KnS agents are colored by the organizational affiliation of their initial location on the KnS-scape.

Knowledge-Creation Opportunity

Each location on the KnS-scape, represented by an (x, y) coordinate, has a knowledge-creation event or opportunity. KnS agents interacting with their environment and with other KnS agents create knowledge. One of the ways a KnS agent interacts with the environment is by moving to a location and then acquiring the knowledge associated with a knowledge-creation event. When an agent acquires the knowledge at a given location, the knowledge is depleted (value = 0) until another knowledge creation event occurs. The value of the knowledge creation event is increased on each cycle of the simulation until the maximum value for that location is achieved. The amount of increase on each cycle is controlled by the "alpha" parameter, described later.

KnS Environment State

Each location on the KnS-scape has a KnS environment state. The states are as follows:

1. **Barrier:** KnS environment has a negative impact on KnS behavior.
2. **Neutral:** KnS environment has no or minimum impact on KnS behavior.
3. **Enabled:** KnS environment enables KnS behavior.
4. **Encouraged:** KnS environment encourages KnS behavior.
5. **Aligned:** KnS environment positively influences KnS behavior.

KnS Organization View

Each location on the KnS-scape, represented by an (x, y) coordinate, has an organizational identifier. When an agent enters the KnS-scape, it is given the organizational identifier of the location where it is placed. The organizational identifier is used in group KnS behaviors.

Interactions: Acquiring and Sharing Knowledge

The KnS agent interacts with the KnS-scape and with other KnS agents. As previously described, each KnS agent comes to the KnS-scape with a vision that allows it to see knowledge-creation opportunities. During each simulation cycle, an agent looks out over the KnS-scape and determines the location of the best knowledge-creation opportunity. It then moves there and acquires the knowledge. If the KnS agent acquires enough knowledge to share, the KnS agent then chooses to share or not to share. The KnS agent can participate in six types of KnS behaviors: individual tacit, individual explicit, group tacit, group explicit, enterprise tacit and enterprise explicit. The impact of each of these KnS interactions is briefly described as follows:

1. Tacit individual: Results in the "current knowledge" attribute of the recipient KnS agent being increased. The physical vicinity of KnS agents restricts this interaction.
2. Tacit group: Results in the "knowledge acquired" attribute of the recipient KnS agents being increased. The "current knowledge" attribute restricts this interaction.
3. Tacit enterprise: Results in the "current knowledge" attribute of all KnS agents being increased. The "organizational affiliation" attribute restricts this interaction.
4. Explicit individual: Results in the "current knowledge" attribute of the recipient KnS agent being increased.
5. Explicit group: Results in an increase of knowledge in the organizational or group repository.
6. Explicit enterprise: Results in an increase of knowledge in the enterprise repository.

The most important aspect of "ba" is interaction. Important to this research is that knowledge is created by the individual knowledge worker as a result of interactions with other knowledge workers and with the environment.

Rules for the KnS-Scape

Eptein and Axtell (1996) describe three types of rules: agent-environment rule, environment-environment rule, and agent-agent rule. There are three types of similar rules in the KnS-scape model:

1. Agent movement rule;
2. Generation of new knowledge creation events rule;
3. KnS rule.

A brief description of each rule is provided here:

- **Agent movement rule:** The KnS agent uses the movement rule to move around the KnS-scape. The movement rule processes local information about the KnS-scape and returns rank ordering of the state according to some criteria. The rules and functions used by the agents are the same for all agents. The values of the parameters change based on the attributes of the agent and the state of the environment. A summary of the movement rule is as follows:
  1. Look out as far as vision (an agent attribute) permits and identify the unoccupied site(s) that best satisfies the knowledge acquisition goal.
  2. If goals can be satisfied by multiple sites, select the closest site.
  3. Move to the site.
  4. Collect the knowledge associated with the knowledge-creation opportunity of the new position.
- **Generation of New Knowledge Creation Events:** A knowledge creation event has a knowledge value. After the knowledge is collected from the site on the KnS-scape, the value goes to zero (it no longer exists). The frequency of new events is driven by the "alpha" parameter. At the end of each cycle, each location on the KnS-scape is incremented by the "alpha" value until it reaches its maximum value.
- **KnS Rule:** After an agent completes the move to the new location and acquires the knowledge there, the KnS rule is executed. The decision to share and the type of knowledge to share is dependent on the KnS behavior of other agents, the KnS environment state, and the "level of knowledge acquired" attribute.
- **E-KnSMOD—Simulation of Enterprise Knowledge Sharing:** Enterprise knowledge sharing is simulated by the e-KnSMOD. Enterprise knowledge sharing is measured by the number of KnS agents participating

in one of the six KnS behaviors, the percent of KnS agents that share, the frequency that KnS agents share, and the number of items deposited into the group or enterprise repositories.

Initializing the e-KnSMOD environment properly is important here. E-KnSMOD, built using the Swarm tool set, has two basic components: the Observer Swarm, and the Model Swarm. Swarms are objects that implement memory allocation and event scheduling. Upon execution of the e-KnSMOD, two probes and a program control panel are displayed. The observer (ObserverSwarm) and model (ModelSwarm) probes consist of default parameters that are modifiable by the user. After the parameters for the Observer Swarm and Model Swarm are processed, the e-KnSMOD environment is established by creating the Observer and Model objects and building the Scheduler. The Observer objects consist of the windows used to display the KnS-scape and KnS agents and other graphs specified by the user. The Model objects consist of the KnS-scape and the KnS agents. These steps are described next:

1. **Creation of the KnS-scape:** The KnS-scape, a 50 x 50 lattice, represents the KnS enterprise environment. Each location (x,y)

on the KnS-scape has a knowledge creation opportunity, an organization identifier, and a KnS environment state. The KnS_event, organization, and KnS_environment data-files (specified in the ModelSwarm probe) are used to build the characteristics of each (x,y) location, respectively. The knowledge creation events, which have a value of 1 through 5, are observable by the user of the KnS model from the KnS-scape window. The value of a knowledge creation (KC) event is distinguishable by color as represented in the KnS-scape window illustrated in *Figure 6.*

2. **Creation of the KnS Agents:** After the KnS-scape is created, the KnS agents are created and randomly placed on the KnS-scape. The "KnSnumAgents" parameter is used to determine how many KnS agents are created. The model creates a heterogeneous population of KnS agents. Some of the attributes are randomly generated, and others are based on where the agent is placed on the KnS-scape. The agents organizational affiliation is determined by the organization associated with the (x, y) coordinate at which the agent is placed. The initial value of current knowledge is based on the vision, which is randomly generated.

Figure 6. Knowledge creation (KC) events on the KnS-scape

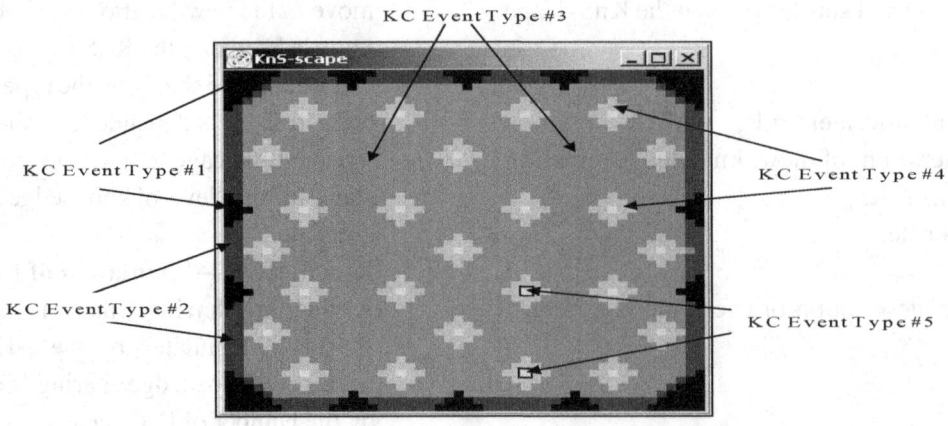

3. **Creation of the Scheduler:** The Observer Swarm and the Model Swarm create a schedule for activities to be performed during each cycle of the model. The Model Swarm schedules the actions to be performed by the KnS agents and the actions to be performed on the KnS-scape. The actions include:

   1. KnS Agent: Move and acquire knowledge.
   2. KnS Agent: Execute KnS behavior rule.
   3. KnS-scape: Update KnS-scape (Knowledge Creation Event View).
   4. KnS Repositories: Update group and enterprise repositories.
   5. Display: Update KnS-scape display window.
   6. Display: Update knowledge distribution graph.
   7. Display: Update KnS attributes over time.
   8. Summary File: Update KnS summary (metrics) file.

4. **Model Output:** The e-KnSMOD has three primary output windows that are updated after each cycle. The windows include: KnS Agent Attributes Over Time, Agent Knowledge Distribution, and the KnS-scape. Additionally, the model maintains a KnS summary data file that captures the KnS metrics of the KnS agents. This data file is used for additional data analysis outside the e-KnSMOD environment. The following KnS metrics are captured by the model: the number of KnS agents that shared, the number of agents that shared by organization, the average amount of knowledge acquired, the number of items contributed to a group repository, and the number of items contributed to an enterprise repository.

Figure 7. Example run – recurring rate for KC events = 1

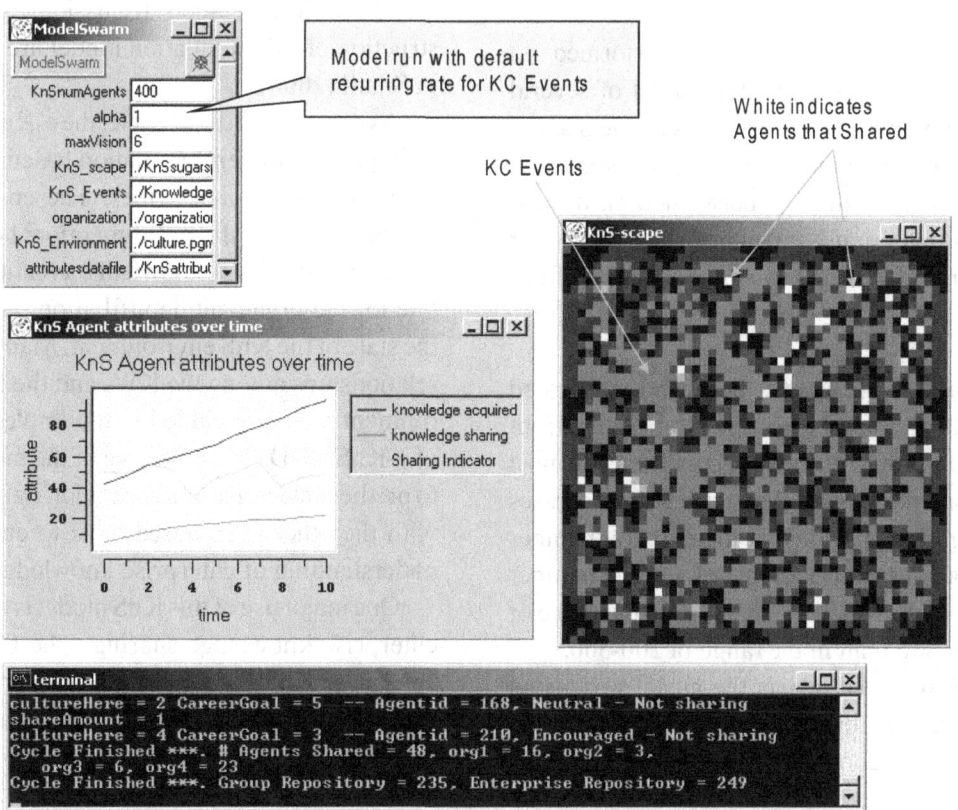

The e-KnSMOD is designed to allow the user to explore possible improvements in enterprise knowledge sharing by observing the impact of KnS influences. The influences identified in the enterprise sharing influence diagram, shown in *Figure 1*, are: KnS environment, KnS behavior of other knowledge workers, and attributes of the knowledge workers. *Figure 7* shows the results of a 10-cycle run using the default "alpha" value (alpha = 1), which causes a depleted KC event to increase one unit per cycle until it reaches its maximum capacity. Examination of the KnS Agent Attributes Over Time window shows that an average number of KnS agents sharing during each cycle is approximately 50, with a steady increase of knowledge acquired. By changing the "alpha" parameter to zero (0), for example, the user can examine what the impact of the KC event not reoccurring has on KnS behavior. Here, the results of a 10-cycle run show that the number of KnS agents sharing began to drop until no sharing occurred. The resulting KnS-scape window shows that there are no KC events.

Sensitivity analysis may be performed on e-KnSMOD by executing the model of several varying conditions in order to determine if small changes to the parameters resulted in unexpected results. Analysis may be performed on the parameters that are used in either the KnS rule or the environment rules. A summary of the findings are:

1. **Number of agents:** the model was tested with the number of agents ranging from 100 to 500 with varying conditions. In most cases, the percent of agents sharing increases slightly (< 1.5%) as one increases the number of agents in increments of 50. The number of agents was more sensitive in the range of 100-300 than in the range of 200-500.

2. **Behavior influence:** the model was tested by setting this parameter to 0 and 1. In all the tests conducted the percent of agents sharing decreased in the range of 1.7 to 4.0 percent when the parameter was changed from 0 to 1.

3. **Max vision:** The maximum vision was tested with the values 7, 14 and 28. In most cases, as the vision increased (7 to 14 to 28) the resulting knowledge sharing increased ~ 1 %. However, the percent was higher when the knowledge creation events with high value (part of the KnS_scape) were further apart.

4. **KnS_scape:**– the percent of agents sharing is impacted most by this parameter. The KnS agents acquire knowledge from the KnS_scape and if the agent does not have knowledge, it does not share.

5. **KnS_environment:** the percent of agents sharing is impacted greatly by this parameter. A difference of one state (i.e., barrier to neutral or neutral to enable) can change the percent of agent sharing from 5 % to 14 %.

Much more detailed discussions of the construction of this simulation model are presented in Small (2006).

As described in this article, the e-KnSMOD, is a simple multi-agent simulation based on simple environment and KnS rules. The environment is represented by three 2-dimentional (50 by 50) lattices: one for the knowledge creation events, one for the organization affiliation, and one for the state of the KnS environment. Many complex relationships among the KWs and the KnS environment are not included in the implementation of e-KnSMOD. The objective of the model is not to predict enterprise KnS behavior, but to be used with the other CAS-based tools to enhance the understanding of enterprise knowledge sharing.

One major use of this KnS model is to improve enterprise knowledge sharing. The CAS-based enterprise KnS model can assist enterprise KM leadership, managers, practitioners, and others involved in KM implementation to characterize the current KnS environment, identify influences of KnS behavior, and better understand the impact of KnS interventions. This model can be applied to enterprises that are about to embark on KnS initiatives, as well as those that have a rich KnS portfolio.

The CAS-based characterization instruments allow a practitioner to characterize enterprise KnS from the perspective of the KW and from that of KM Leadership. Both instruments characterize the frequency of KnS behaviors, the extent of influence of KnS influences and barriers, and the state of the KnS environment. The data gathered using these instruments provide the information needed to characterize and model an enterprise from a CAS perspective.

The KW Profiling Questionnaire is a critical element here. The purpose of the KW Profiling Questionnaire is to determine, from an individual knowledge worker perspective, the answers to four questions:

1.   What are your attributes?
2.   What is your KnS behavior?

3.   What influences your KnS behavior?
4.   What is the state of the KnS environment?

The answers to these questions allow a KM practitioner to investigate the extent of KnS influences on the heterogeneous knowledge worker populations. Addressing the attributes of the knowledge worker is a critical aspect of this CAS-based methodology.

The focus of the KM Leadership Characterization Questionnaire is to determine, from the perspective of KM leadership and implementers, the answers to the following four questions:

• Part I: What is the understanding of the KM Leadership Team regarding the KnS needs (mission perspective) and KnS behavior within the organization?

• Part II: What are the KnS influences and the extent of the influences within your enterprise?

• Part III: What is the state of the KnS enablers/influences within your enterprise?

Figure 8. KM leadership characterization and the CAS-based KnS framework

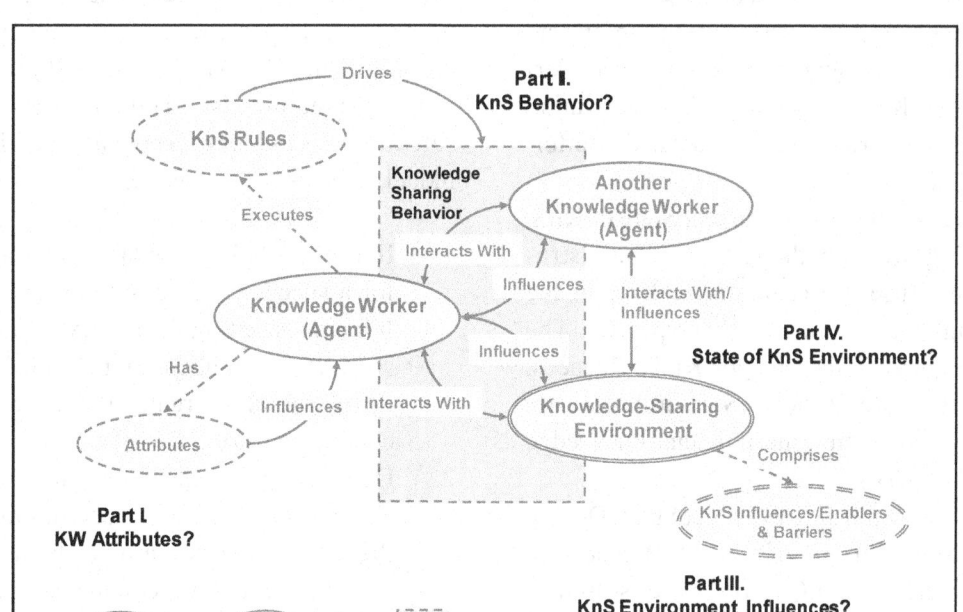

- Part IV: What is the KnS Strategy for Improvement?

Part I and Part IV of the KM leadership characterization instrument relates to the KnS improvement strategy. Part I addresses the importance of KnS to support mission needs, and whether KnS is occurring at the right level (individual, group, enterprise) and frequency. Part IV addresses the KnS strategy, which includes areas of improvement and the priority for achievement. The relationships of these questions to the CAS-based KnS framework are depicted in *Figure 8*.

The CAS-based KnS improvement methodology can be used by either an enterprise about to embark on KnS improvement activities for the first time (Initial Stage) or an enterprise that has a KnS strategy and robust KnS portfolio (Learning Stage). The tools described here can be used to identify and prioritize KnS improvement courses of action. The CAS-based methodology consists of five primary steps:

1. Step 1: Determine KnS Needs in Context of Mission Effectiveness. During this step, the KM practitioner determines the importance of KnS to the organization and assesses whether KnS is occurring at the appropriate frequency to support mission needs. Part I of the KM Leadership Characterization Questionnaire is used to gather this information.

2. Step 2: Characterize Current State of KnS. During this step, the KW profiling instrument is used to characterize KnS in the organization from a CAS perspective. The frequency of KnS behavior, KnS influences, and the state of the KnS environment are characterized from the individual knowledge worker perspective.

3. Step 3: Establish KnS Target State. During this step, Part III of the KM Leadership Characterization Questionnaire is used to capture the target state of the KnS environ-

ment, identify factors in the KnS environment that need improvement, and to establish priority of their implementation.

4. Step 4: Perform CAS-based Analysis. During this step, population analysis is performed based on KW attributes of interest to the organization. A gap analysis is performed on areas targeted for improvement against the extent of influence of the KnS factors identified by the KWs.

5. Step 5: Develop KnS Improvement Strategy. During this step, the results of the CAS-based analysis are used to develop or align the KnS strategy. The current state of the KnS environment (KW perspective), the target state of KnS environment (KM leadership), and the extent of KnS influence (KW perspective) are used to identify areas of improvement and their priority. The CAS-based simulation model can be used to model the planned improvements to gain insight into the possible impacts on KnS behavior.

The steps of the CAS-based KnS methodology should be integrated into the organizational improvement framework. We describe the CAS-based KnS improvement methodology in the context of the IDEAL^{SM} (SEI, 1996) model, an improvement process originally designed for software process improvement. The IDEAL^{SM} model consists of five phases:

1. Initiating: This phase lays the groundwork for a successful KnS improvement effort. It includes setting the context and sponsorship, and establishing the improvement infrastructure (organizations). Step 1 is conducted during this phase.

2. Diagnosing: Assessing the current state of KnS in the enterprise and determining where the organization is relative to the target state. Step 2, 3, and 4 are conducted during this phase.

Figure 9. CAS-based methodology: An IDEAL^{SM} perspective

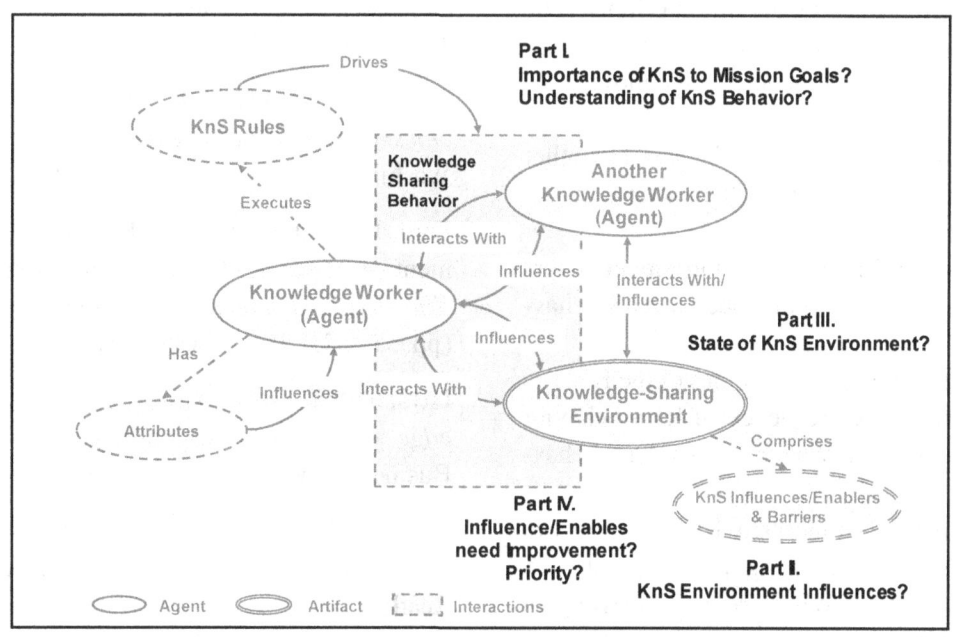

3. Establishing: Developing strategies and plans for achieving the KnS target state. Step 5 is conducted during this phase.
4. Acting: Executing the plan to improve KnS.
5. Learning: Learning from the KnS experience and feedback from mission stakeholders, KM leadership, and knowledge workers.

As shown in *Figure 9*, Step 1 occurs during the Initiating phase. Step 2, 3, and 4 occur during the Diagnosing phase, and Step 5 concurs during the Establishing phase.

SUMMARY

A CAS-based enterprise KnS model is described in this article. The model was evaluated for validity and effectiveness in two case studies. The premise of our research was that modeling enterprise knowledge sharing from a complex adaptive systems (CAS) perspective can provide

KM leadership and practitioners with an enhanced understanding of KnS behavior within their organization. This research found that the CAS-based enterprise KnS model and methodology provides KM leadership with an enhanced understanding of KnS behavior and the KnS influences. In the two case studies conducted in operational environments, members of the KM leadership teams indicated that they had gained a better understanding because of the CAS-based modeling approach. Enhanced understanding of the following was indicated: KnS behavior in their organization; KnS influences in their organization; and the extent of the KnS influences within their organization. KM leadership also indicated that because of the CAS-based modeling, they would either change the target KnS state of the KnS environment or the priority for achieving that state.

The CAS-based enterprise KnS model developed as part of this research was found to be valid. The CAS-based enterprise KnS model was exercised in two case studies. The results of the case studies (Small, 2006) provided support for the validity of the assumptions on which the CAS-based enterprise KnS model was developed. The claims associated with the validity of the CAS-based enterprise KnS model are as follows:

1. Claim 1 (C1): The KnS behavior of other KWs is a significant influence on KnS behavior.
2. Claim 2 (C2): The KnS environment factors are a significant influence on KnS behavior.
3. Claim 3 (C3): The attributes of the KW are related to the frequency of KnS behavior (how often a KW engages in a KnS behavior).
4. Claim 4 (C4): Enterprise KnS behavior can be characterized using a multi-agent CAS model, with a few basic rules that drive agent behavior.

REFERENCES

American Productivity & Quality Center (APQC). (1999). *Creating a knowledge-sharing culture.* Consortium Benchmarking Study -- Best-Practice Report.

American Productivity & Quality Center (APQC). (2000). *Successfully implementing knowledge management.* Consortium Benchmarking Study -- Final Report, 2000.

Anderson, P. (1999). Complexity theory and organization science. *Organization Science, 10*(3), 216-232.

Axelrod, R. (1997). *The complexity of cooperation: Agent-based models of competition and collaboration.* New Jersey: Princeton University Press.

Axelrod, R., & Cohen, M. (1999). *Harnessing complexity: Organizational implications of a scientific frontier.* New York: The Free Press.

Bukowitz, W., & Williams, R. (1999). *The knowledge management fieldbook.* London: Financial Times Prentice Hall.

Epstein, M., & Axtell, R. (1996). *Growing artificial societies: Social science from the bottom up.* Washington D.C.: The Brookings Institution.

Ford, D. (2003). Trust and knowledge management: The seeds of success. In *Handbook on Knowledge Management 1: Knowledge Matters* (pp. 553-575). Heidelberg: Springer-Verlag.

HBSP. (1998). *Harvard Business review on knowledge management.* Cambridge, MA: Harvard Business School Press.

Holland, J. (1995). *Hidden order: How adaptation builds complexity.* MA: Perseus Books Reading.

Holsapple, C.W. (Ed.). (2003). *Handbook on knowledge management 1: Knowledge matters.* Heidelberg: Springer-Verlag.

Holsapple, C.W. (Ed). (2003). *Handbook on knowledge management 2: Knowledge directions.* Heidelberg: Springer-Verlag.

Ipe, M. (2003). *The praxis of knowledge sharing in organizations: A case study.* Doctoral dissertation. University Microfilms.

Ives, W., Torrey, B., & Gordon, C. (2000). Knowledge sharing is human behavior. In *Knowledge management: Classic and contemporary works* (pp. 99-129).

Johnson, P., & Lancaster, A. (2000). *Swarm users guide.* Swarm Development Group.

Leonard, D., & Straus, S. (1998). Putting your company's whole brain to work. In *Harvard Business Review on Knowledge Management* (pp. 109-136). Boston: Harvard Business School Press.

Liao, S., Chang, J., Shih-chieh, C., & Chia-mei, K. (2004). Employee relationship and knowledge sharing: A case study of a Taiwanese finance and securities firm. *Knowledge Management Research & Practice*, 2, 24-34.

Minar, N., Burkhar, R., Langton C., & Askemnazi, M. (1996). *The Swarm Simulation System: A toolkit for building multi-agent simulations*. Retrieved from http://alumni.media.mit.edu/~nelson/research/ swarm/.

Murray, S. (2003). *A quantitative examination to determine if knowledge sharing activities, given the appropriate richness lead to knowledge transfer, and if implementation factors influence the use of these knowledge sharing activities*. Doctoral dissertation. University Microfilms.

Nonaka, I., & Konno, N. (1998). The concept of Ba: Building a foundation for knowledge creation. *California Management Review (Special Issue on Knowledge and the Firm)*, 40(3), 40-54.

Nonaka, I., & Takeuchi, H. (1995). *The knowledge-creating company: How Japanese companies create the dynamics of innovation*. New York: Oxford University Press.

O'Dell, C. (2008). *Web 2.0 and knowledge management*. American Productivity & Quality Center (APQC). Retrieved from http://www.apqc.org.

Small, C. (2006). *An enterprise knowledge-sharing model: A complex adaptive systems perspective on improvement in knowledge sharing*. Doctoral dissertation, George Mason University. University Microfilms.

Small, C., & Sage, A. (2006). Knowledge management and knowledge sharing: A review. *Information, Knowledge, and Systems Management*, 5(3), 153-169.

Software Engineering Institute (SEI). (1996). *IDEAL^{SM}: A user's guide for software process improvement* (Handbook CMU/SEI-96-HB-001). Pittsburgh, PA: Software Engineering Institute, Carnegie Mellon University.

Swarm Development Group. (2004). *Chris Langton, Glen Ropella*. Retrieved from http://wiki.swarm.org/.

This work was previously published in International Journal of Information Technologies and Systems Approach, Vol. 1, Issue 2, edited by M. Mora; D. Paradice, pp. 38-56, copyright 2008 by IGI Publishing (an imprint of IGI Global).

Chapter 10
A Conceptual Descriptive–Comparative Study of Models and Standards of Processes in SE, SwE, and IT Disciplines Using the Theory of Systems

Manuel Mora
Autonomous University of Aguascalientes, México

Ovsei Gelman
Universidad Nacional Autónoma de Mexico, Mexico

Rory O'Connor
Dublin City University, Ireland

Francisco Alvarez
Autonomous University of Aguascalientes, Mexico

Jorge Macías-Lúevano
Autonomous University of Aguascalientes, Mexico

ABSTRACT

The increasing design, manufacturing, and provision complexity of high-quality, cost-efficient and trustworthy products and services has demanded the exchange of best organizational practices in worldwide organizations. While that such a realization has been available to organizations via models and standards of processes, the myriad of them and their heavy conceptual density has obscured their comprehension and practitioners are confused in their correct organizational selection, evaluation, and deployment tasks. Thus, with the ultimate aim to improve the task understanding of such schemes by reducing its business process understanding complexity, in this article we use a conceptual systemic model of a generic business organization derived from the theory of systems to describe and compare two main models (CMMI/SE/SwE, 2002; ITIL V.3, 2007) and four main standards (ISO/IEC 15288, 2002; ISO/IEC 12207, 1995; ISO/IEC 15504, 2005; ISO/IEC 20000, 2006) of processes. Description and comparison are realized through a mapping of them onto the systemic model.

INTRODUCTION

Competitive market pressures in worldwide business firms, because of an accelerated scientific, technological, and human-development progress[1] (Bar-Yam et al., 2004) have fostered the consumer' demands for better and cheaper products and services (e.g., designed with more functional capabilities and offered in more market competitive prices). Consequently, in order to design and manufacture, as well as provision and operate competitive high-quality technical, cost-efficient and trustworthy products and services, worldwide business firms are faced with the intra and inter organizational need to integrate multiple engineering and managerial systems and business processes (Sage & Cupan, 2001).

Such a demanded intra and inter business process integration, in turn, has introduced an engineering and managerial *business process performance complexity* in organizations (but experimented by technical and business managers), and an engineering and managerial *business process understanding complexity* in practitioners (experimented by technical and business managers as well as business process consultants). A *business process performance complexity* in this context is defined as the structural[2] and/or dynamic system's complexity (Sterman, 1999) that confronts technical and business managers to achieve the system organizational performance goals (e.g., efficiency, efficacy, and effectiveness organizational metrics). In similar mode, a *business process understanding complexity* is defined as the structural and/or dynamic system's complexity that confronts technical and business managers (and business consultants) to acquire

a holistic view of such a system under a learning focus.

Manifestations of such raising *business process performance* and *business process understanding complexities* are: (i) critical failures (by cancellations, interruptions, partial use, or early disposal) of enterprises information systems implementations (Standish Group, 2003; CIO UK, 2007); (ii) the apparition (and necessary retirement in the market) of defective products[3] (as tires, toys, software); and (iii) system downtimes and/or low efficiency and effectiveness in critical services such as electricity, nuclear plants, health services, and governmental services (Bar-Yam, 2003).

Consequently, some researchers have proposed the notion of complex system of systems (SoS) (Manthorpe, 1996; Carlock & Fenton, 2001; Sage & Cuppan, 2001) and others have helped to organize such a novel construct (Keating et al., 2003; Bar-Yam et al., 2004), as a conceptual tool to cope with that we call a *business process performance complexity* and a *business process understanding complexity*. Worldwide business firms, then, can be considered SoS and, as such, are comprised of a large variety of self-purposeful internal and external system components and forward and backward system interactions that generate unexpected emergent behaviors in multiple scales. Also, as SoS, the design/engineering and manufacturing/provision complexity of products/services is manifested by the variety of processes, machines/tools, materials, and system-component designs, as well as for the high-quality, cost-efficiency relationships, and value expectations demanded from the competitive worldwide markets. In turn, managerial process complexity is manifested by the disparate business internal and external process to be coordinated to meet the time to market, competitive prices, market-sharing, distribution scope and environmental and ethical organizational objectives, between other financial and strategic organizational objectives to meet (Farr & Buede, 2003). Furthermore, other authors have introduced the notion of complex software-intensive systems (Boehm & Lane, 2006) and complex IT-based organizational systems (Mora et al., 2008) which are characterized by having: *"(i) many heterogeneous ICT (client and server hardware, operating systems, middleware, network and telecommunication equipment, and business systems applications), (ii) a large variety of specialized human resources for their engineering, management and operation, (iii) a worldwide scope, (iv) geographically distributed operational and managerial users, (v) core business processes supported, (vi) a huge financial budget for organizational deployment, and (vii) a critical interdependence on ICT."* And, because such CITOS are critical-mission systems for large-scale organizations and, according to Gartner's consultants Hunter and Blosch (2003, quoted in Mora et al., 2008), these CITOS *"no longer merely depend on information systems ... [but] the systems are the business,"* the need for a better engineering and management process practices based in IT becomes critical in present times.

Under this new business and engineering context, global and large-scale business firms have fostered the development of best organizational practices (Arnold & Lawson, 2004). The purpose is to improve the definition, coordination and execution of business processes and to avoid critical failures in the manufacturing of products and the provision of services. Best practices have been documented (via a deep re-design, analysis, discussion, evaluation, authorization and updating of organizational activities) through models and/or standards of processes by international organizations for the disciplines of systems engineering (SE), software engineering (SwE) and information systems (IS). Some models and standards come

from organizations with a global scope (like ISO: International Organization for Standardization in Switzerland), but others limit their influences in some countries or regions (like SEI-CMU in USA, Canada, and Australia, or British Standard Office in UK). While both types of organizations can differ in their geographic scopes, both keep a similar efficacy purpose: to make available to them a set of generic business processes (technical, managerial, support, and enterprise) which come from the best international practices to correct and improve their organizational process, with the expected outcome to hold, correct, and improve the quality, value, and cost-efficiency issues of the generated products and services.

However, because of (i) the available myriad of models and standards reported in these three disciplines, (ii) the planned convergence for SE and SwE models and standards, and (iii) the critical role played by emergent CITOS in organizations in nowadays, we argue that a correct understanding and organizational deployment of such standards and models of process has been obscured by an inherent *business process complexity understanding* of the engineering and managerial process to be coordinated and the standards and models to be used for such an aim. *Business process understanding complexity* is manifested by a high density of concepts and interrelationships in the models and standards (Roedler, 2006) and by a lack of an integrated/holistic SE, SwE, and IS view of them (Mora et al., 2007a). According to a SEI (2006) statement that points out which *"... in the current marketplace, there are maturity models, standards, methodologies, and guidelines that can help an organization improve the way it does business. However, most available improvement approaches focus on a specific part of the business and do not take a systemic approach to the problems that most organizations are facing,"* and, with the ultimate aim to improve their *business process understanding complexity*, in this article, we report the development and application of a systemic model to describe and compare standards and models of process based in the theory of systems (Ackoff, 1971; Gelman & Garcia, 1989; Mora et al., 2003) by using a conceptual design research approach (Glass et al., 2004; Hevner et al., 2004; Mora & Gelman, 2008). The study's research purpose is limited to access the *business process completeness* and the *business process balance* levels, which are introduced as a guidance of indicators for the selection and evaluation of standards and models of processes. The empirical assessment of the *business process understanding complexity* construct is planned for a subsequent study.

Usefulness of this systemic model is illustrated with the description and comparison of two main models [CMMI/SE/SwE:2002 (SEI, 2002), ITIL V.3:2007 (OGC, 2007)] and four main standards [ISO/IEC 15288:2002 (ISO, 2002), ISO/IEC 12207:1995 (ISO, 1995), ISO/IEC 15504:2005 (ISO, 2005), ISO/IEC 20000:2006 (ISO, 2006a, 2006b)]. The remainder of this article continues as follows: firstly, a general overview of the conceptual design research approach and the face validation process conducted by a panel of experts are reported. Secondly, the rationale of the systemic concepts, which are used in the design of the pro formas to systemically describe and compare the standards and models, is reported. Finally, the application of the systemic descriptive-comparison model is presented and their main findings are discussed. Findings suggest the adequacy of the systems approach for such an aim.

The Conceptual Research Method

Conceptual research has been extensively used in the disciplines of IS and SwE as a non-empirical research method (Glass et al., 2004). Nevertheless, its principles and methods have been implicitly used and its scientific value has been obscured when is compared with empirical research methods which address tangible subjects and objects of study. In a recent systemic (Checkland, 2000)

taxonomy of research methods (Mora & Gelman, 2008), where are related the situational areas under study (A's), the knowledge known on such situations (F's) and the known knowledge on methodological issues (M's) to study the A's, two criteria are used to classify them: (i) the conceptual vs. reality dimension and (ii) the natural/behavioral vs. purposeful design dimension. Both criteria divide the spectrum of research methods in the following four quadrants: (Q1) the conceptual behavioral research, (Q2) the conceptual design research, (Q3) the empirical behavioral research, and (Q4) the empirical design research.

The conceptual dimension accounts for the organized and verifiable/falsifiable subsystem of concepts (e.g., knowledge) on the reality and of itself. The reality dimension (Bhaskar, 1975; Mingers, 2000) accounts for the stratified domains of: (i) observable and not observable events (the empirical and actual domains), and the (ii) broader reality domain of physical and social product-producer generative structures and mechanisms. The scientific knowledge (e.g., the conceptual domain) is socially generated by human beings in concordance with the reality (the truth criteria) and is temporal and relative (Bhaskar, 1975). However, reality existence is independent of human beings from a critical realism philosophical stance. Thus, when we conduct conceptual research we address knowledge objects mapped to a reality and when we perform reality-based research (e.g., empirical) we address real subjects or objects. On the other hand, both conceptual and real entities generated by the nature and social structures and mechanisms can be studied without or with an intervening or modifying purpose. In the former case, we explore, describe, predict, explain, or evaluate conceptual or real entities, and, in the latter, we purposely design, build, and test conceptual or real artifacts (Hevner et al., 2004). This article can be classified both as a *conceptual design research* (Q2) by the design of a systemic model to describe and compare standards and models of processes,

Figure 1. Conceptual research framework

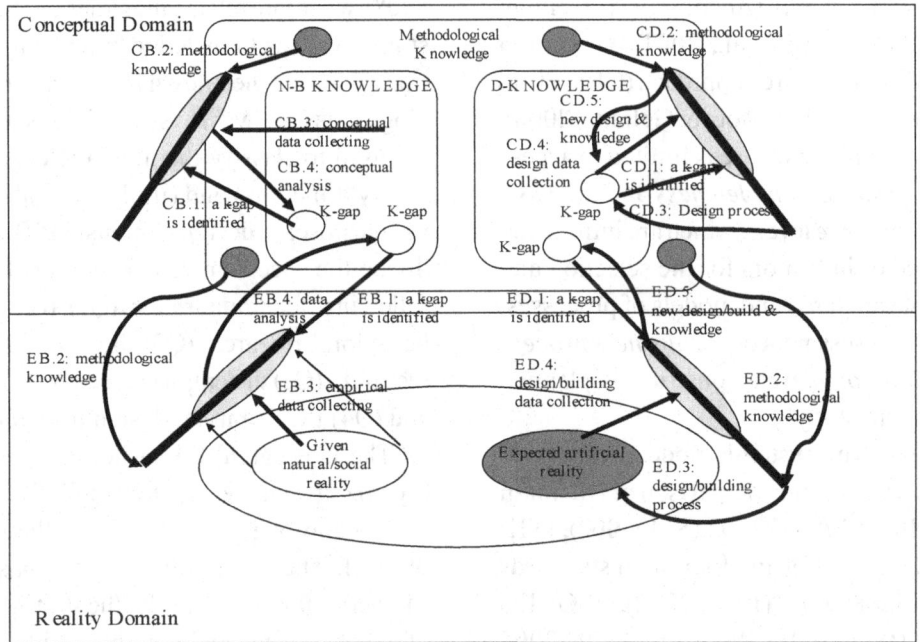

and as a *conceptual behavioral research* (Q1) by the utilization of such a model to describe the schemes. *Figure 1* illustrates the general research methodological framework.

In Mora et al. (2007b, 2007c) the systemic model was designed by applying the following four activities of Q2: CD.1 knowledge gap identification, CD.2 methodological knowledge (conceptual purposeful design), CD.3 conceptual design, CD.4 design data collection, and CD.5 analysis and synthesis where a new conceptual artifact outcome is generated [e.g., a construct, framework/model/theory, method, or system/component (not instanced in a real object)]. Validation is exercised in all five steps: a relevance validity assessment of the knowledge gap in CD.1 and CD.2, a methodological validity assessment in CD.3, CD.4, and CD.5 through a face validity instrument used with two schemes (ISO/IEC 15288 and CMMI/SE).

In contrast to empirical research methods, the validation procedures used in conceptual research can be one of the following: numerical mathematical analysis, mathematical/theorem proof, logical argumentation, or a face validation by a panel of experts. Model validation used in the conceptual design approach was face validation. A panel of four experts participated in the validation. Two experts own an academic joint expertise of 10 years of teaching graduate courses related to standards and models of processes in software engineering. The other two evaluators were invited for their practical knowledge in systems engineering and IT projects with an approximate 30-year joint expertise in IT and SE consulting activities. Because no specific instrument was located in the literature to conduct a model face validation, an instrument previously used to validate conceptual models in several M.Sc. theses was used. Model validation was tested with the description and comparison of the CMMI/SE model and the ISO/IEC 15288

Table 1. Model face validation in conceptual research

CONCEPTUAL INSTRUMENT[1] FOR MODEL FACE VALIDATION						Panel of International Experts					
	Total disagreement			Total agreement		Academic 01	Academic 02	Consultant 01	Consultant 02	Mean	Desv.Std.
I.1 The designed conceptual model is supported by core theoretical foundations regarding the topic under study.	1	2	3	4	5	5	5	5	4	4.75	0.50
I.2 The theoretical foundations used for developing the designed conceptual model are relevant to the topic under study.	1	2	3	4	5	5	5	5	4	4.75	0.50
I.3 There are no critical omissions in the literature used for developing the designed conceptual model.	1	2	3	4	5	5	5	5	4	4.75	0.50
I.4 The designed conceptual model is logically coherent to the purpose to the reality of study.	1	2	3	4	5	4	5	5	5	4.75	0.50
I.5 The designed conceptual model is adequate to the purpose of study.	1	2	3	4	5	4	5	5	5	4.75	0.50
I.6 The outcome (i.e. the designed conceptual model) is congruent with the underlying epistemological philosophy used for its development among positivist, interpretative, critical or critical realism.	1	2	3	4	5	5	4	5	4	4.50	0.58
I.7 The designed conceptual model reports original findings and contributes to the knowledge discipline.	1	2	3	4	5	5	5	5	4	4.75	0.50
I.8 The designed conceptual model is reported using an appropriate scientific style of writing.	1	2	3	4	5	5	4	5	4	4.50	0.58
Mean						4.75	4.75	5.00	4.25	4.67	
Desv.Std.						0.46	0.46	0.00	0.46	0.47	

standard. *Table 1* reports the items used in the validation step and their scores.

In this study, then, we apply the four activities of Q1: CB.1 knowledge gap identification, CB.2 methodological knowledge (e.g., conceptual exploratory review, conceptual descriptive-comparative review or conceptual tutorial review), CB.3 conceptual data collecting, and CB.4 conceptual analysis and synthesis where an exploratory, descriptive-comparative, or tutorial conceptual outcome is generated. Q1 was used for a descriptive/comparative purpose.

Knowledge gaps are reported in the related work section as well as in the introduction section. Methodological knowledge is realized through the utilization of a conceptual descriptive-compara-

tive review approach. Conceptual data collecting was conducted by a systematic reading of the original documents of the three models (CMMI/SE:2002, CMMI/SwE:2002, ITIL V.3:2007) and the three standards (ISO/IEC 15288:2002, ISO/IEC 12207:1995, ISO/IEC 20000:2006) and by an identification of the items required in the systemic model. Finally, the conceptual descriptive-comparative analysis and synthesis of findings was conducted by the two lead authors, broadly reviewed by a third co-author and validated by the remainder two co-authors. The joint-academic expertise of the full research team in systems approach is about 40 years, and 20 years in standards and models of processes.

RELATED WORK

The systems approach has been implicitly used to study organizations as general systems but few papers have reported formal or semi-formal definitions of such constructs (Ackoff, 1971; Feigenbaum, 1968; Wand & Woo, 1991; Gelman & Negroe, 1991; Mora et al., 2003). In the case of models and standards of processes, these have been studied individually (Gray, 1996; Garcia, 1998; Humphrey, 1998; Arnold & Lawson, 2004; Curtis, Phillips, & Weszka, 2001; Menezes, 2002) and comparatively (Sheard & Lake, 1998; Johnson & Dindo, 1998; Wright, 1998; Paulk, 1995, 1998, 1999; Halvorsen & Conrado, 2000; Minnich, 2002; Boehm & Vasili, 2005). While both kinds of studies on standards and models of processes have been useful to describe the main categories of processes, contrast directly two or more schemes, identify their focus of application, strengths and weaknesses, similarities and differences, and their fitness with a particular SE or SwE development approach, all of them have not used a normative-generic systemic model of a worldwide organization to estimate their *process completeness* and *process balance* constructs, neither to estimate their inherent *business process understanding complexity* in practitioners.

For instance, other descriptive and/or comparative studies on standards and models of processes (Sheard & Lake, 1998; Minnich, 2002) have identified core similarities and differences between such schemes. Main similarities are: (i) both provide a map of generic processes from the best international practices, (ii) both establish what and must be instructions rather than how specific procedures, and (iii) both do not impose a mandatory life-cycle of processes but suggest a demonstrative one that is usually taken as a basement. Thus, implementers must complement such recommendations with detailed procedures and profiles of the deliverables. In the case of main differences: (i) the models (at least the early reported) have been focused on process improvement efforts (and consequently include a capability maturity level assessment such as CMMI), while the standards are focused on an overall complain/not complain general assessment (e.g., ISO/IEC 12207), (ii) the models are used under an agreement between companies to legitimate their industrial acceptance (e.g., CMMI in the Americas), while the standards are used under a usually obligatory implicit country-based agreement (e.g., ISO/IEC 15504 in Europe), and (iii) the models can be originated from any organization, while the standards are strongly endorsed by nations.

Our study enhances previous ones through the introduction of a normative-generic systemic model of a business organization that is used to describe and compare the *business process completeness* and *business process balance* of standards and models of processes, as well as the next research goal to assess the *understanding complexity* on such schemes by potential practitioners. *Business process completeness* is defined as the extent of a standard or model fulfills the business process of the organizational subsystems of the generic systemic organization. The categorical scale used is very weak, weak, moderate, strong, and very strong business process completeness. *Business process balance* is defined as the extent of a standard or model provides an equilibrated support for all organizational subsystems of the generic systemic organization. The categorical scale used is very weak, weak, moderate, strong, and very strong business process balance. A high *business process completeness* does not imply a high *business process balance* for a standard or model and vice versa. In the former case, a standard or model could to have a high support for all organizational subsystems but some of them could be redundant. In the latter case, a standard or model could provide similar support for all organizational subsystems but for some organizational subsystems this could be insufficient (e.g., low value). The *business process understanding complexity* construct empirical assessment is planned for a further research.

DESCRIPTION AND COMPARISON OF MODELS AND STANDARDS OF PROCESSES

The Rationale of the Systemic Building-Blocks Constructs of the Normative-Generic Model of an Organization

According to Mora et al. (2007b), the ISO 9000:2000 series of standards (ISO, 2007) contains two principles (Principle 4 and 5) which endorse respectively the process approach and the systems approach as critical management paradigms. Principle 4's rationale states that the resources and activities are managed as processes. In turn, the Principle 5's rationale sets forth that the process be organized via a systems view. Furthermore, the ISO 9000:2000 standard remarks that while "... *the way in which the organization manage its processes is obviously to affect its final (quality of) product*" (ISO, 2007), these standards "... *concerns the way an organization goes about its work ... concern processes not products – at least not directly*" (ISO, 2006). Hence, the concepts of process, system, and product/service and their conceptual interrelationships become critical for understanding the different standards and models under study. In Mora et al. (2007c) are reported three appendices. First appendix

reports the systemic definition of the concepts system, subsystem, component and suprasystem/ entourage. These concepts are used in the second appendix to define the concepts of organization, organizational subsystem, business process and subprocess, business activity, product and service. Finally, in the third appendix, previous concepts are used to define a pro forma of a generic organization as a system. The latter definitions are rooted in the classic cybernetic paradigm (Gelman & Negroe, 1982) and extended to include the information systems subsystem concept (Mora et al., 2003). *Tables 2* and *3* update the definitions reported in the first and second appendices aforementioned. *Table 4* illustrates the cybernetic organizational model mapped to the Porter and Millar (1985) business process model where the IT service processes are explicitly added to the original model.

Definitions in *Table 2* (Mora et al., 2007b, 2007c) are rooted in theory of systems (Ackoff, 1971) and are based in formal definitions reported in Gelman and Garcia (1989) and Mora et al. (2003), and other semiformal definitions (Gelman et al., 2005; Mora et al., 2008). Concepts in *Table 3* (Mora et al., 2007b, 2007c) emerge from an analysis of relationships between the concepts of process, service and system in the context of standards and models of process.

Despite multiple definitions of process, main shared attributes can be identified: (i) an overall

Table 2. Definitions of core system concepts

ID	CONCEPT	CONCEPTUAL DEFINITION
R1	S: system	is a whole into a wider <SS: suprasystem> or <ENT: entourage> that can be modeled with mandatory <A: attributes: a1,a2,a3,a4,a5> (where <a1: purpose>, <a2: function>, <a3: inputs>, <a4: outputs> and <a5:outcomes>) that are co-produced by at least two parts called <sB: subsystems> and the <R: relationships: R1, R2, ...> between this whole, their parts, attributes and/or its suprasystem.
	sB: subsystem	is a <S: system> that is part of a <S: system> and that is decomposable in at least two or more <sB: subsystem> or <C: components>.
R3	C: component	is a constituent of a <sB: subsystem> that is not decomposable (from a modeling viewpoint).
R4	SS: suprasystem	is a <S: system> that contains to the system of interest under observation.
R4'	ENT: entourage	is the supra-system without the system under study.
R4''	W: world	is the entourage of the suprasystem.

Table 3. Definitions of organizational concepts as systems

ID	CONCEPT	CONCEPTUAL DEFINITION
R5	O: organization	is a <S: system> composed of three <OsB: organizational subsystems: driver, driven and IS subsystems>, into in a wider <OSS: organization suprasystem>, and with the generic attribute of <a1:purpose: "to provide valued outcomes for external systems"> additionally to other attributes.
R6	OsB: organizational subsystem	is a <sB: subsystem> composed of three subsystems called <BP: business process: control, operational and informational>.
R7	BP: business process	is a <sB: subsystem> of an <OsB: organizational subsystem> composed of at least two or more subsystems called <BsP: business subprocess> or components called <BA: business activities>, and with the additional mandatory attributes <a6: mechanisms> and <a7: controls>.
R8	BsP: business subprocess	is a <:BP: business process> into a <BP: business process>.
R9	BA: business activity	is a <C: component> into a <BP: business process> or <BsP: business subprocess> with the additional mandatory attributes <a6: tasks>, <a5:7personnel>, <a8: tools & infrastructure>, <a9: methods & procedures> and <a10: socio-political mechanisms & structures>.
R10	Sv: service	is an intangible, and time-continuously but period-limited <a4: people-oriented valued outcomes> from <a3: outputs: acts> of a <BA: business activity>, a <BP: business process>, an < OsB: organizational subsystem> or an <O: organization>.
R11	Pr: product	is a tangible, and discrete <a4: machine-oriented valued outcome> from <a3: outputs: matter> of a <BA: business activity>, a <BP: business process>, an <OsB: organizational sub-system> or an <O: organization>.

Table 4. Mapping of the Porter-Millar business process model onto the systemic model

SYSTEMIC MODEL OF A GENERIC ORGANIZATION		PORTER-MILLAR BUSINESS PROCESS MODEL OF A GENERIC ORGANIZATION	
[<OsB1: driver-organizational subsystem>]	[<OBP1: control business process >]	<STRATEGIC PROCESS>	SUPPORT PROCESSES
		<FINANCIAL PROCESS>	
	[<OBP2: operational business process >]	<HUMAN RESOURCES PROCESS>	
		<ADMINISTRATIVE – LEGAL PROCES>	
	[<OBP3: informational business process>]	<IT SERVICE for MANAGEMENT PROCESS>	
[<OsB2: driven-organizational subsystem>]	[<OBP1: control business process >]	<IN PUT LOGISTIC PROCESS>	PRIMARY PROCESSES
		<OUTPUT LOGISTIC PROCESS>	
	[<OBP2: operational business process >]	<OPERATION PROCESS>	
	[<OBP3: informational business process >]	<IT SERVICE for OPERATION PROCESS>	
[<OsB3: IS-organizational subsystem>]	[<OBP1: control business process >]	<IT SERVICE MANAGEMENT PROCESS>	IT SERVICE PROCESSES
	[<OBP2: operational business process >]	<IT SERVICE ENGINEERING PROCESS>	
	[<OBP3: informational business process >]	<IT SUPPORT PROCESS>	

purpose (transform inputs in outputs), (ii) interrelated activities, and (iii) the utilization of human and material resources, procedures, and methods. Similarly, even though there is no one standard definition of service, several shared attributes can be also identified: (i) intangibility, (ii) non-storable, (iii) ongoing realization, and (iv) a mandatory participation of people to determine the value attribute. We argue that only the human beings can assess a value scale on services (even though such services can usually include machine-based metrics), while that automated processes (by using artificial devices) can assess the quality attributes of products (e.g., to fit some agreed physical specifications). Then, main distinctions between a product and a service are: (i) the tangibility-intangibility dichotomy which leads to the quality (e.g., the attributes expected in the product) versus the value (e.g., the benefits to the quality-prices rate perceived from a customers' perspective), and (ii) the time-discrete utilization of products versus the ongoing experience of services (Teboul, 2007). Concepts reported in *Tables 2* and *3*, then, help to dissolve the conceptual omission of the responsible entity that generates a service: a process or a system. We argue that the concept of system (Gelman & Garcia, 1989) is the logical concept to link process and service/product constructs. Similar conceptualizations are being developed also in the SSME's research stream under the notion of service systems (Spohrer et al., 2007). Hence, we claim that these concepts can be used as conceptual building blocks to describe and compare standards and models of processes.

The Systemic Normative-Generic Model of an Organization

For applying the conceptual building blocks and their interrelationships, we define a set of pro formas (Andoh-Baidoo et al., 2004) for each concept. Pro formas for the concepts system, supra-system, subsystem, component, entourage, and world, as well as for organization, organizational subsystem, business process sub-process and business activity are reported in the *Appendices A* and *B*. Pro formas and the systemic definitions enable us to develop a multi-scale systemic comparison of the standards and models of processes. Because the generic model is mapped onto a very strong and validated business process model (Porter & Millar, 1985), we claim this strategy is better than a direct comparison between them because there is a common normative model against to each standard or model can be compared and because this is useful to estimate an absolute *process completeness* and *process balance* levels. In the opposite case, the assessment would be relative against the considered best model or standard.

The Systemic Description and Comparison of Standards and Models of Processes

In this article, we report the description and comparison of two models (CMMI/SE:2002, CMMI/SwE:2002, ITIL V.3:2007) and four standards (ISO/IEC 15288:2002, ISO/IEC 12207:1995, ISO/IEC 15504:2005, and ISO/IEC 20000:2006) of processes. Description and comparison details are reported in the *Appendix C* but a summary of them is reported in *Table 5*. The symbols: ●, ◕, ◉, ◔, and ○, corresponds directly to the categories of very strong, strong, moderate, weak and very weak.

Assessments reported in *Table 5* are based in the conceptual analysis conducted by the two lead authors and validated by the other three co-authors on the data reported in *Appendix C*. Such descriptions and comparisons are conducted in the organization level of the cybernetic organizational model with initial descriptions and comparisons in the organizational subsystem level (e.g., the driver, the driven and the information organizational subsystems). The analysis was conducted under the premise of an organization interested to deploy a standard or model to manufacture and

Table 5. Business process completeness and balance assessment summary

SYSTEMIC MODEL	PORTER & MILLAR BUSINESS PROCESS MODEL	CMMI/SE/SwE: 2002 Models	ISO/IEC 15288:2002 Standard	ISO/IEC 12207:1995 Standard	ISO/IEC 15504:2006 Standard	ISO/IEC 20000:2005 Standard	ITIL V3 : 2007 Model
[<OsB1: driver-organizational subsystem>]	<STRATEGIC MGT>	⊙	◉	○	⊙	⊙	●
	<FINANCIAL MGT>	○	⊙	○	◉	⊙	●
	<HR MGT>	⊙	◉	⊙	●	◉	◉
	<ADM-LEGAL MGT>	⊙	⊙	⊙	⊙	●	●
	<ITSfM>	○	◉	○	○	⊙	●
	BUSINESS PROCESS COMPLETENESS	◉	◉	○	⊙	⊙	⊙
[<OsB2: driven-org. subsystem>]	<INPUT LOGISTIC>	⊙	●	⊙	⊙	⊙	●
	<OPERATIONS>	⊙	●	⊙	⊙	⊙	⊙
	<OUTPUT LOGISTIC>	⊙	⊙	⊙	⊙	●	●
	<ITSfO>	○	◉	○	○	●	●
	BUSINESS PROCESS COMPLETENESS	⊙	⊙	⊙	⊙	⊙	●
[<OsB3: is-org subsystem>]	<IT SERVICE MANAGEMENT>	○	○	○	○	⊙	●
	<IT SERVICE ENGINEERING>	○	○	○	○	⊙	⊙
	<IT SERVICE SUPPORT >	○	○	○	○	●	●
	BUSINESS PROCESS COMPLETENESS	○	○	○	○	⊙	●
M1	**BUSINESS PROCESS COMPLETENESS WITHOUT OsB3**	⊙	⊙	◉	⊙	⊙	⊙
		Strong	Strong	Moderated	Strong	Strong	Strong
M1'	**OVERALL BUSINESS PROCESS COMPLETENESS**	◉	◉	○	◉	⊙	⊙
		Moderated	Moderated	Weak	Moderated	Strong	Strong
M2'	**BUSINESS PROCESS BALANCE WITHOUT OsB3**	⊙	⊙	◉	⊙	⊙	⊙
		Strong	Strong	Moderated	Strong	Strong	Strong
M2	**OVERALL BUSINESS PROCESS BALANCE**	◉	◉	○	◉	⊙	⊙
		Moderated	Moderated	Weak	Moderated	Strong	Strong

provision products and services strongly based in IT. Furthermore, CMMI, ISO/IEC 15288 and ISO/IEC 15504 claim to be a model/standard for any kind of system/product. Through the generation of the systemic pro formas and their interpretation by the two lead authors, and the additional validation of the validation team, we can summarize the following core findings as follows:

- **Business process completeness on the Porter-Millar's support process:** The six schemes are focused on the core processes related to the lifecycle of man-made systems and related support process. Furthermore, all of them claim to be useful for guiding the design and manufacturing/provision of any kind of system or product/service where software or IT be a core component. However, while this aim is worthy, its overall extent of business process completeness when the whole organization is considered is not so strong in some standards/models. For instance, the ISO/IEC 12207:1995 standard while mainly focused on software products or services also addresses systems that contain software, so its overall completeness should at least be strong. Futhermore, by using the combined systemic and classic process-based organization model (Porter & Millar, 1985), the core strategic management and financial processes are not included or moderately included in the ISO/IEC 12207:1995 and ISO/IEC 15288:2002 schemes. In contrast, others explicitly address such aims through the organizational alignment and financial management processes. Best explicit addressing is realized for the ISO/IEC 20000:2005 and ITIL V.3:2007 schemes. While the strategic process and its links with the remainder process are not considered, the business value of standards and models of process and its full and correct deployment can be obfuscated. For the case of financial management process, two of the

oldest schemes (CMMI/SE/SwE and ISO/IEC 12207:1995) do not explicitly treat it. In contrast, the other four schemes address this important process. Best addressing is from ITIL V.3:2007 followed of ISO/IEC 20000:2006 and ISO/IEC 15288:2002. Latter scheme treats this as the investment management process. Regarding the human resources process, while all of them consider the topic of training and competent human resources (e.g., moderate completeness), only the ISO/IEC 15504:2006 addresses explicitly and adds the KM process. Other worthy effort is considered by CMMI/SE/SwE:2002 model, which assigns to organizational training a strategic focus. The existence of the CMM-People is a proof of this strategic aim but its incorporation into CMMI/SE/SwE:2002 model is not implicit. The completeness on the administrative-legal process is strong for the first four schemes (CMMI/SE/SwE:2002, ISO/IEC 15288:2002, ISO/IEC 12207:1995, ISO/IEC 15504:2006) and very strong in the service-oriented new schemes (ISO/IEC 20000:2005 and ITIL V.3:2007). This happens because the existence of an explicit service level management process in both standards with strong legal considerations. Finally, the IT service for management process is not explicitly addressed in all standards except for the ISO/IEC 20000:2005, and the ITIL V.3:2007, given their aim. However, ISO/IEC 15288:2002 standard considers a general information management process, and the others should address it given the relevance of the IT services process for the modern business firms. Hence, the business process completeness metric for the Porter-Millar support process is strong for ITIL V.3:2007 model, the ISO/IEC 20000:2005, and ISO/IEC 15504:2006 standards, moderated in the CMMI/SE/SwE model, and ISO/IEC 15288:2002 standard, and weak in ISO/IEC

12207:1995 standard by the lack of strategic and financial management processes.

- **Business process completeness on the Porter-Millar's primary process:** Being the six schemes focused on the core processes are related to the lifecycle of man-made systems, it is not an unexpected result a strong completeness assessment in almost all schemes (five of them). ITIL V.3:2007 model is the most complete (e.g., very strong). However, despite such a high assessment for ITIL V.3:2007 model, and the existence of the service release and deployment management process, being this one the core engineering process where the service is built, its general treatment into the high density of the remainder of processes is obfuscated. The relationships of this process with the service design process are critical for a final high-quality, cost-efficient, and trustworthy service, and should be clearly established in the standard. Similarly to its antecessor model (e.g., ITIL V.2, which is enhanced in the new ISO/IEC 20000:2005 standard), this process is weakly elaborated from a systems engineering view. Regarding other processes, the input and output logistic ones, are also strongly completed. The existence of specific process to treat with suppliers or performing as such ones reinforces both processes. CMMI/SE/SwE does not distinguish between suppliers and customers' agreement process. The remainder schemes consider both views: when the organization buys products/services and when it sells them. ITIL V.3:2007 model and ISO/IEC 20000:2005 standard are the most completed schemes by introducing specific service level management and business customers' relationships processes to manage the output logistic process, as well as the supplier management and business supplier relationships to treat with the input logistic process. Regarding the IT service

for operations process, the completeness assessed is similar to the ITSfM process: these ones are not explicitly addressed except for ISO/IEC 20000:2005 standard, and ITIL V.3:2007 model. ISO/IEC 15288:2002 standard considers also a general information management process into the project management category. Hence, the business process completeness metric for the Porter-Millar primary process is strong for five schemes and very strong for ITIL V.3:2007 model.

- **Business process completeness on the Porter-Millar's IT support process:** Our analysis reveals the explicit lack of IT service management, IT service engineering, and IT service support process as a mandatory and relevant component of the standards and models of processes, except for the two designed for such an aim (e.g., ISO/IEC 20000:2005 and ITIL V.3:2007). We consider that under the new business environment characterized by a strong competitive pressure for high quality, cost-efficient, and trustworthy products and services, and the increasing engineering and managerial complexity for achieving them, as well as the increasing dependency of IT services, such a kind of process becomes relevant to be included in updated versions of the models and standards. Hence, the business process completeness metric for the extended Porter-Millar IT service process is strong ISO/IEC 20000:2005, very strong for ITIL V.3:2007 model, and weak for the remainder schemes. The well-structured lifecycle view with design, transition and operation, guided by the strategic and continual improvement service process of ITIL.V3:2007, enhances its antecessor ITIL V.2:2000 model, which is the underlying framework for the ISO/IEC 20000:2005 standard.

- **Overall business process completeness:** Based in the previous assessments, and the fact of the lack of explicit IT service process in most schemes, it is adequate to divide the overall evaluation without and with the OsB3 (e.g., the IS-organizational subsystem). For the first case, five of the six schemes are considered with strong business process completeness and one with a moderated assessment (for ISO/IEC 12207:1995 standard). For the second case, when the OsB3 organizational subsystem is included in the evaluation, the two IT service-oriented schemes keep a strong assessment, but the others reduce it to a moderate assessment (CMMI/SE/SwE model, and ISO/IEC 15228:2002, ISO/IEC 15504:2005 standards) and an overall weak business process completeness assessment (ISO/IEC 12207:1995 standard).

- **Overall business process balance:** Similarly to the business process completeness, the assessment can be divided without and with the OsB3 subsystem. In the former case, five schemes qualify with a strong balance and only ISO/IEC 12207:1995 standard is assessed as moderated. In the latter case, the process balance assessment is reduced to moderate in three schemes: CMMI/SE/SwE model, and ISO/IEC 15288:2002, ISO/IEC 15504:2005 standards. ISO/IEC 12207:1995 standard balance process is assessed as weak. The two IT service-oriented schemes keep a strong assessment. These results are not unexpected. ITIL-based models and standards are of the most updated (e.g., 2005 and 2007 years) and both are based in the new business philosophy of service science, engineering, and management (Spohrer et al., 2007). We consider that the remainder standards and models will follow this approach in short time. For instance, the new planned CMMI-SVC model is being designed for such an aim. In turn, the low scores for ISO/IEC 12207:1995 can explain the two core amendments published in 2001

and 2004. Improvements in the ISO/IEC 12207:1995 standard are clearly exhibited in ISO/IEC 15504:2005:Part 5 standard, which uses the new ISO/IEC 12207:2004 version as an exemplary model for assessment. The problem is the lack of a full document of this standard where all amendments are seamlessly integrated in the previous knowledge. We estimate (by anecdotic but academic sources given the textbook literature on the topic) that main organizational deployments are still using ISO/IEC 12207:1995 version.

- **Implications for IS discipline.** Space and time limitations preclude a deep discussion. Our general and core observation is that, in order for the standards and models studied in this paper to be used and deployed jointly with ITIL-based models and standards, a deep managerial effort will be required to harmonize them. Another core observation is the necessary inclusion in the graduate IS/IT programs of the models/standards topics as mandatory. In the meanwhile, IS/IT practitioners have been alerted to be cautious, given the large economical, human, and organizational resources required to implement successfully such standards and models.

CONCLUSION

We have argued that modern firms are complex systems of systems (SoS) regarding to the engineering and management of their processes to deliver cost-effective, trustworthy, and high-quality products and services. Consequently, the organizations have developed and fostered the exchange of "best practices" through the concepts of standards and models of processes. However, the myriad of them is causing a *business process understanding complexity* that obfuscates their correct deployment. Then, we have posed the utilization of the theory of systems for treating such an understanding problematic situation. Our plausible realization was illustrated with the definition of a systemic model of organization, organizational subsystem and business process, and the model was applied to describe and compare four standards and two models of process. We consider that our systemic model is useful to acquire a holistic view of such schemes through a high-level mapping of the supported organizational processes. This task allows us to assess a *business process completeness* and *business process balance* metrics that can be used as guidance indicators for the selection and evaluation of such schemes. We will continue this research with: (i) studies on specific models/standards under a more fine-granularity level of analysis and with (ii) studies on the semi-automation of such an analysis through ontologies and reasoning computer-based tools.

ACKNOWLEDGMENT

This research is developed with the financial support of the Autonomous University of Aguascalientes, Mexico (www.uaa.mx) (Project PIINF-06-8), and a national grant (Project P49135-Y) provided by the Mexican National Council of Science and Technology (CONACYT, www.conacyt.mx).

REFERENCES

Ackoff, R. (1971). Towards a system of systems concepts. *Management Science, 17*(11), 661-671.

Arnold, S., & Lawson, H. (2004). Viewing systems from a business management perspective: The ISO/IEC 15288 Standard. *Systems Engineering, 7*(3), 229-242.

Andoh-Baidoo, F., White, E., & Kasper, G. (2004). Information systems' cumulative research tradition: A review of research activities and outputs using pro forma abstracts. In *Proceedings of the Tenth Americas Conference on Information Systems*, New York, NY (pp. 4195-4202).

Bar-Yam, Y. (2003). When systems engineering fails -- toward complex systems engineering. In Proceedings of the *International Conference on Systems, Man & Cybernetics* (Vol. 2, pp. 2021-2028). Piscataway, NJ: IEEE Press, 2021- 2028.

Bar-Yam, Y. et al. (2004). The characteristics and emerging behaviors of system of systems. In *NECSI: Complex Physical, Biological and Social Systems Project* (pp. 1-16). Cambridge, MA: New England Complex Systems Institute.

Bhaskar, R. (1975). *A realist theory of science.* Sussex: Harvester Press.

Boehm, B., & Basili, V. (2005). *The CeBASE framework for strategic software development and evolution.* EDSER-3 Position Paper. Retrieved from http://sunset.usc.edu/csse/TECHRPTS/2001/usccse2001-503/usccse2001-503.pdf.

Boehm, B., & Lane, J. (2006, May). 21st century processes for acquiring 21st century software-intensive systems of systems. *Crosstalk: The Journal of Defense Software Engineering*, 1-9.

Carlock, P., & Fenton, R. (2001). System of systems (SoS) enterprise system engineering for information-intensive organizations. *Systems Engineering, 4*(4), 242-261.

CIO UK. (2007). Late IT projects equals lower profits. Retrieved from http://www.cio.co.uk/concern/resources/news/index.cfm?articleid=1563.

Checkland, P. (2000). Soft systems methodology: A 30-year retrospective. In P. Checkland, *Systems thinking, systems practice* (pp. A1-A65). Chichester: Wiley.

Curtis, P., Phillips, M., & Weszka, J. (2002). CMMI – the evolution continues. *Systems Engineering Journal, 5*(1), 7-12.

Farr, J., & Buede, D. (2003). Systems engineering and engineering management: Keys to the efficient development of products and services. *Engineering Management Journal, 15*(3), 3-9.

Feingenbaum, D. (1968). The engineering and management of an effective system. *Management Science, 14*(12), 721-730.

Garcia, S. (1998). Evolving improvement paradigms: Capability maturity models & ISO/IEC 15504 (PDTR). *Software Process Improvement and Practice, 3*(1), 1-11.

Gelman O., & Negroe, G. (1982). Planning as a conduction process. *Engineering National Academy Review, 1*(4), 253-270.

Gelman, O., & Garcia, J. (1989). Formulation and axiomatization of the concept of general system. *Outlet IMPOS (Mexican Institute of Planning and Systems Operation), 19*(92), 1-81.

Glass, R., Armes, V., & Vessey, I. (2004). An analysis of research in computing disciplines. *Communications of the ACM, 47*(6), 89-94.

Gray, L. (1996). ISO/IEC 12207 software lifecycle processes. *Crosstalk: The Journal of Defense Software Engineering, 8*, 1-11.

Halvorsen, C., & Conrado, R. (2000). A taxiomatic attempt at comparing SPI frameworks. In *Proceedings of the Norsk Informatikk Konferanse '2000 (NIK'2000)*, Bodø (pp. 101-116).

Hevner, A.R., March, S.T., Park, J., & Ram, S. (2004). Design science in information systems research. *MIS Quarterly, 28*(1), 75-105.

Humphrey, W. (1998, February). Three dimensions of process improvement. Part I: Process maturity. *Crosstalk: The Journal of Defense Software Engineering*, 1-7.

ISO. (1995). *ISO/IEC 12207: Information technology – software life cycle processes.* Geneva: ISO/IEC.

ISO. (2002). *ISO/IEC 15288: Information technology – systems life cycle processes.* Geneva: ISO/IEC.

ISO. (2005a). *ISO/IEC 20000-1: Information technology – service management. Part 1: Specification.* Geneva: ISO/IEC.

ISO. (2005b). *ISO/IEC 20000-1: Information technology – service management. Part 2: Code of practice.* Geneva: ISO/IEC.

ISO. (2006a). *ISO/IEC 15504-5 information technology – process assessment. Part 5: An exemplar process assessment model.* Geneva: ISO/IEC.

ISO. (2006b). *ISO 9000 and ISO 14000 in plain language.* Retrieved from www.iso.org.

ISO. (2007). *Quality management principles.* Retrieved from www.iso.org.

Johnson, K., & Dindo, J. (1998). Expanding the focus of software process improvement to include systems engineering. *Crosstalk: The Journal of Defense Software Engineering*, 1-13.

Keating, C. et al. (2003). System of systems engineering. *Engineering Management Journal, 15*(3), 36-45.

Manthorpe, W. (1996). The emerging joint system of systems: A systems engineering challenge and opportunity for APL. *John Hopkins APL Technical Digest, 17*(3), 305-313.

Menezes, W. (2002, February). To CMMI or not to CMMI: Issues to think about. *Crosstalk: The Journal of Defense Software Engineering*, 1-3.

Mingers, J. (2000). The contributions of critical realism as an underpinning philosophy for OR/MS and systems. *Journal of the Operational Research Society, 51*, 1256-1270.

Minnich, H. (2002). EIA IS 731 compared to CMMI-SE/SW. *Systems Engineering Journal, 5*(1), 62-72.

Mora, M., Gelman, O., Cervantes, F., Mejia, M., & Weitzenfeld, A. (2003). A systemic approach for the formalization of the information system concept: Why information systems are systems? In J. Cano (Ed.), *Critical reflections of information systems: A systemic approach* (pp. 1-29). Hershey, PA: Idea Group Publishing.

Mora, M., Gelman, O., O'Connor, R., Alvarez, F., & Macías, J. (2007a). On models and standards of processes in SE, SwE and IT&S disciplines: Toward a comparative framework using the systems approach. In K. Dhanda & R. Hackney (Eds.), *Proceedings of the ISOneWorld 2007 Conference,*, Engaging Academia and Enterprise Agendas, Las Vegas, USA (pp. 49/1-18).

Mora, M., Gelman, O., O'Connor, R., Alvarez, F., & Macías, J. (2007b). A systemic model for the description and comparison of models and standards of processes in SE, SwE, and IT disciplines. In *E-Proceedings of the International Conference on Complex Systems 2007*, NECSI, Boston, MA, USA (pp. 1-8).

Mora, M., Gelman, O., O'Connor, R., Alvarez, F., & Macías, J. (2007c). An overview of models and standards of processes in SE, SwE, and IS disciplines In A. Cater-Steel (Ed.), *Information technology governance and service management: Frameworks and adaptations* (pp. 1-20). Hershey, PA: IGI Global. 1-20.

Mora, M., Gelman, O., Frank, M., Cervantes, F., & Forgionne, G. (2008). Toward an interdisciplinary engineering and management of complex IT-intensive organizational systems: A systems view. *International Journal of Information Technologies and the Systems Approach, 1*(1), 1-24.

Mora, M., & Gelman, O. (2008). The case for conceptual research in information systems. Paper submitted to the 2008 International Conference on Information Resources Management (Conf-IRM), Niagara Falls, Ontario, Canada (pp. 1-10).

OGC. (2007). *The official introduction to the ITIL service lifecycle*. London: TSO.

Paulk, M. (1995, January). How ISO 9001 compares with CMM. *IEEE Software*, 74-83.

Paulk, M. (1998). *ISO 12207, ISO 15504, SW-CMM v1.1, SW-CMM v2 Draft C Mapping 1*. SEI/CMU.

Paulk, M. (1999). Analyzing the conceptual relationship between ISO/IEC 15504 (software process assessment) and the capability maturity model for software. In *Proceedings of the 1999 International Conference on Software Quality*, Cambridge, MA (pp. 1-11).

Porter, M., & Millar, V. (1985). How information gives you a competitive advantage. *Harvard Business Review, 63*(4), 149-160.

Roedler, G. (2006). *ISO/IEC JTC1/SC7: Status and plans of alignment of ISO/IEC 15288 and ISO/IEC 12207*. Retrieved from www.15288.com.

Sage, A., & Cuppan, C. (2001). On the systems engineering and management of systems of systems and federations of systems. *Information, Knowledge, Systems Management, 2*, 325-345.

Software Engineering Institute. (2002). CMMI for systems engineering and software engineering. (CMU/SEI-2002-TR-001). Retrieved from www.sei.edu.

Sheard, S., & Lake, J. (1998). Systems engineering and models compared. Retrieved from www.software.org.

Spohrer, J., Maglio, P., Bailey, J., & Gruhl, D. (2007, January). Steps toward a science of service systems. *IEEE Computer*, 71-77.

Sommerville, I. (1998). Systems engineering for software engineers. *Annals of Software Engineering, 6,* 111-129.

Standish Group International. (2003). *The Extreme CHAOS Report.* Retrieved from www.standish-group.com.

Sterman, J. (2001). Systems dynamic modeling: Tools for learning in a complex world. *California Management Review, 43*(4), 8-25.

Teboul, J. (2007). Service is front stage: Positioning services for value advantage. Paris: INSEAD Business Press.

Wand, Y., & Woo, C. (1991). An approach to formalizing organizational open systems concepts. *ACM SIGOIS Bulletin, 12*(2-3), 141-146. Retrieved from www.acm.org.

Wright, R. (1998, October). Process standards and capability models for engineering software-intensive systems. *Crosstalk: The Journal of Defense Software Engineering,* 1-10.

ENDNOTES

[1] At least in well-developed economies and partially in emergent ones.

[2] A complex entity or situation is structurally complex by the large number of relevant elements and interrelationships that affect its behavior and/or dynamically complex by the non-trivial (non lineal and not deterministic ones) forward and backward interactions between their (few or many) elements (Sterman, 1999).

[3] Documented in several internacional news and TV programs.

APPENDIX A. PRO FORMAS OF THE CORE CONCEPTUAL BUILDING-BLOCKS TO STUDY ENTITIES AS SYSTEMS.

CONCEPT	DEFAULT VALUE	DESCRIPTION
[<S: system>]	= [S(X)]	**The X thing that is modeled as a system.**
[<SS: supra-system>]	= [SS(S(X))]	The next up system called supra-system that contains to the modeled S(X) under study.
[<ENT: entourage>]	= [ENT(S(X))]	The supra-system without the modeled S(X) under study.
[<W: world>]	= [W(S(X))] = [ENT (SS(S(X))]	The most up system to be considered in the study without the supra-system of the system under study.
[<A: attributes>]	= [a1+a2+a3+ a4 + a5 + (a6 + a7 + ...)]	The attributes that are defining the system.
[<a1: purpose>]	= [<a1: "to achieve its outcomes" >]	The effectiveness mission of the system.
[<a2: function>]	= [<a2: "to achieve efficiently its outputs">]	The efficacy mission of the system.
[<a3: inputs>]	= [<a3: [{ energy-matter | information-knowledge | acts }^{n}]>]	The system's input flows.
[<a4: outputs>]	= [<a4: [{ energy-matter | information-knowledge | acts }^{n}]>]	The system's output flows.
[<a5: outcomes>]	= [<a5: [{ PoV} | MoV }^{n}] >]	The expected consequences to be generated by the system's outputs. PoV and MoV are respectively people-oriented and machine-oriented valued features.
• ...	...	Other possible attributes.
[[<sB: subsystems>] | [<C: components>]]	= [[sB(X1) | C(X1)] + [sB(X2) | C(X2)] + ([sB(X3) | C(X3)] + ...)]	The main constituents of the system.
[[sB1 | C1]]	= [sB(X1) | C(X1)]	The first constituent of the system.
[[sB2 | C2]]	= [sB(X2) | C(X2)]	The second constituent of the system.
...	...	Other system's constituents.
[<R: relationships>]	= [R1 + (R2 + ...)]	Relationships between the system's parts, attributes and/or its supra-system and entourage.

CONCEPT	DEFAULT INSTANCE	DESCRIPTION
[<sB: subsystem>]	= [sB(X?)]	**The subsystem to be modeled.**
[<S: system>]	= [S(X)]	The owner system of the subsystem.
[<A: attributes>]	= [a1+a2+a3+ a4 + a5 + (a6 + a7 + ...)]	The attributes that are defining the subsystem.
[<a1: purpose>]	= [<a1: "to achieve its outcomes" >]	The effectiveness mission of the subsystem.
[<a2: function>]	= [<a2: "to achieve efficiently its outputs">]	The efficacy mission of the subsystem.
[<a3: inputs>]	= [<a3: [{ energy-matter | information-knowledge | acts }^{n}]>]	The subsystem's input flows.
[<a4: outputs>]	= [<a4: [{ energy-matter | information-knowledge | acts }^{n}]>]	The subsystem's output flows.
[<a5: outcomes>]	= [<a5: [{ PoV} | MoV }^{n}] >]	The expected consequences to be generated by the subsystem's outputs. PoV and MoV are respectively people-oriented and machine-oriented valued features.
• ...	...	Other possible attributes.
[[<sB: subsystems>] | [<C: components>]]	= [[sB(X1) | C(X1)] + [sB(X2) | C(X2)] + ([sB(X3) | C(X3)] + ...)]	The main constituents of the subsystem.
[[sB1 | C1]]	= [sB(X1) | C(X1)]	The first constituent of the subsystem.
[[sB2 | C2]]	= [sB(X2) | C(X2)]	The second constituent of the subsystem.
...	...	Other subsystem's constituents.
[<R: relationships>]	= [R1 + (R2 + ...)]	Relationships between the system's parts, attributes and/or its supra-system and entourage.

CONCEPT	DEFAULT INSTANCE	DESCRIPTION
[<C: component>]	= [C(X?)]	**The component to be modeled.**
[<sB: subsystem> | <S: system>]	= [sB(X?) | S(X)]	The owner subsystem or system that contains to the component.
[<A: attributes>]	= [a1+a2+a3+ a4 + a5 + (a6 + a7 + …)]	The attributes that are defining the component.
[<a1: purpose>]	= [<a1: "to achieve its outcomes" >]	The effectiveness mission of the component.
[<a2: function>]	= [<a2: "to achieve efficiently its outputs">]	The efficacy mission of the component.
[<a3: inputs>]	= [<a3: [{ energy-matter | information-knowledge | acts }^{n}]>]	The component's input flows
[<a4: outputs>]	= [<a4: [{ energy-matter | information-knowledge | acts }^{n}]>]	The component's output flows
[<a5: outcomes>]	= [<a5: [{ PoV} | MoV }^{n}] >]	The expected consequences to be generated by the component's outputs. PoV and MoV are respectively people-oriented and machine-oriented valued features.
• …		Other possible attributes.
[<R: relationships>]	= [R1 + (R2 + …)]	Relationships between the component's attributes and its wider system.

CONCEPT	DEFAULT VALUE	DESCRIPTION
[<SS: suprasystem>]	= [SS(S(X))]	**The next up system that contains to the modeled system under study.**
[<S: system>]	= [S(X)]	The system under study that is a constituent of the suprasystem.
[<ENT: entourage>]	= [ENT(SS(S(X)))] = [W(S(X)]	The supra-system without the modeled S(X) under study.
[<W: world>]	= [W(S(X))] = [ENT (SS(S(X))]	The most up system to be considered in the study without the supra-system of the system under study.
[<A: attributes>]	= [a1+a2+a3+ a4 + a5 + (a6 + a7 + …)]	The attributes that are defining the supra-system.
[<a1: purpose>]	= [<a1: "to achieve its outcomes" >]	The effectiveness mission of the supra-system.
[<a2: function>]	= [<a2: "to achieve efficiently its outputs">]	The efficacy mission of the supra-system.
[<a3: inputs>]	= [<a3: [{ energy-matter | information-knowledge | acts }^{n}]>]	The supra-system's input flows.
[<a4: outputs>]	= [<a4: [{ energy-matter | information-knowledge | acts }^{n}]>]	The supra-system's output flows.
[<a5: outcomes>]	= [<a5: [{ PoV} | MoV }^{n}] >]	The expected consequences to be generated by the supra-system's outputs. PoV and MoV are respectively people-oriented and machine-oriented valued features.
• …		Other possible attributes.
[[sB: <subsystems>] | [C: <components>]]	= [[sB(X1)] + [sB(X2) | C(X2)] + ([sB(X3) | C(X3)] + …)]	The main constituents of the supra-system.
[sB1]	= [sB(X1)] = [S(X)]	The system S is the first constituent of the supra-system.
[[sB2 | C2]]	= [sB(X2) | C(X2)]	The second constituent.
…	…	Other supra-system's constituents.
[<R: relationships>]	= [R1 + (R2 + …)]	Relationships between the supra-system's parts, attributes and its wider system.

CONCEPT	DEFAULT VALUE	DESCRIPTION
[<W: world>]	= [W(S(X))]	**The most up system to be considered in the study without the supra-system of the system under study.**
[<S: system>]	= [S(X)]	The system under study that is a constituent of the suprasystem into the world.
[<SS: supra-system>]	= [SS(S(X))]	The next up system called supra-system that contains to the modeled S(X) under study.
[<ENT: entourage>]	= [ENT(S(X))]	The supra-system without the modeled S(X) under study.
[<A: attributes>]	= [a1 (+ a2+ …)]	The attributes that are defining the world.

CONCEPT	DEFAULT VALUE	DESCRIPTION
[<a1: purpose>]	= [<a1: "to be a system" >]	The effectiveness mission of the world.
• ...		Other possible attributes.
[[sB: <subsystems>] | [C: <components>]]	= [[sB(X1)] + [sB(X2) | C(X2)] + ([sB(X3) | C(X3)] + ...)]	The main constituents of the world.
[sB1]	= [sB(X1)] = [SS(S(X))]	The supra-system SS(S(X) is the first constituent of the world that is modeled as a closed system.
[[sB2 | C2]]	= [sB(X2) | C(X2)]	The second constituent.
...	...	Other world's constituents.
[<R: relationships>]	= [R1 + (R2 + ...)]	Relationships between the world's parts and attributes.

APPENDIX B. PRO FORMAS OF THE SYSTEMIC CONCEPTUAL BUILDING-BLOCKS FOR MODELING AN ORGANIZATION

CONCEPT	GENERIC VALUE	DESCRIPTION
[<O: organization>]	= [O(X)]	**The X thing to be modeled as a systemic organization.**
[<OOS: organizational supra-system>]	= [OSS(O(X))]	The next up system called supra-system that contains to the modeled O(X) under study.
[<OENT: organizational entourage>]	= [OENT(O(X))]	The supra-system without the modeled O(X) under study.
[<OW: organizational world>]	= [OW(O(X))]	The most up system to be considered in the study without the supra-system of the system under study.
[<A: attributes>]	= [a1+a2+a3+ a4 + a5 + (a6 + ...)]	The attributes that are defining the organization.
[<a1: purpose>]	= [<a1: "to provide valued outcomes">]	The effectiveness mission of the organization.
[<a2: function>]	= [<a2: "to achieve efficiently its outputs">]	The efficacy mission of the organization.
[<a3: inputs>]	= [<a3: [{ energy-matter(utilities, artifacts, money) | information-knowledge | acts } ^{n}] >]	The organization's input flows.
[<a4: outputs>]	= [<a4: [{ energy-matter(utilities, artifacts, money) | information-knowledge | acts } ^{n}] >]	The organization's output flows.
[<a5: outcomes>]	= [<a5: [{ <PoV: service> } | <MoV: product >} ^{n}] >]	The expected consequences to be generated by the organizational system's outputs. PoV and MoV are respectively people-oriented and machine-oriented valued features.
• ...		Other possible attributes.
[[sB: <subsystems>] | [C: <components>]] = [<OsB: organizational subsystem>]	= [OsB(X1)] + [OsB(X2)] + [OsB(X3)]	The main constituents of the organization.
[<OsB1: driver-organizational subsystem>]	= [<OsB(X1): [strategic management + financial management + human resources management + administrative-legal management + IT service for management] >]	The organizational subsystem responsible to perform the support business processes. In the Porter-Miller organizational model, this subsystem corresponds to the following support processes: **strategic management, financial management, human resources management, administrative & legal management, and IT service for management.**
[<OsB2: driver-organizational subsystem>]	= [<OsB(X2): [input logistic + operations + output logistic + IT service for operations] >]	The organizational subsystem responsible to perform the primary business processes. In the Porter-Miller organizational model, this subsystems corresponds to the following primary processes: **input logistic, operations, output logistic and IT service for operations.**
[<OsB3: informational-organizational subsystem>]	= [<OsB(X3): [IT service management and engineering] >]	The organizational subsystem responsible to support the informational business processes. In the Porter-Miller organizational model, this is not reported explicitly. We call it the **IT service management and engineering processes (ITSM&E).**

CONCEPT	GENERIC VALUE	DESCRIPTION
[<R: relationships>]	= [R1 + (R2 + …)]	Relationships between the organizational parts, attributes, and/or its supra-system and world.

CONCEPT	DEFAULT INSTANCE	DESCRIPTION
[<OsB: organizational subsystem>]	= [OsB(X1) | OsB(X2) | OsB(X3)]	**The organizational subsystem to be modeled.**
[<O: organization>]	= [O(X)]	The organization to which belongs the organizational subsystem.
[<A: attributes>]	= [a1+a2+a3+ a4 + a5 + (a6 + …)]	The attributes that are defining the organizational subsystem.
[<a1: purpose>]	= [<a1: "to provide valued outcomes">]	The effectiveness mission of the organization.
[<a2: function>]	= [<a2: "to achieve efficiently its outputs">]	The efficacy mission of the organizational subsystem.
[<a3: inputs>]	= [<a3: [{ energy-matter(utilities, artifacts, money) | information-knowledge | acts } ^{n}] >]	The organizational subsystem's input flows.
[<a4: outputs>]	= [<a4: [{ energy-matter(utilities, artifacts, money) | information-knowledge | acts } ^{n}] >]	The organizational subsystem's output flows.
[<a5: outcomes>]	= [<a5: [{ <PoV: service> } | <MoV: product >} ^{n}] >]	The expected consequences to be generated by the organizational subsystem's outputs. PoV and MoV are respectively people-oriented and machine-oriented valued features.
• …	…	Other possible attributes.
[<BP: organizational business processes>]	= [BP1] + [BP2] + [BP3]	The main constituents of the organizational subsystem.
[BP1]	= [<BP1: control business processes>]	The business process responsible for controlling the operational processes into an organizational subsystem.
[BP2]	= [<BP2: operational business processes>]	The business process responsible for doing the core activities into an organizational subsystem
[BP3]	= [<BP3: informational business processes>]	The business process responsible for providing the informational support into an organizational subsystem.
[<R: relationships>]	= [R1 + (R2 + …)]	Relationships between the organizational subsystem parts, attributes and/ or its wider system.

CONCEPT	DEFAULT INSTANCE	DESCRIPTION
[[<BP: business process>] | [<BsP: business subprocess>]]	= [BP1 | BsP1]	**The business process or subprocess to be modeled.**
[[<OsB: organizational subsystem>] | [<BP: business process>]]	= [OsB | BP]	The owner organizational subsystem or business process of the BP or BsP that is being modeled.
[<A: attributes>]	= [a1+a2+a3+ a4 + a5 + a6 + a7 + (a8+ …)]	The attributes that are defining the business process or subprocess.
[<a1: purpose>]	= [<a1: "to provide valued outcomes">]	The effectiveness mission of the organization.
[<a2: function>]	= [<a2: "to achieve efficiently its outputs">]	The efficacy mission of the business process or subprocess.
[<a3: inputs>]	= [<a3: [{ energy-matter(utilities, artifacts, money) | information-knowledge | acts }^{n}] >]	The organizational business process or subprocess' input flows.
[<a4: outputs>]	= [<a4: [{ energy-matter(utilities, artifacts, money) | information-knowledge | acts }^{n}] >]	The organizational business process or subprocess' output flows.

CONCEPT	DEFAULT INSTANCE	DESCRIPTION
[<a5: outcomes>]	= [<a5: [{ <PoV: service> } | <MoV: product >} ^{n}] >]	The expected consequences to be generated by the organizational business process or subprocess' outputs. PoV and MoV are respectively people-oriented and machine-oriented valued features.
[<a6: mechanisms>]	= [<a6: [{ [people | tools | machines] }^{n}]>]	The organizational process' resources used for generating the outputs.
[<a7: controls>]	= [<a7: [{ [information | knowledge}^{n}]>]	The organizational process' resources used for controlling the generation of outputs.
...	...	Other possible attributes.
[[<BsP: business subprocesses>] | [<BA: business activities>]]	= [BsP1 | BA1] + [BsP2 | BA2] + ([BP3 | BA3] + ...)	The main constituents of the organizational business process or subprocess.
[BsP1 | BA1]	= [BsP1 | BA1]	The first business subprocess or activity.
[BsP2 | BA2]	= [BsP2 | BA2]	The second business subprocess or activity.
...	...	Other possible business subprocess or activity.
[<R: relationships>]	= [R1 + (R2 + ...)]	Relationships between the business process' parts, attributes and/or its wider system.

CONCEPT	DEFAULT INSTANCE	DESCRIPTION
[<BA: business activity>]	= [BA]	**The business activity to be modeled.**
[[<BP: business process>] | [<BsP: business subprocess>]]	= [BP | BsP]	The owner organizational business process or subprocess of the BA that is being modeled.
[<A: attributes>]	= [a1+a2+a3+ a4 + a5 + a6 + a7 + (a8+ ...)]	The attributes that are defining the business activity.
[<a1: purpose>]	= [<a1: "to provide valued outcomes">]	The effectiveness mission of the business activity.
[<a2: function>]	= [<a2: "to achieve efficiently its outputs">]	The efficacy mission of the business activity.
[<a3: inputs>]	= [<a3: [{ energy-matter(utilities, artifacts, money) | information-knowledge | acts }^{n}] >]	The organizational business activity's input flows.
[<a4: outputs>]	= [<a4: [{ energy-matter(utilities, artifacts, money) | information-knowledge | acts }^{n}] >]	The organizational business activity's output flows.
[<a5: outcomes>]	= [<a5: [{ <PoV: service> } | <MoV: product >} ^{n}] >]	The expected consequences to be generated by the organizational business activity's outputs. PoV and MoV are respectively people-oriented and machine-oriented valued features.
[<a6: tasks>]	= [t1 + t2 + (...)]	The logical unitary workloads required to complete the BA. At least two are required.
[<a7: personnel>]	= [p1 + (...)]	The people required for that the BA be performed. At least one person is required.
[<a8: tools & infrastructure>]	= [t&i1 + (...)]	The tools and physical infrastructure required for that the BA be performed.
[<a9: methods & procedures>]	= [m&p1 + (...)]	The methods and procedures about how the BA must be performed.
[<a10: socio-political mechanisms & structures>]	= [spm&s1 + (...)]	The socio-political influences (modeled as socio-political norms, values and beliefs) that affect the BA execution.
[<R: relationships>]	= [R1 + (R2 + ...)]	Relationships between the business activity's attributes and/or its wider system.

APPENDIX C. SYSTEMIC DESCRIPTION AND COMPARISON OF THE MODELS AND STANDARDS OF PROCESSES.

Table C.1 Description and comparison of models and standards in the organizational level.

SYSTEMIC CONCEPT	Systemic Map of the CMMI/SE/SwE: 2002 Models	Systemic Map of the ISO/IEC 15288:2002 Standard	Systemic Map of the ISO/IEC 12207:1995 Standard	Systemic Map of the ISO/IEC 15504:2006 Standard	Systemic Map of the ISO/IEC 20000:2005 Standard	Systemic Map of the ITIL V.3 : 2007 Model
[<O: organization>]	[<O: *"is typically an administrative structure in which people collectively manage one or more projects as a whole, and whose projects share a senior manager and operate under the same policies"*>]	[<O: *"a group of people and facilities with an arrangement of responsibilities, authorities and relationships"* >]	[<O: *"is a body of persons organized for some specific purpose, as a club, union, corporation, or society"* and is called a "party" when enters into a contract*>]	[<O: *"an organizational unit deploys one or more processes that have a coherent process context and operates within a coherent set of business goals"*>]	[<O: *"a service provider is the organization aiming to achieve ISO/IEC 20000"*>]	[<O: *"a company, legal entity or other institution ... any entity that has People, Resources and Budgets"* | **"Business unit:** *a segment of the business that has its own Plans, Metrics, Incomes and Costs ... owns Assets and uses these to create value for Customers in the form of goods and services"*>]
[<OSS: organizational supra-system>]	[<OSS: **"Enterprise:** *the full composition of companies"* that belongs the O>]	[<OSS: **"Enterprise:** *the part of an organization with responsibility to acquire and to supply products and/or services according to agreements"*>]	[<OSS: **"Enterprise:** *a system of at least two parties"*>]	[<OSS: *"larger organization: the organization that contains to the organizational unit"*>]	[<OSS: *"business: the organization that that receives the provided services of the service provider "*>]	[<OSS: *"business: an overall corporate entity or Organization formed of a number of Business Unit "*>]
[<a1: purpose:| "to provide valued outcomes"]>]	[<a1: *"to help to deliver products or services through ensuring stable, capable, and mature processes"*>]	[<a1: *"... establishes a common framework for describing the life cycle of systems created by humans ... with the ultimate goal of achieving customer satisfaction"*>]	[<a1: *"... establishes a common framework for software life cycle processes ... applied during the acquisition of a system that contains software, a stand-alone software product, and software service, and during the supply, development, operation, and maintenance of software products"* >]	[<a1: *"... provides a framework for the assessment of process capability"* + "understanding of the state of process" + "process improvement"* >]	[<a1: *"to provide an industry consensus on quality standards for IT service management processes ... (that) deliver the best possible service to meet a customer's business needs within agreed resource levels, i.e. service that is professional, cost-effective and with risks which are understood and managed"* >]	[<a1: *"the objective of the ITIL Service Management practice framework is to provide services to business customers that are fit for purpose, stable and that are so reliable, the business view them as a trusted utility"* >]
[<a2: function:| "to achieve efficiently its outputs"]>]	[<a2: *"to manage the development, acquisition, and maintenance of products or services"*>]	[<2: *"... managing and performing the stages of a man-based system's life cycle"* >]	[<2: *" ... providing a process ... for defining, controlling, and improving software life cycle processes "* >]	[<a2: *" ... planning, managing, monitoring, controlling and improving the acquisition, supply, development, operation, evolution and support of products and services "* >]	[<a2: *"to provide process of management system requirements + service management planning + new or changed services planning & implementing + service delivering + relationships + release + resolution + control "* >]	[<a2: *"to provide robust, mature and time-tested practices into process of service strategy + service design + service transition + service operation + continual service improvement "* >]

continued on following page

[<a3: inputs flows>]	[<a3: [{ energy-matter(utilities, artifacts, money) | information-knowledge | acts }^{n}]>]					
[<a4: outputs flows>]	[<a4: [{ energy-matter(utilities, artifacts, money) | information-knowledge | acts }^{n}]>]					
[<a5: outcomes>]	[<a5: [{ [<PoV1: IT-based services> + <PoV2: capability process profile>] | [<MoV1: IT-based products>] }^{n}]>]	[<a5: [{ [<PoV1: IT-based services> + <PoV2: capability process profile>] | [<MoV1: IT-based products>] }^{n}]>]	[<a5: [{ [<PoV1: IT-based services> + <PoV2: complain-not-complain process profile>] | [<MoV1: IT-based products>]}^{n}]>]	[<a5: [{ [<PoV1: IT-based services> + <PoV2: capability process profile>] | [<MoV1: IT-based products>] }^{n}]>]	[<a5: [{ [<PoV1: IT-based services> + <PoV2: complain-not-complain process profile>] | [<MoV1: IT-based products>]}^{n}]>]	[<a5: [{ [<PoV1: IT-based services> + <PoV2: capability process profile>] | [<MoV1: IT-based products>] }^{n}]>]
[<OsB1: driver-organizational subsystem>]	<STRATEGIC MGT: <Process Mgt: [OPF + OID]>>	<STRATEGIC MGT: <Enterprise P.: [SLCP.MGT + QUA. MGT]>>	<STRATEGIC MGT: <Organizational Life Cycle P.: not defined >>	<STRATEGIC MGT: <Management P. : [ORG.ALIG, QUA. MGT]>, <P. Improvement P.: [PRO.IMPROV]>>	<STRATEGIC MGT: <*Mgt. System Reqs.>, <*Service Mgt. P&I>>	<STRATEGIC MGT: <*Service Strategy: [STRAT.GEN]>>
	<FINANCIAL MGT: <Process Mgt: not defined>>	<FINANCIAL MGT: <Enterprise P.: [INV.MGT]>>	<FINANCIAL MGT: <Organizational Life Cycle P.: [INFRASTR]>>	<FINANCIAL MGT: <Resource & Infst. P.: [INFRASTR]>, <Reuse P.: [ASSET.MGT]>>	<FINANCIAL MGT: <*Service Delivering: [BUDGT.ACCT]>>	<FINANCIAL MGT: <*Service Strategy: [FIN.MGT] >>
	<HR MGT: <Process Mgt : [OT]>>	<HR MGT: <Enterprise P.: [RES.MGT]>>	<HR MGT: <Organizational Life Cycle P.: [TRAINING]>>	<HR MGT: <Resource & Infst. P.: [HR, TRAINING, KM]>>	<HR MGT: <*Mgt. System Reqs.: [Competence, awareness & training]>>	<HR MGT: <*Org. Development>>
	<ADM-LEGAL MGT: <Process Mgt : [OPP + OPD]>>	<ADM-LEGAL MGT: <Enterprise P.: [RES.MGT+ EENV.MGT]>, <Project P.: [INF.MGT]>>	<ADM-LEGAL MGT: <Organizational Life Cycle P.: [MGT.PROC + IM-PROV.PROC]>>	<ADM-LEGAL MGT: <Management P. : [ORG.MGT + MEASRMNT]>, <P. Improvement P.: [PRO.ESTBLSH + PRO.ASSMT]>>	<ADM-LEGAL MGT: <*Service Delivering: [SvL.MGT + Sv.REP] >>	<ADM-LEGAL MGT: <*Service Design: [SvL.MGT]>, <*Continual Service Improvement: [Sv. REP]>>
	<ITSfM: not defined>	<ITSfM: <Project P. : [INF.MGT]>>	<ITSfM: not defined>	<ITSfM: not defined>	<ITSfM: <*Service Delivering: [SvL.MGT + Sv.REP] >>	<ITSfM: <*Continual Service Improvement: [Sv.MEASRMNT + Sv.ANLYS + Sv.REP + Sv.IMPROV]>>
[<OsB2: driven-organizational subsystem>]	<INPUT LOGISTIC: <Project Mgt: [PP+SAM+IPM+ RSKM+QPM]>	<INPUT LOGISTIC: <Agreement P. : [ACQ.PROC]>, <Project P.: [PROJ.PLAN, RSK. MGT]>>	<INPUT LOGISTIC: <Primary Life Cycle P. : [ACQ.PROC]>	<INPUT LOGISTIC: <Management P.: [RSK.MGT + PROJ.MGT]>	<INPUT LOGISTIC: <*Relationships: [SUPPLY.REL. MGT]>	<INPUT LOGISTIC: <*Service Design: [SUPPLY.MGT]>, <*Service Transition: [TRANS.PLAN.SUP + CHNG.MGT + Sv.ASSET.CM + Sv.KM]>>

continued on following page

[<OsB2: driven-organizational subsystem>]	<OPERATIONS: <Engineering: [REQM +CRD + TS + PI + VER + VAL |>, <Support: [CM + PPQA + M&A + DAR + CAR |>>	<OPERATIONS: <Technical P. : [REQ.DEV + REQ.ANLYS + ARCH.DSGN + IMPLMNT + INTGRT + VERIF + TRANSITION + VALID + OPERAT + MANTNC + DISPOSAL |>, <Project P. : [PROJ.CTRL + DEC.MAK + CM + INF.MGT |>>	<OPERATIONS: <Primary Life Cycle P.: [DEV.PROC] >, <Supporting Life Cycle P.: [DOC+ CM+ QA+ VERIF+ VALID+ JOINT.REV+ AUD+ PROB.RES] >>	<OPERATIONS: <Primary Life Cycle P.: [REQ.ELIC + SYS.REQA + SYS.ARCH.DSGN + Sw.REQA + Sw.DSGN + Sw.CNST + Sw.INTGRT + Sw.TEST + SYS.INTGRT + SYS.TEST + Sw.INST + Sw.SYS.MANTNC] >, <Supporting Life Cycle P.: [QA+VERIF+ VALID+JOINT. REV+ AUD+ PRO. EVAL+ USAB+ DOC+ CM+ PROB. RES.MGT+ CHNG. MGT] >, <Reuse P.: [REU.PRO, DOM. ENG] >>	<OPERATIONS: <*Resolution: [INCDNT.MGT + PROB.MGT |>, <*Control: [CM + CHNG.MGT]>, <*New/Changed Services P&I>>	<OPERATIONS: <*Service Operation: [EVENT.MGT + REQST.FULLMT + INCDNT.MGT + PROB.MGT + ACCS.MGT + F.Sv.DESK + F.TECH.MGT + F.IT.OPER.MGT + F.APPLIC.MGT + F.MON.CTRL |>>
	<OUTPUT LOGISTIC: <Project Mgt: [PMC+ IPM+ RSKM+ QPM]>>	<OUTPUT LOGISTIC: <Agreement P.: [SUP.PROC |>, <Project P.: [PROC.ASSMT + PROC.CTRL+ RSK.MGT+ INF. MGT]>>	<OUTPUT LOGISTIC: <Primary Life Cycle P.: [SUPPORT + OPERAT + MANTC]>>>	<OUTPUT LOGISTIC: <Primary Life Cycle P.: [SUPPORT + OPERAT]>>>	<OUTPUT LOGISTIC: <*Service Delivering: [CAPC.MGT + Sv.CONT.AVL. MGT + INF.SEC.MGT + SvL.MGT + Sv.REP] >, <*Release: [RLS.MGT |>, <*Relationships: [BUSS.REL.MGT |>>	<OUTPUT LOGISTIC: <*Service Transition: [VALID.TEST.MGT + REL.DEPLOY. MGT + EVAL.MGT + Sv.KM |>, <*Service Design: [SvL.MGT + Sv.CTLG.MGT + CAPC.MGT + AVL.MGT + INF.SEC.MGT + IT.Sv.CONT. MGT|>
	<ITSfO: not defined>	<ITSfO: <Project P. : [INF.MGT]>>	<ITSfO: not defined>	<ITSfO: not defined>	<ITSfO: embedded in the other processes >	<ITSfO: embedded in the other processes >
[<OsB3: is-org subsystem.>]	<IT SERVICE MANAGEMENT: not defined>	<IT SERVICE MANAGEMENT: not defined>				<IT SERVICE MANAGEMENT: <* Service Strategy >, <*Service Transition>, <*Service Design>, <Continual Service Improvement>>
	<IT SERVICE ENGINEERING: not defined>	<IT SERVICE ENGINEERING: not defined>				<IT SERVICE ENGINEERING: <*Service Transition>, <*Service Design>, <*Service Operation>>
	<IT SERVICE SUPPORT: not defined>	<IT SERVICE SUPPORT: not defined>				<IT SERVICE SUPPORT: <*Service Transition>, <*Service Operation>>

This work was previously published in International Journal of Information Technologies and Systems Approach, Vol. 1, Issue 2, edited by M. Mora; D. Paradice, pp. 57-85, copyright 2008 by IGI Publishing (an imprint of IGI Global).

Chapter 11
Integrating the Fragmented Pieces of IS Research Paradigms and Frameworks:
A Systems Approach

Manuel Mora
Autonomous University of Aguascalientes, Mexico

Ovsei Gelman
National Autonomous University of Mexico, Mexico

Guisseppi Forgionne
Maryland University, Baltimore County, USA

Doncho Petkov
Eastern State Connecticut University, USA

Jeimy Cano
Los Andes University, Colombia

ABSTRACT

A formal conceptualization of the original concept of system and related concepts—from the original systems approach movement—can facilitate the understanding of information systems (IS). This article develops a critique integrative of the main IS research paradigms and frameworks reported in the IS literature using a systems approach. The effort seeks to reduce or dissolve some current research conflicts on the foci and the underlying paradigms of the IS discipline.

INTRODUCTION

The concept of management information systems (MIS) in particular, or information systems (IS) in general, has been studied intensively since the 1950s (Adam & Fitzgerald, 2000). These investigations have been conducted largely by behavioral-trained scientists to study the emergent phenomena caused by the deployment and utilization of computers in organizations.

This discipline, from its conception as a potential scientific field, has been driven by a dual research perspective: technical (design engineering oriented) or social (behavioral focused). This duality of man-made non-living (hardware, software, data, and procedures) and living systems (human-beings, teams, organizations, and societies), the multiple interrelationships among these elements, and the socio-cultural-economic-politic and physical-natural environment, make IS a complex field of inquiry.

The complexity of the IS field has attracted researchers from disparate disciplines—operations research, accounting, organizational behavior, management, and computer science, among others. This disciplinary disparity has generated the utilization of several isolated research paradigms and lenses (e.g., positivist, interpretative, or critical-based underlying research methodologies). The result has been the lack of a generally accepted IS research framework or broad theory (Hirchheim & Klein, 2003) and has produced: (i) a vast body of disconnected micro-theories (Barkhi & Sheetz, 2001); (ii) multiple self-identities perceived by the different stakeholders (*e.g.*, IS researchers, IS practitioners, and IS users); and (iii) partial, disparate and incomplete IS conceptualizations (Benbazat & Zmud, 2003; Galliers, 2004; Orlikowski & Iacono, 2001).

Despite scholastic indicators[1] of maturity, IS, then, has been assessed as: (1) highly fragmented (Larsen & Levine, 2005), (2) with little cumulative tradition (Weber, 1987), (3) deficient of a formal and standard set of fundamental well-defined and accepted concepts (Alter, 2001, p. 3; Banville & Landry, 1989, p. 56; Wand & Weber, 1990, p. 1282) and (4) with an informal, conflicting and ambiguous communicational system (Banville & Landry, 1989; Hirschheim & Klein, 2003). Such findings provide insights for a plausible explanation of the delayed maturation of the field and the conflictive current perspectives on information systems (Farhoomand, 1987; Wand & Weber, 1990).

This article illustrates how systems theory can be used to alleviate the difficulties. First, there is a review of basic *system* and related concepts relevant to information systems (Ackoff, 1960; Bertalanffy, 1950, 1968, 1972; Boulding, 1956; Checkland, 1983; Forrester, 1958; Jackson, 2000; Klir, 1969; Midgley, 1996; Mingers, 2000, 2001; Rapoport, 1968). Next, these systems approach concepts are used to formulate a critique integrative of the main paradigms and frameworks suggested for IS research. Then, a theoretical scheme is developed to integrate holistically and coherently the fragmented pieces of IS research paradigms and frameworks. To end, this article presents future research directions on potential conflictive conclusions presented.

THE SYSTEMS APPROACH: PRINCIPLES AND PARADIGMS

The Principles of the Systems Approach

The systems approach is an intellectual movement originated by the biologist Ludwig von Bertalanffy[2] (1950, 1968, 1972), the economist Kenneth Boulding (1956), and the mathematicians Anatoly Rapoport (1968) and George Klir (1969) that proposes a complementary paradigm (e.g., a worldview and a framework of ideas, methodologies, and tools) to study complex natural, artificial, and socio-politic cultural phenomena.

Lazlo and Lazlo (1997) interpret the modern conceptualization of the systems approach as a worldview shift from chaos to an organized complexity. Boulding (1956) argues that the systems approach—labeled as general systems theory (GST)—is about an adequate trade-off between the scope and confidence in valid theories from several disciplines. In the former case the greater the level of scope the lesser the level of confidence and vice versa. For Rapoport (1968), the systems approach should be conceptualized as a philosophical strategy or direction for doing science. Klir (1969), in turn, considers that GST should contain general methodological principles for all systems as well as particular principles for specific types of systems. Bertalanffy (1972) quotes himself (Bertalanffy, 1950 (reprinted in Bertalanffy, 1968, p. 32)) to explain that GST's "...task is the formulation and derivation of those general principles that are applicable to systems in general."

According to these systems thinkers and additional seminal contributors to this intellectual movement (Ackoff[3] in particular, 1960, 1973, 1981), the systems approach complements the reductionism, analytic, and mechanic worldview with an expansionist, synthetic, and teleological view.

Reductionism implies that the phenomena are isolated or disconnected from wider systems or the environment, while expansionism claims that each phenomenon can be delimited—objectively, subjectively, or coercively—into a central object of interest (e.g., the system under study) and its wider system and/or environment. The analytic view holds that we need only investigate the internal parts and their interrelationships of the phenomenon to understand its behavior. A synthetic view accepts and uses the analytical view but incorporates the interrelationships between the whole and its environment. Furthermore, the synthetic view holds that some events and attributes of the parts are lost when these are not part of the whole and vice versa (e.g., events and

attributes emerge in the whole but are not owned by the parts). The mechanist view holds that the phenomena happen by the occurrence of disconnected and simple linear cause-effect networks, and the systems approach complements this view through a teleological perspective that claims that the phenomena happens via a complex interaction of connected non-linear feed-back networks. Causes or independent constructs are affected by lagged effects.

Under the systems approach, the *systems* own core general properties: *wholeness, purposefulness, emergence, organization, hierarchical order, interconnectedness, competence, information-based controllability, progressive mechanization,* and *centralization. Wholeness* refers to the unitary functional view and existence of a system. *Purposefulness* refers to the extent of a system has predefined or self-generated goals as well as the set of intentional behaviors to reach these targets. *Emergence* involves the actions and/or properties owned solely by the whole and not by their parts. The property *organization* implies a non-random arrangement of its components and *hierarchical order* the existence of multi-level layers of components. *Interconnectedness* accounts for the degree of interdependence effects of components on other components and subgroups. *Competence* implies that the inflows of energy, material, and information toward the system will be distributed in the parts in a competitive manner, and this property also accounts for the conflicts between system, subsystems, and suprasystem's objectives. Finally, the *information-based controllability, progressive mechanization,* and *centralization* are properties which involve the transference of information and control fluxes between components that are required to regulate and govern the relationships. In particular, *progressive mechanization* refers to the extent to which the parts of a system act independently and *centralization* to the extent to which changes in the system result from a particular component.

Research Paradigms of the Systems Approach

Many researchers have shaped the systems approach. These researchers include the hard/functionalist/positivist stream (Forrester, 1958, 1991) supported by a positivist/pragmatist philosophy (Jackson, 2000), the soft/interpretative stream (Checkland, 1983, 2000) linked to Husserl's phenomenology and appreciative philosophy (Checkland, 2000), the critical/emancipative stream (Flood, Norman, & Romm, 1996; Jackson, 2000) underpinned in a critical philosophy from Habermas (referenced by Jackson, 2000) and the emergent critical realism systemic stance (Mingers, 2000, 2002) endorsed by Bhaskar's philosophy (Bhaskar, 1975, quoted by Mingers).

These four main streams can be associated respectively to the following general philosophical principles:

P.1   *the intelligible world is an organized complexity comprised of a variety of natural, man-made, and social systems that own a real existence*

P.2   *the intelligible world can be studied freely through systemic lenses and under an intersubjective social construction*

P.3   *the intelligible world can be uniquely understood when it is studied freely from restrictive social human relationships and a variety of theoretically coherent systemic lenses are used.*

P.4   *the world is intelligible[4a] for human beings because of its stratified hierarchy of organized complexities—the widest container is the* <u>real domain</u> *that comprises a multi-strata of natural, man-made and social structures[4b] as well as of event-generative processes that are manifested in the* <u>actual domain</u> *that in turn contains to the* <u>empirical domain</u> *where the generated events can or cannot be dectected-*

The hard/functionalist/positivist systems approach is based on P.1. The soft/interpretative approach rejects P.1 but supports P.2. The critical/emancipative approach is neutral to P.1, rejects P.2, and endorses P.3. Finally, the emergent critical realism systems approach endorses P.4 and automatically includes P.1 through P.3.

The first three systems paradigms have been extensively studied and applied, However, according to several authors (Dobson, 2003; Mingers, 2001; Mora, Gelman, Forgionne, & Cervantes, 2004), Bhaskar's critical realism has emerged to dissolve theoretical contradictions in the different systems approaches and offer an original expected holistic view of the discipline. Critical realism has been suggested as a common underlying philosophy for management sciences/operations research (Mingers, 2000, 2003) and also recently for information systems research (Carlsson, 2003; Dobson, 2001, 2002; Mingers, 2002). According to Mingers (2002):

Critical realism does not have a commitment to a single form of research, rather it involves particular attitudes toward its purpose and practice. First, the critical realist is never content just with description, whether it is qualitative or quantitative. No matter how complex a statistical analysis, or rich an ethnographic interpretation, this is only the first step . CR wants to get beneath the surface to understand and explain why things are as they are, to hypothesize the structures and mechanisms that shape observable events. Second, CR recognizes the existence of a variety of objects of knowledge—material, conceptual, social, psychological—each of which requires different research methods to come to understand them. And, CR emphasizes the holistic interaction of these different objects. Thus it is to be expected that understanding in any particular situation will require a variety of research methods (multimethodology [Mingers 2001]), both extensive and intensive. Third, CR recognizes the inevitable fallibility of observation, especially in the social

Table 1. Systems research paradigms

	Hard/Positivist/ Functionalist System Paradigm	Soft/Interpretative System Paradigm	Critical System Paradigm	Emergent Critical Realism System Paradigm
Framework of Ideas, Theories, Theoretical Problems and Models	- systems and their underlying structures exist in reality (a systemic ontology) and can be studied, predicted and controlled using systems approach (a systemic epistemology)	- the reality and their underlying structures can be studied systemically (a systemic epistemology) but systems do not have a real existence (non systemic ontology)	- the social reality and their underlying structures can be studied systemically when restrictive power relationships are uncovered (a systemic epistemology) but systems can be contingently considered to be real (a conditioned systemic ontology)	- all reality (natural, designed and social) has real underlying mechanisms and structures (real domain) that generate events observed (empirical domain) as well as non observed (the actual domain including the empirical one), but social reality (concepts, meanings, categories) are socially built and thereby rely on the existence of human beings and a communication language (a systemic ontology) - research involves the underlying mechanisms and structures of the observed events (a systemic epistemology)
Methodology (research strategies and methods)	- systems methods are used with an isolated (just a sole research method is required), an imperialist (just a sole research method is required and superior to others but some features of latter can be added) or a pragmatism (any research tool perceived as useful can be used and combined with others despite theoretical inconsistencies) strategy		- systems methods are used in a pluralist and complementarist view with theoretical and practical coherency	
Areas of Study in the Reality	- natural and designed systems	- social systems (including human activity systems)	- social systems	- all systems
Research Purposes — Response inquiries (understand reality)	- to predict the behavior of the phenomena of interest	- to formulate interpretative theories and models on the phenomena of interest	- to formulate critical theories and models on the phenomena of interest	- to know the correct underlying mechanisms and structures that causes the actual domain
Research Purposes — Solve, resolve or dissolve problems (intervene and modify reality)	-to control the behavior of the phenomena of interest	- to achieve a shared and mutual understanding of conflictive views of the phenomena	- to foster the emancipation of human beings from restrictive power work, life and societal relationships	- to use knowledge to intervene in the phenomena of interest

world, and therefore requires the researcher to be particularly aware of the assumptions and limitation of their research. (p. 302)

Based on Checkland (2000), Jackson's (2000) interpretations of Checkland (1981), Ackoff, Gupta, and Minas (1962), Ackoff (1981), and Midgley (1996), a systemic view of the problem can be articulated with three essential components and five purposes. These components are: (1) the framework **F** of ideas, initial theories, theoretical problems, and models that compose a discipline, (2) the set of philosophical research paradigms and methodologies **M** that define the ontological definitions of the world to be studied as well as the epistemological principles and tools regarding how it can be studied, and (3) the situational area **A** of the reality that contains well-defined or messy situations of interest.

According to Midgley[5] (1996), a science can have the following purposes: (1) to predict and control well-defined objects or situations of study as in the hard/positivist/functionalist systems paradigm, (2) to increase a shared and mutual understanding of messy real situations as in the soft/interpretative systems paradigm, and (3) to increase the quality of work and life of human beings in organizations and societies through an emancipation of power relations between dominant and dominated groups as in the critical systems paradigm. Ackoff et al. (1962) and Ackoff (1981) suggest two main purposes for science: (1) to respond to inquiries and (2) to resolve, solve or dissolve problems.

The integration of these core concepts of the systems research paradigms and the underlying philosophies and research strategies (adapted from Gregory, 1996) leads to the holistic proposal presented in Table 1.

REVIEW AND DISCUSSION OF A CRITICAL REALIST INTEGRATION OF IS RESEARCH PARADIGMS AND FRAMEWORKS

Systemic Integration of the Information Systems Research Paradigms

Six IS research paradigms are reviewed in this section: Weber (1987), Orlikowski and Iacono (2001), Benbazat and Zmud (2003), Hirschheim and Klein (2003), Galliers (2004), and Larsen and Levine (2005), and arguments are articulated for a systemic integration of them.

Weber (1987) critiques the proliferation of research frameworks that have lead to a random and non-selective set of worthy research questions (e.g., a hypothesis generator). Then novel researchers could infer that every relationship is useful to be studied. Weber also asserts that technology-driven research can produce a fragile discipline with a lack of sound theoretical principles. A paradigm is proposed with three required conditions: (i) a set of objects of interest that other disciplines cannot study adequately, (ii) the objects must exhibit an observable behavior, and (iii) a possible underlying order is associated with the object's behaviors. For Weber, two sets of objects are candidates: objects that externally interact with an information system and objects that internally compose the system. The behaviors of interest are performance variables and interrelationships of the two set of objects. Weber claims that an internal order of the second set of objects can and must be assumed to pursue research based on the paradigm. No argument is reported for the first set of objects. Weber also suggests that the IS discipline can have several paradigms. He proposes *static, comparative static,* and *dynamic* paradigms. The articulated paradigm is not the same as a research framework where a definitive set of variables is fixed: "*instead, it provides a way of thinking about the world of IS behavior and the types of research that might be done*" (ibid, p. 16). With such a paradigm, a piecemeal, methodological dominant-oriented and event-day driven research can be avoided.

Orlikowski & Iacono (2001) suggest that IS research should focus on the information technology (IT) artifact as much as its context, effects, and capabilities. According to their study, IT artifacts have been analyzed only as monolithic black-boxes or disconnected from their context, effects, or capabilities. IT artifacts are defined in five different modes: as a *tool*, as a *proxy*, as *an ensemble*, as a *computational resource*, and as a *nominal concept*. The IT artifact can be studied as a *tool* for labor substitution, productivity enhancement, information processing, or to alter social relationships. As a *proxy*, the IT artifact refers to the study of some essential attributes such as individual perceptions, diffusion rates and money spent. As an *ensemble*, the IT artifact is associated with development projects, embedded systems, social structures, and production networks. IT artifacts as *computational resources* can be algorithms or models, and then the interactions of the IT artifact with its social context or its effects on dependent variables are not of interest. Finally, IT artifacts as *nominal concepts* imply that no specific technology is referenced, for example, the IT artifact is omitted in such studies. This *nominal* view was found most common next to the *computational* view (e.g., a computer science-oriented perspective of the IT artifact). Next common view is of IT artifacts as *tools* that affect dependent variables. The *ensemble* view was the least frequently reported. According to the authors, the researchers' original research paradigms or lenses bias the IT artifact conceptualization. *Nominal, tool,* or *proxy* views are used for management and social scientists, while computer scientists consider the *computational* view.

Such disparate views indicate a need to develop conceptualizations and theories on the IT artifacts that could be used in every IS study. Otherwise, IS research will be a fragmented field where its core object is not a "*major player in its own playing field*" (ibid, p. 130). However, for Orlikowski & Iacono (2001), the development of a single grand theory for IT artifacts that accommodates all their context-specificities is not adequate.

Benbazat and Zmud's (2003) suggest that the IS discipline's central identity is ambiguous due to an *under-investigation* of core IS issues and *over-investigation* of related and potentially relevant organizational or technical issues. These authors use Aldrich's (1999) theory of formation of organizations to explain that the IS discipline will be considered a mature discipline when a learning/cognitive and sociopolitical legitimacy is achieved. For this maturity to occur, methodological and theoretical rigor and relevance must in turn be achieved. A dominant design, for example, a central core of properties of what must be studied in the IS phenomena, is suggested to accommodate the topical diversity. For Benbazat and Zmud, this dominant identifying design for the IS discipline does not preclude the utilization of an interdisciplinary effort. In this view, the central character for the IS discipline is defined as the composition of the IT artifact that enables/supports some task(s), into structures and later into contexts, and its *nomological* network of IT managerial, technological, methodological, operational, and behavioral capabilities and practices of the pre- and post-core activities to the existence of some IT artifacts. Like Orlokowski and Iacono (2001), Benbazat and Zmud (2003) reject the IS research based in the black-box IT concept.

Hirschheim and Klein's (2003) thesis is that the IS discipline is fragmented with structural deficiencies manifested in a missing and generally accepted body of knowledge and in internal and external communication gaps. These authors build on Habermas' theory of communication (and of knowledge) to pose that any inquiry has two cognitive purposes[6]: a rationale for IS design and the communication for mutual understanding and agreement of disparate perceptions and interpretations (called *technical* and *practical* originally by Habermas). Hirschheim and Klein accept that the technical purpose seeks the prediction and control of the law-based IS phenomena, while the practical seeks the accommodation of disparate viewpoints underpinned in different norms, values, and contexts. They also agree that IS frameworks, called *categorization schemes*, are useful to start a shared body of knowledge (BoK) but fail to indicate how the IS knowledge as a whole—for example, as a system—can be articulated. Also, they accept theoretical and methodological diversity in the discipline. Like others, Hirschheim and Klein suggest that the lack of a shared core set of underlying knowledge weakens the IS discipline. They identify four challenges for the IS community: to accept and understand through clear communication the theoretical and methodological pluralist status, to develop a common general theoretical base, and to conduct research with methodological rigor and relevancy for IS research stakeholders. For the second challenge, they encourage the development of studies that take fragmented pieces of evidence and put them in a broader theoretical framework, as in Ives, Hamilton, and Davis study (1980). In their view, the IS BoK should be integrated with four types of knowledge with similar relevance: theoretical, technical, ethical, and applicative.

Galliers (2004) disagrees with the current evaluation of the IS discipline as a field in crisis. According to Galliers, Kuhn's ideas on paradigms can be interpreted as an *evolution* rather than a *revolution*. Core ideas, then, should not be abandoned but complemented, as in the organizational sciences (Gellner, 1993). For Galliers, the concepts of information and information system must be uncovered to understand the IS discipline. Supporting Land and Kennedy-McGregor's view (1987, quoted by Galliers, 2004), Galliers considers IS as a system that comprises formal and informal human, computer and external systems. In Galliers' view, Benbazat and Zmud's IT artifact is not conceptually sufficient to embrace these elements and thereby the *"essentially human activity of data interpretation and communication, and knowledge sharing and creation"* (2003, p. 342) could be diminished. In addition, Galliers rejects the notion of an IS as solely a generic social system with a strong technological component. Instead, Galliers considers IS a complex, multi-faceted and multi-leveled phenomenon that requires a trans-disciplinary research effort. Gallier's thesis for a mature discipline is the acceptance of a dynamic and evolutionary field in research focus, boundaries, and diversity/pluralism versus a prefixed set of core concepts with a monolithic, and dominant, perspective of the discipline. As a tentative strategy, Galliers uses Checkland's (1981) definition of a system's boundary and environment and its dependence on the observer's research purposes.

In summary, Galliers disagrees with the limited concept of the IT artifact and notes *"inclusion errors"* resulting from the closed boundaries of the IT artifact. Galliers also notes the IT artifact concept ignores relevant topics such as EDI, inter-organizational IS, knowledge management systems, and the digital divide concept.

For Larsen and Levine (2005), the crisis in the field has been over-assessed. While these authors accept the lack of coherence, the paucity of a cumulative tradition, and the loss of relevant research, they blame the university education and disciplinary knowledge aggravated by the effects of a rapid evolution of ITs. Based on Kuhn's field concept, Larsen and Levine suggest that the IS discipline can be considered pre-paradigmatic: *"a common set of theories derived from a paradigmatic platform do not exist in MIS"* (p. 361). They suggest that Kuhn's ideas, built on natural sciences, could be inadequate. Instead, they propose the socio-political Frickel and Gross' (2005, referenced by Larsen & Levine, 2005) concept of scientific/intellectual movement (SIM) in which several SIMs tied to multiple research approaches can co-exist under a common umbrella and compete for recognition and status.

Larsen and Levine use a novel co-word analysis technique (Monarch, 2000; quoted by authors) to identify networks of associated concepts, represented by *leximaps* (Monarch, 2000; quoted by Larsen & Levine), and measure the associative strength of pairs of concepts. Concepts highly connected are considered the center of coherence. A total of 1,325 research articles from five top IS journals in the 1990-2000 period are the dataset. The researchers divide this dataset in two sub-periods: 1990-1994 and 1995-2000. A key finding is that in both sub-periods the center of coherence related to IS generic concepts, like system, information, and management, is present. However, in the *leximap* of 1995-2000, new concepts appear, such as model, process, technology, user, and research. The number of centers of coherence related with identified theories is minimal. Furthermore, four selective pairs of concepts were used to trace a building theory activity. The scarce evidence and the minimal number of centers of coherence for theories are interpreted by Larsen and Levine as a lack of cumulative tradition where innovation is more appreciated than building on the work of others.

The previous studies provide the alternative proposals to establish a framework **F** of ideas, theories, theoretical problems, and models that are suggested to define the distinctive identity of the IS discipline. Table 2 summarizes the key findings from this research.

From the diverse systems paradigms exhibited in Table 1, this article argues that a critical realism systems approach is ontologically and epistemologically valid and comprehensive to integrate with theoretical and pragmatic coherence the shared ideas, theories, theoretical problems, and models from such studies.

Table 3 exhibits a summarized systemic proposal of integration. There are two competitive research paradigms: the IS discipline as a classic Kuhn's imperialist or dominant framework of ideas, theories, and models, and the IS discipline as a post-Kuhnian paradigm, as a dynamic body of knowledge and a diversified intellectual movement under one umbrella. The first includes the approaches of Weber (1987) and Benbazat and Zmud (2003) and partially Orlikowski and Iacono (2001), while second incorporates the work of Galliers (2004) and Larsen and Levine (2005), and partially the paradigm proposed by Hirschheim and Klein (2003).

The critical realism approach claims that the IS discipline can have a framework **F** based in permanent and shared generic knowledge structures on systems, as well as of dynamic or changing concepts that will emerge as in any systemic structure. Also, it supports the utilization of a set of methodologies **M** that are theoretical and pragmatically coherent according to the purpose of a specific research study (Midgley, 1996; Mingers, 2000) and consequently offers a pluralist-complementarist research strategy. The central theme is information systems conceptualized as *systems* (Gelman & Garcia, 1989; Gelman, Mora, Forgionne, & Cervantes, 2005; Mora, Gelman, Cervantes, Mejia, & Weitzenfeld, 2003; Mora, Gelman, Cano, Cervantes, & Forgionne, 2006; Mora, Gelman, Cervantes, & Forgionne, in press). Such an approach incorporates the different components and interrelationships of the system as well as of the lower (subsystems and so on) and upper (suprasystems and wider systems) systemic levels, including the environment. The specific components, attributes, events, and interrelationships and levels of lower and upper systemic layers depend on the specific research purposes, resources, and schedules required.

This critical realism stance can accommodate the two competitive approaches posed for the IS discipline through the acknowledgement of the complexity and diversity of the phenomenon. Weber's (1987) foci of IS research is identified as systemically founded: its components and its environment are based on Miller's living systems

Table 2. Analysis of main IS research paradigms

IS Research Paradigm or Framework	Main weaknesses identified in the IS Discipline	Maturity Criteria	Foci for IS Research	Concept of what an Information System is	Underlying System Theories Used
Weber (1987)	- lack of a research paradigm - little cumulative tradition - lack of grand stream direction - fashion event-day driven research	- existence of at least one Kuhn's Paradigm (used as a grand theory) - pattern of literature citations in the field	- set of objects that interacts with an information system - set of objects that comprises an information system	- not reported	- Milller's Living Systems Theory - Simon's concept of Complex Systems
Orlikowski & Iacono (2001)	- not engaged with the central object for IS: the IT artifact - thus the IT artifact is not studied per se but is studied within its context or as it affects the dependent variable - IT as a monolithic black-box or even absent - IT artifacts are conceptualized in multiple ways by management, social and computer scientists - lack of theories on IT artifacts	- IT artifact is included in every IS research in any of its multiple views	- the IT artifact, its context, effects and capabilities	- IT artifact as a software-hardware package with cultural and material properties - IT artifacts are not natural, neutral, universal and given - IT artifacts are embedded in some time, place, discourse and community - IT artifacts involve an arrangement of components - IT artifacts are not fixed, static, or independent from context	- no explicit theory of systems - pluralism and multi-methodology is encouraged
Benbazat & Zmud (2003)	- lack of a core collective identity - errors of inclusion by doing research on non IS issues and of omission by not studying core IS issues	- central character, claimed distinctiveness and claimed temporal continuity (based on the central identify concept for organizations of Albert & Whetten, 1985) - cognitive legitimacy	- the IT artifact (any application of IT to support tasks, embedded in structures and latter in contexts) and its nomological network (IT managerial, technological, methodological, operational and behavioral capabilities) and practices for pre and post core activities of an IT existence	- IT artifacts related with tasks, inserted in structures and latter in contexts	- no explicit theory of systems - interdisciplinary research is encouraged

model (1978). Also, Benbazat and Zmud's (2003) IT artifact and its nomological network can be accommodated in a systemic structure. As Galliers (2004) suggests implicitly, the nomological network should be considered a dynamic rather than static set of concepts. Additional research could extend the inside and outside elements, at-

tributes, events, and interrelationships according their specific purposes. The systems approach provides the methodological tools for this extended analysis. Since Orlikowski and Iacono's (2001) framework of ideas is a subset of Benbazat and Zmud (2003), the previous arguments apply also for this framework.

Table 2. continued

IS Research Paradigm or Framework	Main weaknesses identified in the IS Discipline	Maturity Criteria	Foci for IS Research	Concept of what an Information System is	Underlying System Theories Used
Hirschheim & Klein (2003)	- IS field is fragmented - internal and external communication gap - intellectually rigid and lack of fruitful communication - disagreement about the nature of IS field includes - lack of a shared core set of underlying knowledge - a high strategic task uncertainty	- an accepted IS body of knowledge is available	- the IS body of knowledge	- not defined but implicitly inferred as instruments for process and organizational efficiency and effectiveness	- partial use of Habermas' philosophy - no other systems theory but it is accepted that IS are systems (pp. 282)
Galliers (2004)	- change and new challenges as opportunities for field evolution instead of a crisis status	- a field that evolves dynamically in research focus and boundaries - a trans-disciplinary criteria - practice improved through research - IS interaction with other disciplines	- organizations, individuals and information systems	- an IS is composed of six elements: formal and informal human, computer and external systems (based on Land and Kennedy-McGregory, 1981)	- based on Checkland's soft systems view - trans-disciplinary holistic-systemic approach encouraged - Asbhy's Law of Variety of Requisite - methodological, theoretical and topical diversity and pluralism encouraged
Larsen & Levine (2005)	- lack of cumulative tradition - weak coherence in the field - affected by the current intellectual anxiety on the role of university education and disciplinary knowledge and augmented by the rapid change of IT - a Kuhnian pre-paradigm status - ambiguity, fragmentation and change patterns as most frequent	- Kuhn's field in its own right manifested by an shared exemplar study as base, an image of the subject matter, theories, and methods and instruments or - scientific/intellectual movement umbrella (based on Frickel & Gross, 2005)	- centers of coherence linked to the concepts of system, information, management, process, model, user, research, technology mainly.	(Lee's (1999) definitions supported) - MIS includes an information system and its organizational context - MIS includes information technology and its instantiation - MIS includes the activities of a corporative function	- the system of systems (SoS) concept is subtly endorsed from Systems Engineering discipline

Galliers' (2004) conceptualization for the IS discipline focused on information systems, organizations and their individuals; that is, philosophically supported by the soft/interpretative systemic approach (Checkland, 2000), can be also accommodated in the critical realism stance

Table 3. Systemic integration of main IS research paradigms.

Key Systemic Issues for Integrating IS Research Paradigms	IS discipline toward a well-defined Kuhnian's Paradigm	IS discipline toward a Kuhnian's Evolution of the discipline	IS discipline toward a holistic and critical realism integration
F: *Framework of fundamental concepts*	- a central character manifested by a core set of concepts linked to the IT artifact and nomological network based in contexts, effects and capabilities (inferred from Weber (1987); Orlikowski & Iacono (2001); Benbazat & Zmud (2003))	- a dynamic IS body of knowledge – with centers of coherence varying in time that considers the formal and informal human, organizational and technical components of IS (inferred from Hirschheim & Klein (2003), Galliers (2004) and Larsen & Levine (2005)	- a broad systems view of the field with permanent and shared generic constructs and dynamic and emergent system properties
M: *Methodological research tools*	- monolithic research approach (hard/ quantitative or soft/interpretative tools)	- multi-methodology research approach but limited to some dominant tools/lenses	- real multi-methodology research approach not limited to dominant tools/lenses (all variety of hard/quantitative, soft/interpretative or critical/intervention tools) (inferred from Midgley (1996), Mingers (2000))
A: *Areas/situations of study in the world*	- IT artifact (internal view) and its nomological network based on contexts, effects and capabilities (inferred from Weber (1987); Orlikowski & Iacono (2001); Benbazat & Zmud (2003))	- Information systems, individual, organizations and society and future centers of coherence (inferred from Galliers (2004) and Larsen & Levine (2005))	- Information systems as *systems* (and automatically includes both perspectives) (inferred from Gelman & Garcia (1989) and Mora et al (2003))
Research Strategy	- an imperialist paradigm that integrates other paradigms that strengthen the dominant paradigm (inferred from Weber (1987), Benbazat & Zmud (2003))	- pluralist paradigm under the concept of scientific/intellectual movement umbrellas (inferred from Orlikowski & Iacono (2001), Galliers (2004) and Larsen & Levine (2005). Hirschheim & Klein (2003) holds a neutral position.	- pluralist paradigm with theoretical and practical coherence between different philosophical paradigms (Mingers, 2001, 2002).
Philosophy of Science	Kuhn's imperialist or paradigmatic stance	Post-Kuhnian imperialist or paradigmatic stance	- Critical realism: all reality (natural, designed and social) has real underlying mechanisms and structures (real domain) that generate observable events (empirical domain) as well as non observable events (the actual domain including the empirical one), but social reality (concepts, meanings, categories) are socially built and thereby rely in the existence of human beings and a communication language (a systemic ontology).

by the arguments reported in Table 1. Larsen & Levine's (2005) framework of ideas, based in a subtle concept of *system of systems* and the empirical evidence to keep as center of coherence the concept of information systems with some dynamic concepts, also can be accommodated in the critical realism stance as follows: the underlying mechanisms and structures of the real domain become the permanent center of coherence to be searched and the dynamic elements of knowledge are located in the empirical domain of the events observed. Then, according to the specific research purposes, tools, resources, and schedules, some events generated in the actual domain will be observed in the empirical domain. In this way, the permanent and dynamic central themes are linked to a critical realism view of the world.

Hirschheim and Klein (2003) admit the usefulness of a broad underlying structure for the IS discipline that can organize the fragmented pieces of the IS knowledge in a coherent whole (e.g., a conceptual *system)*. This IS body of knowledge initiative relies on a systemic approach. Furthermore, its philosophical support, based in Habermas' theory of knowledge, links their ideas automatically with the systems intellectual movement.

Hence, we claim that competitive and conflictive perspectives posed for the IS discipline can be dissolved under a critical realism view as articulated in Table 3.

A Systemic Integration of the Information Systems Research Frameworks

Ives, Hamilton, and Davis' study (1980) can be considered the first effort to develop a comprehensive IS research framework. According to Ives et al. (1980), the previous five similar studies were dimensionally incomplete (e.g. capture a partial view of the IS field). These previous studies do not account for the overall processes and environments to develop, operate, and evolve the IS

artifact, are focused on specific types of IS, or omit the organizational environment except by the type of managerial levels related with the IS artifact. Based on Mora et al. (2006), Ives et al.'s (1980) IS research framework contributes to the integration of the disparate dimensions and provides a structured framework to organize and classify IS research. However, Mora et al. (2006) suggest that Ives et al. (1980) do not articulate a correct systemic organization (e.g. a hierarchical definition of <system, subsystems, environment> and the conceptual differentiation of system's outcomes with systems elements in the model), and the concept of <organization> and <organizational environment> are not well differentiated.

A second IS research framework is reported by Nolan and Wetherbe (1980). This framework also draws on the same five past studies analyzed by Ives et al (1980). However, Nolan and Wetherbe build on a more fundamental conceptualization of the theory of systems (Boulding, 1956). As result, the IS research framework is more congruent with the formal concept of *system*. According to Mora et al. (2006), this framework is composed of

a <MIS Technology System> that is part of an < Organization> and it of its < Organizational Environment >. The <MIS Technology system> is conceptualized as a system composed of the following subsystems: <hardware>, <software>, <data base>, <personnel> and <procedures>. In turn, the <Organization>, as the wider system for the <MIS Technology system> is conceptualized in five subsystems: <goals and values>, <psychosocial>, <structural>, <technical> and <managerial>.

Nolan and Wetherbe's contribution can be identified as a more coherent articulation of the main theory of system elements of interest to be studied in the IS discipline. Nonetheless, Mora et al. (2006, p. 3) report the following deficiencies: (1) the outputs of the <MIS Technology system> are only conceptualized in terms of types of IS, omitting other outcomes that it can generate such as <IT audits>, <IT proposal assessments> and <IT services> in general; (2) the model does not conceptualize the interactions between the systems considered as wholes and the systems considered as a set of components—e.g. the system type I and type II views respectively defined in Gelman and García (1989) and updated in Mora, Gelman et al (2003)- and then influences like <IT suppliers>, <IT trends> or the conceptualization of an <Inter-organizational IS> cannot be modeled; and (3) the time dimension that is critical for some of the 33 cases reported—e.g., on system's evolutions- is implicitly assumed and not related with the state $\omega(t)$ of the system, subsystem or environment.

In a third IS framework (Silver & Markus, 1995), the researchers quote an MBA student's claim: "*I understand the pieces but I don't see how to fit together*" (p. 361). Based on Bertlanffy's (1951) ideas and using Ackoff's (1993) recommendations to study a phenomenon from a systems view, the researchers recognize that the study of an IS as a *system* implies the need to identify its supersystem—e.g. its suprasystem—as well as its subsystems. The Silver and Markus' model places the IS as the central object of learning into a supersystem: the organization, and this in its wider environment. In the organization as a *system*, the following elements are identified: firm strategy, business processes, structure and culture, and the IT infrastructure. Also, additional elements for the model are included: IS features and IS implementation process. For each category, a list of sub-elements are also identified. Yet the three levels of systems, suprasystem, and subsystems are inconsistently structured from a formal systemic perspective (Johnson, Kast, & Rosenzweig, 1964). Conceptual categories for subsystems are mixed with system outcomes, actions, and attributes. For example, firm strategy can be categorized as a system's outcome instead of a subsystem, and the IS implementation process is disconnected from the subsystem of business processes.

Furthermore, the initial systemic views for IS and for organization—exhibited in Figures 1 and 2 (ibid, p. 364)—are disconnected from the final model. The IT infrastructure element —viewed as a subsystem of the organization —affects the component, but it is not part of the IS system, and the people component—also an initial subsystem of the organization— is lost or transformed in the structure and culture element. Then the formal utilization of the Systems Approach is incomplete.

In a fourth IS research framework (Bacon & Fitzgerald, 2001), the researchers contrast arguments on the advantages and disadvantages of frameworks and conclude that the potential benefits exceed the potential limitations. The researchers also support empirically the academic and practical need to have and use frameworks for the IS discipline. This evidence is based on a survey of 52 prominent IS individuals from 15 representative countries in North-America, Europe and Oceania. However, given the current philosophical and methodological debate, it is noted that a general IS research framework could be not totally possible but it is highly encouraged to be pursued (Bacon & Fitzgerald, 2001, pp. 51). According to the researchers, previous related studies fail to describe a holistic—e.g.integrated, overall, and systemic (ibid, pp. 47)—view of the discipline.

Through a grounded theory research method and after an extensive review of concepts from the literature, IS syllabus, IS curricula proposals and opinions of IS academicians, Bacon and Fitzgerald induce five categories for IS research framework: (a) information for knowledge work, customer satisfaction, and business performance, (b) IS development, acquisition, and support activities, (c) information and communication technologies, (d) operations and network management activities, and (e) people and organizations issues.

Despite four references to recognized systems researchers (Checkland & Howell, 1995; Mason & Mitroff, 1973; Stowell & Mingers, 1997; van

Gich & Pipino, 1983), no specific systems model or approach is used to structure the conceptual system posed. Then its framework is not systemically articulated. A formal systemic analysis reveals that this systemic model lacks: (a) a coherent set of subsystems, (b) a description of its subsystems as systems, and (c) an environment for the system.

Hence the four comprehensive IS research frameworks posited, despite their theoretical and practical contributions, are incomplete and non-comprehensive from a formal systems-based view. A framework with systemic theoretical soundness that is able to integrate holistically all dimensions considered in past frameworks and the few dimensions omitted is still required. Mora et al. (2006) report a framework that can be useful for such purposes. This systems-founded IS research framework is based on formal definitions of *systems*[7] (Gelman & Garcia, 1989) and formal definitions of *organization, business process* and *information systems* (Gelman et al., 2005; Mora et al., 2003, 2006, in press).

According to Gelman and Garcia (1989), Gelman et al. (2005), and Mora et al. (2003, 2006, in press), to define an object of study as a **system-I** implies to specify it as a whole composed by *attributes, events,* and *domains* for *attributes.* For the case of **system-II**, the formal definition offers

the classic view of a *system* as a set of interrelated components. Furthermore, the definition used here also considers the *output/input* relationships between any *subsystem* and the whole *system.* In turn, the auxiliary definitions — reported in Mora et al. (2006, in press)—help to support the expansionist systemic perspective that indicates that every *system* always belongs to another larger *system* (Ackoff, 1971). Figure 1—borrowed from Mora et al. (in press)—exhibits a graphical interpretation of the *system-II* view. In turn, Figure 2 exhibits a diagram of the concept *organization O(X)* as *system* with its *high-level, low-level,* and *socio-political business processes* as *subsystems.*

Mora et al.'s (2006) framework uses an integrative cybernetic, as well as interpretative socio-political systemic paradigm theoretically coherent through a critical realism stance, where $S_{II}(X.1)$ and $S_{II}(X.2)$ are conceptualized as a *driving-org-subsystem* and a *driven-org-subsystem,* respectively, $S_{II}(X.3) = HLBP(X.3)$ as a *information-org-subsystem,* and $S_{II}(X.4) = SSBP(X.4)$ as a *socio-political-org-subsystem.* The Figure 3[8]—also borrowed from Mora et al. (2006)—exhibits the systemic articulation of the concepts: *organization* and *information systems,* as well as of its wider *systems* and *subsystems.*

Figure 1. A diagram of the multilevel layers of the concept system and related terms

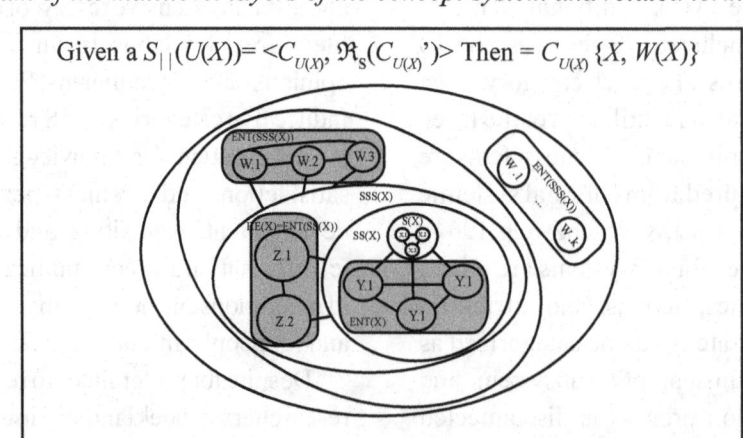

Figure 2. A schematic view of an organization as a system

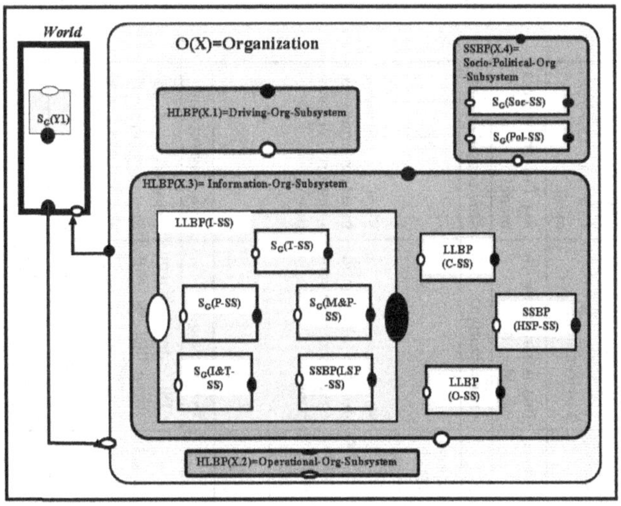

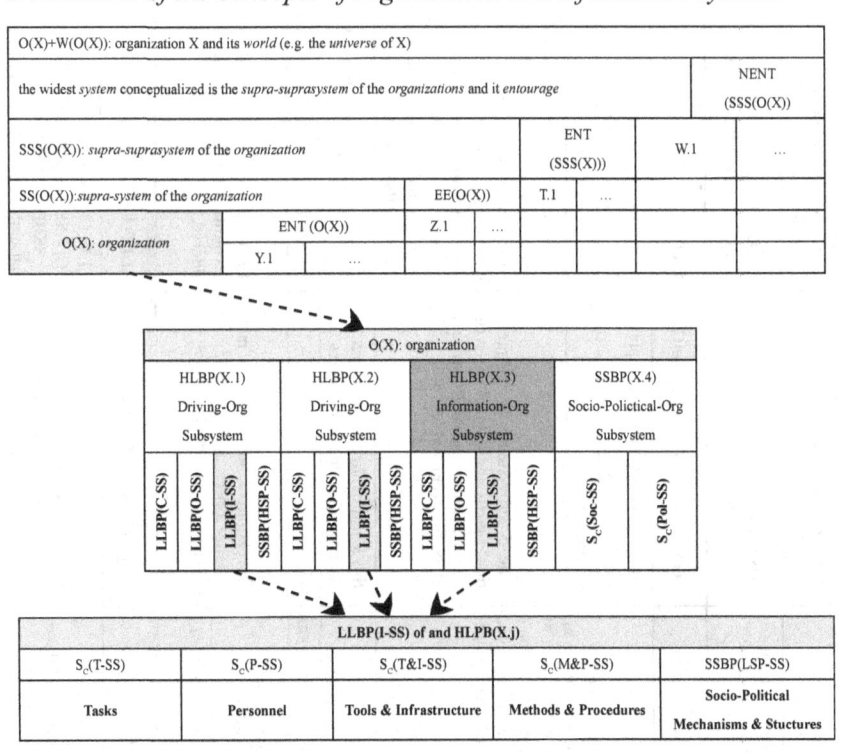

Figure 3. The articulation of the concepts of organization and information systems

Table 4. Systemic map of the concepts for IS Research in the five frameworks

The Critical Integrative IS Research Framework				
Ives, Hamilton & Davis' Framework	Nolan & Wetherbe's Framework	Mora, Gelman, Cano, Cervantes & Forgionne's Framework	Bacon & Fitzgerald's Framework	Silver & Markus' Framework
<external environment> : legal social, cultural, economic, educational, resource and industry/trade systems	this is not explicitly considered	EE(O(X)) : (the systems to be modeled at this level of analysis are determined by the researcher)	this is not explicitly considered	this is not explicitly considered
<organizational environment>	<environment of the organization>	ENT(O(X))	Environment	External Environment
	competitors, government, suppliers, customers, etc.	(the systems to be modeled at this level of analysis are determined by the researcher)	this is not explicitly considered	five Porter's forces model
The concept of <organizational environment> used by Ives et al, does not consider the environment of an organization, but the organization's attributes per se such as: goals, tasks, structure, volatility and management philosophy/style	<organization>	O(X)	<organization>	<the organization>
	The 5 subsystems of <organization>: goals and value SS, psychosocial SS, managerial SS, structural SS and technical SS	Goals & Value SS, psychosocial SS, and structural SS, are considered in the SSBP(X.4) and the SSBP(HSP-SS) for any HLBP(X.j). The managerial and technical SS respectively corresponds to the HLBP(X.1) (e.g. the driving-org subsystem) and to HLBP(X.2) (e.g. the driven-org subsystem). Ives' et al such as: goals, tasks, volatility and management philosophy/style, as well as <organization> issues of others can be also modeled.	<organization & people> are considered in the same component	Firm strategy, business processes, structure & culture
<user environment> + <use process>	this is not explicitly considered but is implicit in the 5 subsystems	Any HLBP(X.j) and SSBP(X.4) of the O(X)		
IS as function is implicit in the framework		HLBP(X.3) = The IS function as system	IS Function	IT infrastructure
<IS development environment> + <IS development process>	Development, operations and maintenance aspects are considered in this concept of <MIS Technology>	LLBP(O-SS) and SSBP(HSP-SS) of the HLBP(X.3) accounts for any IS operation and process to be considered.	<IS development, acquisition & support>	Resources of the firm to generate IT applications and <Implementation Process>
<IS operations environment> + <IS operations process>			<Operations & Network management>	
IS as <properties and effects>	IS as <<MIS Technology>>	LLBP(I-SS) = The IS artifact as system	IS as ICT + Impacts	IS as core element
<Information Subsystem> (attributes of content, presentation, time, etc)	(hardware, software, data base, procedures and personnel)	LLBP(I-SS) of any HLBP(X.j) comprised of the SG(T-SS) +SG(P-SS) + SG(T&I-SS) + SG(M&P-SS) + SSBP(LSP-SS) (e.g. the low-level business process that corresponds to the <informational subsystem> of any high-level business process in an O(X))	<Information for KW, CS & BP> + <ICT>	(hardware, software, procedures, data and people)

Finally, Table 4 exhibits a mapping of the concepts posited in the previous four frameworks onto the systemic framework. It can be inferred from the formal definitions of *system-I, system-II,* and *general system, organization, suprasystem, supra-suprasystem, envelop, entourage* and *world, high-level process, low-level high process, socio-political process,* and *information systems* that previous frameworks are systemically incomplete.

CONCLUSION

We have reviewed the main IS research paradigms and frameworks reported in the IS literature by using a Systems Approach. This review has identified that previous studies have been developed using no, informal or few, systemic concepts from the formal spectrum of systemic concepts developed by the systems approach intellectual movement. Then, through the acceptance of a critical realist view, an IS research framework has been developed to integrate theoretically these disparate and conflictive views of IS as objects of study as well as a discipline.

We claim that this systemic framework: (1) is congruent with formal definitions of *system;* (2) permits the modeling of all variables reported in previous IS research frameworks as sub-systems or attributes and relationships of sub-systems as well as of the wider systems; (3) permits the study of static or dynamic IS phenomenon through the consideration of the concept of state of the *system*; (4) integrates theoretically the different positivist, interpretative, and emancipative paradigms through a critical realism stance; and (5) provides a systemic-holistic backbone and main ramifications to start the building of the required IS BoK.

We admit that this framework must be considered a research starting point rather than an end point in the long-term aim to have a non-fragmented discipline with a strong cumulative tradition. Davis' (1974) seminal ideas for IS were related with general systems theory. Back to the basics could be useful to coherently organize the discipline.

REFERENCES

Ackoff, R. (1960). Systems, organizations and interdisciplinary research. *General System Yearbook, 5,* 1-8.

Ackoff, R., Gupta, S., & Minas, J. (1962). *Scientific method: Optimizing applied research decisions.* New York: Wiley.

Ackoff, R. (1971). Towards a system of systems concepts. *Management Science, 17*(11), 661-671.

Ackoff, R. (1973). Science in the systems age: Beyond IE, OR and MS. *Operations Research, 21*(3), 661-671.

Ackoff, R. (1981). The art and science of mess management. *Interfaces, 11*(1), 20-26.

Ackoff, R. (1993, November). From mechanistic to social systems thinking. In *Proceedings of Systems Thinking Action Conference*, Cambridge, MA.

Adam, F., & Fitzgerald, B. (2000). The status of the information systems field: Historical perspective and practical orientation. *Information Research, 5*(4), 1-16.

Aldrich, H. (1999). *Organizations evolving.* Thousand Oaks, CA: Sage.

Alter, S. (2001). Are the fundamental concepts of information systems mostly about work systems? *CAIS, 5*(11), 1-67.

Alter, S. (2003). 18 reasons why IT-reliant work systems should replace "The IT Artifact" as the core subject matter of the IS field. *CAIS, 12*(23), 366-395.

Bacon, J., & Fitzgerald, B. (2001) A systemic Framework for the Field of Informaiton Systems. *The DATA BASE for Adavances in Informaion Systems, 32(2)*, 46-67.

Banville, C., & Landry, M. (1989). Can the field of MIS be disciplined? *Communications of the ACM, 32*(1), 48-60.

Barkhi, R., & Sheetz, S. (2001). The state of theoretical diversity of information systems. *CAIS, 7*(6), 1-19.

Benbazat, I., & Zmud, R. (2003). The crisis identity within the IS discipline: Defining and communicating the discipline's core properties. *MIS Quarterly, 27*(2), 187-194.

Beer, S. (1966). *Decision and control.* Chichester: Wiley.

Bertalanffy, L. von. (1950). An outline of general systems theory. *British Journal of the Philosophy of Science, 1*, 134-164 (reprinted in Bertalanffy (1968)).

Bertalanffy, L. von. (1951). The theory of open systems in physics and biology. *Science, 111*, 23-29.

Bertalanffy, L. von. (1968). *General systems theory–foundations, developments, applications.* New York: G. Brazillier.

Bertalanffy, L. von. (1972). The history and status of general systems theory. *Academy of Management Journal, December*, 407-426.

Bhaskar, R. (1975). *A realist theory of science.* Sussex: Harvester Press.

Boulding, K. (1956). General systems theory–the skeleton of the science. *Management Science, 2*(3), 197-208.

Carlsson, S. (2003, June 16-21). Critical realism: A way forward in IS research. In *Proceedings of the ECIS 2003 Conference* Naples, Italy.

Checkland, P. (1981). *Systems thinking, systems practice.* Chichester: John Wiley.

Checkland, P. (1983). O.R. and the systems movement: mappings and conflicts. *Journal of the Operational Research Society, 34*(8), pp. 661-675.

Checkland, P. (2000). Soft systems methodology: A thirty year retrospective. *Systems Research and Behavioral Science, 17*, S11–S58.

Checkland, P., & Holwell, S. (1995). Information systems: What's the big idea?. *Systemist, 7*(1), 7-13.

Davis, G. (1974). *Management information systems: Conceptual foundations, structure and development.* New York: McGraw-Hill.

Dobson, P. (2001). The philosophy of critical realism -- An opportunity for information systems research. *Information Systems Frontiers, 3*(2), 199-210.

Dobson, P. (2002). Critical realism and information systems research: Why bother with philosophy? *Information Systems Research, 7*(2). Retrieved from http://InformationR.net/ir/7-2/paper124,html.

Dobson, P. (2003). The SoSM revisited – A critical realist perspective. In Cano, J. (Ed). *Critical reflections of information systems: A systemic approach* (pp. 122-135). Hershey, PA: Idea Group Publishing.

Farhoomand, A. (1987). Scientific progress of management information systems. *Database, Summer*, 48-57.

Farhoomand, A., & Drury, D. (2001). Diversity and scientific progress in the information systems discipline. *CAIS, 5*(12), 1-22.

Flood, R., & Jackson, M. (1991). *Creative problem solving: Total systems intervention.* New York: Wiley.

Flood, R., & Romm, N. (Eds). (1996) *Critical Systems Thinking*. New York: Plenum Press.

Forrester, J. (1958). Industrial dynamics – A major breakthrough for decision makers. *Harvard Business Review, 36*, 37-66.

Forrester, J. (1991). *Systems dynamics and the lessons of 35 years*. (Tech. Rep. D-4224-4). Retrieved from http://sysdyn.mit.edu/sd-group/home.html

Frickel, S., & Gross, N. (2005). A general theory of scientific/intellectual movements. *American Sociological Review, 70*, 204-232.

Galliers, R. (2004). Change as crisis or growth? Toward a trans-disciplinary view of information systems as a field of study: A response to Benbasat and Zmud's call for returning to the IT artifact. *JAIS, 4*(7), 337-351.

Gellner, E. (1993). What do we need now? Social anthropology and its new global context. *The Times Literary Supplement, 16*, July, 3-4.

Gelman, O., & Garcia, J. (1989). Formulation and axiomatization of the concept of general system. *Outlet IMPOS (Mexican Institute of Planning and Systems Operation), 19*(92), 1-81.

Gelman, O., Mora, M., Forgionne, G., & Cervantes, F. (2005). M. Khosrow-Pour (Ed.) Information Systems and Systems Theory. In *Encyclopedia of Information Science and Technology* (1491-1496). Hershey, PA: IGR.

Gigch, van J., & Pipino, L. (1986). In search of a paradigm for the discipline of information systems. *Future Computer Systems, 1*(1), 71-97.

Gregory, W. (1996). Dealing with diversity. In R. Flood & N. Romm (Eds.), *Critical Systems Thinking* (37-61). New York: Plenum Press.

Habermas, J. (1978). Knowledge and human interests. London: Heinemann.

Hirschheim, R., & Klein, H. (2003). Crisis in the IS field? A critical reflection on the state of the discipline. *JAIS, 4*(10), 237-293.

Hoffer, J., George, J., & Valachi, J. (1996). *Modern systems analysis and design*. Menlo Park, CA: Benjamin/Cummings.

Ives, B., Hamilton, S., & Davis, G. (1980). A framework for research in computer-based management information systems. *Management Science, 26*(9), 910-934.

Jackson, M. (2000). *Systems approaches to management*. New York: Kluwer.

Johnson, R., Kast, F., & Rosenzweig, J. (1964). Systems theory and management. *Management Science, 10*(2), 367-384.

Klir, G. (1969). *An approach to general systems theory*. New York: Van Nostrand.

Kuhn, T. (1970). *The structure of scientific revolutions*. Chicago: Chicago University Press.

Land, F., & Kennedy-McGregor, M. (1987). Information and information systems: Concepts and perspectives. In R. Galliers (Ed.), *Information analysis: Selected readings* (pp. 63-91). Sydney: Wesely.

Larson, T., & Levine, J. (2005). Serching for Management Information Systems: Coherence and Change in the Discipline. *Information Systems Journal, 15*, pp. 357-381.

Lazlo, E., & Krippner, S. (1998). In J.S. Jordan (Ed.), *Systems theories and a priori aspects of perception* (47-74). Amsterdam: Elsevier Science.

Lazlo, E., & Lazlo, A. (1997). The contribution of the systems sciences to the humanities. *Systems Research & Behavioral Science, 14*(1), 5-19.

Leavitt, H., & Whisler, T. (1958). Management in the 80's. *Harvard Business Review, 36*(6), 41-48.

Mason, R., & Mitroff, I. (1973). A program of research on MIS. *Management Science, 19*(5), 475-485.

Midgley, G. (1996). What is this thing called CST? In R. Flood & N. Romm (Eds.), *Critical systems thinking* (11-24). New York: Plenum Press.

Miller, J. (1978). *Living systems.* New York: McGraw-Hill.

Mingers, J. (2000). The contributions of critical realism as an underpinning philosophy for OR/MS and systems. *Journal of the Operational Research Society 51,*1256-1270.

Mingers, J. (2001). Combining IS research methods: Towards a pluralist methodology. *Information Systems Research, 12*(3), 240-253.

Mingers, J. (2002). Realizing information systems: Critical realism as an underpinning philosophy for information systems. In *Proceedings of the 23rd International Conference in Information Systems, 295-303.*

Mingers, J. (2003). A classification of the philosophical assumptions of management science methodologies. *Journal of the Operational Research Society, 54*(6), 559-570.

Monarch, I. (2000). Information science and information systems: Converging or diverging? In *Proceedings of the 28th Annual Conference of the Canadian Association in Information Systems,* Alberta.

Mora, M., Gelman, O., Cano, J., Cervantes, F., & Forgionne, G. (2006, July 9-14). Theory of systems and information systems research frameworks. In *Proceedings of the International Society for the Systems Sciences 50th Annual Conference,* Rohnert Park, CA.

Mora, M., Gelman, O., Cervantes, F., Mejia, M., & Weitzenfeld, A. (2003). A systemic approach for the formalization of the information system concept: why information systems are systems? In J. Cano (Ed), *Critical reflections of information systems: A systemic approach* (1-29). Hershey, PA: Idea Group Publishing.

Mora, M., Gelman, O., Forgionne, G., & Cervantes, F. (2004, May 19-21). Integrating the soft and the hard systems approaches: A critical realism based methodology for studying soft systems dynamics (CRM-SSD). In *Proceedings of the 3rd. International Conference on Systems Thinking in Management (ICSTM 2004),* Philadelphia, PA.

Mora, M., Gelman, O., Forgionne, G., & Cervantes, F. (in press). Information Systems: a systemic view. M. Khosrow-Pour (Ed.) In *Encyclopedia of information science and technology, 2nd ed.* Hershey, PA: IGR.

Nolan, R., & Wetherbe J. (1980). Toward a comprehensive framework for MIS research. *MIS Quarterly, June,* 1, 1-20.

Orlikowski, W., & Iacono, S. (2001). Desperately Seeking the IT in IT research. *Information Systems Research, 7*(4), 400-408.

Rapoport, A. (1968). Systems Analysis: General systems theory. *International Encyclopedia of the Social Sciences, 14,* 452-458.

Silver, M., Markus, M., & Beath, C. (1995) The Information Techonlogy Interation Model: a Foundation of the MBA Course. *MIS Quarterly, 19(3),* pp. 361-369.

Stowell, F., & Mingers, J. (1997). Introduction. In J. Mingers, & F. Stowell (Eds.), *Information systems: An emerging discipline?* London: McGraw-Hill.

Vickers, G. (1965). *The art of judgment.* London: Chapman & Hall.

Wand, Y., & Weber, R. (1990). An ontological model of an information system. *IEEE Transactions on Software Engineering, 16*(11), 1282-1292.

Weber, R. (1987). Toward a theory of artifacts: A paradigmatic basis for information systems research. *Journal of Information Systems, 2*, 3-19.

ENDNOTES

[1]   Such as: (1) a vast set of undergraduate, master, and doctoral programs; (2) a network of research centers focused on IS topics; (3) 100 relevant specialized conferences and journals; and (4) the existence of professional and academic associations.

[2]   According to Lazlo and Lazlo (1997), Bertalanffy's ideas were also influenced by the mathematician Alfred Whitehead and the also biologist Paul Weiss.

[3]   Ackoff (1973) describes the machine age as useful for some kind of problems but not sufficient for studying the complex phenomena of the present age, hence the emergence of a systems age.

[4a]   Bhaskar (1975, p. 30) explains that "*it is not the character of science that imposes a determinate pattern or order in the world; but the order of the world that, under certain determinate conditions, makes possible the cluster of activities we call science.*"

[4b]   For Bhaskar (1975), the reality exists per se independently of the existence of human beings: "*a law governed world independently of man*" p. 26. However, the social structures and mechanisms are conditioned to the existence of human beings at first and then these have a real existence that can be studied and intervened.

[5]   Midgley reports his interpretation from Flood and Jackson 's ideas (1991) on Habermas' theory of knowledge.

[6]   The emancipative Habermas' third purpose is not considered by the authors.

[7]   Gelman and Garcia analyzed the formal definitions of the concept *system* from Ackoff, Arbib, Bertalanffy, Kalman, Lange, Mesarovic, Rapoport and Zadeh.

[8]   Interactions between subsystems are not diagrammed.

This work was previously published in Information Resources Management Journal, Vol. 20, Issue 2, edited by M. Khosrow-Pour, pp. 1-22, copyright 2007 by IGI Publishing (an imprint of IGI Global).

Chapter 12
System-of-Systems Cost Estimation:
Analysis of Lead System Integrator Engineering Activities

Jo Ann Lane
University of Southern California, USA

Barry Boehm
University of Southern California, USA

ABSTRACT

As organizations strive to expand system capabilities through the development of system-of-systems (SoS) architectures, they want to know "how much effort" and "how long" to implement the SoS. In order to answer these questions, it is important to first understand the types of activities performed in SoS architecture development and integration and how these vary across different SoS implementations. This article provides results of research conducted to determine types of SoS lead system integrator (LSI) activities and how these differ from the more traditional system engineering activities described in Electronic Industries Alliance (EIA) 632 ("Processes for Engineering a System"). This research further analyzed effort and schedule issues on "very large" SoS programs to more clearly identify and profile the types of activities performed by the typical LSI and to determine organizational characteristics that significantly impact overall success and productivity of the LSI effort. The results of this effort have been captured in a reduced-parameter version of the constructive SoS integration cost model (COSOSIMO) that estimates LSI SoS engineering (SoSE) effort.

INTRODUCTION

As organizations strive to expand system capabilities through the development of system-of-systems (SoS) architectures, they want to know "how much effort" and "how long" to implement the SoS.

Efforts are currently underway at the University of Southern California (USC) Center for Systems and Software Engineering (CSSE) to develop a cost model to estimate the effort associated with SoS lead system integrator (LSI) activities. The research described in this article is in support of the development of this cost model, the constructive SoS integration cost model (COSOSIMO). Research conducted to date in this area has focused more on technical characteristics of the SoS. However, feedback from USC CSSE industry affiliates indicates that the extreme complexity typically associated with SoS architectures and political issues between participating organizations have a major impact on the LSI effort. This is also supported by surveys of system acquisition managers (Blanchette, 2005) and studies of failed programs (Pressman & Wildavsky, 1973). The focus of this current research is to further investigate effort and schedule issues on "very large" SoS programs and to determine key activities in the development of SoSs and organizational characteristics that significantly impact overall success and productivity of the program.

This article first describes the context for the COSOSIMO cost model, then presents a conceptual view of the cost model that has been developed using expert judgment, describes the methodology being used to develop the model, and summarizes conclusions reached to date.

COSOSIMO CONTEXT

We are seeing a growing trend in industry and the government agencies to "quickly" incorporate new technologies and expand the capabilities of legacy systems by integrating them with other legacy systems, commercial-off-the-shelf (COTS) products, and new systems into a system of systems, generally with the intent to share information from related systems and to create new, emergent capabilities that are not possible with the existing stove-piped systems. With this development approach, we see new activities being performed to define the new architecture, identify sources to either supply or develop the required components, and then to integrate and test these high level components. Along with this

"system-of-systems" development approach, we have seen a new role in the development process evolve to perform these activities: that of the LSI. A recent Air Force study (United States Air Force Scientific Advisory Board, 2005) clearly states that the SoS Engineering (SoSE) effort and focus related to LSI activities is considerably different from the more traditional system development projects. According to this report, key areas where LSI activities are more complex or different than traditional systems engineering are the system architecting, especially in the areas of system interoperability and system "ilities;" acquisition and management; and anticipation of needs.

Key to developing a cost model such as CO-SOSIMO is understanding what a "system-of-systems" is. Early literature research (Jamshidi, 2005) showed that the term "system-of-systems" can mean many things across different organizations. For the purposes of the COSOSIMO cost model development, the research team has focused on the SoS definitions provided in Maier (1999) and Sage and Cuppan (2001): an evolutionary net-centric architecture that allows geographically distributed component systems to exchange information and perform tasks within the framework that they are not capable of performing on their own outside of the framework. This is often referred to as "emergent behaviors." Key issues in developing an SoS are the security of information shared between the various component systems, how to get the right information to the right destinations efficiently without overwhelming users with unnecessary or obsolete information, and how to maintain dynamic networks so that component system "nodes" can enter and leave the SoS.

Today, there are fairly mature tools to support the estimation of the effort and schedule associated with the lower-level SoS component systems (Boehm, Valerdi, Lane, & Brown 2005). However, none of these models supports the estimation of LSI SoSE activities. COSOSIMO, shown in Figure 1, is a parametric model currently under develop-

Figure 1. COSOSIMO model structure

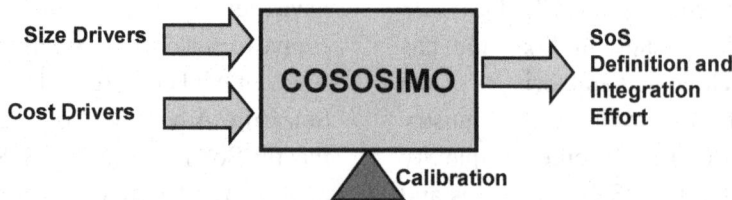

ment to compute just this effort. The goal is to support activities for estimating the LSI effort in a way that allows users to develop initial estimates and then conduct tradeoffs based on architecture and development process alternatives.

Recent LSI research conducted by reviewing LSI statements of work identifies the following typical LSI activities:

- Concurrent engineering of requirements, architecture, and plans
- Identification and evaluation of technologies to be integrated
- Source selection of vendors and suppliers
- Management and coordination of supplier activities

- Validation and feasibility assessment of SoS architecture
- Continual integration and test of SoS-level capabilities
- SoS-level implementation planning, preparation, and execution
- On-going change management at the SoS level and across the SoS-related integrated product teams to support the evolution of requirements, interfaces and technology.

With the addition of this new cost model to the constructive cost model (COCOMO) suite of cost models, one can easily develop more comprehensive estimates for the total SoS development, as shown in Figure 2.

Figure 2. Architecture-based SoS cost estimation

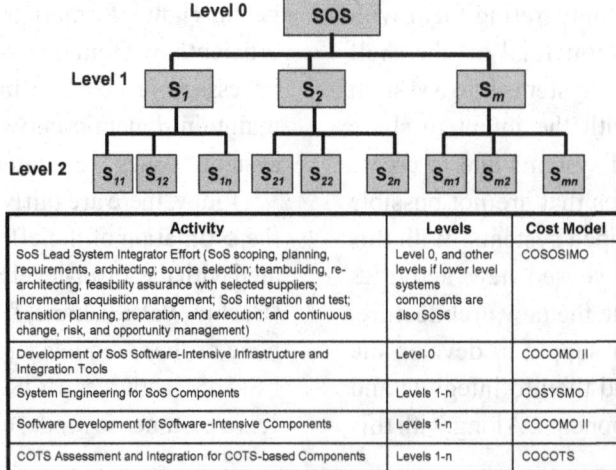

Activity	Levels	Cost Model
SoS Lead System Integrator Effort (SoS scoping, planning, requirements, architecting; source selection; teambuilding, re-architecting, feasibility assurance with selected suppliers; incremental acquisition management; SoS integration and test; transition planning, preparation, and execution; and continuous change, risk, and opportunity management)	Level 0, and other levels if lower level systems components are also SoSs	COSOSIMO
Development of SoS Software-Intensive Infrastructure and Integration Tools	Level 0	COCOMO II
System Engineering for SoS Components	Levels 1-n	COSYSMO
Software Development for Software-Intensive Components	Levels 1-n	COCOMO II
COTS Assessment and Integration for COTS-based Components	Levels 1-n	COCOTS

LSI EFFORT ESTIMATION APPROACH

As mentioned above, key to an LSI effort estimation model is having a clear understanding of the SoSE activities performed by the organization as well as which activities require the most effort. In addition, it is important to understand how these SoSE activities differ from the more traditional systems engineering activities. Analysis presented in Lane (2005) describes how the typical LSI SoSE activities differ from the more traditional system engineering activities identified in EIA 632 (Electronic Industries Alliance, 1999) and the Software Engineering Institute (SEI) Capability Maturity Model Integration (CMMI) (Software Engineering Institute, 2001). Subsequently, Delphi surveys conducted with USC CSSE industry affiliates have identified key size drivers and cost drivers for LSI effort and are shown in Table 1.

Because there are concerns about the availability of effort data from a sufficient number of SoS programs to support model calibration and validation, current efforts are focusing on defining a "reduced parameter set" cost model or ways to estimate parts of the LSI effort using fewer, but more specific, parameters. The following paragraphs present the results of this recent research.

Further observations of LSI organizations indicate that the LSI activities can be grouped into three areas: 1) planning, requirements management, and architecting (PRA), 2) source selection and supplier oversight (SS), and 3) SoS integration and testing (I&T). There are typically different parts of the LSI organization that are responsible for these three areas. Figure 3 illustrates, conceptually, how efforts for these three areas are distributed across the SoS development life cycle phases of inception, elaboration, construction, and transition for a given increment or evolution of SoS development.

Planning, requirements, and architecting begin early in the life cycle. As the requirements are refined and the SoS architecture is defined and matured, source selection activities can begin to identify component system vendors and to issue contracts to incorporate the necessary SoS-enabling capabilities. With a mature SoS architecture and the identification of a set of component systems for the current increment, the integration team can begin the integration and test planning activities. Once an area ramps up, it continues through the transition phase at some nominal level to ensure as smooth a transition as possible and to capture lessons learned to support activities and plans for the next increment. Boehm and Lane (2006) describe how some of these activities directly support the current plan-driven SoS development

Table 1. COSOSIMO cost model parameters

Size Drivers	Cost Drivers
• # SoS-related requirements • # SoS interface protocols • # independent component system organizations • # SoS scenarios • # unique component systems	• Requirements understanding • Architecture maturity • Level of service requirements • Stakeholder team cohesion • SoS team capability • Maturity of LSI processes • Tool support • Cost/schedule compatibility • SoS Risk Resolution • Component system maturity and stability • Component system readiness

Figure 3. Conceptual LSI effort profile

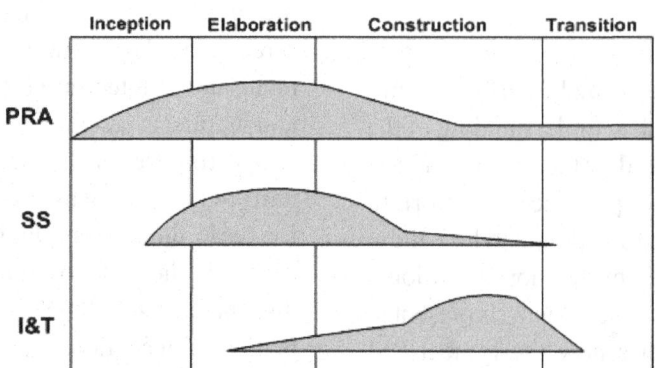

effort while others are more agile, forward looking, trying to anticipate and resolve problems before they become huge impacts. The goal is to stabilize development for the current increment while deferring as much change as possible to future increments. For example, the planning/requirements/architecture group continues to manage the requirements change traffic that seems to be so common in these large systems, only applying those changes to the current increment that are absolutely necessary and deferring the rest to future increments. The architecture team also monitors current increment activities in order to make necessary adjustments to the architecture to handle cross-cutting technology issues that arise during the component system supplier construction activities. Likewise, the supplier oversight group continues to monitor the suppliers for risks, cost, and schedule issues that arise out of SoS conflicts with the component system stakeholder needs and desires. As the effort ramps down in the transition phase, efforts are typically ramping up for the next increment or evolution.

By decomposing the COSOSIMO cost model into three components that correspond to the three primary areas of LSI SoSE effort, the parameter set for each COSOSIMO component can be reduced from the full set and the applicable cost drivers made more specific to the target area. Table 2 shows the resulting set of size and cost drivers for each of the three primary areas. This approach allows

the model developers to calibrate and validate the model components with fewer parameters and data sets. It also allows the collection of data sets from organizations that are only responsible for a part of the LSI SoSE activities. Finally, this approach to LSI SoSE effort estimation allows the cost model to provide estimates for the three areas, as well as a total estimate—a key request from USC CSSE industry affiliates supporting this research effort.

Detailed definitions and proposed ratings for these parameters may be found in Lane (2006). The following provides a brief description of each of the COSOSIMO parameters. Note that several of the COSOSIMO parameters are similar to those defined for the constructive systems engineering cost model (COSYSMO) and are identified in the descriptions below.

COSOSIMO Size Drivers

Number of SoS-Related Requirements[1]

This driver represents the number of requirements for the SoS of interest at the SoS level. Requirements may be functional, performance, feature, or service-oriented in nature depending on the methodology used for specification. They may also be defined by the customer or contractor. SoS requirements can typically be quantified by counting the number of applicable shalls,

Table 2. COSOSIMO parameters by SoSE area

COSOSIMO Component	Associated Size Drivers	Associated Cost Drivers
PRA	• # SoS-related requirements • # SoS interface protocols	• Requirements understanding • Level of service requirements • Stakeholder team cohesion • SoS PRA capability • Maturity of LSI PRA processes • PRA tool support • Cost/schedule compatibility with PRA processes • SoS PRA risk resolution
SS	• # independent component system organizations	• Requirements understanding • Architecture maturity • Level of service requirements • SoS SS capability • Maturity of LSI SS processes • SS tool support • Cost/schedule compatibility with SS activities • SoS SS risk resolution
I&T	• # SoS interface protocols • # SoS scenarios • # unique component systems	• Requirements understanding • Architecture maturity • Level of service requirements • SoS I&T capability • Maturity of LSI I&T processes • I&T tool support • Cost/schedule compatibility with I&T activities • SoS I&T risk resolution • Component system maturity and stability • Component system readiness

wills, shoulds, and mays in the SoS or marketing specification. Note: Some work may be required to decompose requirements to a consistent level so that they may be counted accurately for the appropriate SoS-of-interest.

Number of SoS Interface Protocols

The number of distinct net-centric interface protocols to be provided/supported by the SoS framework. Note: This does NOT include interfaces internal to the SoS component systems, but it does include interfaces external to the SoS and between the SoS component systems. Also note that this is not a count of total interfaces (in many SoSs, the total number of interfaces may be very dynamic as component systems come and go in the SoS environment —in addition, there may be multiple instances of a given type of component system), but rather a count of distinct protocols at the SoS level.

Number of Independent Component System Organizations

The number of organizations managed by the LSI that are providing SoS component systems.

Number of Operational Scenarios[1]

This driver represents the number of operational scenarios that an SoS must satisfy. Such scenarios include both the nominal stimulus-response thread plus all of the off-nominal threads resulting from bad or missing data, unavailable processes, network connections, or other exception-handling cases. The number of scenarios can typically be quantified by counting the number of SoS states, modes, and configurations defined in the SoS concept of operations or by counting the number of "sea-level" use cases (Cockburn, 2001), including off-nominal extensions, developed as part of the operational architecture.

Number of Unique Component Systems

The number of types of component systems that are planned to operate within the SoS framework. If there are multiple versions of a given type that have different interfaces, then the different versions should also be included in the count of component systems.

COSOSIMO Cost Drivers

Requirements Understanding[1]

This cost driver rates the level of understanding of the SoS requirements by all of the affected organizations. For the PRA sub-model, it includes the PRA team as well as the SoS customers and sponsors, SoS PRA team members, component system owners, users, and so forth. For the SS sub-model, it is the understanding level between the LSI and the component system suppliers/vendors. For the I&T sub-model, it is the level of understanding between all of the SoS stakeholders with emphasis on the SoS I&T team members.

Level of Service Requirements[1]

This cost driver rates the difficulty and criticality of satisfying the ensemble of level of service requirements or key performance parameters (KPPs), such as security, safety, transaction speed, communication latency, interoperability, flexibility/adaptability, and reliability. This parameter should be evaluated with respect to the scope of the sub-model to which it pertains.

Team Cohesion[1]

Represents a multi-attribute parameter, which includes leadership, shared vision, diversity of stakeholders, approval cycles, group dynamics, integrated product team (IPT) framework, team dynamics, trust, and amount of change in responsibilities. It further represents the heterogeneity in stakeholder community of the end users, customers, implementers, and development team. For each sub-model, this parameter should be evaluated with respect to the appropriate LSI team (e.g., PRA, SS, or I&T).

Team Capability

Represents the anticipated level of team cooperation and cohesion, personnel capability, and continuity, as well as LSI personnel experience with the relevant domains, applications, language, and tools. For each sub-model, this parameter should be evaluated with respect to the appropriate LSI team (e.g., PRA, SS, or I&T).

Process Maturity

A parameter that rates the maturity level and completeness of the LSI's processes and plans. For each sub-model, this parameter should be evaluated with respect to the appropriate LSI team processes (e.g., PRA, SS, or I&T).

Tool Support[1]

Indicates the coverage, integration, and maturity of the tools in the SoS engineering and management environments. For each sub-model, this parameter should be evaluated with respect to the tool support available to appropriate LSI team (e.g., PRA, SS, or I&T).

Cost/Schedule Compatibility

The extent of business or political pressures to reduce the cost and schedule associated with the LSI's activities and processes. For each sub-model, this parameter should be evaluated with respect to the cost/schedule compatibility for appropriate LSI team activities (e.g., PRA, SS, or I&T).

Risk Resolution

A multi-attribute parameter that represents the number of major SoS/LSI risk items, the maturity of the associated risk management and mitigation plan, compatibility of schedules and budgets, expert availability, tool support, and level of uncertainty in the risk areas. For each sub-model, this parameter should be evaluated with respect to the risk resolution activities for the associated LSI team (e.g., PRA, SS, or I&T).

Architecture Maturity

A parameter that represents the level of maturity of the SoS architecture. It includes the level of detail of the interface protocols and the level of understanding of the performance of the protocols in the SoS framework. Two COSOSIMO sub-models use this parameter, and it should be evaluated in each case with respect to the LSI activities covered by the sub-model of interest.

Component System Maturity and Stability

A multi-attribute parameter that indicates the maturity level of the component systems (number of new component systems versus number of component systems currently operational in

other environments), overall compatibility of the component systems with each other and the SoS interface protocols, the number of major component system changes being implemented in parallel with the SoS framework changes, and the anticipated change in the component systems during SoS integration activities.

Component System Readiness

This indicates readiness of component systems for integration. User evaluates level of verification and validation (V&V) that has/will be performed prior to integration and the level of subsystem integration activities that will be performed prior to integration into the SoS integration lab.

COSOSIMO COST MODEL DEVELOPMENT METHODOLOGY

The COSOSIMO cost model is being developed using the proven cost model development methodology developed over the last several years at the USC CSSE. This methodology, described in (Boehm, Abts, Brown, Chulani, Clark, & et al., 2002), is illustrated in Figure 4.

For COSOSIMO, the *literature review* has focused on the definitions of SoSs and SoSE; the role and scope of activities typically performed

Figure 4. USC CSE cost model development methodology

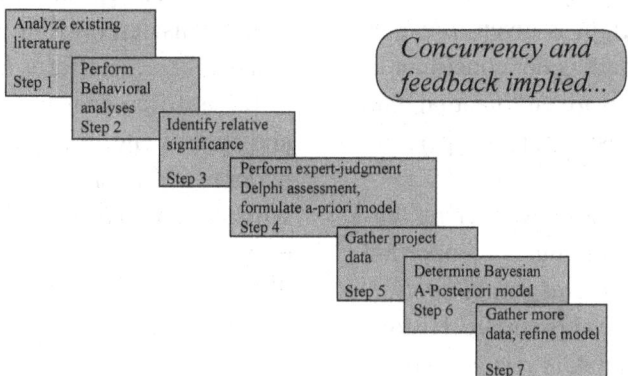

by LSIs; and analysis of cost factors used in related software, systems engineering, and COTS integration cost models, as well as related system dynamics models that investigate candidate SoSE cost factors.

The *behavioral analyses* determine the potential range of values for the candidate cost drivers and the relative impact that each has on the overall effort associated with the relevant SOSE activities. For example, if the stakeholder team cohesion is very high, what is the impact on the PRA effort? Likewise, if the stakeholder team cohesion is very low, what is the resulting impact on PRA effort? The results of the behavioral analyses are then used to *develop a preliminary model form.* The parameters include a set of one or more size drivers, a set of exponential scale factors, and a set of effort multipliers. Cost drivers that are related to economies/diseconomies of scale as size is increased are combined into an exponential factor. Other cost drivers that have a more linear behavior with respect to size drivers are combined into an effort multiplier.

Next, the model parameters, definitions, range of values, rating scales, and behaviors are reviewed with industry and research experts using *a wideband Delphi process.* The consensus of the experts is used to update the preliminary model. In addition to expert judgement, *actual effort data* is collected from successful projects covering the LSI activities of interest. A second model, based on actual data fitting, is then developed. Finally, the expert judgment and actual data *models are combined using Bayesian techniques.* In this process, more weight is given to expert judgement when actual data is not consistent or sparse, and more weight is given to actual data when the data is fairly consistent and experts do not strongly agree.

Since technologies and engineering approaches are constantly evolving, it is important to *continue data collection and model analysis* and update the model when appropriate. Historically, this has led to parameters related to older technologies being dropped and new parameters added. In the case of COSOSIMO, it will be important to track the evolution of SoS architectures and integration approaches and the development of convergence protocols.

For COSOSIMO, each of the sub-models will go through this development process. Once the sub-models are calibrated and validated, they may be combined to estimate the total LSI effort for a proposed SoS development program. To date, several expert judgment surveys have been conducted and actual data collection is in process.

CONCLUSION

LSI organizations are realizing that if more traditional processes are used to architect and integrate SoSs, it will take too long and too much effort to find optimal solutions and build them. Preliminary analysis of LSI activities show that while many of the LSI activities are similar to those described in EIA 632 and the SEI's CMMI, LSIs are identifying ways to combine agile processes with traditional processes to increase concurrency, reduce risk, and further compress overall schedules. In addition, effort profiles for the key LSI activities (the up-front effort associated with SoS abstraction, architecting, source selection, systems acquisition, and supplier and vendor oversight during development, as well as the effort associated with the later activities of integration, test, and change management) show that the percentage of time spent on key activities differs considerably from the more traditional system engineering efforts. By capturing the effects of these differences in organizational structure and system engineering processes in a reduced parameter version of COSOSIMO, management will have a tool that will better predict LSI SoSE effort and to conduct "what if" comparisons of different development strategies.

REFERENCES

Blanchette, S. (2005). *U.S. Army acquisition – The program executive officer perspective*, (Special Report CMU/SEI-2005-SR-002). Pittsburgh, PA: Software Engineering Institute.

Boehm, B., Abt, C., Brown, A., Chulani, S., Clark, & et al. (2000). *Software cost estimation with COCOMO II*. Upper Saddle River, NJ: Prentice Hall.

Boehm, B., Valerdi, R., Lane, J., & Brown, A. (2005). COCOMO suite methodology and evolution. *CrossTalk, 18*(4), 20-25.

Boehm, B., & Lane, J. (2006). 21st century processes for acquiring 21st century systems of systems. *CrossTalk, 19*(5), 4-9.

Cockburn, A. (2001). *Writing effective use cases*. Boston: Addison-Wesley.

Electronic Industries Alliance. (1999). EIA Standard 632: Processes for engineering a system.

Jamshidi, M. (2005). System-of-systems engineering - A definition. *Proceedings of IEEE System, Man, and Cybernetics (SMC) Conference*. Retrieved January 29, 2005 from http://ieeesmc2005.unm.edu/SoSE_Defn.htm

Lane, J. (2005). *System of systems lead system integrators: Where do they spend their time and what makes them more/less efficient*. (Tech. Rep. No. 2005-508.) University of Southern California Center for Systems and Software Engineering, Los Angeles, CA.

Lane, J. (2006). *COSOSIMO Parameter Definitions*. (Tech. Rep. No. 2006-606). University of Southern California Center for Systems and Software Engineering, Los Angeles, CA.

Maier, M. (1998). Architecting principles for systems-of-systems. *Systems Engineering, 1*(4), 267-284.

Pressman, J., & Wildavsky, A. (1973). *Implementation: How great expectations in Washington are dashed in Oakland*. Oakland, CA: University of California Press.

Sage, A., and Cuppan, C. (2001). On the systems engineering and management of systems of systems and federations of systems. *Information, Knowledge, and Systems Management 2*, 325-345.

Software Engineering Institute (2001). *Capability maturity model integration* (CMMI) (Special report CMU/SEI-2002-TR-001). Pittsburgh, PA: Software Engineering Institute.

United States Air Force Scientific Advisory Board (2005). *Report on system-of-systems engineering for Air Force capability development*. (Public Release SAB-TR-05-04). Washington, DC: HQUSAF/SB.

Valerdi, R (2005). *The constructive systems engineering cost model (COSYSMO)*. Unpublished doctoral dissertation, University of Southern California, Los Angeles.

ENDNOTE

[1] Adapted to SoS environment from COSYSMO (Valerdi, 2005).

This work was previously published in Information Resources Management Journal, Vol. 20, Issue 2, edited by M. Khosrow-Pour, pp. 23-32, copyright 2007 by IGI Publishing (an imprint of IGI Global).

Chapter 13
Could the Work System Method Embrace Systems Concepts More Fully?

Steven Alter
University of San Francisco, USA

ABSTRACT

*The **work system method** was developed iteratively with the overarching goal of helping business professionals understand IT-reliant systems in organizations. It uses general systems concepts selectively, and sometimes implicitly. For example, a work system has a boundary, but its inputs are treated implicitly rather than explicitly. This chapter asks whether the further development of the work system method might benefit from integrating general systems concepts more completely. After summarizing aspects of the work system method, it dissects some of the underlying ideas and questions how thoroughly even basic systems concepts are applied. It also asks whether and how additional systems concepts might be incorporated beneficially. The inquiry about how to use additional system ideas is of potential interest to people who study systems in general and information systems in particular because it deals with bridging the gap between highly abstract concepts and practical applications.*

BACKGROUND

The idea of using the concept of **work system** as the core of a systems analysis method for business professionals was first published in Alter (2002), although the ideas had percolated for over a decade. Experience as vice president of a manufacturing software company in the 1980s

convinced me that many business professionals need a simple, yet organized approach for thinking about systems without getting swamped in details. Such an approach would have helped our customers gain greater benefits from our software and consulting, and would have helped us serve them more effectively across our entire relationship. A return to academia and production of an

IS textbook provided an impetus to develop a set of ideas that might help. Starting in the mid-1990s I required employed MBA and EMBA students to use the ideas in an introductory IS course to do a preliminary analysis of an information system in their own organizations. The main goal was to consolidate their learning; a secondary benefit was insight into whether the course content could actually help people understand systems in business organizations.

To date over 300 group and individual papers have contributed to the development of the **work system method** (WSM). At each stage, the papers attempted to use the then current version of WSM. With each succeeding semester and each succeeding cycle of papers, I tried to identify which confusions and omissions were the students' fault and which were mine because I had not expressed the ideas completely or clearly enough.

Around 1997 I suddenly realized that I, the professor and textbook author, had been confused about what system the students should be analyzing. Unless they are focusing on software or hardware details, business professionals thinking about information systems should not start by describing or analyzing the information system or the technology it uses. Instead, they should start by identifying the work system and summarizing its performance gaps, opportunities, and goals for improvement. Their analysis should focus on improving work system performance, not on fixing information systems. The necessary changes in the information system would emerge from the analysis, as would other work system changes separate from the information system but necessary before information system improvements could have the desired impact.

After additional publications (available for download at www.stevenalter.com) helped develop various aspects of WSM, the overall approach became mature enough to warrant publication of a book (Alter, 2006) that combines and extends the main ideas from the various papers, creating a

coherent approach that is organized, flexible, and based on well-defined concepts. Use to date by MBA and EMBA students (early career business professionals) indicates that WSM might be quite useful in practice. Recent developments motivated by widespread interest and concern about services and the service economy led to an attempt to extend the work system approach to incorporate the unique characteristics of services. The main products to date of those efforts are the service value chain framework and service responsibility tables. (Alter, 2007, 2008) Further development of WSM might proceed in many directions, including improving the concepts, testing specific versions in real world settings, and developing online tools that make WSM easier to use and more valuable.

WSM uses system concepts, but the priority in developing WSM always focused on practicality. System concepts and system-related methods that seemed awkward or difficult to apply were not included in WSM. For example, WSM might have incorporated certain aspects of soft system methodology (SSM) developed over several decades by the British researcher Peter Checkland (1999). An area of similarity is SSM's identification of 6 key aspects of a "**human activity system**." Those include customers, actors, transformations, worldview, owner, and environment. Based on an unproven belief that SSM is too abstract and too philosophical to be used effectively by most (American) MBA and EMBA students, WSM was designed to be very flexible but also much more prescriptive than SSM and much more direct about suggesting topics and issues that are often relevant for understanding IT-reliant work systems.

At this point in the development of WSM it is worthwhile to ask whether additional systems concepts might be incorporated beneficially and might contribute to its value for practitioners. Searching for possibilities is a bit awkward because there is very little agreement about what constitutes general systems theory and general systems thinking.

General Systems Theory (GST) integrates a broad range of special system theories by naming and identifying patterns and processes common to all of them. By use of an overarching terminology, it tries to explain their origin, stability and evolution. While special systems theory explains the particular system, GST explains the systemness itself, regardless of class or level. (Skyttner, 1996)

A **system** is not something presented to the observer, it is something to be recognized by him. Most often the word does not refer to existing things in the real world but rather to a way of organizing our thoughts about the real world." ... 'A system is anything unitary enough to deserve a name.' (Weiss, 1971) ... 'A system is anything that is not chaos' (Boulding, 1964) ...[A] system is 'a structure that has organized components.' (Churchman, 1979)." (Skyttner, 2001)

One of the problems in trying to incorporate general system ideas is that so many different types of systems fit under the GST umbrella (Larses and El-khoury, 2005):

- Concrete (living, non-living), conceptual, or abstract
- Open, closed, or isolated
- Decomposable, near-decomposable, or non-decomposable
- Static or dynamic
- Black, gray, or white box

This chapter dissects some of the ideas underlying WSM and questions how faithfully even basic systems concepts are incorporated into it. This chapter proceeds as follows. After summarizing WSM, it summarizes four work systems to illustrate the range of systems that WSM addresses (and conversely, the types of systems it does not address). Building on this clarification of the context, the chapter looks at typical concepts from writings about the systems approach or general systems. In each case it discusses whether those ideas already appear in WSM and whether they might be incorporated to a greater extent. The goal is two-fold, to find directions for improving WSM and to reflect on whether typical general systems ideas are truly useful for understanding information systems and other work systems from a business professional's viewpoint.

THE WORK SYSTEM METHOD

WSM focuses on work systems rather than the information systems that support them and often overlap with them. WSM is designed to produce shared understandings that can lead to better technical specifications needed to develop software. It does not produce the type of specification that might be converted mechanically into software. Although WSM can be used for totally new systems, its basic form assumes that a set of problems or opportunities motivate the analysis of an existing work system. The structure and content of WSM attempt to provide both conceptual and procedural knowledge in a readily usable form, and try to express that knowledge in everyday business language.

- **Definition of work system:** A **work system** is a system in which human participants and/or machines perform work using information, technology, and other resources to produce products and/or services for internal or external customers. Typical business organizations contain work systems that procure materials from suppliers, produce products, deliver products to customers, find customers, create financial reports, hire employees, coordinate work across departments, and perform many other functions. Almost all significant work systems in business and governmental organizations employing more than a few people cannot operate efficiently or effectively without using IT. Most practical IS research is about the development, operation, and maintenance of such systems and their components. In effect, the IS field is basically about IT-reliant work systems. (Alter, 2003)

- **Work system framework:** The nine elements of the **work system framework** (Alter, 2003, 2006) are the basis for describing and analyzing an IT-reliant work system in an organization. Even a rudimentary understanding of a work system requires awareness of each of the nine elements. Four of these elements (work practices, participants, information, and technologies) constitute the work system. The other five elements that fill out a basic understanding of the work system include the products and services produced, customers, environment, infrastructure, and strategies (see Figure 1).

- **Work system life cycle model:** WSM's other basic framework describes how work systems change over time. Unlike the system development life cycle (SDLC), which

is basically a project model rather than a system life cycle, the work system life cycle model (WSLC) is an iterative model based on the assumption that a work system evolves through a combination of planned and unplanned changes. (Alter, 2003, 2006, 2008) Consistent with Markus and Mao's (2004) emphasis on the distinction between system development and system implementation, the planned changes occur through formal projects with initiation, development, and implementation phases. The unplanned changes are ongoing adaptations and experimentation that adjust details of the work system without performing formal projects. In contrast to control-oriented versions of the SDLC, the WSLC treats unplanned changes as part of a work system's natural evolution. Ideas in WSM can be used by any business or IT professional at any point in the WSLC. The steps in WSM (summarized later) are most pertinent in the initiation phase as

Figure 1. The Work System Framework, slightly updated (Source: S. Alter, The Work System Method: Connecting People, Processes, and IT for Business Results, Larkspur, CA: Work System Press, 2006). All rights reserved.

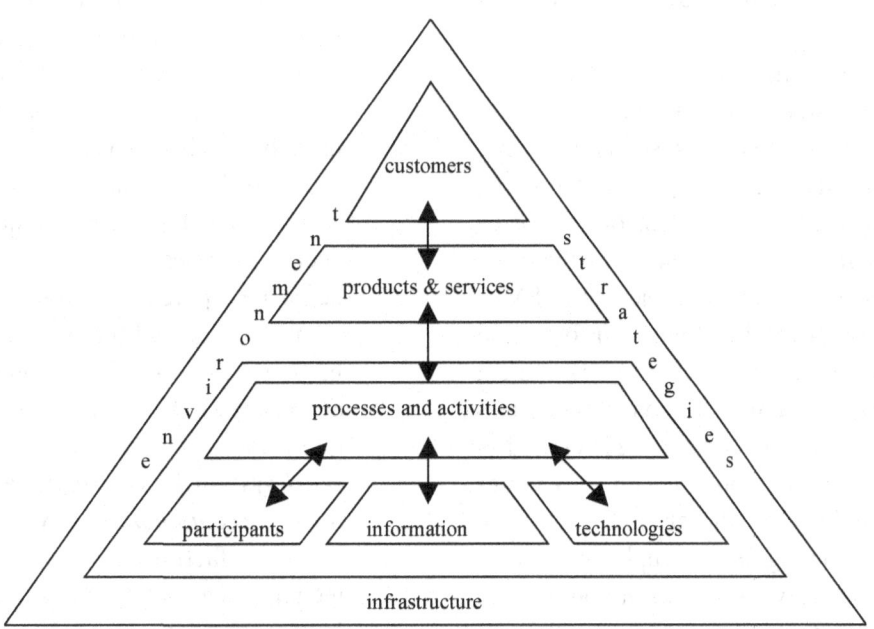

individuals think about the situation and as the project team negotiates the project's scope and goals.

- **Information systems as a special case of work systems:** Work system is a general case of systems operating within or across organizations. Special cases of work systems include **information systems**, projects, supply chains, e-commerce web sites, and totally automated work systems. For example, an information system is a work system whose work practices are devoted to processing information, i.e., capturing, transmitting, storing, retrieving, manipulating, and displaying information. Similarly, a project is a work system designed to produce a product and then go out of existence. The relationship between the general case and the special cases is useful because it implies that the special cases should inherit vocabulary and other properties of work systems in general. (Alter, 2005) Based on this hierarchy of cases, an analysis method that applies to work systems in general should also apply to information systems and projects.

Examples of Information Systems

A common problem in reading general discussions of information systems is the lack of clarity about which types of information systems are being discussed and which are being ignored. Similarly, discussions of general systems theory are often unclear about the types of systems for which specific concepts or principles are relevant. WSM is designed for situations in which an information system is viewed as an IT-reliant work system devoted to processing information. WSM is less applicable if the information system is viewed as a technical artifact that operates on a computer and is used by "users" who are external to the system. The following four examples from Alter (2006) illustrate the types of information systems to which WSM applies:

- *Work system #1: How a bank approves commercial loans:* A large bank's executives believe that its current methods for approving commercial loans have resulted in a substandard loan portfolio. They are under pressure to increase the bank's productivity and profitability. The work system for approving loan applications from new clients starts when a loan officer helps identify a prospect's financing needs. The loan officer helps the client compile a loan application including financial history and projections. A credit analyst prepares a "loan write-up" summarizing the applicant's financial history, projecting sources of funds for loan payments, and discussing market conditions and the applicant's reputation. Each loan is ranked for riskiness based on history and projections. Senior credit officers approve or deny loans of less than $400,000; a loan committee or executive loan committee approves larger loans. The loan officer informs the loan applicant of the decision.

- *Work system #2: How a software vendor tries to find and qualify sales prospects:* A software vendor sells HR software to small and medium sized enterprises. It receives initial expressions of interest through inquiries from magazine ads, web advertising, and other sources. A specialized sales group contacts leads from other sources and asks questions to qualify them as potential clients. A separate outside sales force contacts qualified prospects, discusses software capabilities, and negotiates a purchase or usage deal. Management is concerned that the sales process is inefficient, that it misses many good leads, and that the outside sales group receives too many unqualified prospects.

- *Work system #3: How consumers buy gifts using an ecommerce web site:* The web site of a manufacturer of informal clothing for teenagers has not produced the anticipated level of sales. Surveys and logs of web site usage reveal that customers who know exactly what they want quickly find the product on the web site and make the purchase. Customers who are not sure what they want, such as parents buying gifts for teenagers, often find it awkward to use the site, often leave without making a purchase, and have a high rate of after purchase returns. The company wants to extend existing sales channels. Its managers want to improve the level of sales by improving the customer experience.

- *Work system #4: How an IT group develops software*: The IT group buys commercial application software whenever possible, but also produces home-grown software when necessary. Many of the IT group's software projects miss schedule deadlines, go over budget, and/or fail to produce what their internal customers want. Software developers often complain that users can't say exactly what they want and often change their minds after programming has begun. Users complain that the programmers are arrogant and unresponsive. Use of the company's computer aided software engineering (CASE) software is uneven. Enthusiasts think it is helpful, but other programmers think it interferes with creativity. The IT group's managers believe that failure to attain greater success within several years could result in outsourcing much of the group's work.

Several features of the four examples should be noted. First, each example concerns an IT-reliant work system, and each of these work systems is an information system. (Yes, software development, work system #4, is basically an information system). In addition, as the analysis proceeds in each case, the system to be analyzed will be defined based on the problem or opportunity posed by management. The system will not be defined based on the software that happens to be used as a part of the system.

Three Levels for Using the Work System Method

The current version of WSM (Alter, 2006) consists of three problem-solving steps (SP, AP, and RJ) related to systems in organizations:

- **SP - Identify the System and Problems:** Identify the work system that has the problems that launched the analysis. The system's size and scope depend on the purpose of the analysis.
- **AP - Analyze the system and identify Possibilities:** Understand current issues and find possibilities for improving the work system.
- **RJ - Recommend and Justify changes:** Specify proposed changes and sanity-check the recommendation.

Recognizing the varied nature of analysis situations and goals, WSM can be used at three levels of detail and depth. The level to use depends on the user's particular situation:

- **Level One (Define):** Be sure to remember the three main steps when thinking about a system in an organization.
- **Level Two (Probe):** Within each main step, ask questions that are typically important. These questions include:
  - *Five SP questions* (What is the system? What is the problem? What are the constraints? And so on)
  - *Ten AP questions* (One question about how well each work system element is performing, and one question about the work system as a whole)
  - *Ten RJ questions* (What is the recommendation? How does it compare to an ideal system? Was the original problem solved? What new problems will the recommendation cause? How favorable is the balance of costs and benefits? And so on).

- **Level Three (Drill Down):** For each question within each step, apply guidelines, concepts, and checklists that are often useful.

The most recent version of WSM uses an electronic questionnaire, the first page of which is basically a Level One outline of the executive summary of a typical business analysis and recommendation. The next pages present the 25 questions in Level Two, with space to fill in answers. Vocabulary and concepts identified throughout Alter (2006) and arrayed in checklists, templates, and tabular forms provide additional (Level Three) support for the analysis. These tools help in identifying common topics that might be considered, providing hints about common issues, and providing blank tables that might be used to summarize specific topics or perspectives.

WSM is built on the assumption that an organized structure combining flexibility with considerable depth can be effective in helping business professionals pursue whatever amount of detail and depth is appropriate. Because WSM is designed to support a business professional's analysis, even Level Three does not approach the amount of detail or technical content that must be analyzed and documented to produce computerized information systems.

WORK SYSTEM METHOD AS A SYSTEMS APPROACH

At a superficial level WSM surely represents a systems approach because it describes a situation as a system consisting of interacting components that operate together to accomplish a purpose. A closer look is worthwhile, however, because some aspects of WSM use systems concepts in an idiosyncratic manner. In particular, a careful look at the four examples (beyond the scope of this article) would show that each is a system but that describing each as a set of interacting components operating together to accomplish a purpose might

miss some insights related to the type of system WSM studies. To reflect on how WSM applies a systems approach, it is possible to look at the form and prominence of basic system concepts within the current version of WSM:

- ***Identification of the system*:** WSM users start their analysis by defining the work system. As a general guideline, the system is the smallest work system that has the problem or opportunity that launched the analysis. The system's scope is revealed by identifying the work practices (typically a business process, but possibly other activities as well) and participants who perform the work.
- ***The observer*:** Systems thinking recognizes that system is a mental construct imposed on a situation in order to understand it. Different observers have different system views of the same situation. Part of WSM's value is as a way to help people come to agreement about what system they are trying to improve.
- ***Boundary and environment*:** The identification of the work system in the initial part of WSM automatically sets the boundaries. The work system framework includes some elements that are part of the work system and some that are outside of the work system. The four elements inside the work system are work practices, people, information, and technologies. The other five elements are not part of the work system but are included in the framework because it tries to identify the components of even a rudimentary understanding of a work system.

Mora et al (2002) noted several logical problems with treatment of the system concept *context* in an earlier version of the work system framework that was used in 2001. (*Context* appeared where *environment* now appears.) They also noted that the customers are not included in the box called context (now environment). These observations are accurate from a definitional viewpoint, but the priority in creating WSM is to combine system-related ideas in a way that makes WSM as useful as possible for typical business professionals. Explicitly saying that a work system exists to produce products and services for customers encourages the WSM user to pay special attention to the products and services and customers. Various aspects of the environment such as culture and political issues may matter greatly in some situations and may be unimportant in other situations, but products and services and customers are always important for understanding work systems (including information systems) in organizations.

The term **infrastructure** is also problematic in relation to boundaries. Real world work systems could not operate without infrastructure owned and managed by the surrounding organization and external organizations. Infrastructure is included as one of the elements for understanding a work system because ignoring external infrastructure may be disastrous. However, it is awkward to treat computer networks, programming languages, IT personnel, and other shared resources as internal parts of the work systems they serve. If these components of infrastructure were treated as internal components of the work system, even small work systems involving a few people and several activities would become gigantic. They would be like an iceberg, with visible aspects of the work system above the water and an enormous mass of shared infrastructure largely invisible below the waterline. To discourage unnecessary attention to distinctions between technology and infrastructure early in the analysis, WSM users creating a 'work system snapshot' for summarizing a work system should assume that the difference between technology and technical infrastructure is unimportant for the initial summary. The distinction should be explored only if it is important for understanding the work system in greater depth.

Inputs and Outputs

The work system framework contains neither the term *inputs* nor the term *outputs*. The term output is not used because it sounds too mechanistic and is associated too much with computer programs. In terms of logic and structure, there is no problem in calling a work system's outputs products and services. Also, that terminology helps focus attention on the work system's goal of providing products and services customers want, rather than just producing whatever outputs it is programmed to produce. Even the terms products and services are occasionally problematic. For example, consider a work system that produces entertainment. The product might be described as a temporally sequenced information flow that is sensed and interpreted by viewers (the customers). It also might be viewed as the customer's stimulation, peace of mind, or enjoyment. At minimum there is question about whether the customer plays an active role as a participant in the work system. As the nature of the product becomes more ambiguous, the boundary of the system also becomes more ambiguous.

The work system framework ignores inputs altogether because it assumes that important inputs will be understood implicitly. The first step in the work practices typically will describe something about receiving, transforming, or responding to something that comes from outside of the work system. (If important inputs are not mentioned anywhere in the work practices, it is likely that the summary of work practices will be insufficient.) Also, information from external sources is often listed under the information used or produced by the work system. And what about other inputs, such as the air the participants breathe, the food they metabolize as they do their work, or the skills they received from a training course last year? It is easy to say in general that systems have inputs, but substantially more difficult to identify which inputs are worth mentioning in an analysis. The work system framework lacks a slot for inputs because it is easier to infer important inputs from the steps listed in the work practices.

Transformations

The term transformation is particularly meaningful in physical systems such as assembling a set of tangible components to produce an automobile. Transformation is less meaningful for various aspects of each of the work system examples mentioned earlier. For example, it doesn't feel natural to say that the loan approval system transforms loan applications into approvals or denials; nor does it feel natural to say that the ecommerce web site transforms customer desires into purchase decisions. Thus, the term transformation is often unsatisfactory for summarizing the activities that occur within the work system. During the development of WSM, alternatives to the term transformation included activities, actions, business process, and work practices. Work practices was selected because it includes business process and other perspectives for thinking about activities, such as communication, decision making, and coordination.

Goals, Controls, and Feedback

Most observers say that purposive systems contain control mechanisms that help the system stay on track or help the system move toward equilibrium (as with a thermostat). A thermostat-like goal and feedback metaphor is appropriate for heating a house, but doesn't fit well for many information systems. For example, the role of feedback control in the use of the ecommerce web site is not apparent. Similarly, the software development system may or may not have formal feedback mechanisms. Those mechanisms will be more apparent in a highly structured software development environment, and much less apparent in an agile development environment that proceeds through a series of incremental changes that receive individual feedback but may or may not lead to a larger goal.

WSM treats control as one of the perspectives for thinking about work practices. The basic question is whether controls are built into existing or proposed work practices, and whether a different type or different amount of control effort would likely generate better results.

Wholeness

One of the major premises of the system approach is that systems should be treated as wholes, not just as a set of components. The structure of WSM is designed to recognize systemic issues, but WSM certainly doesn't favor wholeness over analysis of components. The structure of WSM calls for looking at each element separately and drilling down to understand the elements in enough depth to spot problems within each element. Simultaneously, however, the work system framework contains explicit links between elements, showing the main routes through which they interact in the operation of the work system as a whole.

An interesting issue with wholeness is that many components of work systems are not wholly dedicated to those work systems. The work practices are the activities within the work system, but the participants may be involved in many other work systems. Their activities within a particular work system may absorb only an hour a day or an hour a week. Similarly, the information and technologies may be used in other work systems. In the real world, the wholeness of the work system is often challenged when work system participants feel torn about their responsibilities in multiple work systems.

Emergent Properties

Users of WSM automatically observe emergent properties when they look at how a system operates as a whole. However, the level of detail and broad-brush modeling used in WSM is usually insufficient to reveal the types of counterintuitive system behaviors (Forrester, 1971) that are sometimes revealed and understood by system modelers using techniques such as systems dynamics.

Hierarchy, Subsystems and Supersystems

Systems in organizations are typically viewed as subsystems of larger systems. Relationships between information systems and the work systems they support have changed over recent decades. Before real time computing, computerized tracking systems and transaction processing systems were often separate from the manual processes they served or reported. As real time computing became commonplace, information systems became an integral part of the work systems they served. Remove the work system and the information system has no meaning. Turn off the information system and the work system grinds to a halt. All four of the work systems mentioned earlier have some aspect of this feature. All were selected as information systems that are somewhat independent of other work systems, yet major failures in any of these systems would have significant impacts on larger systems or organizations that they serve.

System Elements

Even the idea of system elements can be called into question. The nine elements of the work system framework are different types of things. For example, work practices are different from people (participants) and are different from information. Compare that view of work system elements to Ackoff's (1981) definition of a system as a set of two or more elements that satisfies the following three conditions:

- The behavior of each element has an effect on the behavior of the whole.
- The behavior of the elements and their effects on the whole are interdependent.

- However subgroups of the elements are formed, all have an effect on the behavior of the whole but none has an independent effect on it(cited by Skyttner, 2001).

In Ackoff's view, each element is a separate component that has the capability of behaving. In contrast, the work system framework says that the work practices are the behaviors and the work system participants and in some highly automated cases the technologies have the capability of behaving. In other words, each work system is basically a separate element in Ackoff's terms. The implication for future development of the work system method is that it might be possible to include explicit forms of interaction between work systems. Currently the only form of interaction included in WSM is the potential use of one work system's products and services in the work practices of another work system.

System Evolution

General system theory is primarily concerned with how systems operate, and somewhat less concerned with how they evolve over time. Patterns through which work systems evolve through planned and unplanned change are extremely important in WSM because the justification of a proposed system change includes preliminary ideas about how the system can be converted from its current configuration to a desired future configuration. As explained in substantial detail in Alter (2006), the work system life cycle model says much more about the evolution of a work system than is implied by most general system discussions. As WSM develops further it will surely absorb more ideas and principles related to system change, but most of these will probably come from the IS literature and the organizational behavior and innovation literatures rather than from the general systems literature. On the other hand, it is possible that some aspect of Beer's (1981) viable systems model might be incorporated in future explanations or explorations of the work system life cycle model.

Chaos, Complexity, Entropy, Self-Organization

Concepts such as chaos, complexity, entropy, and self-organization are often part of sophisticated discussions of systems. Although these ideas are sometimes tossed around at a rather non-specific, metaphoric level (the chaos of everyday management, the complexity of our lives, etc.), bringing these concepts into WSM while maintaining their deeper, more precise meanings in sophisticated discussions of systems seems impractical at this point. Attaining insightful analysis when using the existing WSM vocabulary is challenging enough.

The use of the term complexity in WSM illustrates the challenge of moving toward more advanced concepts. Within the current version of WSM, complexity is applied with its every day meaning and is treated as one of many strategy choices for work practices. Simpler work practices deal with fewer variables and are easier to understand and control; the opposite is true for complex work practices. Even with that simple definition, MBA and EMBA students have shown little inclination to use that term when evaluating a work system (i.e., it is too simple or too complex) and little inclination to use it to describe proposed improvements. If they shy away from relatively simple usage of that type, it is unlikely that they would be willing or able to apply advanced understandings of complexity that require insight at a very abstract level.

CONCLUSION

This reflection on WSM's use of general system concepts is highly subjective because different authors use different definitions of terms and have different views of which terms belong under the umbrella of general systems concepts. Evaluation of WSM in relation to general systems theory is all the more difficult because WSM was not developed as an application of general systems theory. It was developed to provide a set of ideas and tools that business professionals can use when trying to understand and analyze systems from a business viewpoint. At every stage in its development, every choice between maximizing ease of use and maximizing conceptual purity was decided in favor of ease of use.

The review of the relationship between typical system concepts and concepts within WSM showed that WSM uses a system approach and system concepts, but sometimes uses those terms idiosyncratically. Consistent with its practical goal of helping business professionals understand and analyze IT-reliant work systems, WSM adapts system concepts within a framework that is easier to understand and apply than any of the frameworks typically associated with general system theory (at least in my opinion).

Real world examples were introduced in this article as a reality check because general systems theory tends to include under one umbrella many different types of systems at vastly different levels (e.g., Miller's (1978) inclusion of cells, organs, organisms, groups, organizations, communities, societies, and supranational systems within the category of living systems). Potential changes in WSM concepts and process should be tested against realistic examples of IT-reliant work systems. If a change would make typical examples clearer to typical business professionals, then it might be appropriate within the spirit of WSM, especially if it could also co-exist with the rest of WSM or if it would make an existing part of WSM unnecessary.

It is unclear whether a detailed review of general system theory and its sophisticated extensions related to concepts such as chaos, complexity, entropy and self-organization might lead to useful improvements in WSM. Although this is a possibility, the path would probably be long. The first step would involve finding real situations in which sophisticated use of these concepts would help in evaluating, analyzing, and designing the types of systems that WSM addresses. It seems likely that sophisticated applications of concepts such as chaos, complexity, entropy, and self-organization are less pertinent to in typical work systems and more pertinent to physical and mathematical systems whose components and component interactions are more amenable to mathematical analysis.

The original question was "Could the work system method embrace systems concepts more fully?" At this point the answer to that question is a weak maybe. WSM is mature enough that its value to business and IT professionals can be tested in a number of different settings. Informal results thus far show that many users find it useful, and imply that at least some future users will suggest ways to make it easier to use in general or easier to apply to specific types of situations.

If I had to guess, I would say that the suggestions most directly associated with general systems concepts would be related to subset/ superset relationships and supplier/customer (output/input) relationships between separate work systems. The current form of WSM focuses on a single work system and says that it may be convenient to subdivide one work system into several or combine several work systems into one. An effective way to handle relationships between separate work systems without making the analysis too awkward probably would be a very useful extension of the current version of WSM.

At a more theoretical level, it also would be interesting to look at general systems concepts and principles at much greater depth than was possible in this brief chapter. For example, Skyttner (2001, pp. 61-64) lists 39 different "widely known laws, principles, theorems, and hypotheses). It would be interesting to look at each in turn and to decide whether it says anything that is both non-obvious about IT-reliant work systems and useful in understanding them in real world situations.

REFERENCES

Ackoff, R. (1981). *Creating the corporate future*. New York: John Wiley & Sons.

Alter, S. (2002). The work system method for understanding information systems and information system research. *Communications of the AIS*, *9* (6), 90-104.

Alter, S. (2003). 18 reasons why IT-reliant work systems should replace the IT artifact as the core subject matter of the IS field. *Communications of the AIS*, *12* (23), 365-394.

Alter, S. (2005). Architecture of Sysperanto - A model-based ontology of the IS field. *Communications of the AIS*, *15* (1), 1-40.

Alter, S (2006). *The work system method: Connecting people, processes, and IT for business results*. Larkspur CA: Work System Press.

Alter, S. (2007). *Service responsibility tables: A new tool for analyzing and designing systems*. Paper presented at the 13th Americas Conference on Information Systems, Keystone, CO.

Alter, S. (2008). Service system fundamentals: work system, value chain, and life cycle. *IBM Systems Journal*, 47(1), 2008, 71-85. Available at http://www.research.ibm.com/journal/sj/471/alter.html

Beer, S. (1981). *Brain of the firm*, 2nd ed. Chichester, UK and New York: John Wiley.

Boulding, K. (1964). General systems as a point of view. In J. Mesarovic (Ed), *Views on general systems theory*. New York: John Wiley.

Checkland, P. (1999). *Systems thinking, systems practice* (Includes a 30-year retrospective). Chichester, UK: John Wiley.

Churchman, C. W. (1979). *The design of inquiring systems: Basic concepts of systems and organizations*. New York, Basic Books.

Forrester, J. (1971). Counterintuitive Behavior of Social Systems. *Technology Review, 73* (3), January.

Larses, O., and El-Khoury, J. (2005). *Review of Skyttner (2001) in O. Larses and J. El-Khoury, "Views on General Systems Theory."* Technical Report TRITA-MMK 2005:10, Royal Institute of Technology, Stockholm, Sweden. Retrieved June 30, 2006 on the World Wide Web: http://apps.md.kth.se/publication_item/web.phtml?ss_brand=MMKResearchPublications&department_id='Damek'

Markus, M.L. and Mao, J.Y. (2004). Participation in development and implementation – updating an old, tired concept for today's IS contexts. *Journal of the Association for Information Systems, 5* (11: 14).

Miller, J. G. (1978). *Living systems.* New York: McGraw-Hill.

Mora, M., Gelman, O., Cervantes, F., Mejía, M., and Weitzenfeld, A. (2002). A Systemic Approach for the Formalization of the Information Systems Concept: Why Information Systems are Systems. In J.J. Cano, (Ed.), *Critical reflections on information systems: A systemic approach.* Hershey, PA: Idea Group Publishing, 1-29.

Skyttner, L. (2001). *General systems theory.* Singapore: World Scientific Publishing

Skyttner, L, (1996). General systems theory: Origin and hallmarks. *Kybernetes, 25* (6), 16, 7 pp.

Weiss, P. (1971). *Hierarchically organized systems in theory and practice.* New York: Hafner.

Chapter 14
Information and Knowledge Perspectives in Systems Engineering and Management for Innovation and Productivity through Enterprise Resource Planning

Stephen V. Stephenson
Dell Computer Corporation, USA

Andrew P. Sage
George Mason University, USA

ABSTRACT

This article provides an overview of perspectives associated with information and knowledge resource management in systems engineering and systems management in accomplishing enterprise resource planning for enhanced innovation and productivity. Accordingly, we discuss economic concepts involving information and knowledge, and the important role of network effects and path dependencies in influencing enterprise transformation through enterprise resource planning.

INTRODUCTION

Many have been concerned with the role of information and knowledge and the role of this in enhancing systems engineering and management (Sage, 1995; Sage & Rouse, 1999) principles, practices, and perspectives. Major contemporary attention is being paid to enterprise transformation (Rouse, 2005, 2006) through these efforts. The purpose of this work is to discuss many of these efforts and their role in supporting the definition, development, and deployment of an enterprise resource plan (ERP) that will enhance transformation of existing enterprises and development of new and innovative enterprises.

Economic Concepts Involving Information and Knowledge

Much recent research has been conducted in the general area of information networks and the new economy. Professors Hal R. Varian and Carl Shapiro have published many papers and a seminal text, addressing new economic concepts as they apply to contemporary information networks. These efforts generally illustrate how new economic concepts challenge the traditional model, prevalent during the Industrial Revolution and taught throughout industry and academia over the years. In particular, the book *Information Rules* (Shapiro & Varian, 1999) provides a comprehensive overview of the new economic principles as they relate to today's information and network economy. The book addresses the following key principles:

- Recognizing and exploiting the dynamics of positive feedback
- Understanding the strategic implications of lock-in and switching costs
- Evaluating compatibility choices and standardization efforts
- Developing value-maximizing pricing strategies
- Planning product lines of information goods
- Managing intellectual property rights
- Factoring government policy and regulation into strategy

These concepts have proven their effectiveness in the new information economy and have been fundamental to the success of many information technology enterprises introducing new ideas and innovations into the marketplace. Paramount to an enterprise's success in reaching critical mass

for its new product offering is the understanding and implementation of these new economic concepts.

Economides (1996) has also been much concerned with the economics of networks. He and Himmelberg (1994) describe conditions under which a critical mass point exists for a network good. They characterize the existence of critical mass points under various market structures for both durable and non-durable goods. They illustrate how, in the presence of network externalities and high marginal costs, the size of the network is zero until costs eventually decrease sufficiently, thereby causing the network size to increase abruptly. Initially, the network increases to a positive and significant size, and thereafter it continues to increase gradually as costs continue to decline. Odlyzko (2001) expands on the concept of critical mass and describes both the current and future growth rate of the Internet and how proper planning, network budgeting, and engineering are each required. He emphasizes the need for accurate forecasting, since poor planning can lead to poor choices in technology and unnecessary costs.

Economides and White (1996) introduce important concepts with respect to networks and compatibility. They distinguish between direct and indirect externalities, and explore the implications of networks and compatibility for antitrust and regulatory policy in three areas: mergers, joint ventures, and vertical restraints. They also discuss how compatibility and complementarity are linked to provide a framework for analyzing antitrust issues. Strong arguments are made for the beneficial nature of most compatibility and network arrangements, with respect to vertical relationships, and policies are set forth to curb anti-competitive practices and arrangement. Farrell and Katz (2001) introduce concepts of policy formulation in preventing anti-competitive practices and, in addition, explore the logic of predation and rules designed to prevent this in markets that are subject to network effects. This work discusses how the imposition of the leading proposals for rules against predatory pricing may lower or raise consumer welfare, depending on conditions that may be difficult to identify in practice.

Research conducted on these economic concepts establishes a solid foundation and baseline for further research in the area of enterprise resource planning and new technology innovations (Langenwalter, 2000). In this work, he extends the traditional enterprise resource planning (ERP) model to incorporate a total enterprise integration (TEI) framework. He describes TEI as a superset of ERP and also describes how it establishes the communications foundation between customer, manufacturer, and supplier. Each entity is linked internally and externally, allowing the TEI system to enhance performance and to provide process efficiencies that reduce lead times and waste throughout the supply chain. This work illustrates how ERP is uniquely integrated with customers and suppliers into the supply chain using TEI and how it significantly improves customer-driven performance. The model for this includes five major components: executive support, customer integration, engineering integration, manufacturing integration, and support services integration. These components are essential for integrating all information and actions required to fully support a manufacturing company and its supply chain. TEI presents a strategic advantage to an enterprise, rather than just improving operating efficiencies. The TEI framework provides the enterprise a competitive edge by:

- Maximizing speed and throughput of information and materials
- Minimizing response time to customers, suppliers, and decision makers
- Pushing decisions to the appropriate levels of the organization
- Maximizing the information made available to the decision-makers
- Providing direct integration into the supply chain

In addition to the technology, TEI also incorporates stakeholders. People are empowered at all levels of the enterprise to improve the quality of their decision-making. One result of this is MRP II (Manufacturing Resources Planning) systems. MRP II evolved from MRP (Material Requirements Planning), which was a method for materials and capacity planning in a manufacturing environment. Manufacturing plants, to plan and procure the right materials in the right quantities at the right time, used this method. MRP became the core planning module for MRP II and ERP. MRP was later replaced by MRP II, which expanded the MRP component to include integrated material planning, accounting, purchasing of materials for production, and the shop floor. MRP II integrated other functional areas such as order entry, customer service, and cost control. Eventually, MRP II evolved into enterprise resource planning (ERP), integrating even more organizational entities and functions such as human resources, quality management, sales support, and field services. ERPs became richer in functionality and involved a higher degree of integration than their predecessors MRP and MRP II.

Another very well-known contributor to the field of enterprise resource planning is Thomas H. Davenport (2000). In *Mission Critical: Realizing the Promise of Enterprise Systems*, the need to take a customer or product focus when selecting an operational strategy is emphasized. To enable this, a direct connection should exist between the daily operations and the strategic objectives of the enterprise. This is made possible through the use of operational data, that is used to enhance the operational effectiveness of the enterprise. Operational data is defined by the organization seeking to measure the operational effectiveness of its environment. Operational data may be defined in terms of various parameters such as cycle time (CT), customer response time (CRT), or MTTR (mean time to repair). These are only a few of the parameters, and they are contingent on the operational strategy the organization is seeking to adopt. For example, an organization that seeks to reduce cycle time (CT) for processing orders in order to minimize cost may look to capture CT in its operational data. This data is captured over time as process efficiencies are instituted within the existing order process. Operational effectiveness is then determined by comparing the future CT state of the order process with that of its initial CT benchmark. For example, if cycle time to process an order was originally 15 minutes, and after the process efficiencies were instituted, CT was then 5 minutes, then operational effectiveness improved by 10 minutes. Now it takes fewer resources to process orders, thus reducing operational costs.

Davenport (2000) introduces a data-oriented culture and conveys the need for data analysis, data integrity, data synthesis, data completeness, and timely extracts of data. Data is used across organizational boundaries and shared between the various entities in an effort to enhance operational effectiveness. For example, transaction data must be integrated with data from other sources, such as third-party vendors, to support effective management decision-making. One's ability to interpret and analyze data can effect the decisions that are made and the confidence management has in pursuing particular ongoing decisions. Davenport believes that a combination of strategy, technology, data (data that is relevant to the organization), organization, culture, skills and knowledge assist with developing an organization's capabilities for data analysis. When performing data analysis, various organizations may have similar results, but with different meanings. He indicates that a typical corporation may have divisions that have a need to store customer data in different customer profile schemes. Therefore, a common shared master file between the divisions may not be feasible. This approach takes on more of a distributed approach versus a centralized approach to data management. The operational effectiveness of each of these divisions will vary based on the

benchmarks and target improvements they have set for themselves.

Christopher Koch (2006) supports Davenport's data concept and elaborates on the value of an ERP and how it can improve the business performance of an enterprise. He demonstrates the value of an ERP by integrating the functions of each organization to serve the needs of all stakeholders. The associated framework attempts to integrate all organizational entities across an enterprise onto a single-systems ERP platform that will serve the needs of the various entities. This single platform replaces the standalone systems prevalent in most functional organizations such as human resources, finance, engineering, and manufacturing, thereby allowing people in the various organizations to access information not only in the most useful manner but also from their own perspectives. This information may be the same shared data used between the organizations or may vary, based on the need of each of the organizations. Each organization in the enterprise and its stakeholders will have their own set of requirements for accessing, viewing and manipulating their data. Data management may even take on a hybrid of a centralized and distributed approach. Some organizations may need a view of the same data, while others may have their own unique data requirements. Koch (2006) indicates that there are five major reasons why an enterprise adopts an ERP strategy:

1. Integrate financial information
2. Integrate customer order information
3. Standardize and speed up manufacturing processes
4. Reduce inventory
5. Standardize human resources (HR) information

Each organization within an enterprise has its own requirements for an ERP. They may share the same ERP solution; however, the ERP may be designed to support the specific business need of each organization. Some organizations may have a need to view the same data. For example, a sales and customer care-focused organization may need to view the same customer profile data to access customer contact information. In comparison, a human resources-focused organization may not need to be privy to this same information. They may be more interested in accessing internal employee personnel records for employee performance monitoring. The senior executive level of an enterprise will also have its own unique data requirements in order to make key strategic and tactical decisions. This executive level may need the capability to access data from each of the organizational units in order to effectively manage the operations of the business. The organizations, within an enterprise each have their own instances of an ERP with respect to accessing data and implementing processes. Some organizations may share a common process such as the order fulfillment process. For example, this process may be shared between organizational entities such as sales, operations, and revenue assurance. Sales would complete a service order, operations would deliver the service, and revenue assurance would bill the customer. However, there are processes that are only unique to a particular organization. For example, the marketing organization may not be interested in the escalation process used by operations to resolve customer issues. This process is unique to operations and, as a result, the ERP would be designed for such uniqueness. The design of an ERP should, of course, take organizational data and process requirements into account and support management of the enterprise and its inter-workings in a transdisciplinary and transinstitutional fashion (Sage 2000, 2006).

William B. Rouse had produced very relevant and important popular work surrounding new technology innovation with respect to the enterprise. In *Strategies for Innovation*, Rouse (1992) addresses four central themes to introduce strategies for innovation in technology-based enterprises. Rouse discusses the importance of strategic thinking and how some enterprises fail to plan long term. This is based on the notion that "while people may want to think strategically, they actually do not know how(p. 3)." He emphasizes the need for stakeholders to understand the solutions offered as a result of new innovation, and how strategies are critical for ensuring successful products and systems. Most importantly, these strategies must also create a successful enterprise for developing, marketing, delivering, and servicing solutions, thus leading to the need for human-centered planning, organization, and control. These are among the approaches needed to stimulate innovation in products and services (Kaufman & Woodhead, 2006).

Rouse (1992) describes the need for applying a human-centered design methodology to the problem of enhancing people's abilities and overcoming their limitations. In the process of planning, organizing, and controlling an enterprise, he illustrates how technology-based enterprises differentiate themselves from each other based on their core product technologies. This strategic strength is based on the unique value that the core product can provide to the marketplace. He indicates that the enterprise should continuously analyze the market and measure core product value to determine the benefits that can be provided. Assessing and balancing the stakeholders' interests will be necessary to ensure success of the core product. Stakeholders consist of both producers and consumers. Each may have a stake in the conceptualization, development, marketing, sales, delivery, servicing, and use of the product. The three key processes highlighted in this work are: strategic planning, operational management, and the engineering/administration, vehicles used by the enterprise to assist stakeholders with pursuing the mission of the enterprise.

Rouse further addresses strategic approaches to innovation in another one of his books. In *Essential Challenges of Strategic Management* (Rouse, 2001), he illustrates the strategic management challenges faced by all enterprises and introduces best practices for addressing these challenges. He disaggregates the process of strategically managing an enterprise into seven fundamental challenges. The essential challenges he describes, which most enterprises are confronted with, are: growth, value, focus, change, future, knowledge, and time. Growth is critical to gaining share in saturated and declining markets and essential to the long-term well-being of an enterprise. A lack of growth results in declining revenues and profits, and, in the case of a new enterprise, there is the possibility of collapse. He describes value as the foundation for growth, the reason an enterprise exists. Matching stakeholders' needs and desires to the competencies of the enterprise, when identifying high-value offerings, will justify the investments needed to bring these offerings to market. While value enhances the relationships of processes to benefits and costs, focus will provide the path for an enterprise to provide value and growth. Focus involves pursuing opportunities and avoiding diversions, that is, making decisions to add value in particular ways and not in others are often involved. For example, allocating too few resources among many projects may lead to inadequate results or possible failure.

The focus path is followed by another path called change. An enterprise challenged with organizational re-engineering, downsizing, and rightsizing often takes this change path. The enterprise will continue to compete creatively while maintaining continuity in its evolution. As the nature of an organization changes rapidly during an enterprise's evolution, managing change becomes an art. According to Rouse (2001), investing in the future involves investing in inherently unpredictable outcomes. He describes the future as uncertain. The intriguing question is, "If we could buy an option on the future, how would we determine what this option is worth(p. 6)?" A new enterprise will be faced with this challange when coming into the marketplace.

The challenge of knowledge is transformation of information from value-driven insights to strategic programs of action. Determining what knowledge would make an impact, and in what ways, is required. This understanding should facilitate determining what information is essential and should provide further elaboration on how it is to be processed and how its use will be supported. The most significant challenge identified is that of time. A lack of time is the most significant challenge facing best use of human resources. Most people spend too much time being reactive and responding to emergencies, attending endless meetings, and addressing an overwhelming number of e-mails, all of which cannibalize time. As a result, there is little time for addressing strategic challenges. Carefully allocating the scarcest resource of an organization is vital to the future of an enterprise. Some of the best practices Rouse (2001) has presented in addressing the seven strategic challenges may be described as follows.

- **Growth:** Buying growth via strategic acquisitions and mergers; fostering growth from existing market offerings via enhanced productivity; and creating growth through innovative new products and brand extensions.
- **Value:** Addressing the nature of value in the market; using market forces in determining the most appropriate business process; and designing cost accounting system to align budgets and expenditures with value streams.
- **Focus:** Deciding what things to invest in and those things to be avoided or stopped; and linking decisions or choices to organizational goals, strategies, and plans.
- **Change:** Instituting cross-functional teams for planning and implementing significant changes; and redesigning incentive and reward systems in order to ensure that people align their behaviors with desired new directions.

- **Future:** Employing formal and quantitative investment decision processes; and creating mechanisms for recognizing and exploiting unpredictable outcomes.
- **Knowledge:** Ensuring that knowledge acquisition and sharing are driven by business issues in which knowledge has been determined to make a difference; using competitive intelligence and market/customer modeling to provide a valuable means for identifying and compiling knowledge.
- **Time:** Committing top management to devoting time to challenges; and improving time management, executive training, and development programs, in addition to providing increased strategic thinking opportunities.

Gardner (2000) takes a complementary approach to the enterprise and to innovation by focusing on the valuation of information technology. He addresses the difficulties of defining the value of new technologies for company shareholders using integrated analytical techniques in his book *The Valuation of Information Technology*. Gardner presents methodologies for new enterprise business development initiatives and presents techniques for improving investment decisions in new technologies. This 21st-century approach to valuation avoids making investment decisions on an emotional basis only, in favor of predicting shareholder value created by an information technology system before it is built. Determining the contribution an information technology system makes to a company's shareholder value is often challenging and requires a valuation model. Gardner suggests that the primary objective of information technology systems development in business is to increase the wealth of shareholders by adding to the growth premium of their stock. The objective of maximizing shareholder wealth consists of maximizing the value of cash flow generated by operations. This is accomplished by generating future investment in information

technology systems. As an example, this could be a state-of-the-art enterprise resource planning system, which could easily maximize what we will call operational velocity and, as a result, maximize shareholder wealth. The process that Gardner suggests using would be to first identify the target opportunity, align the information technology system to provide the features the customer wants in a cost-effective manner, and then to accurately measure the economic value that can be captured through this.

Some of the techniques Gardner uses to compute economic value are net present value (NPV), rate of return (ROR), weighted average cost of capital (WACC), cost of equity, and intrinsic value to shareholders of a system. Each of these techniques may be used to determine aspects of the shareholder value of an information technology system. The results from computing these values will assist an enterprise with making the right decisions with respect to its operations. For example, if the rate of return on capital is high, then time schedule delays in deploying an information technology system can destroy enormous value. Time to market becomes critical in this scenario. Gardner suggests that it may be in the best interest of the company to deploy the system early by mitigating the potential risk and capitalizing on the high rate of return. A risk assessment must be performed to ensure that the customer relationship is not compromised at the expense of implementing the system early. If the primary functionality of the system is ready, then the risk would be minimal, and the other functional capabilities of the system may be phased in at a later time.

If the rate of return is low, however, schedule delays will have a lesser effect on value and deployment of a system does not immediately become crucial to the success of the enterprise. This approach to predicting value takes a rational approach to decision making by weighing the rewards and risks involved with an information technology system investment. The author suggests moving away from the more intuitive approach of valuation often practiced in the high-tech industry, which is said to be very optimistic, spotty, and driven by unreasonable expectations from management. Gardner describes this intuitive practice as a non-analytical approach to assessing the economic viability of an information technology system. This practice primarily ignores the bare essentials that management must consider in assessing whether the economics of an information technology system are attractive. Gardner has established an analytical framework for analyzing the economics of information technology systems. His process is comprised of the three following steps:

1. Identify the target customer opportunity.
2. Align the information technology system to cost-effectively provide the features the customer wants.
3. Measure the economic value that can be captured.

The result of utilizing the framework is the quantification of the shareholder value created by an information technology system.

Boer (1999) also has much discussion on the subject of valuation in his work on *The Valuation of Technology*. He illustrates links between research and development (R&D) activity and shareholder value. In addition, he identifies the languages and tools used between business executives, scientists, and engineers. The business and scientific/engineering communities are very different environments and are divided by diverse knowledge and interest levels. Bridging the gap between these communities is made possible through the process of valuation, which fosters collaboration and communication between both communities. Boer identifies the link between strategy and value and addresses the mutual relationship between corporate strategy and technology strategy. He introduces tools and approaches used to quantify the link between technological

research and commercial payoff within the value model of an enterprise.

This value model is comprised of four elements: operations, financial structure, management, and opportunities. The opportunity element is most critical to the future growth of an enterprise. The options value of an enterprise and how it is addressed strategically will determine the fate of an emerging enterprise. Boer illustrates how productive research and development creates options for the enterprise to grow in profitability and size. He views R&D as a component of operations, since this is the point at which new technology is translated into commercial production. In the competitive marketplace, the enterprise evolves in order to generate opportunity and growth. R&D serves as the vehicle for converting cash into value options for the enterprise. Boer introduces R&D stages (conceptual research, feasibility, development, early commercialization), where the level of risk, spending, and personnel skills vary. Each stage of the R&D process allows management to make effective decisions regarding the technology opportunity and perform levels of risk mitigation. R&D can be instrumental in decreasing capital requirements with results of a very high rate of return on the R&D investment. The art of minimizing capital requirements requires good and effective communication between the scientific/engineering and business communities. This will allow both communities to share their views and foster the need for driving this essential objective.

Some of the methods Boer uses for asset valuation are similar to Rouse's methods. Boer uses discounted cash flow (DCF), NPV, cost of money, weighted average cost of capital, cost of equity, risk-weighted hurdle rates for R&D, and terminal value methods for assessing valuation. In accelerated growth situations, as in the case of an emerging enterprise, Boer emphasizes that the economic value is likely to be derived from the terminal value of the project, not from short-term cash flows. A lack of understanding of terminal value can compromise the analysis of an R&D project. R&D can be a cash drain, and the outcomes are difficult to predict. Boer's techniques provide a vehicle for converting cash into opportunity and creating options for the enterprise.

Another work that addresses valuation is entitled *The Real Options Solution: Finding Total Value in a High-Risk World* (Boer, 2002). Here, the author presents a new approach to the valuation of business and technologies based on options theory. This innovative approach, known as the total value model, applies real options analysis to assessing the validity of a business plan. All business plans are viewed as options. These plans are subject to both unique and market risks. While business plans seem to create no value on a cash flow basis, they do become more appealing once the full merit of all management options is recognized. Since management has much flexibility in execution, the model offers a quantifiable approach to the challenge of determining the strategic premium of a particular business plan. Boer defines total value as "the sum of economic value and the strategic premium created by real options (p. vii)." He presents a six-step method for applying this model in a high-risk environment for evaluating enterprises, R&D-intensive companies, bellwether companies, capital investments, and hypothetical business problems. His method reveals how changes in total value are driven by three major factors: risk, diminishing returns, and innovation.

Boer's option theory efforts provide the enterprise with a vehicle for computing the strategic premium to obtain total value. This six-step method to calculate total value is comprised of:

1. Calculation of the economic value of the enterprise,
2. Framing the basic business option,
3. Determining the option premium,
4. Determining the value of the pro forma business plan,
5. Calculating the option value, and
6. Calculating total value.

Options theory approached to valuation leverage on elements of uncertainty such as these afford enterprise managers major investment opportunities. This was not common using more traditional valuation methods such as NPV- and internal rate of return (IRR)-based calculations. As Boer (2002) illustrates, the new options theory emphasizes the link between options, time, and information. Boer states: "Options buy time. Time produces information. Information will eventually validate or invalidate the plan. And information is virtual (p. 106)." This theory and its extensions (Boer, 2004) may well pave the way for a new generation of enterprise evolution and enterprise innovation.

Rouse (2005, 2006) is concerned with the majority of these issues in his development of systems engineering and management approaches to enterprise transformation. According to Rouse, enterprise transformation concerns change, not just routine change but fundamental change that substantially alters an organization's relationships with one or more of its key constituencies: customers, employees, suppliers, and investors. Enterprise transformation can take many forms. It can involve new value propositions in terms of products and services and how the enterprise should be organized to provide these offerings and to support them. Generally, existing or anticipated value deficiencies drive these initiatives. Enterprise transformation initiatives involve addressing the work undertaken by an enterprise and how the work is accomplished. Other important elements of the enterprise that influence this may include market advantage, brand image, employee and customer satisfaction, and many others.

Rouse suggests that enterprise transformation is driven by perceived value deficiencies due to existing or expected downside losses of value; existing or expected failures to meet promised or anticipated gains in value; or desire to achieve new, improved value levels through marketing and/or technological initiatives. He suggests three ways to approach value deficiencies: improve how work is currently performed; perform current work differently; and/or perform different types of work. Central to this work is the notion that enterprise transformation is driven by value deficiencies and is fundamentally associated with investigation and change of current work processes such as to improve the future states of the enterprise. Potential impacts on enterprise states are assessed in terms of value consequences.

Many of the well-known contributors in the field of enterprise resource planning presented had developed their own unique model. Each had established a strategy to address the evolution and growth of the enterprise. Differences between the models varied based on the challenge presented and the final objective to be achieved by the enterprise. A comparison of the ERP models presented is illustrated in Table 1.

Fundamentally, system engineering and system management are inherently transdisciplinary in attempting to find integrated solutions to problems that are of a large scale and scope (Sage, 2000). Enterprise transformation involves fundamental change in terms of reengineering of organizational processes and is also clearly transdisciplinary as that success necessarily requires involvement of management, computing, and engineering, as well as behavioral and social sciences. Enterprises and associated transformation are among the complex systems addressed by systems engineering and management. Rouse's efforts (2005, 2006) provide a foundation for addressing these issues and the transdisciplinary perspective of systems engineering and management provide many potentially competitive advantages to deal with these complex problems and systems.

Network Effects and Their Role in Enterprise Resource Planning

In today's information economy, introducing new technologies into the marketplace has become a significant challenge. The information economy is not driven by the traditional economies of scale and diminishing returns to scale that are prevalent among large traditional production companies. It has been replaced by the existence of network effects (also known as network externalities), increasing returns to scale and path dependence. This is the core economic reality, and not at all a philosophy, which has revolutionized traditional economic theories and practices, resulting in a new approach to economic theory as it pertains to the information economy.

There are a number of market dynamics or external variables that impact the success of any new technology entering the market. The most common variable is the element of network effects. A product exhibits network effects when its value to one user depends on the number of other users. Liebowitz and Margolis (1994) define network effects as the existence of many products for which the utility that a user of them derives from their consumption increases with the number of other agents that are also utilizing the product, and where the utility that a user derives from a product depends upon the number of other users of the product who are in the same network. Network effects are separated into two distinct parts, relative to the value received by the consumer. Liebowitz and Margolis (1994) denote the first component as the autarky value of a technology product, the value generated by the product minus the other users of the network. The second component is the synchronization value, the value associated when interacting with other users of the product. The social value derived from synchronization is far greater than the private value from autarky. This social value leads the way to increasing returns to scale, by creating path dependence (also known as positive

Table 1. Comparison of ERP models

ERP Models				
Contributor	**Model**	**Strategy**	**Challenge**	**Objective**
Gary A. Langenwalter	Total enterprise integration (TEI) framework	• Integrates customer, manufacturer, and supplier • Provides competitive edge by: maximizing speed of information, minimizing response time, pushing decisions to the correct organizational level, maximizing information available to decision-makers, and direct integration of supply chains	• Establishing seamless communication • Multi-functional integration	• Incorporate all stakeholders • Empower people at all levels of the organization • Improve quality of decision-making
Thomas H. Davenport	Operational data model	• Introduces data-oriented culture • Supports a customer and product focus • Uses operational data to measure operational effectiveness	• Defining organizational boundaries • Enhancing operational effectiveness	• Define operational performance parameters • Measure operational effectiveness • Support effective decision-making
Christopher Koch	Business performance framework	• Supports data sharing • Integrates financial information • Integrates customer order information • Standardizes manufacturing process • Reduces inventory • Standardizes HR information	• Centralized and distributed approach to data management • Establishing requirements for accessing, viewing, and manipulating data	• Integrate all organizational entities across a single systems platform • Manage enterprise in transdisciplinary and transinstitutional fashions
William B. Rouse	Strategic innovation model	• Introduces a strategic approach to innovation • Focuses on the need for human-centered planning, organization, and control • Differentiating from the competition based on core product technologies	• Enhancing people's abilities and overcoming their limitations • Essential challenges: growth, value, focus, change, future, knowledge, and time	• Support strategic planning, operational management, and engineering • Ensure the successful innovation of products and systems
Christopher Gardner	Valuation model	• Presents methodologies for new enterprise business development initiatives • Determines the contribution an enterprise system makes to a company's shareholder value	• Defining the value of new technologies • Mitigating the potential risk and capitalizing on the high rate of return	• Increase shareholder wealth • Maximize the value of cash flow generated by operations
Peter F. Boer	Options model	• Bridges the gap between the business and scientific/engineering communities • Introduces research and development that creates options for the enterprise to grow in profitability and size	• Identifying the link between corporate strategy and technology strategy • Minimizing capital requirements • Understanding terminal value of a project	• Introduce research and development stages for assessing technology opportunities • Determine strategic premium created by real options

feedback) and influencing the outcome for network goods. These efforts and others are nicely summarized in Liebowitz (2002) and Liebowitz and Margolis (2002).

Path dependence is essential for a company to reach critical mass when introducing new technologies into the market. As the installed customer base grows, more customers find adoption of a new product or technology of value, resulting in an increase in the number of consumers or users. Consumer choices exhibit path dependence for new products as others realize their value, eventually leading to critical mass. Path dependence is simply an effect whereby the present position is a result of what has happened in the past. The path dependence theory demonstrates that there are a number of stable alternatives, one of which will arise based on the particular initial conditions. Path dependence is evident when there is at least persistence or durability in consumer decision-making. Decisions made by early adopters can exhibit a controlling influence over future decisions or allocations made by late adopters. These product decisions are often based on the individual arbitrary choices of consumers, persistence of certain choices, preferences, states of knowledge, endowments, and compatibility. The outcome may depend on the order in which certain actions occur based on these behavioral determinants.

Network effects, increasing returns, and path dependence can be better illustrated when applied to the concept of a virtual network. The virtual network has similar properties to a physical or real network, such as a communications network. In such networks, there are nodes and links that connect the nodes to each other. In a physical network, such as a hard-wire communications network, the nodes are switching platforms and the links are circuits or telephone wires. Conversely, the virtual network nodes may represent consumers and transparent links represent paths, as driven by network effects and path dependence, that impact consumer behavior. The value of connecting to the

network of Microsoft Office users is predicated on the number of people already connected to this virtual network. The strength of the linkages to the virtual network and its future expansion is based on the number of users who will use the same office applications and share files.

Path dependence can easily generate market dominance by a single firm introducing a new technology. This occurs when late adopters latch onto a particular virtual network, because the majority of users already reside on this infrastructure and have accepted the new technology. As more consumers connect to the virtual network, it becomes more valuable to each individual consumer. Consumers benefit from each other as they connect to the infrastructure. The larger network becomes more attractive to the other consumers who eventually become integrated. A communications network can best illustrate this concept. For example, additional users who purchase telephones and connect to a communications infrastructure bring value to the other users on the network, who can now communicate with the newly integrated users. This same concept applies to the virtual network and has the same impact. Real and virtual networks share many of the same properties and, over time, are destined to reach a critical mass of users.

New and emerging startup enterprises seeking to take advantage of network effects and path dependence when launching a new technology or innovation in the marketplace must have a reliable and operationally efficient enterprise resource planning (ERP) solution in place. The ERP solution must be capable of attaining operational velocity to address market demands. Miller and Morris (1999) indicate that traditional methods of managing innovation are no longer adequate. They suggest that as we make the transition to fourth generation R&D, appropriate complex timing for innovations remains a significant challenge. These authors assert that as new technologies and new markets emerge, management must deal with complexity, enormous discontinuities, increasing volatility, and the rapid evolution of industries. The challenge becomes that of linking emerging technologies with emerging markets through methods such as an ERP solution to bridge this link and to allow new emerging enterprises, or established mature enterprises seeking to transform themselves, to adapt quickly to the dynamics of the marketplace. The solution supports both continuous and discontinuous innovation as defined by Miller and Morris (1999). This form of innovation works well when customer needs in a competitive environment can be met within existing organizational structures.

In contrast to this, discontinuous innovation may bring forth conditions emanating from fundamentally different new knowledge in one or more dimensions of a product or service, and offer significantly different performance attributes. Discontinuous change potentially brings about change in a deep and systematic way. It offers a potential lifestyle change to customers that can be dramatic. Miller and Morris (1999) note, for example, the transition from typewriters to personal computers for producing written documents. In part, this occurred because customers no longer were satisfied with the existing framework of capability offered by the typewriter. New knowledge, organizational capabilities, tools, technology, and processes changed the behavior and desires of the customer. In addition to this change was also the change resulting in supporting infrastructure. Miller and Morris (1999) emphasize that discontinuous innovation affects not only products and services but also the infrastructures integral to their use, as well as extensive chains of distribution that may involve a plethora of affiliated and competing organizations.

As the threat of unexpected competition surrounds any new enterprise entering the market, the risk associated with technology shifts and the compression of the sales cycle make successfully managing discontinuous innovation a necessary challenge for success. We must be able to gauge how the market is evolving and what

organizational capabilities must exist to sustain competitiveness as a result of this evolution. Because innovation usually requires large capital infusions, decreasing the time for appearance of a positive revenue stream is critical to the success of the enterprise. This decrease in time is made possible through operational velocity attainment, which requires changes in existing implementation strategies and organizational capabilities. This requires a collaborative effort between the various involved organizations to understand what is needed to support new innovations. Responsibility for supporting new innovation is not only supported by internal organizations but by such external organizations as suppliers, customers, and partners. Organizational structure, capabilities, and processes are fundamental to an evolutionary ERP model and serve as the framework for supporting new technology adoption in the marketplace.

The information economy is driven by network effects (also termed demand-side economies of scale or network externalities). Network effects support path dependence and are predicated on Metcalfe's Law, which suggests that the value of a network goes up as the square of the number of users (Shapiro & Varian, 1999), or on recent suggested modifications to this (Briscoe, Odlyzko, & Tilly, 2006). Positive effects occur when the value of one unit increases with an increase in the number of the same unit shared by others. Based on this premise, it is possible to create an enterprise resource planning model that influences positive feedback from human behavior in adopting new technologies and accelerates critical mass early in the deployment phase of the product development lifecycle, by attaining operational velocity. Operational velocity is defined in terms of speed in delivering products or services to market, meeting all customer expectations in a timely manner, and decreasing the time for appearance of a positive revenue stream as much as possible. This ERP model would support the integration of data, standardization of processes, order fulfillment, inventory control, supply-chain management, and customer relationship management (CRM) as critical drivers to result in enterprise transformation.

William B. Rouse, in his work *Strategies for Innovation* (Rouse, 1992), states "A prerequisite for innovation is strategies for making stakeholders aware of enabling technology solutions, delivery of theses solutions in a timely fashion, and providing services that assure the solutions will be successful. These strategies must not only result in successful products or systems, they must also create a successful organization—an enterprise—for developing, marketing, delivering, and serving solutions" (p. 2). His philosophy encompasses the human-centered design approach that takes into account the concerns, values, and perceptions of all stakeholders during a design initiative. This approach entertains the views of all the stakeholders, balancing all human considerations during the design effort.

Traditionally, when designing an enterprise resource planning solution, very few enterprises are easily able to think strategically. Most are only concerned with today's products and services and the financial profits and revenue growth realized in the short term. They often fail to properly forecast future growth and to properly scale their ERP in order to meet the potential consumer demands of the future. An enterprise must be able to plan for and respond to future demands by analyzing the market and evaluating the impact that their core product technologies will have in the marketplace. Market demand will drive consumer needs and desire for these core product technologies, as well as the type of ERP that will be used to support these products. An effective ERP must be capable of assessing and balancing all stakeholders' interests consciously and carefully. The market share that an enterprise is able to acquire for its core product technologies can be tied to how well an ERP is developed, deployed, and implemented in order to provide the operational support infrastructure needed. Many of the traditional success factors

for an enterprise have been their position in the marketplace, achievements as innovators, productivity, liquidity and cash flow, and profitability. In order for an enterprise to grow and mature, it must be able to respond to market demand in a timely manner. Responding to market demand includes timely delivery of products and services, immediate attention to customer problem/resolution, and continuous process improvements. Operational velocity attainment becomes the focus and the critical success factor in the execution of an evolutionary ERP strategy, thus supporting the long-term vision of the enterprise by ensuring a strategic advantage for the enterprise.

A well-thought-out ERP strategy will require advanced planning to determine how each of the organizations will be integrated in supporting the long-term objective. Critical to the success of an enterprise is how well its associated organizations can adapt to organizational change, as the company begins to mature and demand increases for the new innovative products and services. Change may include the type of culture that is fostered, tools used, and level of knowledgeable resources required to make the organizational transitions. Most importantly, customer experiences becomes the focus. How fast an enterprise can service customers to meet their expectations may determine how soon it meets revenue expectations. The quality of on-time customer service could impact the number of future sales. A good product or service, combined with excellent customer service, may drive more business to the enterprise, decreasing the time taken to meet revenue forecasts. The mechanism used to drive customer on-time service becomes what we call an evolutionary ERP model. In order for new core technology products to become acceptable to a newly installed base of customers, service delivery and customer response times must be minimized as much as possible. True enterprise growth and profitability can be made possible through this model for emerging enterprises delivering new innovations to the marketplace. The model takes into account the long-term vision of the enterprise, which is a key to its consistent success. Rouse (1992) states this well when he says that many technology-based startup companies are very attracted to learning about new technologies, using these to creating new products, and hiring appropriate staff to accomplish these. Such activities may get the product, resulting from the enterprise vision, into the marketplace. Initial sales and profit goals may be achieved. He appropriately notes that without a long-term vision, plans for getting there, and an appropriate culture; no amount of short-term oriented activity will yield consistent long-term success.

The strategic advantages that a well-defined, developed, and deployed ERP brings to the enterprise are: integration across the enterprise, communication, operating efficiencies, modeling, and supply chain management. These effective strategies assist with bridging the overall corporate strategies with the organizational objectives. Integration across the enterprise supports the following organizational objectives:

- Maximization of speed and throughput of information,
- Minimization of customer response times,
- Minimization of supplier and partner response times,
- Minimization of senior management response times,
- Decision-making authority pushed to the appropriate levels within the organization, using work flow management,
- Maximization of information to senior management,
- Direct integration of the supply chain,
- Reduction of inventories,
- Reduction in order-to-ship time,
- Reduction in customer lead times, and
- Total quality integration.

Communication links the enterprise to both the suppliers and the customers. Good communication between supplier and the enterprise can help reduce design errors, foster good supplier and enterprise relationships, reduce enormous costs, reduce the supplier's time to respond to the enterprise, and improve performance and market adoption of a new core technology product.

Langenwalter (2000) indicates in his work on enterprise resource planning that integrating the design process with customers can surface customer responses with respect to their true needs. He emphasizes the voice of the customer (VOC) as a proven methodology that addresses the true needs and expectations of the customer. VOC serves as basic input to the evolutionary ERP model. Key customer considerations in achieving operational velocity using this model are ranked customer expectations, performance metrics, and customer perceptions of performance.

In *The Valuation of Technology*, Boer (1999) is also concerned with these customer considerations by including the concept of the value proposition from the customer's viewpoint. He emphasizes that stakeholders must find useful ways to determine the value added in every step of the business process from the viewpoint of the customer. The enterprise must exist to deliver value to the extent that it improves operational performance and/or lower costs through new or enhanced products, processes and services. For example, the operations of an enterprise will focus on procuring equipment and materials from vendors and suppliers to produce products on time and within budget. The operations objective is to meet customer demand through scheduling, procurement, implementation, and support, to meet the ever-changing needs of the customer environment. These changes must be measured so that the operations of the enterprise may be able to meet the needs of the marketplace. Such flexibility of operations in the marketplace is essential in keeping up with the dynamic needs of the customer.

In the new technology age, markets are moving much faster than traditional predictive systems suggest. Flexibility therfore becomes an essential and necessary element in achieving operational velocity. To achieve this, Langenwalter (2000) introduces a new measurement system that recognizes the ever-changing dynamics of products, customers, workers, and processes. His approach is based on the assumption that all products have life cycles and should have four key metrics: profitability, time, quality, and company spirit. Encompassing this approach would be the execution of a continuous process improvement initiative, with respect to the operational component of the product lifecycle. He proposes that the enterprise measure each organizational contribution to profit for the entire lifecycle of the product. An ERP can effectively measure the contribution to margin that a sales organization may make on new product releases. Unprofitable products can be immediately identified and retired. In comparison, an ERP can also track the total lifecycle cost that a manufacturing organization incurs when producing a product. Total profit and loss (P&L) responsibilities can be tracked and material procurement and cost strategies can be evaluated to enhance profitability to the extent possible. Other organizational facets such as engineering and marketing can increase profits, by accessing customer profile information from an ERP and trending product demand for various new features and functionality. Incorporating new design considerations in future product releases may also increase potential profitability, as more high-end products are released.

The element of time is an important metric and is truly a measure of process, although process efficiencies can also translate into cost savings. Langenwalter (2000) describes three key time dimensions: time to market, time to customer, and velocity. Each is a component of operational velocity. In achieving operational velocity, time to market is critical for new technology adoption. It is crucial for new enterprises to launch their core technology product(s) on time, in order to sustain long-term product profitability. This is especially

true if new technology is involved. Langenwalter (2000) indicates that a study performed by the McKinsey Consulting Group reflects that a six-month delay in entering a market results in a 33% reduction in after-tax profit over the life of the product. In addition, the six-month delay is five times more costly than a 50% development-cost overrun and approximately 30% more costly than having production costs 10% over budget.

An ERP should be capable of monitoring product development and manufacturing processes to ensure timely delivery of products to market. Such items as customer requirements, technical viability, manufacturing costs, production volumes, staffing levels, work order priorities, material requirements, and capacity requirements can be accessible via the ERP, and allow both the engineering and manufacturing components in an organization to respond to product demands quickly. The ERP supports time to market in that these two organizations are able to ensure efficient product development manufacturing processes and organizational communication in launching new products to market. The ERP, so enabled, becomes the common domain and communications intermediary between engineering and manufacturing.

Time to customer is the next most critical dimension, or aspect, of time as described by Langenwalter (2000). This time dimension is focused on reducing lead times to customers. For example, manufacturers look to reduce the lead-time it takes to produce a product, component, or assembly. Although it may have taken weeks to produce a particular component, improved manufacturing capabilities may now enable this process in only two days. This may have been accomplished through the use of an ERP, which made it possible to track performance metrics of the various manufacturing processes. As a result of isolating various inhibiting manufacturing processes and improving these processes, time to customer was reduced significantly, thus supporting the operational velocity objective of the enterprise.

Another good example is customer care, achieved by responding to a product fault scenario and providing technical support capability to the customer for fault resolution. Response to a customer call may have originally taken 72 hours to resolve the problem due to the lack of an effective scheduling tool for the timely dispatching of technical support field resources. With the integration of a resource-scheduling tool within ERP, customer care can now respond perhaps within four hours and provide timely customer support. Velocity, the final dimension that Langenwalter presents, is defined as the total elapsed time consumed by a process divided by the actual value-added time contributed by the same process.

The quality metric of the product life cycle, as described by Langenwalter, focuses on continuous improvement. Quality metrics are very much tied to what may be called an evolutionary enterprise resource planning architecture framework. Operational velocity is only as good as the product and the service that is delivered. Any compromise in quality may translate to potential customer attrition and/or the degradation of market share. A good ERP should be capable of tracking product component failure rates and product design flaws, so that immediate action may be exercised on the part of the enterprise. Speed without quality only becomes a formula for failure. Product failures are not the only inhibitors of quality. A lack of knowledgeable and skilled resources can compromise quality, and this describes Langenwalter's last critical metric – company spirit. He emphasizes the fact that people are the ones who develop relationships with customers and suppliers, eventually leading to new products and processes. This metric goes outside much traditional thinking. However, during the enterprise startup technology revolution, company spirit is generally the most important element of survival and success among enterprises. This leads to a greater

sense of ownership and responsibility among the people involved. An enterprise without a healthy team spirit and aggressive workforce has little chance of success.

Rouse (1992) introduces yet another interesting growth strategy that further supports the concept of operational velocity for new technology adoption. He describes a strategy for growth via enhancing productivity through process improvement and information technology. This approach leads to higher quality and lower cost of products and services and, eventually, to greater market share and profits. Enterprise performance is not as visible as product performance, so the money and time saved on process refinements often go unnoticed. Each approach has its own value. Rouse describes product value as the foundation for growth and indicates that the challenge of value concerns matching stakeholders' needs and desires to the enterprise's competencies in the process of identifying high-value offerings that will justify investments needed to bring these to market. Value to the customer is dependent on the particular market domain. The most noticeable form of value comes in the form of new innovations that meet a customer's economics or needs. Customers quickly realize the benefits of a new technology product; however, the real value is determined at the enterprise level, where customer support becomes critical. Technology products are sophisticated and require a high level of customer support when potential problems arise. After the sale of the product, the relative performance of the enterprise becomes the focus of the customer. Lack of timely and quality support can erode consumer confidence and eventually erode market share for an enterprise.

After the launch of its first product, an enterprise is immediately under the scrutiny of the public. Often, early adopters of new technologies can either make or break an emerging enterprise. Early adopters will assess the enterprise on product quality, delivery, and customer support. If the product is reliable and performs well, then delivery and customer support become the two most critical criteria that a customer will evaluate. It is usually the shortfalls in these two areas that diminish consumer confidence and challenge the credibility of a new enterprise. An enterprise that has an ERP strategy to address these criteria is better positioned for success. If the ERP is designed well, it will allow the enterprise to ensure quality delivery and customer support to the end users. The true value to the customer is realized in enterprise performance as opposed to product performance. Historically, customers have been prone to pursue other vendors because of lack of customer support, moreso than with average product performance. The result of a well-executed ERP strategy enables the enterprise to react immediately and consistently, enabling the organizational components to focus their human and financial capital in the right areas.

Rouse describes the challenge of focus as deciding the path whereby the enterprise will provide value and grow. Rouse (2001) introduces some common challenges in and impediments to an organization's decision making, including:

- Assumptions made,
- Lack of information,
- Prolonged waiting for consensus,
- Lack of decision-making mechanisms,
- Key stakeholders not involved, and
- Decisions made but not implemented.

An enterprise is capable of addressing these challenges if it institutes an ERP solution during its evolution. The ERP solution will bridge many of the communication gaps common among enterprises that are often organizationally disconnected. A good ERP solution will support information sharing, track performance metrics, and archive information, thus providing methods and tools in supporting rapid decision making and furthering the concept of operational velocity. Many times, senior management is unable to focus on key areas due to lack of information

and decision-making tools. This problem can be overcome by integrating these capabilities with the ERP. An ERP can scale easily to meet the business needs. The enterprise that plans for growth through its evolution can scale more easily and adapt to change.

Rouse (2001) states that "given a goal (growth), a foundation (value), and a path (focus), the next challenges concern designing an organization to follow this path, provide this value, and achieve this goal" (pp. 5-6). The climate of the enterprise changes rapidly and dramatically throughout its evolution. As new core technology products are launched, the environment is subject to change. Enterprises find ways to scale their infrastructures to meet growth, fend off competition, restructure, reengineer, and support virtual organizations. The objective of change is to improve quality, delivery, speed, and customer service. All of this is made possible through a well-integrated ERP. An ERP capable of facilitating change allows the enterprise to foster new opportunities for growth and reward. As an enterprise evolves over time into a major corporation, business practices change and a paradigm shift occurs over several phases of maturation. The ERP can assist an enterprise in transitioning new business philosophies and practices and to help pave the way for future growth. There is a major need to anticipate future opportunities and threats, plan for contingencies, and evolve the design of the enterprise so that the plans are successful.

The value of the future is difficult to estimate; this realization has lead to another interesting concept, the option value of an enterprise. As previously mentioned, Boer (2002) is a major proponent of options value as applied to the enterprise. This concept explores investment decisions based on buying an option on the future and determining what that option is worth. An enterprise must plan for future growth and weigh the various investment alternatives available. These include looking at the following:

- Strategic fit,
- Financial payoff,
- Project risk and probability of success,
- Market timing, and
- Technological capability of the enterprise.

The above factors weigh into the decisions made to invest in the future. It is through investments in education, training and organizational development that the enterprise is enabled to meet future objectives through resource allocation.

Other investments in research and development technology make decision-making much more complex. However, they may yield promising future results if planned well and integrated with other decisions taken. Investments in R&D require knowledgeable resources that can influence the abilities of an enterprise to provide value. Knowledge management becomes a key element in the overall ERP strategy. Rouse (2001) indicates that knowledge management and knowledge sharing (Small & Sage, 2006) will promote an integrated approach to identifying, capturing, retrieving, sharing, and evaluating an enterprise's information assets. This may be achieved by applying knowledge management concepts to the ERP strategy. A sound return on investment (ROI) model for an ERP should assess the dynamics of the enterprise, changes needed, and projected savings from these changes. The changes themselves should be measurable. An ERP must be planned carefully and, most importantly, well-executed with all resource considerations made during its evolution. The benefits derived from a well-executed ERP should reveal improvements in task management, automation, information sharing, and process workflow. Each of these components improves the most scarce resources that people face within the enterprise, that of time.

Time is a key ingredient for gaining organizational control. An ERP system with integrated tools and methods for communicating and modeling assists human resources with time management. Time management can be a critical problem

and human resources can easily find themselves becoming reactive versus proactive in their day-to-day activities. Rouse (2001) emphasizes that it is important to increase the priority given to various long-term strategic tasks, especially since they too often suffer from demands for time from the many near-term operational tasks. A well-integrated ERP supports time management and allows human resources to gain control of their time and allocate it across the appropriate tasks. It further supports the need for long-term planning by supplying various tools and methods for enhancing strategic thinking. The tools and methods integrated within the ERP should improve both the efficiency and effectiveness of time allocation to tasks. An ERP that is incapable of handling the challenge of time diminishes the true value of the ERP. Time management is a crucial component in achieving operational velocity and must be controlled, in order for the enterprise to respond quickly to customer demands.

The seven challenges to strategic management of Rouse (2001) are all critical elements that need to be considered when designing an ERP. A well-designed ERP helps position the enterprise well in the market and gives it a strategic advantage. The true gauges of success of an enterprise, with a successfully executed ERP, will be reflected in how it is positioned in the marketplace. Rouse (1992) has identified five gauges of success:

1. Standings in your markets,
2. Achievements as an innovator,
3. Productivity,
4. Liquidity and cash flow, and
5. Profitability.

Each of these gauges of success is tied to shareholder value. Gardner (2000) also raises a major consideration about designing an ERP in his book *The Valuation of Information Technology*. He asks the questions:

What contributions will an information technology system make to a company's shareholder value?

How can an information technology system be constructed to create shareholder value? In other words, not just determine the effect of a system on shareholder value but guide the activities involved in its construction in the first place. (p. 63)

He emphasizes the need to predict the shareholder value that will be created by an information system before it is actually built. In the context of an ERP, the objective is to increase the wealth of shareholders by adding premium growth to their stock. An ERP can improve the asset utilization of an enterprise by allowing shareholders to increase their returns on invested capital. The traditional approach to increasing shareholder wealth consists of maximizing the value of the cash flow stream generated by an operational ERP. The cash flow generated from the ERP is allocated among the shareholders and debt holders of the enterprise. Shareholder value is traditionally measured by using the DCF method, which is central to the valuation of assets and the return they generate in the future. Boer (1999) addresses the DCF method well in his book *The Valuation of Technology*. He defines the premise of the DCF method "as a dollar received tomorrow is worth less than one in hand today" (p. 63).The question that arises from this premise is how much should one invest today in order to earn a dollar tomorrow. To address this, Boer presents one of the common DCF methods known as net present value.

The NPV method can be used to compute the value of tomorrow's dollar. Boer properly defines NPV as "the present value of a stream of future cash flow less any initial investment (p. 98)." NPV addresses the time value of money, which is essential for developing an ERP strategy, with the objective of attaining operational velocity. Gardner (2000) illustrates how this has a significant effect on the management of ERP systems. If the rate of return is high, schedule delays in deploying an ERP can erode value, which makes time to market critical; and since short product life generates as much value as long product

life, there should be little resistance in replacing legacy systems. In comparison, if the rate of return is low, delays have little effect on value, and a longer product lifecycle is feasible, thereby allowing for a more thorough systems development effort. Gardner extends the NPV method to an ERP system and illustrates how shareholder value is created by changes in the present value of the cash flow to shareholders due to the use of the ERP system.

The DCF method illustrated here focuses solely on the economic value of the enterprise. Boer (2002) introduces a concept known as the options value of the enterprise in his book *The Real Options Solution: Finding Total Value in a High-Risk World*. The options method is presented as a means to value the strategic capital of an enterprise. This method is known as the total value model and combines the economic value and strategic value of the enterprise, and also takes into account three major drivers that affect value: risk, diminishing returns, and innovation. Enterprises satisfactorily releasing new technologies into the marketplace normally increase their strategic value if consumers adopt these new technologies to meet their needs. New technology adoption in the marketplace can vary based on need, price, standards, and other related factors. Once the need is recognized, operational velocity becomes critical to answering the customer's needs. How fast customers can be served and cared for will drive the strategic value of the enterprise. A well-designed and executed ERP can assist with operational velocity attainment by improving efficiencies, speed, and time to market. Boer's total value model uses a six-step approach to computing the total value of an enterprise. His practical six-step approach encompasses the following:

- **Step 1.** Calculate the economic value of the enterprise, where free cash flow (FCF) is defined as the actual cash flow minus the amount of cash that must be reinvested:

Economic Value = FCF / (Cost of Capital – Growth Rate).

- **Step 2.** Frame the basic business option and identify strategic options. For example, leasing space at another site and expanding the enterprise may yield additional future revenue. Here, investment in an ERP system may yield future revenue, as a result of enhancing operational velocity.
- **Step 3.** Determine the option premium, which is the premium paid or expenditures incurred to make the plan actionable. For example, this may include the option cost of technology, people, partners, financing, systems, and R&D.
- **Step 4.** Determine the value of the pro forma business plan, where NPV is computed to determine valuation of the enterprise
- **Step 5.** Calculate the option value. Here, the Black-Scholes option formula is used using five key elements: value of the underlying security, strike price, time period of the option, volatility, and risk-free rate.
- **Step 6.** Calculate total value according to *Total Value = Economic Value + Strategic Value.*

Boer's model computes the true value of the enterprise taking options thinking into consideration, thus reflecting real life and the strategic payoff that can result if an enterprise is successful. To clarify the concept, Boer makes an interesting analogy by illustrating the strategic value of a common family with a low standard of living. The family's principal economic activities concern the income produced. Costs such as mortgage, utilities, and gas are set against this revenue. Any savings are stored away as additional income. The income and expenses mentioned thus far only reflect the economic value of the family. The potential strategic value lies in the education of its children. Education could pay off in the long term and increase the family's standard of living. However, there are also significant market risks.

Once the children are educated, the marketplace may not demand their skills, or they may not meet the various job requirements of their profession. In comparison, an enterprise may have potential strategic value in a new technology that it develops. The enterprise may have sufficient venture capital to cover R&D expenses for the next few years. Once the technology goes to marketplace for the first time, the enterprise has first mover advantage in the market if it attracts enough early adopters to build market momentum. Critical mass can be achieved as momentum for the product accelerates. However, there could be the risk of competitors with a similar technology that may go to market during the same time frame. In addition, the competitor may have a similar product with different performance standards, which adds to the competitive nature of the situation. This leads to a race for market share and ultimate establishment of the preferred technology standard between the products.

Strategic value is not always predictable, and the dynamics of the market change constantly. A negative impact on strategic value could result in zero return; this results in a loss of venture capital to cover the R&D expenses. There is evidence during the past five years that a number of startup technology enterprises never arrived at fruition in strategic value. The strategic value represents the potential revenue that could be realized if market conditions are ideal for the enterprise. Gardner (2000) estimates the revenue opportunity for an enterprise using *Annual Revenue = ;Annual Market Segment Size x Annual Likelihood of Purchase x Annual Price.* The terms in this relation are time dependent and are critical to new technology adoption in the marketplace. Forecasting potential annual revenue requires understanding the purchasing decisions and patterns customers will make. Decreasing the time for appearance of a positive revenue stream for an enterprise, a new technology into the marketplace is highly desirable. The mechanism for achieving this objective is the evolutionary enterprise resource planning architecture framework, which will accelerate critical mass early in the deployment phase of the product development lifecycle by achieving operational velocity. Thus , the work established by the early pioneers of ERP and technology valuation methods has laid the foundation for a new ERP paradigm to evolve and support operational velocity attainment.

Network Elements Influencing Path Dependence and Network Effects

Consumers who become completely satisfied with a new technology product or innovation realize the value proposition derived from this new creation. For example, the value of a digital subscriber line (DSL) at home brings value to the home PC user who now has high-speed access to the Internet. The home user is no longer confined to the limiting speed capability of a 56 Kbps dial-up modem. As more users adopt DSL, due to its broadband capabilities, increasing returns to scale and path dependence are achieved. The economy has shifted from the supply-side economies of scale, based on the traditional industrial era of mass production driven by unit costs, to increasing returns to scale (also known as demand-side economies of scale) driven by consumer attitudes and expectations. Strategic timing is vital with respect to demand-side economies of scale. First, introducing an immature technology into the marketplace may result in negative feedback from potential consumers. For example, potential de-

sign flaws, functional limitations and constrained feature sets may overshadow the true value of the technology, making it less attractive to potential consumers. In addition, moving too late in the market means not only missing the market entirely but also the opportunity to acquire any significant market share. Moving without an effective ERP strategy compromises new customer acquisition and customer retention.

The marketplace is subject to various network elements that influence path dependence and network effects of new technology adoption. These network elements directly impact consumer decision-making and lead to the formulation of consumer perceptions and expectations of new technology. Network elements can be defined as economic, business, regulatory, market, and technological influences that impact consumer decision making relative to new technology adoption. Understanding what drives consumer behavior and how it can be controlled allows innovators and technologists to achieve better success in launching new products while gaining market acceptance.

In *Information Rules*, Shapiro and Varian (1999) identify 10 primary network elements that influence consumer decision-making. They describe how these network elements impact consumer decision making with respect to new technology adoption. The network elements described are: partnerships, standards, pricing differentials, product differentials, lock-in and switching costs, complementary products, first

Figure 1. Network elements

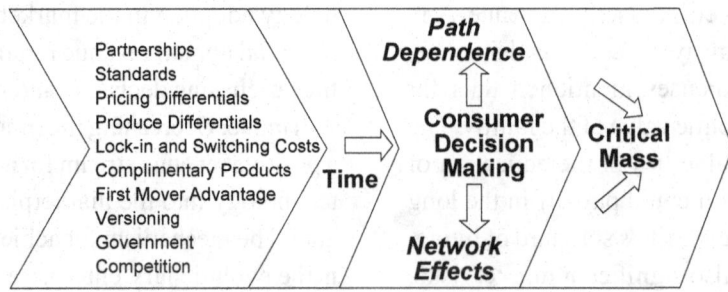

mover advantage, versioning, government, and competition. Figure 1 reflects these 10 primary network elements that influence consumer decision making over time. These network elements will shape consumer choice, based on the degree of consumer confidence, need, desire, satisfaction, and comfort with adopting a new technology. The degree that these human traits will vary among consumers will determine the speed with which a new technology will be adopted. Consumers will most likely fall into three categories of adoption: early, evolving, and late. As a technology becomes popular, consumer decision-making becomes positive with respect to new product acquisition. Early adopters of the technology will begin to generate demand for the product.

Based on the success of the initial product, more consumers will see and understand the value proposition realized by the early adopters. A large number of consumers begin to evolve connecting to the network of users. At this stage, consumer choice begins to exhibit path dependence and network effects. As the network of users begins to accelerate, critical mass is realized. Critical mass occurs when a large enough customer-installed base is established, as a result of positive feedback derived from the growing number of adopters. The network continues to expand until these late adopters eventually interconnect and the product reaches maturity in the marketplace. Network elements are also critical to consumer decision-making and can impact the destiny of a new technology if unrecognized. A good illustration of this was the competition between Beta and VHS in the 1970s. Beta was believed by most to be clearly superior to VHS in quality; however, VHS was the de-facto standard among consumers due to its compatibility. Operational velocity is one of the most fundamental critical success factors influencing adoption of new technology the presence of network elements. Operational velocity is a factor that needs the most attention and the one that can easily be controlled by implementing an effective ERP model. Since

understanding the influence network elements have on achieving critical mass is essential, a narrative follows describing each one of the elements shown in Figure 1.

The first network element reflects partnerships, which provide a strategic advantage. New technology enterprises, possessing a leading-edge niche product in the marketplace, may find that one or more partnerships, with major players offering a complementary product suite, may be the answer to acquiring critical mass early in the game. An emerging enterprise would have the opportunity to immediately sell its new product to the existing installed customer base of its partner. This existing installed customer base may have taken the partner years to establish and grow, thus offering an advantage to a new enterprise, which has not yet established customer relationships or gained brand name recognition. An opportunity to sell into an existing installed base of customers, by gaining the visibility and credibility via a strong strategic partner, can shorten the sales cycle and accelerate critical mass. Alliances can even be established through suppliers and rivals as a means of accelerating critical mass attainment. It would also be advantageous for the enterprise to offer incentives when possible. Consumer confidence may be won, along with new customer acquisitions, by allowing customers who are undecided over a new technology to sample or test the new product.

The next element reflects standards. Standard setting is one of the major determinants when it comes to new customer acquisitions. Consumer expectations become extremely important when achieving critical mass, especially as each competitor claims they have the leading standard. Standards organizations try to dispel any notions or perceptions as to which company drives the predominant standard; however, most of these standards groups are comprised of industry players, each of whom attempts to market their own agendas. Most will try to influence the direction of standards setting for their own best interests.

Standards are necessary for the average consumer, who wants to reduce potential product uncertainties and lock-in (defined as consumers forced to use a non-standard proprietary product). The product that consumers expect to be the standard will eventually become such, as standards organizations and large industry players begin to shape and mold consumer expectations. Standards increase the value of the virtual network and build credibility for new technologies introduced into the market.

One strategy often used among new and aggressive companies in order to gain market momentum is that of pricing differentials. This network element can ignite market momentum by under-pricing competitors and targeting various consumer profiles. Some enterprises may use various pricing strategies to offer incentives to new customers. As a result, this may be an effective strategy, since some customers may be more price sensitive and may not be as influenced by factors such as standards. A common pricing strategy is differential pricing; this may take the form of personalized or group pricing. Personalized pricing takes the form of selling to each consumer at a different price. The focus is in understanding what the consumer wants and tailoring a price to meet the consumer's needs. Group pricing will set targets for various consumer profiles and group them accordingly. This affords flexibility to potential consumers and takes into account various price sensitivities that may impact decision-making. Consumer lock-in may be achieved through pricing strategies by offering incentives such as discounts, promotions, and the absorption of consumer switching costs.

Making product differentials available is another strategy that is very common in the technology industry and that can effectively influence consumer decision-making. Product differentials offer consumers a choice across several product derivatives. By designing a new product from the top down, the company can easily engage any potential competition by introducing the high-end

solution first. Once the high-end consumers have been acquired, a low-end solution can be made available to capture the low end of the market. The low-end product also may be used to position the high-end product, when using an up-selling strategy. When introducing a new technology to the market, the market should be segmented based on several factors such as user interface, delay, image resolution, speed, format, capability, flexibility, and features. These factors help target and span various consumer profiles.

As various pricing schemes, product features, and functionality are offered to the consumer, the fears of lock-in and excessive switching costs enter into the decision-making. This network element is one of the most common ones that can halt adoption of a new technology, especially if consumers only deal with one vendor. Most consumers want to deal with two or more vendors in order to maintain a level of integrity among the suppliers offering the product or service. This alleviates the possibility of lock-in with any one particular vendor, as long as they share the same standard. Consumers who deal with only one supplier may face the possibility of lock-in and high switching costs should they decide to select another vendor later. If the existing supplier has not kept up with standards and new technology trends, the consumer may be bound by old legacy infrastructure, which could result in complications if the consumers can no longer scale their environment to meet their own business needs. Some enterprises may absorb the switching costs of a consumer to win their business, if it is in their own best interest, and also if they need to increase their customer base and market share to gain critical mass. New enterprises gaining minimal market momentum with cutting-edge technology product introductions may be more willing to take this approach.

A common competitive strategy used by many high-technology organizations is the selling of complementary products to their installed base of customers. These complementary product offerings can arrive internally within a company by entering new product domains, or externally by offering a partner's complementary product and leveraging on its core competencies.

One of the most challenging network elements that an enterprise faces is having time to market a new innovation, better known as first-mover advantage. First-mover advantage is the best way to gain both market momentum and brand name recognition as the major provider of this new technology. Microsoft, Sun Microsystems, and Netscape serve as good examples of companies that have succeeded in gaining first mover advantage and that have become leaders in their industries (Economides, 2001). An early presence in the market place has allowed these companies to secure leadership positions throughout the years. We note, however, that Netscape has lost considerable market share to Microsoft's Internet Explorer for reasons that are also explainable by this theory.

Over the years, versioning has become a common practice among technology companies. The network element of versioning offers choices to consumers. Companies will offer information products in different versions for different market segments. The intent is to offer versions tailored to the needs of various consumers and to design them to accommodate the needs of different groups of consumers. This strategy allows the company to optimize profitability among the various market segments and to drive consumer requirements. The features and functions of information products can be adjusted to highlight differences and variations of what consumers demand. Companies can offer versions at various prices that appeal to different groups.

As observed with the Microsoft antitrust legislation proceedings, the government can impact the direction of new technology, whether it attempts to control a monopoly or fuel demand for new technologies (Economides, 2001). This network element can be the most restrictive in achieving critical mass. The government, in efforts intending

to ensure that there are no illegal predatory practices that violate true competition, scrutinizes mergers and acquisitions involving direct competitors. There is every reason to believe that it will continue to focus on controlling genuine monopoly power and take action where necessary. All mergers and acquisitions are subject to review by the Department of Justice and the Federal Trade Commission. In addition, the government can serve as a large and influential buyer of new technologies. It can become a catalyst by financing, endorsing and adopting new technologies in order to accelerate their development, adoption, and use. Federal government IT spending on emerging technologies over the next several years can potentially aid those enterprises that are struggling for business and survival as a result of downturns in the economy.

Another network element that can restrict critical mass attainment is competition. Competition in the marketplace will continue as new enterprises are entering the market and presenting a challenge to some large established companies that are plagued by inflexibility and bureaucratic challenges. Companies will compete on new innovations, features, functionality, pricing, and, more importantly, standards. Information products are costly to produce but inexpensive to reproduce, pushing pricing toward zero. Companies that are challenged with a negative cash flow, and have limited venture capital, will need to devise creative strategies to keep themselves in the game. Margins begin to diminish as pricing reaches zero; a complementary set of products or services may be necessary or required to maintain a level of profitability. Knowing the customer, owning the customer relationship, and staying ahead of the competition are the major keys to survival.

Operational velocity is the critical success factor, making a much more profound impact on revenue and profit than the individual network elements described and illustrated in Figure 1. This critical mass determinant, which is the key to the success of an enterprise, is often given very little attention due to the organizational dynamics that take place. Operational velocity, as defined earlier, is speed in delivering products or services to market, meeting all customer expectations in a timely manner, and decreasing the time for appearance of a positive revenue stream as much as possible. This may appear to be a simple concept; however, it is very difficult to master. Without an evolutionary ERP approach, it will be quite challenging to scale a business to meet aggressive future customer demands. There exists a direct relationship between an effective evolutionary the ERP model and operational velocity attainment that allows an enterprise to scale its business accordingly while meeting customer demand in a timely manner. More importantly, there is a unique organizational process lifecycle and key behavioral influences that are essential to implementing an effective ERP model. Without these, the model becomes ineffective, in that ERP has not been implemented in an appropriate and effective manner.

Many enterprises lack any initial operations plan or back-office infrastructure to support new product launches in the marketplace. This is a major challenge in the commercial world, where time to market is critical and development of an effective ERP may be neglected in favor of seemingly more pressing and immediate needs. The primary focus of a new technology company is to amass customers immediately at minimal cost. Often a number of senior executives hired to manage a new enterprise come from sales backgrounds and have very little experience in running a company from a strategic IT, operations, and financial perspective. They sometimes lack the associated fundamental technical and non-technical skill sets, which can easily compromise the future of the business. This often stems from senior executives who come from large corporations but who lack the entrepreneurial experience necessary to launch new businesses. For example, they may fail to see the value of hiring a chief operating officer (COO) who has the required operations background and who understands how

to run a business in its operational entirety. The importance of the COO role is later recognized, but many times it is too late as much of the infrastructure damage has already occurred.

Many of the chief executive officers (CEO) hired to lead new enterprises are prior senior vice presidents of sales. It is believed that they can bring immediate new business to the enterprise and begin instant revenue-generating activity. The sole focus becomes revenue generation and new customer acquisitions. The common philosophy is that the company will resolve the back-office infrastructure later. This is usually a reactionary approach to developing a back-office versus a proactive approach. The lack of a sound evolutionary approach in developing an ERP from concept to market maturity for new products can result in missed customer opportunities, customer de-bookings, loss of market share, lack of credibility, competitive threats and, most importantly, bankruptcy of the business.

Other potential plaguing factors that can impact implementation of an effective ERP strategy are undefined, or at least under-defined, organizational requirements, sometimes termed business rules, and lack of business process improvement (BPI— also known as workflow management) initiatives and strategies. Organizational requirements and BPI for supporting new product launches should be addressed early in the development phase of the new technology. How a product is supported and the relationship and communication between the respective support organizations will be vital to the success of the product. Quite often, organizational requirements and BPI are lacking due to limited understanding and use of contemporary IT principles and practices. Many of the savvy technologists who have started the enterprise may lack knowledge in formal methods, modeling, systems development, and integration. They may be great internal design engineers who have come across a new innovation or idea; however, they lack infrastructure knowledge for commercializing the new technology. This had been a common problem among a number of new enterprises. Most new enterprises that have succeeded with these challenges have first mover advantage, a positive cash flow to continue hiring unlimited human resources, and, although reacting late in the process, have implemented an infrastructure that could support the business. The infrastructure was a splintered systems environment lending only to a semi-automated environment. The systems migration strategy occured too late in the product launch phase to allow for a seamless automated process.

Another factor that often plagues the enterprise is the lack of IT personnel who have business-specific skills. Personnel in the IT organization who lack business skills in the various vertical markets such as engineering, manufacturing, healthcare, financial, legal, and retail may have a difficult time eliciting internal customer requirements when developing and implementing an ERP. They may also lack the various business skills internally, if they are unfamiliar with the business and technical requirements of the other functional organizational elements such as sales, marketing, finance, operations, engineering, logistics, transportation, manufacturing, human resources, business development, alliances, product development, legal, along with any other relevant enterprise elements.

Finally, not all employees hired into an enterprise come with an entrepreneurial spirit. Some still have a corporate frame of mind and do not become as self-sufficient as is necessary to keep up the pace. They have a tendency to operate in closed groups and do not interact well with other business units. A team philosophy and aggressive work ethic is essential in order to succeed in an enterprise environment.

The approach, suggested here, to achieving operational velocity is to develop an ERP model that meets the following 15 performance criteria:

1. Reduces service delivery intervals;
2. Maintains reliable inventory control;
3. Reduces mean-time-to-repair (mttr);
4. Enhances customer response time;
5. Establishes timely and effective communications mechanism;
6. Automates processes;
7. Creates tracking mechanisms;
8. Maintains continuous business process improvement;
9. Supports fault management through problem detection, diagnosis, and correction;
10. Manages customer profiles;
11. Monitors business performance;
12. Establishes best practices;
13. Creates forecasting tools;
14. Supports supply chain management; and
15. Integrates all systems within the ERP model such as sales tools, order entry, CRM, billing, and fault management.

These performance attributes are ones that companies have adopted to monitor, manage, support, and measure success of their operational environment. Companies are also continuously challenged with developing and implementing an effective model to support these attributes. The challenges stem primarily from a lack of knowledge and limited use of contemporary IT principles and practices. Enterprises must realize the need for appropriate performance metrics in order to measure success criteria and to plan for future growth and expansion.

Of all the network elements impacting the adoption of new technology, operational velocity is the most compelling, since it will influence customer expectations based on how quickly customer needs can be serviced. These needs may consist of rapid customer service response time, product delivery, problem resolution, and maintenance. Operational velocity, like the network elements, will influence consumer decision making on new technology adoption. If a new technology product has long delays in service delivery or lacks customer support, new customer acquisition and retention eventually become compromised. Under these circumstances, it is possible to lose business to the competition, which may be introducing a similar product into the marketplace. Consumers become disappointed, less patient and quickly begin to look for alternatives. The lack of a reliable operational infrastructure would have been the result of a poorly executed ERP. An effective ERP must be automated, capable of tracking, serve as a communications mechanism, and support various tools. If these criteria are recognized and controlled by the core team of an enterprise, the ERP can provide many benefits as the business begins to scale and the product begins to meet customer expectations. Network elements can influence the outcome of a new technology or the destiny of the product. Understanding the impact that the various network elements have on the enterprise can help position the business in taking on the challenges that prevail. The market timing of the product and the influence on customer decision making will determine the end result of critical mass attainment. An enterprise that prepares and develops strategies, and which takes into account the large number of potential network influences, will accordingly realize this end result. There are a number of complex adap-

tive system challenges associated with these, and these must be explored as well.

Many of the enterprise resource planning efforts cited in this article can be traced to the three basic core elements of an ERP: people, process, and systems. Each of these elements were addressed in the various models and frameworks identified by early contributors in the field. As an ERP architecture evolves, each of the ERP elements goes through a maturity state. The evolution of a fully developed and integrated ERP architecture can be inferred from the phases of a basic systems engineering lifecycle. Table 2 illustrates this inference through a framework of key systems engineering concepts that can be applied to the development of an enterprise resource planning architecture.

This suggested framework could be used to develop an enterprise architecture using six key system engineering concepts. To support the ERP development effort, this 6x3 matrix of Table 2 could be used with the six general system engineering concept areas as rows and the three columns depicting the core components of an ERP. This defines the structural framework for

Table 2. Key systems engineering concepts framework

Framework of key systems engineering concepts			
Concept Area	ERP Core Components		
	People	Process	Systems
Requirements definition and management	Organizational requirements elicited	High-level operational processes defined	System functions identified
Systems architecture development	High-level architecture developed by team	Architecture supports organizational processes	Systems defined to address organizational requirements
System, subsystem design	Unique data and functionality criteria addressed for each organization	Operational processes at the organizational level are developed	Organizational system components are designed
Systems integration and interoperability	Shared/segmented data and functionality is designed	Operational processes are fully integrated, seamless, and automated	All organizational system components and interfaces are fully integrated and interoperable
Validation and verification	Organizations benchmark and measure operational performance	Operational process efficiencies and inefficiencies are identified	System response time and performance are benchmarked and measured
System deployment and post deployment	Team launches complete and fully integrated ERP architecture	Operational readiness plan is executed and processes are live	Systems are brought online into production environment and supporting customers

systems engineering concepts and their relevance in developing, designing, and deploying an enterprise architecture. ERP maturity states are represented in each of the quadrants of the 6x3 matrix. As an ERP matures, each of the maturity states is realized and can be directly correlated to its respective systems engineering concept. It can be seen from the framework that the phases of the systems engineering lifecycle can be applied to ERP development. The various ERP models presented in this article revealed that a systems engineering paradigm may be inferred. The SE concept framework clearly illustrates a systems engineering orientation with respect to ERP.

SUMMARY

In this article we have attempted to summarize the very important effects of contemporary issues surrounding information and knowledge management as they influence systems engineering and management strategies for enhanced innovation and productivity through enterprise resource planning. To this end, we have been especially concerned with economic concepts involving information and knowledge and the important role of network effects and path dependencies in determining efficacious enterprise resource planning strategies. A number of contemporary works were cited. We believe that this provides a very useful, in fact, a most-needed, background for information resources management using systems engineering and management approaches.

REFERENCES

Boer, P. F. (1999). *The valuation of technology: business and financial issues in R&D*. Hoboken, NJ: Wiley.

Boer, P. F. (2002). *The real options solution: finding total value in a high-risk world*. Hoboken, NJ: Wiley.

Boer, P. F. (2004). *Technology valuation solutions*. Hoboken, NJ: Wiley.

Briscoe, B., Odlyzko, A., & Tilly, B. (2006). Metcalfe's Law is wrong. *IEEE Spectrum*, July, 26-31.

Davenport, T. H. (2000). *Mission critical: Realizing the promise of enterprise systems*. Boston, MA: Harvard Business School Press.

Economides, N. (1996), The economics of networks. *International Journal of Industrial Organization, 14*(6), 673-699.

Economides, N. (2001). The Microsoft antitrust case. *Journal Of Industry, Competition And Trade: From Theory To Policy, 1*(1), 7-39.

Economides, N., & Himmelberg, C. (1994). Critical mass and network evolution in telecommunications. *Proceedings of Telecommunications Policy Research Conference, 1-25*, Retrieved from http://ww.stern.nyu.edu/networks/site.html.

Economides, N., & White, L., J. (1996). One-way networks, two-way networks, compatibility, and antitrust. In D. Gabel & D. Weiman (Eds.), *The regulation and pricing of access*. Kluwer Academic Press.

Farrell, J., & Katz, M. (2001). Competition or predation? Schumpeterian rivalry in network markets. (Working Paper No. E01-306). University of California at Berkeley. Retrieved from http://129.3.20.41/eps/0201/0201003.pdf

Gardner, C. (2000). *The valuation of information technology: A guide for strategy, development, valuation, and financial planning*. Hoboken, NJ: Wiley.

Kaufman, J. J., & Woodhead, R. (2006). *Stimulating innovation in products and services*. Hoboken, NJ: Wiley.

Koch, C. (2006, January). The ABCs of ERP, *CIO Magazine*.

Langenwalter, G. A. (2000). *Enterprise resource planning and beyond: Integrating your entire organization.* Boca Raton, FL: CRC Press, Taylor and Francis.

Liebowitz, S. J. (2002). *Rethinking the networked economy: The true forces driving the digital marketplace.* New York: Amacom Press.

Liebowitz, S. J., & Margolis, S. E. (1994). *Network externality: An uncommon tragedy. Journal of Economic Perspectives, 19*(2), 219-234.

Liebowitz, S. J., & Margolis, S. E. (2002). *Winners, losers & Microsoft.* Oakland, CA: The Independent Institute.

Miller, W. L., & Morris, L. (1999). *Fourth generation R&D: Managing knowledge, technology and innovation.* Hoboken, NJ: Wiley.

Odlyzko, A. (2001). Internet growth: Myth and reality, use and abuse. *Journal of Computer Resource management, 102*, Spring, pp. 23-27.

Rouse, W. B. (1992). *Strategies for innovation: Creating successful products, systems and organizations.* Hoboken, NJ: Wiley.

Rouse, W. B. (2001). *Essential challenges of strategic management.* Hoboken, NJ: Wiley.

Rouse, W. B. (2005). A theory of enterprise transformation. *Systems Engineering, 8*(4) 279-295.

Rouse, W. B. (Ed.). (2006). *Enterprise transformation: Understanding and enabling fundamental change.* Hoboken, NJ: Wiley.

Sage, A. P. (1995). *Systems management for information technology and software engineering.* Hoboken, NJ: John Wiley & Sons.

Sage, A. P. (2000). Transdisciplinarity perspectives in systems engineering and management. In M. A. Somerville & D. Rapport (Eds.) *Transdisciplinarity: Recreating integrated knowledge* (158-169), Oxford, U.K.: EOLSS Publishers Ltd.

Sage, A. P. (2006). The intellectual basis for and content of systems engineering. *INCOSE INSIGHT, 8*(2) 50-53.

Sage, A. P., & Rouse, W. B. (Eds.). (1999). *Handbook of systems engineering and management.* Hoboken, NJ: John Wiley and Sons.

Shapiro, C., & Varian, H. R. (1999). *Information rules – A strategic guide to the network economy.* Boston, MA: Harvard Business School Press.

Small, C. T., & Sage, A. P. (2006). Knowledge management and knowledge sharing: A review. *Information, knowledge, and systems management. 5*(6) 153-169

This work was previously published in Information Resources Management Journal, Vol. 20, Issue 2, edited by M. Khosrow-Pour, pp. 44-73, copyright 2007 by IGI Publishing (an imprint of IGI Global).

Chapter 15
A Critical Systems View of Power–Ethics Interactions in Information Systems Evaluation

José-Rodrigo Córdoba
University of Hull, UK

ABSTRACT

Current developments in information systems (IS) evaluation emphasise stakeholder participation in order to ensure adequate and beneficial IS investments. It is now common to consider evaluation as a subjective process of interpretation(s), in which people's appreciations are taken into account to guide evaluations. However, the context of power relations in which evaluation takes place, as well as their ethical implications, has not been given full attention. In this article, ideas of critical systems thinking and Michel Foucault's work on power and ethics are used to define a critical systems view of power to support IS evaluation. The article proposes a system of inquiry into power with two main areas: 1) Deployment of evaluation via power relations and 2) Dealing with ethics. The first element addresses how evaluation becomes possible. The second one goes in-depth into how evaluation can proceed as being informed by ethical reflection. The article suggests that inquiry into these relationships should contribute to extend current views on power in IS evaluation practice, and to reflect on the ethics of those involved in the process.

INTRODUCTION

It has been argued extensively in the literature of information systems (IS) evaluation that failures in implementation of information systems occur due to lack of consideration of different (e.g., softer) as-pects that influence information systems adoption (Hirschheim & Smithson, 1999; Irani, 2002; Irani & Fitzgerald, 2002; Irani, Love, Elliman, Jones, & Themistocleus, 2005; Serafeimidis & Smithson, 2003). Among these aspects, the issue of ethics also gains importance, yet few evaluation approaches consider it explicitly (Ballantine, Levy, Munro, & Powell, 2003). When evaluating the implementation of information systems, there is still a need to consider the *context* of human relations within which evaluation takes place (Walsham, 1999), and more specifically, the nature and impacts of *power relations* (Doolin, 2004; Gregory, 2000; Introna, 1997). This consideration has also been noticed in the realm of systems thinking, but there is a dearth of approaches to deal with the complexities of power (Gregory & Jackson, 1992; Jackson, 2000). In IS evaluation, power has been mainly considered as a "contextual," "political," or "external" variable (Serafeimidis & Smithson, 1999), and its impacts in practice (for instance regarding the treatment of ethical issues) are far from clear. Power is often understood as "politics" (Bariff & Galbraith, 1978), "interests playing" or struggle between parties (Walsham, 1993), and is associated with the dynamics of organisational change that are said to be difficult to manage (Lyytinen & Hirschheim, 1987). These connotations could limit a better understanding of the nature of power in IS evaluation and how practitioners can act in relation to it.

Awareness of the nature of power for intervention has been a subject of discussion in *critical systems thinking*, a set of ideas and methodologies that aim to clarify stakeholders' understandings prior to the selection and implementation of intervention methods in situations of social design (Flood & Jackson, 1991b; Jackson, 2000; Midgley, 2000). Using the commitments of critical systems thinking to critical awareness, pluralism, and improvement as well as Michel Foucault's ideas on power and ethics, this article extends current understandings of power to inform IS evaluation. The article proposes a relational view of power that is dynamic, transient, and pervasive, and which influences, and is influenced by, individuals' ethics. With this view, the article defines a "system of inquiry" with two elements of analysis for IS evaluation: (1) Exploring the *deployment* of evaluation via power relations; and (2) Dealing

with ethics. With these areas, different manifestations of power can be accounted for and related in evaluation interventions. In addition, inquiry into these areas enables people involved to reflect on the ethics of their own practices.

The article is structured as follows. Critical systems thinking is introduced in relation to three (3) commitments that can inform systems thinking and practice. Then, information systems (IS) evaluation as *interpretation(s)* is described and reviewed in relation to how the issue of power is currently being addressed. It is argued that a critical, pluralistic and ethically oriented view of power is needed. To build up this view, the article presents the basic tenets of Michel Foucault's work on power and ethics, highlighting implications for IS evaluation. A system of inquiry into power for IS evaluation is defined, and its relevance for evaluation practice discussed.

CRITICAL SYSTEMS THINKING

This article stems from the UK-based systems research and practice, in which there is a variety of systems methodologies that contain principles, ideas, and methods to facilitate intervention for social improvement (Checkland, 1981; Flood & Jackson, 1991b; Flood & Romm, 1996; Jackson, 2000, 2003; Midgley, 2000; Stowell, 1995). The use of systems ideas has also pervaded the information systems (IS) field. Currently, it has been accepted that a *systemic* view of IS practice, one that looks at different elements of activity in organisational, social, and technical domains, can contribute to make sense of a variety of efforts in the IS field (Avison, Wood-Harper, Vidgen, & Wood, 1998; Checkland, 1990; Checkland & Holwell, 1998). This view also shares a common idea with other systems research movements elsewhere that conceive of an information system as part of an organisational system (Mora, Gelman, Cervantes, Mejia, & Weitzenfeld, 2003).

In the UK, the popularity of systems thinking can also be reflected through the use of soft systems methodology (SSM) as a learning tool (Checkland, 1981) and its applications in several areas in information systems. These include information requirements definition (Checkland, 1990; Checkland & Scholes, 1990; Lewis, 1994; Wilson, 1984, 2002), systems development (Avison & Wood-Harper, 1990), intervention methodology (Clarke, 2001; Clarke & Lehaney, 2000; Midgley, 2000; Ormerod, 1996, 2005), and professional practice (Avison et al., 1998; Checkland & Holwell, 1998).

To this popularity, however, it has also been argued that the use of some methodologies like SSM can help in reinforceing the 'status quo' in a situation if it is not used in a more critical and informed manner (Jackson, 1982; Mingers, 1984). Jackson (1992) argues that the practice of information systems can be further developed if systems-based interventions are not only guided by one type of rationality, methodology, or research paradigm, and if assumptions about the 'status quo' in a situation of social design are critically reviewed. Using systems ideas, practitioners should be able to foster creativity, complementarity and social responsibility.

Jackson and others have developed a collection of ideas, methodologies, and approaches under the name of "critical systems thinking" (Flood & Jackson, 1991b; Flood & Romm, 1996; Gregory, 1992; Jackson, 2000, 2003; Midgley, 2000; Mingers, 1992, 2005; Mingers & Gill, 1997; Ulrich, 1983). Critical systems thinking (CST) has been defined as a continuous dialogue between systems practitioners who are concerned with the issue of improvement (Midgley, 1996). As an evolving set of ideas, it contains a variety of notions that aim to foster continuous stakeholders' reflection prior to the selection and implementation of planning and design methods.

In critical systems thinking, Midgley (1996) distinguishes three common and inter-related commitments to guide the efforts of researchers and practitioners. These commitments are: (1) *Critical awareness*, continuous re-examining of taken-for-granted assumptions in a situation (including those inherent to systems methodologies); (2) *Pluralism* (or complementarism), using a variety of ideas and approaches in a coherent manner to tackle the complexity of the situation; and (3) *Improvement*, ensuring that people advance in developing their full potential by freeing them of potential constraints like the operation of power.

The commitments of critical systems thinking have been put into practice in different ways. For instance, there is a system of systems methodologies (Jackson & Keys, 1984) to help those involved in an intervention choose the most adequate system methodologies to tackle a problem situation according to methodologies' own strengths and weaknesses. In addition to methodology choice, creativity can also be fostered when thinking about problem situation with the use of metaphors, and reflection is included to enable learning and understanding through the use of methodologies (Flood & Jackson, 1991a; Flood & Romm, 1996). Recently, systems practice has also been enriched with generic principles to ensure that intervention is guided by continuous critique, the use of different methods and definition of local and temporary improvements (Jackson, 1999, 2003; Midgley, 2000).

An emerging (UK- and non-UK-based) slant on critical systems thinking is that developed by Ulrich (1983; 2003) and Midgley (2000) on boundary critique. According to them, our processes of producing knowledge about a situation are *bounded* by a number of assumptions about purpose(s), clients, theories, methodologies, methods, and other aspects related to an intervention. These assumptions are intimately linked to systems boundaries. Here the idea of a system is that of an intellectual construction that guides analysis and decision-making (Churchman, 1970, 1979). According to Ulrich and Midgley, such boundaries and their underpinning assumptions need to be identified, analysed and debated with people involved in relation to their value content, so that individuals can make more informed decisions regarding the implications of privileging some boundaries at the expense of others.

In line with the above, in critical systems thinking, the issue of *power* has been discussed at length, and it has been argued that power can inhibit individuals' own reflection about the conditions that influence their own improvement (Flood, 1990; Flood & Romm, 1996; Valero-Silva, 1996; Vega-Romero, 1999). Power has not been defined in a unique way. It has been associated with phenomena of coercion, which affects relationships between stakeholders (Gregory & Jackson, 1992; Jackson, 2000). Critique on systems boundaries adopted for analysis and decision making in a social situation has been enhanced with the idea that such boundaries are the result of the operation of *power* and its manifold manifestations (Flood, 1990; Midgley, 1997; Vega-Romero, 1999). Despite acknowledging the importance of power for systems practice, in critical systems thinking there is little about how practitioners can identify and act in relation to power issues in intervention. Although this could be attributed to the diversity of meanings of power (and hence an interpretation of a commitment to pluralism), there is a need to provide further insights into the nature of power and how reflection about it can be developed in practice, if a commitment to improvement in social situations is to be honoured.

In this article, we use the above commitments in critical systems thinking to develop a view of power for intervention. With this view, we generate a "system" (e.g. a "whole") of inquiry into power that aims to follow these commitments. We apply our view and system to the domain of information systems (IS) evaluation in order to provide guidance to practitioners on how to identify and manage power in evaluation practice. In the next section we review the practice of IS evaluation in relation to power.

INFORMATION SYSTEMS EVALUATION

In general terms, information systems (IS) evaluation is about assessing the continuous *value* that systems and communication technologies give to organisations and individuals (Irani & Love, 2001; Parker, Benson, & Trainor, 1988; Piccoli & Ives, 2005; Remenyi & Sherwood-Smith, 1999). IS evaluation is still considered a "wicked" phenomenon (Farbey, Land, & Targett, 1999), a "thorny" and complex process (Irani, 2002; Serafeimidis & Smithson, 2003) that is difficult to carry out given different aspects that affect its outcomes. To date, there are a number of approaches and techniques that are used to support successful evaluation of IS and technology investments prior to, during, or after their implementation, although a strong focus on financial techniques still remains (Irani, 2002; Parker et al., 1988; Serafeimidis & Smithson, 1999).

In IS evaluation, it has also been argued that success depends on the usefulness of evaluation processes and outcomes to inform managerial decision-making. This usefulness has been related to the identification of different issues (i.e., financial, ethical, organisational, and cultural) that affect IS implementation so that these are promptly and adequately addressed (Avison & Horton, 1992; Ballantine et al., 2003; Doherty & King, 2001; Hirschheim & Smithson, 1999; Irani, 2002; Irani & Love, 2001; Symons & Walsham, 1988). With the inclusion of a variety of issues in IS evaluation, a growing concern is the usefulness that evaluation will have for those individuals involved and affected by it (Irani, 2002; Irani & Love, 2001; Serafeimidis & Smithson, 1999). People would like to benefit from being involved in an evaluation or using evaluation outcomes.

Therefore, individual perceptions have become relevant, and researchers have suggested that IS evaluation can be better understood as a continuous and *subjective process of interpretation(s)* (Hirschheim & Smithson, 1999; Smithson &

Tsiavos, 2004; Walsham, 1999). In other words, evaluation is a process of "experiential and subjective judgement, which is grounded in opinion and world views, and therefore challenges the predictive value of traditional [IS] investment methods" (Irani et al., 2005, p. 65) (brackets added). For Walsham (1999), IS evaluation processes are about understanding and learning through stakeholders' perspectives and actions; stakeholder participation can contribute to minimise resistance IS to implementation (Walsham, 1993). The idea of IS evaluation being a subjective process is expanded by Serafeimidis and Smithson (2003) who argue that IS evaluation "is a socially embedded process in which formal procedures entwine with the informal assessments by which actors *make sense* of their situation" (p.253, emphasis added). They provide the following roles of IS evaluation as:

1. *Control*, meaning that evaluation is and becomes embedded in traditional procedures of organisational appraisal. IS evaluation processes adhere to existing hierarchies and accepted ways of assessing and monitoring investments. The aim of IS evaluation is to deliver value to the business. Financial techniques that appraise the contribution of information systems and technologies to business strategies are preferred to any other type of evaluation approach (Serafeimidis & Smithson, 1999). In control-evaluation, traditional channels of communication are used. Participation of stakeholders contributes to minimise the risks related to investments and to ensure commitment. However, those people who benefit from controlling other individuals can use evaluation to advance their own interests.

2. *Sense making*, or clarifying any implications that IS investments and projects could have to stakeholders. Informal communication complements formal communication. In sense-making evaluation, establishing a common language helps those leading the evaluation (evaluators) and those taking part (evaluands) to share their expectations and concerns about IS investments or projects. Sense-making evaluation, though, does not exclude the possibility that the revealing of meanings can be used for political purposes or to advance the evaluators' own interests (Legge, 1984; Weiss, 1970).

3. *Social learning*, or fostering the creation, storing, and exchange of knowledge. Stakeholders can take part in this exchange and contribute so that they reduce any uncertainty about the implementation and success of information systems. In social learning, evaluators facilitate the exchange of knowledge through interactions with stakeholders (for example by promoting conversations about how systems will address people or business-related expectations). The selection of what type of knowledge is relevant for evaluation can become an instrument of political influence (e.g. directed to achieve particular objectives), as well as the ways in which this knowledge can be disseminated or exchanged.

4. An *exploratory* exercise, to help organisations to clarify their strategic direction and promote change. Those involved in IS evaluation develop new ways of appraising and monitoring the value that systems have to organisations. This requires thinking creatively. In doing so, people involved in evaluation can contribute to shift the existing balance of *power*: They can challenge those who advocate evaluation techniques based solely on financial benefits or traditional accounting and reporting techniques.

In each of the above orientations on IS evaluation, the perceptions and actions of stakeholders can be used to reinforce or shift the balances of power, but power has not been defined yet. The wider (non-IS) literature on evaluation suggests situations of disadvantage or conflict can be addressed via more participation or empowerment (Guba & Lincoln, 1989; Mertens, 1999; O'Neill, 1995; Weiss, 1970, 1998). Moreover, it is suggested that evaluators should "sign in" with disadvantaged groups and ensure that their concerns, claims and issues are adequately considered and listened to in the evaluation process (Guba & Lincoln, 1989). However, as Gregory (2000) contends, participative evaluation approaches can easily overlook the operation of power and how it can contribute to generate and maintain the very same conditions that enable or inhibit participation to occur. By trying to address imbalances in participation, evaluators may well be privileging their own power as experts or facilitators, or inadvertently reinforcing the power of those who are in managerial control in a situation (Wray-Bliss, 2003). For Gregory (2000), the problem of participation in evaluation can only be approached through a wider understanding of *power* and its *operation* in practices that prohibit or promote such participation. There needs to be considerations about the context of power in which evaluation is taking place, as well as the role of those being involved in it as part of evaluation practice.

Table 1 contains a summary of four different notions of power that can be related to the IS evaluation roles discussed before. These notions are drawn from existing classifications in the IS literature (Dhillon, 2004; Horton, 2000; Jasperson et al., 2002) and elsewhere (Lukes, 1974; Oliga, 1996). As seen in the table, it can be common to associate power with tangible or distinguishable resources (i.e., information), skills or authority that some people have and use to control others (Bariff & Galbraith, 1978; Horton, 2000). Power can be also associated with institutional structures, so that its use can reinforce, perpetuate, or resist existing organisational hierarchies and "games" (Bloomfield & Coombs, 1992; Dhillon, 2004; Markus, 2002). Or power can be seen as the *influence* that any action of particular individuals have in the behaviour of others (Handy, 1976; Walsham & Waema, 1994). This includes, for instance, the influence that IS experts have over systems users (Horton, 2000), the political skills (Checkland & Scholes, 1990), or the style that managers have to define, implement, and evaluate IS plans (Walsham & Waema, 1994).

The above views presented about power show individual notions, as if power had different but not intersecting manifestations. Nevertheless, power could be an intertwining of capacities, influences, or resources. These views describe

Table 1. power in orientations for is evaluation

IS Evaluation as (Serafeimidis & Smithson, 2003)	Power as	Manifestations
Control	Resources (Bariff & Galbraith, 1978)	Authority, skills, information, use of technology.
Sense-making	Capacity (Markus, 2002)	Structures that facilitate (or inhibit) communication
Social-learning, exploratory	Influence (Checkland & Scholes, 1990; Handy, 1976; Walsham & Waema, 1994)	Expertise and styles used to facilitate (or inhibit) knowledge exchange and change
Relational	All of the above	In the relations between people (Foucault, 1984a), as a backdrop (Horton, 2000) and in the conditions that make evaluation possible.

very little about how power comes to be considered as such, in other words, how power is *deployed* as such in a situation. In IS practice, it has been acknowledged that explicit exercise of power can contribute to systems implementation (Markus, 2002; Serafeimidis & Smithson, 2003; Walsham & Waema, 1994). However, this does not fully consider the often indistinguishable, unintended, contradictory, and complex consequences of power in IS/IT implementations in a context of intervention (Jasperson et al., 2002; Robey & Boudreau, 1999).

Therefore, it can be argued that IS evaluation faces a similar problem to critical systems thinking, that of not providing enough guidance to practitioners on how to identify and act in relation to power as a multifarious and complex issue that affects any action for improvement. It is necessary to consider a *critical* view on power in which power is studied in its deployment (how, why), and not only taking power as a *given*. The view also needs to be *pluralistic* in order to include different manifestations and forms of power, as well as the relationships between them. Moreover, an alternative view of power should help practitioners to explore possibilities for *improvement* in action in relation to power relations. To develop this view in line with the commitments of critical systems thinking, Michel Foucault's ideas on power and ethics are now presented.

Foucault on Power

Michel Foucault's work on the history of Western civilisation provides relevant insights into the problem of the human *subject*, be it individual or collective. For Foucault, the main question in modern society is how human beings are constituted as subjects (Foucault, 1982a, 1982b). His aim is to show connections between what counts as knowledge, the power relations used to make it valid, and the ethical forms that support its deployment. This for Foucault is a way of developing critique in contemporary society (Foucault, 1980b). For Foucault, the meaning of a "subject" is twofold: "someone subject to someone else by control and dependence, and tied to his own identity by a conscience or self-knowledge" (Foucault, 1982a, p.212). Both meanings in the above definition suggest a form of *power*, which subjugates and makes one subject to it (Foucault, 1982a). This suggests that power operates in different ways (targeting individuals and/or groups), influencing the ways in which people relate to themselves and each other.

According to Foucault, the end result of processes of production of knowledge is the potential operation of forms of "normalisation" in society which constrain our behaviour and limit our freedom as individuals. The set of analyses on how people become normalised is called by Foucault "subjectivity" (Foucault, 1977). With his historical analyses, Foucault also shows that the ways individuals define themselves and relate to others have been contingently defined, contested and deployed via *power* relations as "the ways we fashion our subjectivity" (Bernhauer & Mahon, 1994, p. 143). Subjectivity refers to the practices we perform on ourselves, and this includes what we consider ethical, as will be shown later.

In Foucault's analyses, one can find different definitions of power that also show power's dynamic nature in society (Foucault, 1980b). Power can be identified in the relations between people, between actions influencing other actions. Power means *power strategies* through which individual try to define, determine, or guide the conduct of others (Foucault, 1984a). Power also helps deploying some forms of knowledge at a particular moment in time whilst obscuring others, so that certain practices prevail as the valid ones. Power can be seen as a "total structure of actions brought to bear upon possible actions: in incites, it induces, it seduces, it makes easier or difficult" (Foucault, 1982a, p. 220).

For Foucault (1980b, 1984b), power is not an objective issue; it can only be identified in its operation through the relations that it establishes, maintains (including resisting), or creates between individuals. Power is an analytical device that helps us to understand how we have been constituted as the subjects we currently are in the relations with ourselves and others. Such relations are mobile, transient, and dynamic; they target single individuals or entire populations; their operation occurs across institutions and at different levels (micro, macro) in society. New forms of power emerge that reinforce, support, undermine, or resist previous ones, and this happens at any level (e.g., individual, micro and macro). In Foucault's work, power is present where there is freedom and is essential to regulate relations between individuals in society (Foucault, 1984a). Power can be used intentionally, but the consequences of doing so cannot be fully determined (Foucault, 1984a, 1984b).

Foucault's work has been used in the realm of information systems to understand the effects of information systems planning and implementation in managerial practices (Ball & Wilson, 2000; Córdoba & Robson, 2003; Doolin, 2004; Horton, 2000; Introna, 1997). For instance, Introna (1997) suggests that Foucaultivan notions of power helps to identify some "obligatory passage points" in the design and implementation of information systems as sets of relations that determine what types of information and the practices associated with its management count as organisationally accepted. According to Bloomfield and Coombs (1992), such awareness can also help IS practitioners to map and better understand the conditions that enable the implementation of systems in an organisation. For Doolin (2004), the Foucaultvian concept of power can help explain how people can resist or react to existing implementation practices and how implementation is the by-product of many different organisational factors, some of which emerge in opposition to the implementation itself. In these accounts, the issue of ethics has not been explicitly addressed using Foucault's ideas (Burrell, 1988), and this will be revisited later in the article.

From the above discussion, we elaborate a fourth notion of power to support IS evaluation (see last row of Table 1). In this notion, power operates in the *relations* between individuals. It includes different manifestations *as well* as the conditions and relations that make possible the existence and use of power as a resource, structure, or influence in evaluation as previously discussed. These different manifestations of power not only generate potential constraints that inhibit action (including the evaluation itself), but also opportunities that will make action feasible according to the "landscape" of possibilities that individuals are part of (Brocklesby & Cummings, 1996; Foucault, 1980b). As will be seen in the next section, these possibilities can be better defined in relation to the ethics of individuals.

ETHICS

According to Brooke (2002), some authors see Foucault as failing to provide a concrete space within which debate can take place given the ever presence of power even as resistance to it. In particular, Foucault's acceptance of the idea that "Yesterday's resistance can become today's normalisation… which in turn can become the conditions for tomorrow's resistance and/or normalisation" (Darier, 1999, p. 18) is lacking any normative content and thus generates ambiguity or confusion (Rowlinson & Carter, 2002; Taylor, 1984). A question arises about how one can then discern and decide on ethical issues in evaluation (Ballantine et al., 2003). This question gains importance in light of a critical systems-based commitment with *improvement* as mentioned before. To the potential ambiguity of power analyses, more structured ways of dealing with questions of ethics in IS evaluation like the ones presented by Ballantine et al (2003) (based on Habermas) can provide alternative and systematic answers. These alternatives focus on reviewing and developing spaces for equal debate about ethical issues, as well as providing general rules for examining or conducting debate. In contrast to these alternatives, for Foucault it is essential to explore the *conditions* that led debate and inequalities to emerge in the first place. These conditions could be unique in a context of intervention (Brooke, 2002), including those that enable participation in IS evaluation to take place.

To address the above question, there is a still largely unexplored area in Foucault's work that needs to be made more explicit, and that is ethics. Foucault's work is not power but the *human subjects*, how we have been constituted as the individuals we are (Chan & Garrick, 2002; Foucault, 1982a). According to Foucault (1977), any action in relation to power *cannot* be considered exterior to power relations, so that inevitably any debate on issues (including ethical) in evaluation takes place in relation to power relations. Therefore, we need to look at power relations from the inside (Brooke, 2002). Foucault's analyses aim to show how subjects position themselves to situations according to what they think it is *ethical* (Darier, 1998; Foucault, 1977). In his study of the history of sexuality, Foucault says:

Morality [ethics] also refers to the real *behaviour of individuals to the rules and values that are recommended to them...the manner in which they respect or disregard a set of values... (p. 25)... those intentional and voluntary actions by which men not only set themselves rules of conduct, but also seek to transform themselves, to change themselves in their singular being.* (Foucault, 1984b, p. 10)

This means that it is possible for subjects to make strategic use of their freedom (Foucault, 1984a, 1984c) and use it to "no longer being, doing or thinking what we are, do, or think" (Foucault, 1984c, p. 46). Foucault is aware that we need to continuously recognise the limits of our actions, what is no longer necessary (or dangerous) for the

constitution of us as autonomous subjects and *act* accordingly. He says:

The question, in any event, is that of knowing how the use of reason can take the public form that it requires, how the audacity to know can be exercised in broad daylight, while individuals are obeying as scrupulously as possible. (Foucault, 1984c, p. 37)

This means that in the light of power relations in a particular context of intervention, it is possible to develop a reflexive and ethically oriented practice of individual freedom. Ethical practice becomes a way of providing direction to action for improvement, an opportunity to (re) develop forms of ethics within what is possible in relation to power relations. This aspect will be further discussed when proposing a system of inquiry into power for IS evaluation in the next section.

Towards a System of Inquiry into Power for IS Evaluation

From the above discussion on power and ethics, two important implications can be derived to inform the definition of a critical systems view of power for IS evaluation. First, the inclusion of power would require considering it as a backdrop (Horton, 2000) of relations against which any IS evaluation orientation can be studied. Any manifestation of power (as a resource, capacity, structure, or influence) in IS evaluation should be considered the by-product and medium of power relations operating in a context of intervention, with these relations having varied implications (for instance, economic, political, social, and cultural). Identification and analysis of how power relations operate would help those involved in evaluation to reflect on how they become subjects of evaluation activities and what they can do about it. The above does not mean that power should be avoided but its possibilities and constraints used strategically according to what individuals consider relevant to do (Brocklesby & Cummings, 1996) in relation to what has been institutionally unfolded and accepted as IS evaluation (Smithson & Tsiavos, 2004). For those involved in evaluation, analysis of power requires them to reflect on their participation in power relations that make evaluation (im) possible and that facilitate or inhibit unfolding of events. It becomes necessary to explore the origins and deployment of power relations in which IS evaluation has arisen as a process to be carried out.

Secondly, the analysis of power relations as a system requires ethical awareness from those involved about ethical issues that they adopt, debate, or resist in IS evaluation, *and this also includes the ethical issues that are adopted to analyse power.* This requires direct intervention from the "inside" of evaluation. Foucault is proposing to continuously study power in order to see the limitations of its "normalising" ethical systems, and how power can also offer possibilities for action for people as they see them fit (Brocklesby & Cummings, 1996) or ethically appropriate (Vega-Romero, 1999). In other words, Foucault is proposing to study and reflect on the internal conditions that can make ethical action possible in IS evaluation in order to define the "battleground" and possibilities for further action.

Considering the above, the following is the definition of a system of inquiry into power to support IS evaluation as presented in Figure 1. The system is composed of two areas interacting with each other and informing existing role(s) of the evaluation process as described by Serafeimidis and Smithson (2003). This analysis brings together an understanding of evaluation as a series of interpretations as described by Smithson and Tsiavos (2004), and a way of reflecting on ethical issues in IS evaluation as proposed by Ballantine et al (2003), so that those involved in evaluation can reflect on power from their own participation. The areas of inquiry are:

Figure 1. A system of inquiry into is evaluation

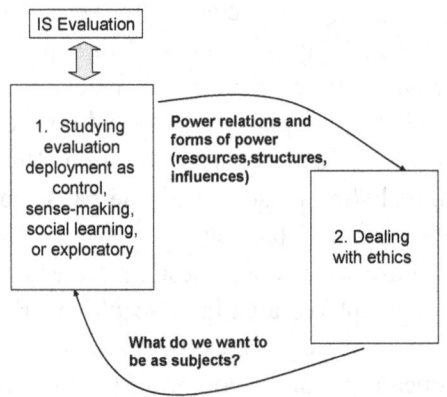

1. *Exploring the deployment of IS evaluation.* Analysis of power in relation to forms of being, knowing, and acting consists of locating how power relations contribute to deploy (implement) or undermine IS evaluation activities. The purpose is to identify how evaluation became possible and accepted as such, and how it progresses. This type of analysis requires unveiling power relations at different levels (for instance, economic, social, political)—as maps of actions influencing other actions—(Foucault, 1984a) that constitute the definition, approval and unfolding of the evaluation under study. A good starting point or "points of entry" to analyse power is to see how it helps in the deployment of accepted evaluation roles (i.e., as control, sense-making, social-learning, and exploration) (Serafeimidis & Smithson, 2003); in other words, to study how these roles came to being, and the wider relations that made them possible and valid. The analysis can then be complemented or developed with the following questions (Córdoba & Robson, 2003): How is that evaluation is defined and approved? How does it engage those involved? What role(s) for evaluation are accepted? Through which mechanisms and justifications? How do activities in evaluation become successful or unsuc-

cessful? How are evaluations institutionally completed or abandoned?

2. *Dealing with ethics.* As said before, for Foucault (1977), one cannot be exterior to the power relations one is analysing or intervening. Therefore, analyses should also show how individual subjects position themselves in situations (Darier, 1998; Foucault, 1977). This consideration should lead those involved in IS evaluation to consider *what is ethical for them to do according to power*, and to go beyond the idea of interpretations. Analysis of power should also yield insights as to what behaviours and actions are ethically acceptable or unacceptable (including the analysis itself as a practice that is guided by ethical values), and what to do about them. Those involved in evaluation can decide to adopt a critical stance and go beyond what is being established, to imagine new forms of being and acting (Foucault, 1984c). This could mean that the purpose and nature of evaluation are re-defined according to what people consider ethical to do in a context of intervention.

Using Foucault's (1984b) elements of analysis of ethics, those involved in evaluation can formulate the following questions to help them decide on how to treat ethical issues: In the dominant role(s) of evaluation, what part of our behaviour (thinking, acting) do we need to be ethically concerned with? Through which evaluation activities (including analysis of power) ought we to show our ethical behaviour? What individual activities do we need to work on to become ethical? Most importantly, what type of ethical subjects do we want to be in relation to existing power? Answers to these questions can yield further insights as to how to define action to carry on with evaluation activities.

Figure 1 shows how these two areas of analysis are related. To the deployment of IS evaluation through power relations, analysis of power (e.g.,

how is evaluation deployed?), could trigger the identification of ethical issues for those involved. Using this analysis, individuals could then identify and reflect upon their ethics and how to develop it by considering what has been deployed as ethically acceptable. This could place people in a better position to define their possibilities and constraints for action according to existing power relations. As new issues of concern emerge in an evaluation process, further analysis of power and forms of ethics is required, as the interactions between the elements of Figure 1 show.

AN EXAMPLE

As an example of how to use the above elements of inquiry, let us consider that in an IS evaluation process, financial control and communication to stakeholders are seen as essential (Irani, 2002; Serafeimidis & Smithson, 2003) in order to guarantee compliance with organisational procedures of auditing. In this context, evaluation can then be seen as control mechanism, more specifically as a way of exerting control over IS investments (or perhaps as a way of enabling financial officers to exert control over the rest of the organisation). The deployment of evaluation as an accepted process can have many manifestations. These could include, for instance, continuous exercise of formal authority (e.g., via established practices of reporting to finance officers), traditional use of financial skills and resources to get evaluation activities "done" (e.g., by an influential chief financial officer), or emerging pertinence of financial matters in IS investment decisions (e.g., a sound business case with "numbers" that now needs to be elaborated before being approved). These manifestations could be the by-product of previous practices (i.e., a history of financial success or failure in the organisation).

With this understanding of evaluation as a deployment of power relations, those involved in evaluation could then proceed to reflect on how a particular issue (e.g., communication) and its treatment can be dealt with. This issue can be then considered "ethical," and the expected behaviours or ways of thinking about it identified. People involved in evaluation could decide not to pay any more attention, for instance, to requests to analyse or communicate (financial) progress to other stakeholders or use existing communications to raise a different set of ethical issues (e.g., confidentiality, quality, etc.) Decisions can follow people's desire to become ethically different (e.g., more professional in their practice) or to be "seen" as ethical (and then using the power available to make themselves known). These decisions need to be examined in the light of potential consequences for individuals and their organisations, and any effect that could be foreseen (for instance, excessive professionalism could then generate a desire for people to become "professionally accredited"). In this particular case, the emergence of new issues to be discussed in evaluation (for example, due to new business practices related to improvement in customer service), or new ways of conducting evaluation in the context of intervention (e.g., those seen as more "professional") can then trigger further analyses on how these elements are being deployed and how they need to be managed. Although this example is brief, it illustrates the type of analysis that can be conducted and the actions that could result to improve the practice of IS evaluation. The example can also prompt evaluation practitioners to reflect on the scope of their analyses by considering manifestations and effects of power at different levels (economic, social, "political", etc).

CONCLUSION

In this article, a review of the issue of power in critical systems thinking and information systems evaluation has been undertaken to define an alternative view about it. It has been found that existing interpretations of power as operating "externally"

from those involved in evaluation leaves individuals with little guidance in relation to how to identify and act about it. Using the commitments of critical systems thinking and Foucault's ideas on power and ethics, the article develops a view of power and a system of inquiry into how it can be analysed in IS evaluation. The system enables practitioners and others involved in evaluation to be *critically aware* of the influence of power to deploy evaluation. It also allows for the inclusion and study of *different* manifestations of power and relations between them. Using this system, practitioners can inquire about how evaluation becomes possible through power relations. Inquiry should lead practitioners to reflect on ethical issues associated with IS evaluation and develop their own actions to *improve* their practice according to what they consider is ethical to do.

In comparison with other perspectives on power, Foucault's ideas can prompt those involved in evaluation to study the *power conditions* of the evaluation itself before establishing any possibilty of dialogue or debate6. This can help them to frame their actions into the possibilities and constraints given by power relations in the context where they are immersed. In evaluation practice, there is still a need to compare the study of power from this perspective with others. We see an opportunity to incorporate the use of the proposed system of inquiry with the use of systems methodologies to promote participative IS evaluation. We hope the view on power developed in this article contributes to open up further opportunities of dialogue and research between critical systems thinking and information systems.

REFERENCES

Avison, D., & Horton, J. (1992). *Evaluation of information systems* (Working Paper). Southampton: University of Southampton, Department of Accounting and Management Science.

Avison, D. E., & Wood-Harper, A. T. (1990). *Multiview: An exploration in information systems development.* Henley on Thames, UK: Alfred Waller (McGraw-Hill Publishing Company).

Avison, D., Wood-Harper, A. T., Vidgen, R. T., & Wood, J. R. G. (1998). A further exploration into information systems development: The evolution of Multiview2. *Information Technology and People, 11*(2), 124-139.

Ball, K., & Wilson, D. (2000). Power, control and computer-based performance monitoring: A subjectivist approach to repertoires and resistance. *Organization Studies, 21*(3), pp. 539-565.

Ballantine, J., Levy, M., Munro, I., & Powell, P. (2003). An ethical perspective on information systems evaluation. *International Journal of Agile Management Systems, 2*(3), 233-241.

Bariff, M., & Galbraith, J. R. (1978). Intraorganizational power considerations for designing information systems. *Accounting, organizations and society, 3*(1), 15-27.

Bernhauer, J., & Mahon, M. (1994). The ethics of Michel Foucault. In G. Gutting (Ed.), *The Cambridge Companion to Foucault* (pp. 141-158). Cambridge, UK: Cambridge University Press.

Bloomfield, B., & Coombs, R. (1992). Information technology, control and power: The centralization and decentralization debate revisited. *Journal of Management Studies, 29*(4), 459-484.

Brocklesby, J., & Cummings, S. (1996). Foucault plays Habermas: An alternative philosophical underpinning for critical systems thinking. *Journal of the Operational Research Society, 47*(6), 741-754.

Brooke, C. (2002). What does it mean to be "critical" in IS research? *Journal of Information Technology, 17*(2), 49-57.

Burrell, G. (1988). Modernism, post modernism and organizational analysis: The contribution of Michel Foucault. *Organization Studies, 9*(2), 221-235.

Chan, A., & Garrick, J. (2002). Organisation theory in turbulent times: The traces of Foucault's ethics. *Organization, 9*(4), 683-701.

Checkland, P. (1981). *Systems thinking, systems practice.* London: John Wiley and Sons.

Checkland, P. (1990). Information systems and systems thinking: Time to unite? In P. Checkland & J. Scholes (Eds.), *Soft systems methodology in action* (pp. 303-315). Chichester, UK: John Wiley & Sons Ltd.

Checkland, P., & Holwell, S. (1998). *Information, systems and information systems: Making sense of the field.* Chichester, UK: John Wiley and Sons.

Checkland, P., & Scholes, P. (1990). *Soft systems methodology in action.* Chichester: John Wiley and Sons.

Churchman, C. W. (1970). Operations research as a profession. *Management Science, 17*, b37-b53.

Churchman, C. W. (1979). *The systems approach and its enemies.* New York: Basic Books.

Clarke, S. (2001). *Information systems strategic management : An integrated approach.* London: Routledge.

Clarke, S., & Lehaney, B. (2000). Mixing methodologies for information systems development and strategy: A higher education case study. *Journal of the Operational Research Society, 51*, 542-566.

Córdoba, J. R., & Robson, W. D. (2003). Making the evaluation of information systems insightful: Understanding the role of power-ethics strategies. *Electronic Journal of Information Systems Evaluation, 6*(2), 55-64.

Darier, E. (1998). Time to be lazy: Work, the environment and modern subjectivities. *Time & Society, 7*(2), 193-208.

Darier, E. (1999). Foucault and the environment: An introduction. In E. Darier (Ed.), *Discourses of the environment* (pp. 1-33). Oxford: Blackwell.

Dhillon, G. (2004). Dimensions of power and IS implementation. *Information & Management, 41*, 635-644.

Doherty, N., & King, M. (2001). An investigation of the factors affecting the successful treatment of organisational issues in systems development. *European Journal of Information Systems, 10*, 147-160.

Doolin, B. (2004). Power and resistance in the implementation of a medical management information system. *Information Systems Journal, 14*(4), 343-362.

Farbey, B., Land, F., & Targett, D. (1999). Moving IS evaluation forward: Learning themes and research issues. *Journal of Strategic Information Systems, 8*(2), 189-207.

Flood, R. L. (1990). *Liberating systems theory.* New York: Plenum Press.

Flood, R. L., & Jackson, M. C. (1991a). Total systems intervention: A practical face to critical systems thinking. *Systems Practice, 4*, 197-213.

Flood, R. L., & Jackson, M. C. (Eds.). (1991b). *Critical systems thinking: Directed readings.* Chichester: John Wiley and Sons.

Flood, R. L., & Romm, N. (1996). *Diversity management: Triple loop learning.* Chichester: John Wiley and Sons.

Foucault, M. (1977). *The history of sexuality volume one: The will to knowledge* (Vol. 1). London: Penguin.

Foucault, M. (1980a). Truth and power. In P. Rabinow (Ed.), *The Foucault reader: An introduction to Foucault's thought* (pp. 51-75). London: Penguin.

Foucault, M. (1980b). Two lectures. In C. Gordon (Ed.), *Power/knowledge: Selected interviews and other writings Michel Foucault* (pp. 78-108). New York: Harvester Wheatsheaf.

Foucault, M. (1982a). Afterword: The subject and power. In H. Dreyfus & P. Rabinow (Eds.), *Michel Foucault: Beyond structuralism and hermeneutics* (pp. 208-226). Brighton: The Harvester Press.

Foucault, M. (1982b). On the genealogy of ethics: An overview of work in progress. In P. Rabinow (Ed.), *The Foucault reader: An introduction to Foucault's thought* (pp. 340-372). London: Penguin.

Foucault, M. (1984a). The ethics of the concern of the self as a practice of freedom (R. e. a. Hurley, Trans.). In P. Rabinow (Ed.), *Michel Foucault: Ethics subjectivity and truth: Essential works of Foucault 1954-1984* (pp. 281-301). London: Penguin.

Foucault, M. (1984b). *The history of sexuality volume two: The use of pleasure.* London: Penguin.

Foucault, M. (1984c). What is enlightenment? (C. Porter, Trans.). In P. Rabinow (Ed.), *The Foucault reader: An introduction to Foucault's thought* (pp. 32-50). London: Penguin.

Gregory, A. (2000). Problematizing participation: A critical review of approaches to participation in evaluation theory. *Evaluation, 6*(2), 179-199.

Gregory, W. J. (1992). *Critical systems thinking and pluralism: A new constellation.* Unpublished doctoral dissertation, City University, London.

Gregory, A., & Jackson, M. C. (1992). Evaluation methodologies: A system for use. *Journal of the Operational Research Society, 43*(1), 19-28.

Guba, E. G., & Lincoln, Y. S. (1989). *Fourth generation evaluation.* Newbury Park, CA: Sage Publications.

Handy, C. (1976). *Understanding organizations.* Aylesbury: Penguin.

Hirschheim, R., & Smithson, S. (1999). Evaluation of information systems: A critical assessment. In L. Willcocks & S. Lester (Eds.), *Beyond the IT productivity paradox* (pp. 381-409). Chichester, UK: John Wiley and Sons.

Horton, K. S. (2000). The exercise of power and information systems strategy: The need for a new perspective. *Proceedings of the 8th European Conference on Information Systems (ECIS),* Vienna.

Introna, L. D. (1997). *Management, information and power: A narrative of the involved manager.* Basingstoke: Macmillan.

Irani, Z. (2002). Information systems evaluation: Navigating through the problem domain. *Information & Management, 40,* 11-24.

Irani, Z., & Fitzgerald, G. (2002). Editorial. *Information Systems Journal, 12*(4), 263-269.

Irani, Z., & Love, P. E. (2001). Information systems evaluation: Past, present and future. *European Journal of Information Systems, 10*(4), 189-203.

Irani, Z., Love, P. E., Elliman, T., Jones, S., & Themistocleus, M. (2005). Evaluating E-government: Learning from the experiences of two UK local authorities. *Information Systems Journal, 15*(1), 61-82.

Jackson, M. C. (1982). The nature of soft systems thinking: The work of Churchman, Ackoff and Checkland. *Journal of Applied Systems Analysis, 9,* 17-29.

Jackson, M. C. (1992). An integrated programme for critical thinking in information systems research. *Information Systems Journal, 2,* 83-95.

Jackson, M. C. (1999). Towards coherent pluralism in management science. *Journal of the Operational Research Society, 50*(1), 12-22.

Jackson, M. C. (2000). *Systems approaches to management.* London: Kluwer Academic/Plenum Publishers.

Jackson, M. C. (2003). *Creative holism: Systems thinking for managers.* Chichester, UK: John Wiley and Sons.

Jackson, M. C., & Keys, P. (1984). Towards a system of system methodologies. *Journal of the Operational Research Society, 35,* 473-486.

Jasperson, J. S., Carte, T., Saunders, C. S., Butler, B. S., Croes, H. J. P., & Zheng, W. (2002). Power and information technology research: A metatriangulation review. *MIS Quarterly, 26*(4), 397-459.

Legge, K. (1984). *Evaluating planned organizational change.* London: Academic Press.

Lewis, P. (1994). *Information systems development: Systems thinking in the field of information systems.* London: Pitman Publishing.

Lukes, S. (1974). *Power: A radical view.* London: Macmillan.

Lyytinen, K., & Hirschheim, R. (1987). Information systems failures - A survey and classification of the empirical literature. *Oxford Surveys in Information Technology, 4,* 257-309.

Markus, M. L. (2002). Power, politics and MIS implementation. In M. Myers & D. Avison (Eds.), *Qualitative research in information systems.* London: Sage.

Mertens, D. (1999). Inclusive evaluation: Implications of transformative theory for evaluation. *American Journal of Evaluation, 20*(1), 1-14.

Midgley, G. (1996). What is this thing called CST? In R. L. Flood & N. Romm (Eds.), *Critical Systems Thinking: Current Research and Practice* (pp. 11-24). New York: Plenum Press.

Midgley, G. (1997). Mixing methods: Developing systemic intervention. In J. Mingers & A. Gill (Eds.), *Multimethodology: The Theory and Practice of Combining Management Science Methodologies.* (pp. 249-290). Chichester, UK: John Wiley and Sons.

Midgley, G. (2000). *Systemic intervention: Philosophy, methodology and practice.* New York: Kluwer Academic/Plenum.

Mingers, J. (1984). Subjectivism and soft systems methodology: A critique. *Journal of Applied Systems Analysis, 11,* 85-113.

Mingers, J. (1992). Technical, practical and critical OR: Past, present and future? In M. Alvesson & H. Willmott (Eds.), *Critical management studies* (pp. 90-112). London: Sage.

Mingers, J. (2005). 'More dangerous than an unanswered question is an unquestioned answer': A contribution to the Ulrich debate. *Journal of the Operational Research Society, 56*(4), 468-474.

Mingers, J., & Gill, A. (1997). *Multimethodology: The theory and practice of combining management science methodologies.* Chichester, UK: John Wiley & Sons Ltd.

Mora, M., Gelman, O., Cervantes, F., Mejia, M., & Weitzenfeld. (2003). A systemic approach for the formalization of information systems concept: Why information systems are systems? In J. Cano (Ed.), *Critical reflections in information systems: A systemic approach* (pp. 1-29). Hershey (PA): Idea Group Publishing.

Oliga, J. (1996). *Power, ideology and control.* New York: Plenum.

O'Neill, T. (1995). Implementation frailties of Guba and Lincoln's fourth generation evaluation theory. *Studies in Educational Evaluation, 21*(1), 5-21.

Ormerod, R. (1996). Information systems strategy development at Sainsbury's supermarket using "soft" OR. *Interfaces, 16*(1), 102-130.

Ormerod, R. (2005). Putting soft OR methods to work: the case of IS strategy development for the UK Parliament. *Journal of the Operational Research Society, 56*(12), 1379-1398.

Parker, M. M., Benson, R., & Trainor, H. E. (1988). *Information economics: Linking business performance to information technology.* Englewood Cliffs, NJ: Prentice Hall.

Piccoli, G., & Ives, B. (2005). IT-dependent strategic initiatives and sustained competitive advantage: A review and synthesis of the literature. *MIS Quarterly, 29*(4), 747-776.

Remenyi, D., & Sherwood-Smith, M. (1999). Maximise information systems value by continuous participative evaluation. *Logistics Information Management, 12*(1/2), 145-156.

Robey, D., & Boudreau, M. (1999). Accounting for the contradictory organizational consequences of information technology: Theoretical directions and methodological implications. *Information Systems Research, 10*(2), 167-185.

Rowlinson, M., & Carter, C. (2002). Foucault and history in organization studies. *Organization, 9*(4), 527-547.

Serafeimidis, V., & Smithson, S. (1999). Rethinking the approaches to information systems evaluation. *Logistics Information Management, 12*(1-2), 94-107.

Serafeimidis, V., & Smithson, S. (2003). Information systems evaluation as an organizational institution - Experience from a case study. *Information Systems Journal, 13,* 251-274.

Smithson, S., & Tsiavos, P. (2004). Re-constructing information systems evaluation. In C. Avgerou, C. Ciborra & F. Land (Eds.), *The social study of information and communication technology: Innovation, actors and contexts* (pp. 207-230). Oxford: Oxford University Press.

Stowell, F. (1995). *Information systems provision: The contribution of soft systems methodology.* London: McGraw-Hill.

Symons, V., & Walsham, G. (1988). The evaluation of information systems: A critique. *Journal of Applied Systems Analysis, 15,* 119-132.

Taylor, C. (1984). Foucault on freedom and truth. *Political Theory, 12*(2), 152-183.

Ulrich, W. (1983). *Critical heuristics of social planning: A new approach to practical philosophy.* Berne: Haupt.

Ulrich, W. (2003). Beyond methodology choice: critical systems thinking as critically systemic discourse. *Journal of the Operational Research Society, 54*(4), 325-342.

Valero-Silva, N. (1996). A Foucauldian reflection on critical systems thinking. In R. L. Flood & N. Romm (Eds.), *Critical systems thinking: Current research and practice.* (pp. 63-79.). London: Plenum.

Vega-Romero, R. (1999). *Care and social justice evaluation: A critical and pluralist approach.* Hull: University of Hull.

Walsham, G. (1993). *Interpreting information systems in organisations.* Chichester, UK: John Wiley and Sons.

Walsham, G. (1999). Interpretive evaluation design for information systems. In L. Willcocks & S. Lester (Eds.), *Beyond the IT productivity paradox* (pp. 363-380). Chichester, UK: John Wiley and Sons.

Walsham, G., & Waema, T. (1994). Information systems strategy and implementation: A case study of a building society. *ACM Transactions on Information Systems, 12*(2), 150-173.

Weiss, C. (1970). The politicization of evaluation research. *Journal of Social Issues, 26*(4), 57-68.

Weiss, C. (1998). Have we learned anything new about the use of evaluation? *American Journal of Evaluation, 19*(1), 21-34.

Wilson, B. (1984). *Systems: Concepts, methodologies, and applications.* Chichester, UK: John Wiley and Sons.

Wilson, B. (2002). *Soft systems methodology: Conceptual model and its contribution.* Chichester, UK: John Wiley and Sons.

Wray-Bliss, E. (2003). Research subjects/research subjections: Exploring the ethics and politics of critical research. *Organization, 10*(2), 307-325.

This work was previously published in Information Resources Management Journal, Vol. 20, Issue 2, edited by M. Khosrow-Pour, pp. 74-89, copyright 2007 by IGI Publishing (an imprint of IGI Global).

Chapter 16
Information Technology Industry Dynamics:
Impact of Disruptive Innovation Strategy

Nicholas C. Georgantzas
Fordham University Business Schools, USA

Evangelos Katsamakas
Fordham University Business Schools, USA

ABSTRACT

This chapter combines disruptive innovation strategy (DIS) theory with the system dynamics (SD) modeling method. It presents a simulation model of the hard-disk (HD) maker population overshoot and collapse dynamics, showing that DIS can crucially affect the dynamics of the IT industry. Data from the HD maker industry help calibrate the parameters of the SD model and replicate the HD makers' overshoot and collapse dynamics, which DIS allegedly caused from 1973 through 1993. SD model analysis entails articulating exactly how the structure of feedback relations among variables in a system determines its performance through time. The HD maker population model analysis shows that, over five distinct time phases, four different feedback loops might have been most prominent in generating the HD maker population dynamics. The chapter shows the benefits of using SD modeling software, such as iThink®, and SD model analysis software, such as Digest®. The latter helps detect exactly how changes in loop polarity and prominence determine system performance through time. Strategic scenarios computed with the model also show the relevance of using SD for information system management and research in areas where dynamic complexity rules.

INTRODUCTION

In challenging business environments, where even the best thought-of and executed strategies can fail dramatically (Raynor, 2007), disruptive innovation is emerging as a mainstream strategy that firms use first to create and subsequently to sustain growth in many industries (Bower & Christensen, 1995; Christensen, 1997; Christensen, et al 2002; Christensen & Raynor, 2003).

Honda's small off-road motorcycles of the 60s, for example, personal computers and Intuit's accounting software initially under-performed established product offers. But such innovations bring new value propositions to new users, who do not need all the performance incumbent firms offer. After establishing themselves in a simple application or user niche, potentially disruptive products (goods or services) improve until they "change the game" (Gharajedaghi, 1999), driving incumbent firms to the sidelines.

Christensen and Raynor (2003) see disruptive innovation strategy (DIS) not as the product of random events, but as a repeatable process that managers can design and replicate with sufficient regularity and success, once they understand the circumstances associated with the genesis and distinct dynamics a DIS entails. Similarly, Christensen et al (2002, p. 42) urge technology managers, adept in developing new business processes, to design robust, replicable DIS for creating and nurturing new growth business areas. In so doing, they must (a) seek to balance resources that sustain short-term profit and investments in high-growth opportunities and (b) use both separate screening processes and separate criteria for judging sustaining and disruptive innovation projects.

DIS can crucially affect the dynamics of IT, causing turbulence and industry shake-outs. Anthony and Christensen (2004) and Christensen *et al.* (2002) argue that is extremely important for technology managers to understand DIS. To help them make it so, this chapter shows a system dynamics (SD) model that replicates Christensen's (1992) data on hard-disk (HD) maker population dynamics. The model draws on archetypal SD overshoot and collapse work (Alfeld & Graham, 1974; Mojtahedzadeh, Andersen & Richardson 2004), which covers SD models in many areas with similarities in the structure of causal processes.

Cast as a methodological IS industry case, the chapter also shows the use and benefits of model analysis with Mojtahedzadeh's (1996) pathway participation metric (PPM), implemented in his *Digest®* software (Mojtahedzadeh et al, 2004). Shown here is a small part of a modeling project that combined DIS theory with SD to answer specific client concerns about the dynamic consequences of implementing disruptive innovation strategies in established high-technology markets, which contain over- and under-served (current and potential) users.

By definition, DIS is a dynamic process. Any model that purports to explain the evolution of a dynamic process also defines a dynamic system either explicitly or implicitly (Repenning, 2002). A crucial aspect of model building in any domain is that any claim a model makes about the nature and structure of relations among variables in a system must follow as a logical consequence of its assumptions about the system. And attaining logical consistency requires checking if the dynamic system the model defines can generate the real-life performance of the dynamic process the model tries to explain.

But most existing DIS models are merely textual and diagrammatic in nature. Given a particular disruptive innovation situation, in order to determine if a prescribed DIS idea can generate superior performance, which only 'systemic leverage' endows (Georgantzas & Ritchie-Dunham, 2003), managers must mentally solve a complex system of differential or difference equations. Alas, relying on intuition for testing logical consistency in dynamic business processes might contrast sharply with the long-certified human cognitive limits (Morecroft, 1985; Paich & Sterman 1993; Sterman 1989; Sastry 1997).

Aware of these limits, the chapter makes multiple contributions. *One* is the culmination of the early disruptive innovation literature into a generic model of the hard-disk makers' overshoot and collapse. Using a generic structure from prior SD overshoot and collapse work, the model contains assumptions common to seemingly diverse theories in economics, epidemiology, marketing and sociology. *Two* is the translation of these seemingly diverse components into a simulation model that allows addressing the specific concerns of a real-life client, by generating the overshoot and collapse dynamics of the hard-disk makers' population.

Furthermore, the chapter aims at expanding the relatively scarce but insightful IS research using the SD modeling method (Dutta & Roy, 2005; Kanungo, 2003; Abdel-Hamid & Madnick 1989). By describing the SD method and demonstrating its value, the chapter encourages a wider adoption of the SD modeling method in information systems research.

The model analysis results show that, over five distinct time phases, four different feedback loops become prominent in generating the HD maker population dynamics from 1973 through 1993. The chapter does not merely translate Christensen's DIS work into a SD model to replicate his results. It dares to ask *how* and *why* the model produces the results it does. With the help of *Digest*®, the chapter ventures beyond *dynamic* and *operational thinking*, seeks insight from system structure and thereby accelerates *circular causality thinking* (Richmond, 1993). *Digest*® helps detect exactly how changes in loop prominence determine system performance.

Following are a review of the disruptive innovation strategy literature and an overview of the SD modeling method. Then the chapter proceeds with model description and discussion of the simulation and model analysis results.

DISRUPTIVE INNOVATION STRATEGY (DIS) THEORY

The management innovation literature has delineated multiple ways to dichotomize innovation, such as radical vs. incremental, competency destroying vs. competency enhancing, and component vs. architectural (Hill & Jones, 1998; Christensen, 1997; Tushman & Anderson, 1986). The DIS theory offers a new dichotomy of innova-

tion: *sustaining* vs. *disruptive* (Christensen, 1997). The defining feature of DIS is that it emphasizes new performance virtues or dimensions, which are not the primary performance dimensions of the mainstream market. Conversely, sustaining innovation emphasizes the improvement of extant product performance dimensions.

Typically, DIS or *disrupter* firms start out small and operate on the fringes of existing markets for a while, growing and establishing a foothold under incumbents' radar screens. At the heart of DIS, with the potential to disrupt a mature industry, perhaps even to overtake and to displace incumbent firms through time, is a technology and a good or service platform that marks a departure from sustaining innovation, in the form of product extensions and add-ons to existing goods and services. Such a technology fills a previously unidentified or unaddressed niche with a value proposition aligned with user needs or 'jobs to do' (Christensen & Raynor, 2003).

A disrupter firm offers new choices in the form of stripped down functionality at a lower price, i.e., a 'less for less' offering. Adapted from Christensen and Raynor (2003, p. 44), Figure 1 shows the low-end and non-consumption markets disrupters exploit. The sustaining innovations of established firms often over-supply users with technological functionality or services that users do not actually need. The broken straight lines of Figure 1 show the trajectories of increasing user requirements for a given good or service. The sustaining innovations solid line on the front panel of Figure 1 is the increasing performance the good or service offers, which is steeper than the user requirements broken line. For example, mainframe and mini-computers in the late 1980s offered users higher levels of performance, features and capability than they could use. This oversupply left a vacuum at the low end of the market for a 'simpler' product offering: the personal computer (PC).

Once introduced, along the solid, low-end disruption line on Figure 1, PCs had offered lower

Figure 1. Low-end and non-consumption market disruption (adapted from Christensen and Raynor, 2003, p.44)

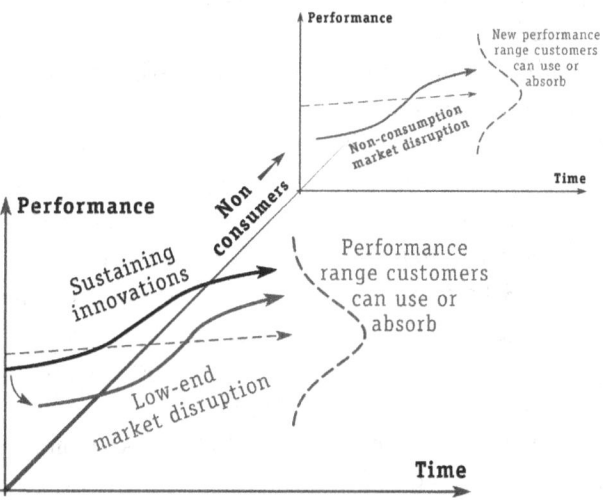

performance to users than mainstream mainframe and mini computers did. But a niche of users valued PCs and, through time, PC technological performance improved along the trajectory of the low-end disruption line. At some point, PC performance equaled that demanded by the average mainstream users of mainframes and minicomputers. So users started to switch, causing a widespread disruption of the established computer market, thereby driving many incumbent firms out of business. Depending on the performance ranges users can use or absorb to get a job done (i.e., to fulfill a need), new goods and services continually improve, usually faster than the average user's requirements, leaving space for new-market disruption waves among non-users on the back panel of Figure 1. Potentially, for example, the fast evolving personal digital assistant (PDA) and related mobile devices might next disrupt the PC market further in the future.

Disrupter firms typically target market segments currently unable to purchase a good or service or to fill a specific need. In effect, they create new markets by addressing non-user needs. Each disrupter firm exploits its ability to appeal to incumbent firms' low-end markets, i.e., over-served users facing a good or a service, with

functionality that far exceeds their needs, at a price they only pay reluctantly for lack of alternatives. Contrary to fitly served users, users in such markets cannot absorb sustaining performance improvements that exceed the range of utility they need or know how to exploit.

Once a disrupter firm becomes successful at penetrating non-consumption and low-end tiers, and has been on the market long enough to improve service delivery, to strengthen core business processes and to achieve a reasonable level of profitability, then the DIS firm is poised for an up-market march. This entails going after incumbent firms' high-end segments with an improved or expanded good/service offering and enhanced functionality at higher price points. The disrupter must be aware that moving up market to contest an incumbent firm's lock-in of lucrative users might trigger a wave of retaliation. So disrupters must ensure sufficient readiness to address the competitive response prior to embarking on their up-market march.

Disrupter firms exploit incumbents' exclusive focus on sustaining innovations and improved presence in the high-end, most profitable market segments (Christensen, 1997). As incumbents pay little attention to new and lower-end markets, they allow DIS entrants to move in and position themselves to eventually move up-market and to begin carving paths into the very markets established players serve. Incumbents begin to compromise long-term growth by allowing disrupters to eat into the lower-end segments and undermine their competitive position. Often, incumbents face a cost disadvantage compared to the typically light DIS cost structure. This limits when and how incumbents can respond to the DIS threat. Taking a long-term view might well suggest retaliating early and with great force.

Disrupters are typically ideally positioned to take advantage of the time lag to retaliation. They strengthen their presence and improve the quality and functionality of their offering and its overall value proposition as they prepare to embark upon an up-market march. A successful up-market march can spell a prolonged period of upset and transformation for entire industries. Old ways of doing business and serving users give way to superior ways of addressing user needs or jobs to do, at a more granular level and at a lower price.

This chapter provides insights in industry transformation, focusing on the effects of DIS on the number of firms operating in IT industry. It shows that DIS may crucially affect the dynamics of the IT industry, causing turbulence and consolidation of firms operating in the industry. The number of firms is a core aspect of industry structure because it affects product price and variety, as well as firm profitability and the value enjoyed by technology users (Tirole, 1988). The chapter also contributes to the emerging IS literature on disruptive innovation (e.g. Lyytinen & Rose, 2003a,b; Katsamakas & Georgantzas, 2008; Georgantzas & Katsamakas, 2008). Much of that literature focuses on the organizational impact of disruptive innovations (e.g., Lyytinen & Rose, 2003a,b), but pays little attention on the explanation of industry dynamics.

THE SYSTEM DYNAMICS (SD) MODELING METHOD

Client-driven, the entire SD modeling process aims at helping managers articulate exactly how the structure of circular feedback relations among variables in the system they manage determines its performance through time (Forrester & Senge, 1980). In the endless hunt for superior performance, SD's basic tenet is that the structure of feedback loop relations in a system gives rise to its dynamics (Meadows, 1989; Sterman, 2000, p. 16).

SD moves beyond mere - (Gharajedaghi, 1999; Senge, et. al, 1994) to systems formal modeling. Pioneered by MIT's Forrester (1961) and influenced by engineering control theory, SD calls for formal simulation modeling that provides a rigorous understanding of system behavior. Formal simulation modeling is an essential tool because "people's intuitive predictions about the dynamics of complex systems are systematically flawed" (Sterman, 1994, p. 178), mostly because of human's bounded rationality. Fontana (2006) sees SD as a most coherent modeling method, with high descriptive ability and theory building potential.

Two types of diagrams help formalize system structure: causal loop diagrams (CLDs) and stock and flow diagrams. CLDs depict relations among variables (e.g., Figure 5b). Arrows show the direction of causality and '+' and '−' signs the polarity of relations, i.e., how an increase in a variable affects change in a related variable. The culmination of all variable relations describes a set of positive or reinforcing and negative or balancing feedback loops characterizing a system.

Complementary to CLDs (Sterman, 2000), stock and flow diagrams depict how flow variables accumulate into stock variables, i.e., how stocks integrate the flows and how the flows differentiate the stocks (e.g., Figure 3). Stock and flow diagrams include causal loops, and provide the system with useful features such as memory and inertia. So they are essential in determining the dynamic behavior of the system under study. Figure 2 shows possible system behavior patterns through time. At the right level of abstraction, SD researchers encounter similar causal processes that underlie *seemingly* highly diverse phenomena (Forrester, 1961).

Model Analysis in the SD Modeling Method

Both as an inquiry and as a coherent problem-solving method, SD can attain its spectacular Darwinian sweep (Atkinson, 2004) as long as it formally links system structure and performance. In order to help academics and managers see exactly what part of system structure affects performance through time, i.e., detect shifting loop polarity and dominance (Richardson, 1995), SD researchers use tools from discrete mathematics and graph theory first to simplify and then to automate model analysis (Gonçalves, Lerpattarapong,

& Hines, 2000; Kampmann, 1996; Mojtahedzadeh, 1996; Mojtahedzadeh, et al 2004; Oliva, 2004; Oliva & Mojtahedzadeh, 2004). Mostly, they build on Nathan Forrester's (1983) idea to link loop strength to system eigenvalues.

Mojtahedzadeh's *Digest®* software plays a crucial role in the analysis of this chapter's model. The pathway participation metric inside *Digest®* detects and displays prominent causal paths and loop structures by computing each selected variable's dynamics from its slope and curvature, i.e., its first and second time derivatives. Without computer simulation, even experienced modelers find it hard to test their intuition about the connection between circular causality and SD (Oliva, 2004; Mojtahedzadeh *et al* 2004). Using *Digest®* is, however, a necessary but insufficient condition for insight. Insightful articulations that link performance to system structure integrate insight from dynamic, operational and feedback loop thinking (Mojtahedzadeh *et al* 2004; Richmond, 1993).

Figure 2. Eight archetypal performance (P) dynamics (i.e., behavior patterns through time) might exist within a single phase of behavior for a single variable (adapted from Mojtahedzadeh et al 2004)

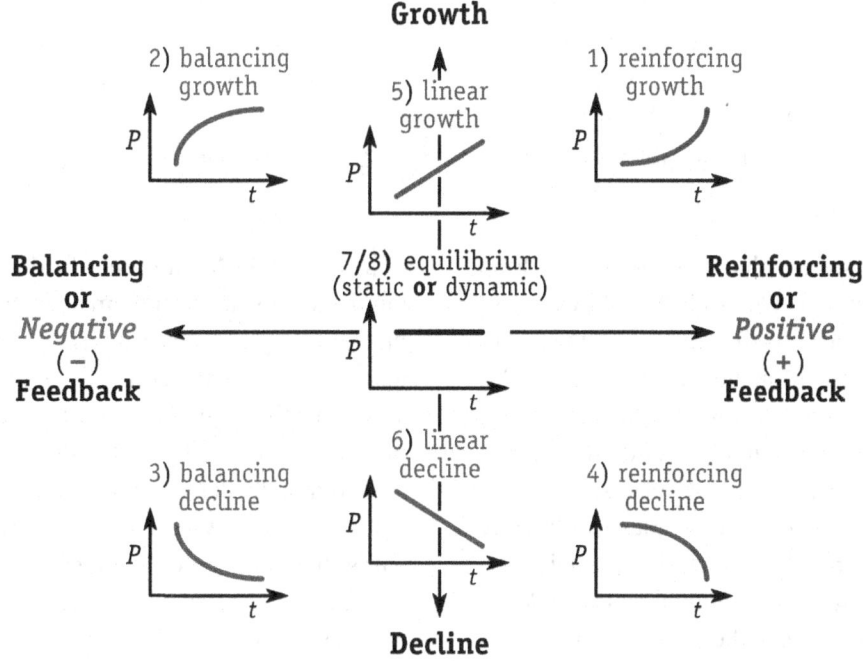

Figure 3. Hard-disk (HD) makers' population and users' jobs to do sector

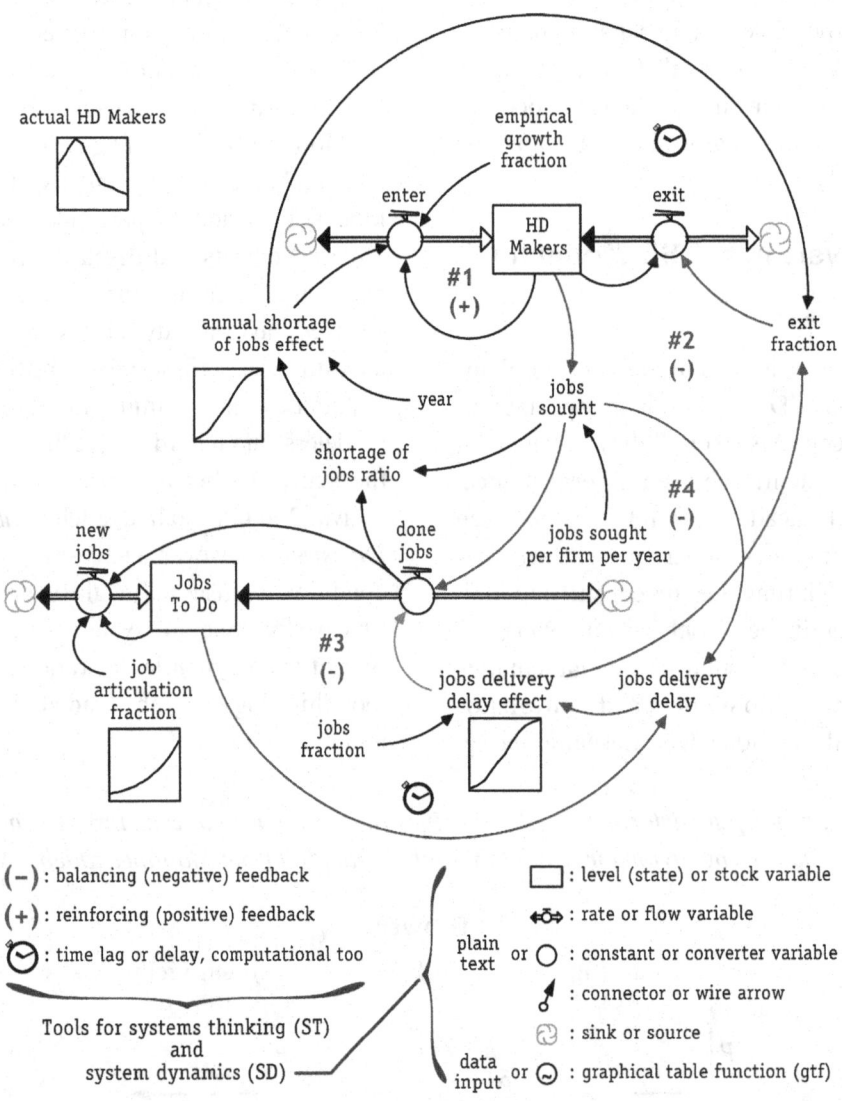

(−) : balancing (negative) feedback

(+) : reinforcing (positive) feedback

⊘ : time lag or delay, computational too

Tools for systems thinking (ST)
and
system dynamics (SD)

☐ : level (state) or stock variable

⬦ : rate or flow variable

plain text or ○ : constant or converter variable

⌁ : connector or wire arrow

⟲ : sink or source

data input or ⊙ : graphical table function (gtf)

Linked to eigenvalue and dominant loop research, Mojtahedzadeh's (1996) PPM is most promising in formally linking performance to system structure. Mojtahedzadeh *et al* (2004) give an extensive overview of PPM that shows its conceptual underpinnings and mathematical definition, exactly how it relates to system eigenvalues and concrete examples to illustrate its merits. Very briefly, the pathway participation metric sees a model's individual causal links or paths among variables as the basic building blocks

of structure. PPM can identify dominant loops, but does not start with them as its basic building blocks. Using a recursive heuristic approach, PPM detects compact structures of chief causal paths and loops that contribute the most to the performance of a selected variable through time.

Mojtahedzadeh *et al* (2004, pp. 7-11) also present *Digest®* software. *Digest®* detects the causal paths that contribute the most to generating the dynamics a selected variable shows. It first slices a selected variable's time path or trajectory into discrete phases, each corresponding to one of eight possible behavior patterns through time (Figure 2). Once the selected variable's time trajectory is cut into phases, PPM decides which pathway is most prominent in generating that variable's performance within each phase. As causal paths combine to form loops, combinations of such circular paths shape the most influential or prominent loops within each phase.

Mojtahedzadeh is testing PPM with a multitude of classic SD models, such as, for example, Alfeld and Graham's (1976) urban dynamics model (*cf* Mojtahedzadeh *et al* 2004). Similarly, Oliva and Mojtahedzadeh (2004) use *Digest*® to show that the shortest independent loop set (SILS), which

Oliva (2004) structurally derived via an algorithm for model partition and automatic calibration, does contain the most influential or prominent causal paths that *Digest*® detects. Methodologically, this chapter contributes to this line of work.

MODEL DESCRIPTION

The SD model consists of two major components or sectors. First, we describe the hard-disk makers' population and user *jobs to do* sector, and then the behavior reproduction testing sector. The SD simulation model was developed using the *iThink*® SD software (Richmond 2006).

Figure 4. Behavior reproduction testing sector

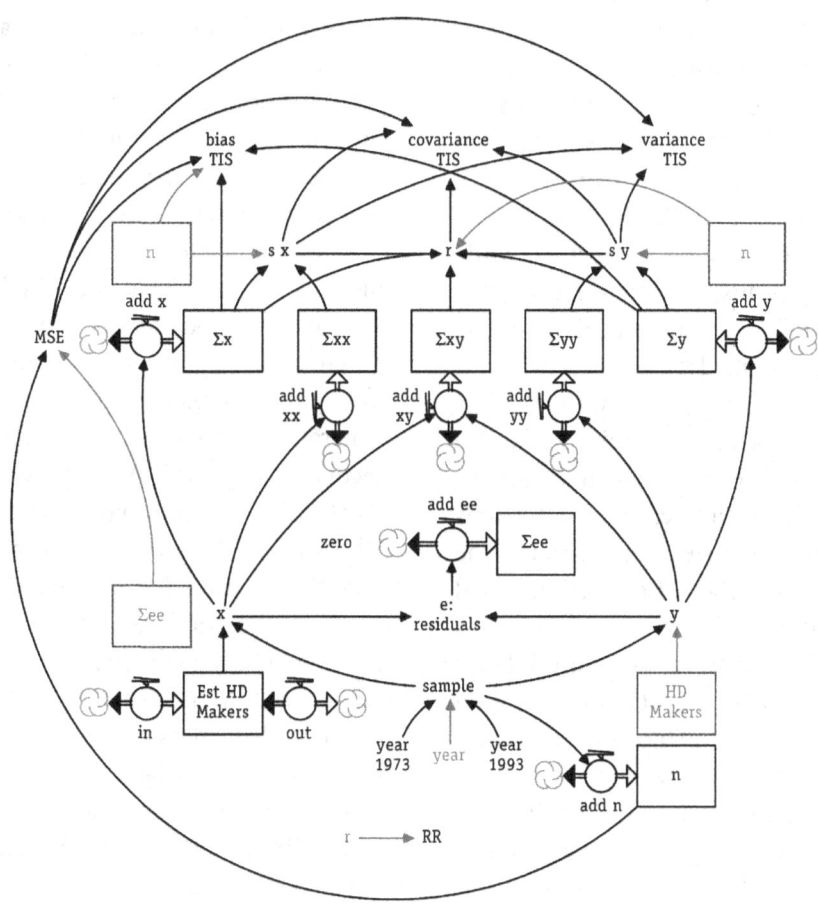

Hard-Disk (HD) Maker Population and User Jobs to do Sector

Figure 4 shows the model's hard-disk (HD) makers' population and user jobs to do sector. Listed in the Appendix, Table 1 shows the equations of the model. There is a one-to-one association between the model diagram of Figure 3 and its equations on Table 1. Building a model entails diagramming system structure and then specifying differential equations and parameter values. The software enforces consistency between model diagrams and equations, while its built-in functions help quantify parameters and variables pertinent to the HD Makers' overshoot and collapse dynamics.

Rectangles represent stocks or level variables that accumulate in SD, such as the population of HD Makers (Figure 3 and Eq. 1, Table 1). Emanating from cloud-like *sources* and ebbing into cloud-like *sinks*, the double-line, pipe-and-valve-like icons that fill and drain the stocks represent flows or rate variables that cause the stocks to change. The exit outflow of Figure 3 and Eq. 5, for example, bleeds the HD Makers stock, initialized (INIT) with 18 hard-disk maker firms (Eq. 1.1, Table 1) per Christensen's (1992) data. Single-line arrows represent information connectors, while circular icons depict auxiliary converters where constants, behavioral relations or decision points convert information into decisions.

The enter inflow (Eq. 4), which fills the HD Makers stock, depends, for example, on the HD Makers population itself, multiplied by the industry's empirical growth fraction, an exogenous auxiliary constant parameter (Eq. 7), and by the annual shortage of jobs effect (Eq. 16), a graphical table function (gtf).

The stock and flow diagram on Figure 3 shows accumulations and flows essential in generating the performance dynamics of the hard-disk maker population. The fate of this population was determined by the disruptive innovation diffusion process (Christensen, 1992). This diagram also tells, with the help of the equations on Tables 1,

what drives the flows in the system. In the context of systems thinking (ST), stock and flow diagrams like the one on Figure 3 help accelerate what Richmond (1993) calls *operational thinking*.

The model on Figure 3 and Table 1 is based on a classic structure that illustrates how the population of firms in a particular industry grows through time until the resources needed to support its growth are depleted (Alfeld & Graham, 1976; Mojtahedzadeh et al 2004). The model captures real-world processes as feedback loops that might cause the performance dynamics of its pertinent variables. Caught in a web of eleven feedback loops, the HD Makers' population, for example, grows when, *ceteris paribus*, new hard-disk makers enter through a reinforcing or positive (+) loop and declines when, again ceteris paribus, they exit through a balancing or negative (–) loop (Figure 3). Once new firms join the hard-disk makers' population, they immediately begin to deplete the users' Jobs To Do stock (Eq. 2), a vital resource for HD Makers to stay in business.

The shortage of jobs ratio (Eq. 12), i.e., the ratio of done jobs (Eq. 3) to jobs sought (Eq. 13), also affects new firm entry and exit indirectly. Last but not least, the users' Jobs To Do stock controls its own depletion rate, i.e., done jobs, by modulating the jobs delivery delay (Eq. 12), i.e., the ratio of Jobs To Do to jobs sought (Eq. 13).

Given its specific set of parameters and initial values, to explain the dynamics the model generates, the question is: which of the eleven feedback loops HD Makers are caught in are most influential or prominent in generating the HD Makers' behavior Christensen (1992) observed. For example, what made the users' Jobs To Do decline rapidly? What drove HD Makers to grow rapidly in the first few years? What part of the structure is responsible for the decline of the hard-disk makers' population followed by its growth? Those familiar with this archetypal model structure might easily explain the growth and declining phases. It might not be as easy, however, to distinguish which part of the model contributes most to the dynamics of HD Makers in the transition from reinforcing (+) growth to a balancing (–) decline. Using *Digest®* allows detecting the most prominent or influential feedback loops as the HD Makers dynamics unfolds.

Behavior Reproduction Testing Sector

To replicate the DIS-caused overshoot and collapse dynamics of the HD Makers' population that Christensen (1992) reports, the model's specific set of parameters and initial values were set to minimize the mean square error (MSE) between actual and simulation data. Shown on Figure 4, Theil's (1966) inequality statistics (TIS) subsequently decompose MSE on Figure 7.

TIS provide an elegant decomposition of the MSE into three components: bias (U^M), unequal variance (U^S) and unequal covariance (U^C), so that $U^M + U^S + U^C = 1$ (Oliva, 1995; Sterman, 1984 and 2000; Theil, 1966). Briefly, bias arises when competing data have different means. Unequal variance implies that the variances of two time series differ. Unequal covariance means imperfectly correlated data that differ point by point. Dividing each component by the MSE gives the MSE fraction due to bias (U^M), due to unequal variance (U^S) and due to unequal covariance (U^C). A large U^M reveals a potentially serious systematic error. U^S errors can be systematic too. When unequal variation dominates the MSE, the data match on average and is highly correlated but the variation in two time series around their common mean differs. One variable is a stretched out version of the other. U^S may be large either because of trend differences, or because the data have the same phasing but different amplitude fluctuations (Sterman, 2000, p. 876). If most of the error is concentrated in unequal covariance, then the data means and trends match but individual data points differ point by point. When U^C is large, then most of the error is unsystematic and, according to Sterman: "a model should not be faulted for

failing to match the random component of the data" (2000, p. 877).

Figure 4 shows the stock and flow diagram of the behavior reproduction testing model sector and Table 2 the sector's equations, complete with explanatory comments included for this TIS implementation. Worth noting, however, on Figure 4 and Table 2 are the estimated hard-disk makers stock (Est HD Makers, Eq. 19), along with its associated in and out flows (Eqs 34 and 35). These last three model components help replicate Christensen's (1992) data exactly, with zero error, using the built-in *STEP* function of *iThink®*. This may seem like a futile exercise at the outset, but it helped convince the client of the much larger modeling project than what is shown here that replicating real-life data does not necessarily produce much insight, nor does it help one appreciate a dynamically complex system.

RESULTS

To be useful, model analysis must create insight via coherent explanations of how influential pieces of system structure give rise to performance through time. Figure 6 shows the simulation results for the hard-disk maker population performance, with time phases and prominent feedback loops. The Est HD Makers behavior faithfully reproduces the actual HD Makers dynamics without error (Figure 5a). But zero error in behavior pattern reproduction can also mean zero insight for appreciating a dynamically complex system. The HD Makers behavior (line #3 on Figure 5a) provides a less impressive data fit, but the feedback loop web behind its dynamics is where insight lives.

The vertical lines on the time domain output of Figure 5a show five distinct time phases in the HD Makers dynamics, which *Digest®* identified by detecting behavior pattern shifts. Phase I of the HD Makers dynamics on Figure 5a shows reinforcing growth (Figure 2), which lasts for about 4 years. During this time, both the slope

Figure 5. Simulation results with time phases and prominent feedback loops

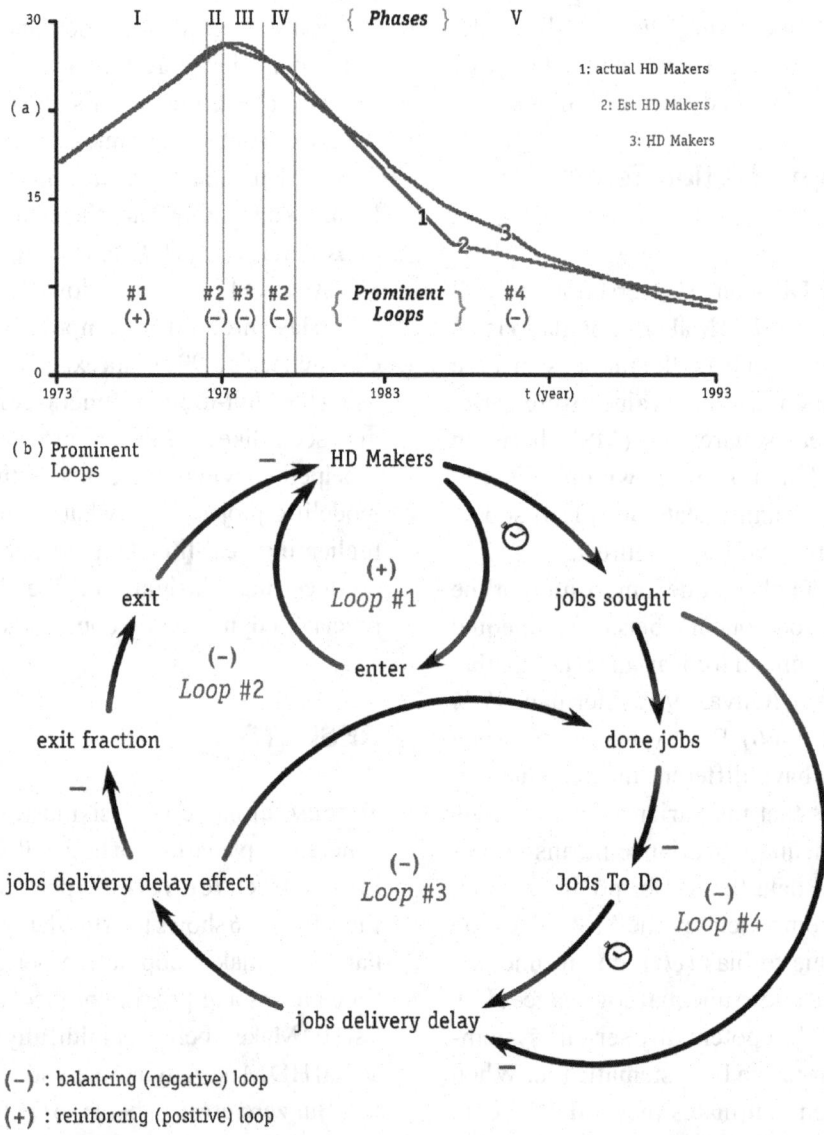

(−) : balancing (negative) loop
(+) : reinforcing (positive) loop
⊘ : time lag or delay, computational too

(first time derivative) and the curvature (second time derivative) of the variable of interest, HD Makers, remain positive. Phase II on Figure 5a shows balancing growth. The slope and curvature of HD Makers have opposite signs in this phase. Phases III and IV show reinforcing decline. And lastly, in its fifth distinct phase (Figure 5a) the HD Makers dynamics shows balancing decline (Figure 2).

In addition to discerning distinct time phases in the dynamics of a variable of interest, *Digest*® also detects and displays the most influential or prominent structures that contribute the most to the selected behavior pattern in each phase. Corresponding to the first phase of the behavior of HD Makers is reinforcing feedback loop #1 of Figure 5b which, according to *Digest*®, is the most prominent loop in generating the reinforc-

Figure 6. Phase plots of relations among pertinent variables

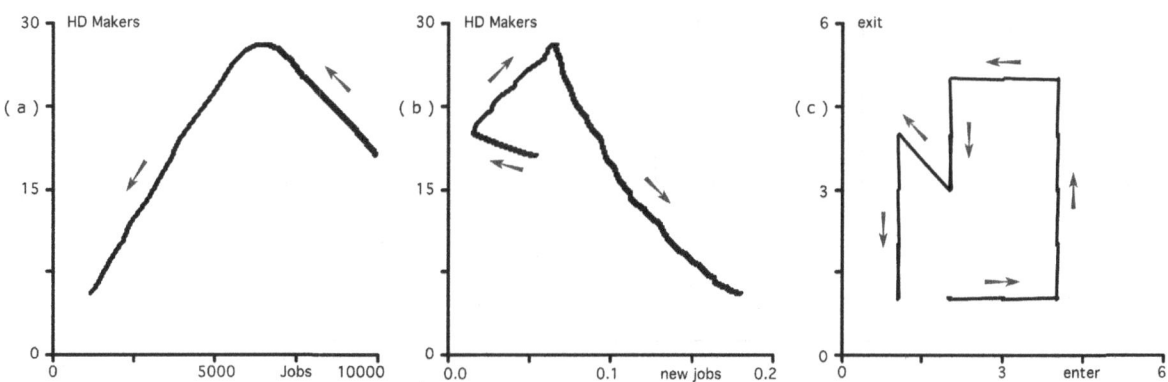

ing growth in the HD maker population. Initially, HD Makers attract new hard-disk makers to enter the industry, increasing HD Makers further. By inspecting the model structure on Figure 3, one could identify eleven feedback loops surrounding HD Makers. Using its pathway participation metric, *Digest*® automatically selects reinforcing feedback loop #1 as the most prominent one among all the other loops in the model.

In phase II of the HD Makers dynamics, system control shifts from reinforcing loop #1 to the most influential structure or balancing feedback loop #2 of Figure 5b, associated with the users' Jobs To Do stock. Initially plenty in phase I, users' Jobs To Do now begin to fall, along the pathway that carries the effect of balancing loop #2 to HD Makers. This same structure is also most prominent in phase IV of HD Makers' dynamics.

In phase III of the HD Makers behavior, balancing loop #3 becomes the most influential structure of Figure 5b, associated with the users' Jobs To Do stock and the delivery delay. By phase III, the large HD Makers population causes the done jobs rate to deplete the users' articulated Jobs To Do faster. And the more job delivery delay decreases because of the–by now–large hard-disk maker population, the more it causes the exit fraction to increase, thereby forcing some HD Makers to exit, while preventing new ones from entry.

In phase IV, prominent loop #2 takes over again, now from loop #3, while keeping the users' Jobs To Do stock in focus. In phase V, however, with HD Makers already dropping, prominent loop #4 bypasses the Jobs to Do stock, increasing the jobs delivery delay directly, indirectly causing the exit fraction, and thereby the exit rate of HD Makers, to slow down. Prominent loop #4 remains most influential until the end of the simulation.

In phase II of Figure 5, while trying to explain why HD Makers is generating a balancing growth, it may be easy to spot the role of the balancing feedback loop that controls the Jobs To Do stock depletion. The users' articulated Jobs To Do is dropping, thereby preventing new hard-disk makers from entry. The subtlety in explaining the behavior of the HD Makers is the subsequent reinforcing decline in HD Makers' dynamics in phase IV. Some novices may even look for reinforcing feedback to explain the reinforcing decline dynamics. But *Digest*® tells that what forces HD Makers to fall faster and faster is exactly the same process that keeps their population at bay. Balancing loop #2, which controls Jobs To Do, prevents new hard-disk makers from entering and, once new entries fall behind those who exit, the HD Makers stock goes into a reinforcing decline.

The relations among select pertinent variables on the phase plots of Figure 6 confirm the above. Polarity changes in all three cases, making it

Figure 7. Dynamic behavior reproduction test results

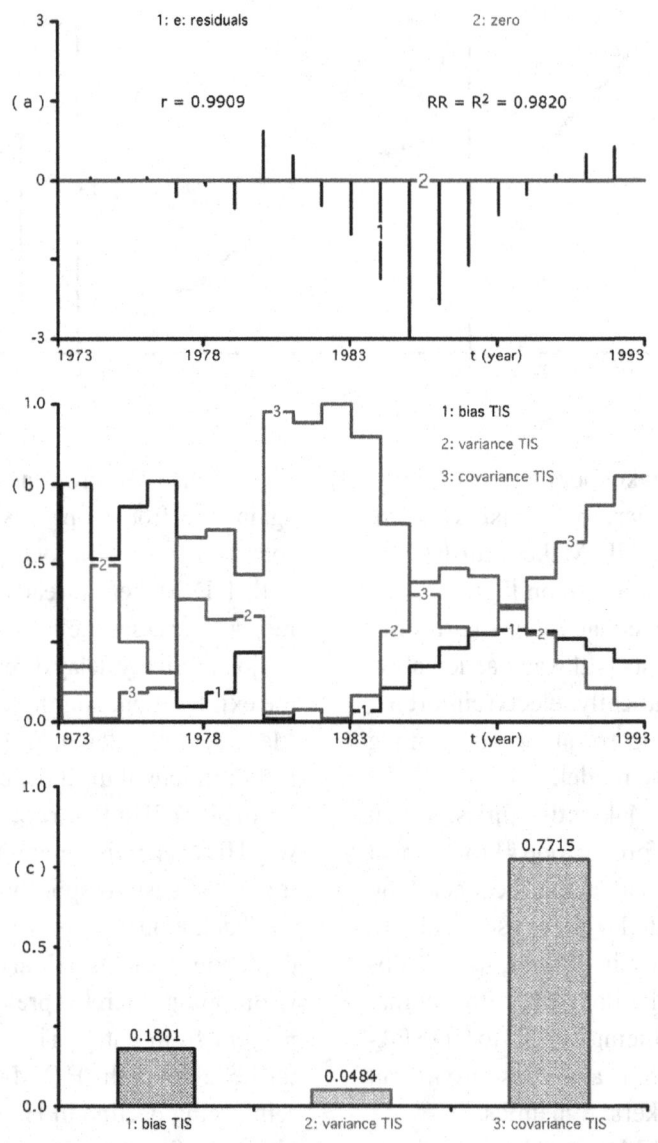

rather impossible to assess such relations with correlation statistics. On Figure 6a, for example, as the users' articulated Jobs To Do decline, HD Makers initially rise and subsequently fall. On Figure 6b, new jobs gradually decrease as the HD Makes stock grows exponentially, then they begin to grow once the HD maker population slows down, i.e., begins to increase at a declining rate but, lastly, they increase even more rapidly once the HD Makes stock decreases.

Behavior Reproduction Test Results

The coefficient of determination, R^2, which measures the variance in the data explained by the model as a dimensionless fraction, is a common statistic used to assess a model's ability to reproduce system behavior. The coefficient of determination is the square of the correlation coefficient, r, which measures the degree to which two series co-vary.

Although widely reported because audiences expect it (Figure 7), R^2 is actually not very useful. Two series with the same error can generate very different R^2 values depending on their trend (Sterman 2000, p. 874). Conversely, Theil's (1966) inequality statistics (TIS) use the mean square error (MSE), which measures the average error between competing data series in the same units as the variable itself and weights large errors more heavily than small ones.

The residual plot of Figure 7a shows an uneven pattern of serially autocorrelated errors, but both the r and the R^2 values are high. And Theil's inequality statistics (Figure 7b and c) do support the model's usefulness. The unequal covariance TIS, U^C, dominates throughout the simulation (Figure 7b), and Figure 7c shows the end TIS values on a vertical bar graph. Most of the MSE fraction is concentrated above U^C, showing that the model captures the mean and trends in the actual data rather well, differing mostly point by point.

Computed Strategic Scenarios

Both academics and managers can benefit from the leading interpretive instruments used in SD model analysis, such as eigenvalues, Theil's (1966) inequality statistics (Oliva, 1995; Sterman, 1984 and 2000) and the pathway participation metric, implemented in the *Digest*® software (Mojtahedzadeh, 1996; Mojtahedzadeh, et al 2004).

But SD models also allow computing strategic scenarios of what might happen in the future as well as of what might have been in the past. Both academics and managers again can benefit from the insight such scenarios provide, with respect to the potential effects that changes in environmental and policy parameters and variables might have or might have had on chosen performance variables of interest.

Back to the time domain of Figure 8, the computed strategic scenarios of Figure 8a show, for example, what might have been the effect of increasing the available Jobs To Do stock, i.e., the HD Makers' market size, on the hard-disk maker population. *Ceteris paribus*, a larger market size back in 1973 might have prolonged the reinforcing growth phase of HD Makers, perhaps giving enough time to some of them to respond more timely to new entrant firm's DIS. But the assumed structure of relations among variables in the system would still render inevitable the balancing growth phase that followed (Phase II, Figure 5a).

Again *all other things being equal*, less greed on HD Maker's behalf, in terms of the sought-jobs-per-firm-per-year policy parameter (Figure 8b), might have similarly prolonged the Phase I reinforcing growth of Figure 5a, giving time to some HD Makers to develop an effective disruptive innovation response strategy (DIRS). The computed strategic scenarios of Figure 8 show but

Figure 8. Strategic scenarios computed with the SD model

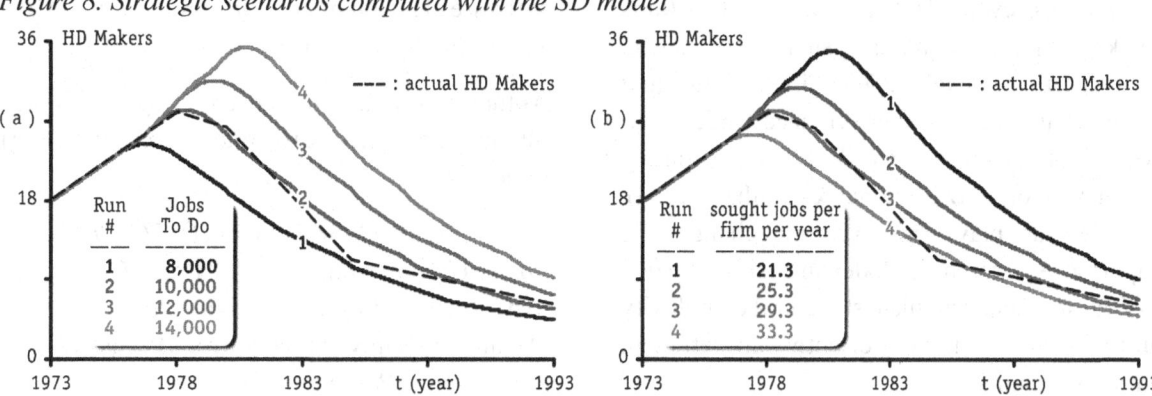

one example of how academics and managers can benefit from the insight SD models can give them, about the strategic leverage of pertinent strategic performance variables and policy parameters or strategy levers. Choosing which lever to push or pull on and when is crucial for both DIS and DIRS design in IS research and practice.

CONCLUSION

In business processes and systems, "randomness is a measure of our ignorance" (Sterman, 2000, p. 127). And Christensen and Raynor (2003) might be right to see disruptive innovation strategies as repeatable processes and not as the products of random events. But have the DIS and DIRS theory proponents used the right tools to help managers understand the circumstances associated with the genesis and distinct dynamics that DIS and DIRS entail?

Purely deterministic, this chapter's SD model is rather useful in explaining the HD Makers dynamics. With four different feedback loops becoming prominent along five distinct time phases, the chapter demonstrates the indispensable role of SD modeling in explaining the hard-disk makers' rise and fall between 1973 and 1993.

It is Mojtahedzadeh's *Digest®*, with its analysis of shifting prominent structure and polarity phases that has helped reveal the model analysis results. Indeed, tools such as PPM can help make sense of the dynamically complex structure of SD models, even if Oliva (2004, p. 331) finds SD keen in understanding system performance, "not structure *per se*", in lieu of its core tenet that system structure causes performance. Undeniably, while looking for *systemic leverage* in strategy making (Georgantzas & Ritchie-Dunham, 2003), modelers do play with structural changes for superior performance. Model analysis tools such as *Digest®* help articulate structural complexity and thereby enable both effective and efficient strategy designs.

The SD model behavior might resemble IS technology industry dynamics beyond the DIS effects in the hard-disk industry. For example, industry overshoot and collapse dynamics have been observed in e-commerce early in this century, when a large number of Internet firms entered the industry and then went out of business (Oliva, Sterman & Giese, 2003). The SD modeling method and tools described here can be extended and used equally well in these diverse contexts and that should be a fruitful direction for future research.

The SD modeling method can provide dynamic leverage insights in the dynamic complexity of information technology markets, and the design, development, implementation and management of IS, DIS and DIRS within organizations. A more extensive adoption of system dynamics method in IS research and practice should be fruitful. To that direction, the chapter authors organized a Workshop with theme "Complex Information System Dynamics" in NYC on June 11 2008. They are also guest-editing a forthcoming *System Dynamics Review* special issue on "Information Systems Dynamics", aiming to fuse a reinforcing feedback loop that will breed needed high-quality research on complex information systems dynamics.

REFERENCES

Abdel-Hamid, T. and Madnick, S. (1989). Lessons learned from modeling the dynamics of software development. *Communications of the ACM*, 32(12), 1426-1455.

Anthony, S. and Christensen, C. (2004). Forging innovation from disruption. *Optimize*, (Aug) issue 24.

Alfeld, L.E. and Graham, A. (1976). *Introduction to Urban Dynamics*. MIT Press, Cambridge MA. Reprinted by Productivity Press: Portland OR and currently available from Pegasus Communications: Waltham, MA.

Atkinson, G. (2004). Common ground for institutional economics and system dynamics modeling. *System Dynamics Review,* 20(4), 275-286.

Bower, J.L. and Christensen, C. (1995). Disruptive technologies: catching the wave. *Harvard Business Review* (Jan-Feb), 43-53.

Christensen, C.M. (1992). *The Innovator's Challenge: Understanding the Influence of Market Environment on Processes of Technology Development in the Rigid Disk Drive Industry.* Ph.D. Dissertation, Harvard Business School: Boston, MA.

Christensen, C.M. (1997). *The Innovator's Dilemma: When New Technologies Cause Great Firms to Fail.* Harvard Business School Press: Cambridge, MA.

Christensen, C.M., Johnson, M. & Dann, J. (2002). Disrupt and prosper. *Optimize* (Nov), 41-48.

Christensen, C.M. & Raynor, M.E. (2003). *The Innovator's Solution: Creating and Sustaining Successful Growth.* Harvard Business School Press: Boston MA.

Christensen, C.M., Raynor, M.E. & Anthony, S.D. (2003). Six Keys to Creating New-Growth Businesses. *Harvard Management Update* (Jan).

Dutta, A. and Roy, R. (2005). Offshore outsourcing: a dynamic causal model of counteracting forces. *Journal of Management Information Systems,* 22(2), 15-35.

Fontana, M. (2006). Simulation in economics: evidence on diffusion and communication. *Journal of Artificial Societies and Social Simulation,* 9(2) (http://jasss.soc.surrey.ac.uk/9/2/8.html).

Forrester, J.W. (1961). *Industrial Dynamics.* MIT Press: Cambridge, MA.

Forrester, J.W. (2003). Dynamic models of economic systems and industrial organizations. *System Dynamics Review,* 19(4), 331-345.

Forrester, J.W. & Senge, P.M. (1980). Tests for building confidence in system dynamics models. In AA Legasto Jr, JW Forrester and JM Lyneis (Eds), *TIMS Studies in the Management Sciences, Vol. 14: System Dynamics.* North-Holland: New York, NY, pp. 209-228.

Forrester, N. (1983). Eigenvalue analysis of dominant feedback loops. In *Plenary Session Papers Proceedings of the 1st International System Dynamics Society Conference,* Paris, France: 178-202.

Georgantzas, N.C. & Ritchie-Dunham, J.L. (2003). Designing high-leverage strategies and tactics. *Human Systems Management,* 22(1), 217-227.

Georgantzas, N.C. & Katsamakas, E. (2008). *Disruptive service-innovation strategy.* Working Paper, Fordham University, New York, NY.

Gharajedaghi, J. (1999). *Systems Thinking: Managing Chaos and Complexity: A Platform for Designing Business Architecture.* Butterworth-Heinemann: Boston MA.

Gonçalves, P., Lerpattarapong, C. and Hines, J.H. (2000). Implementing formal model analysis. In *Proceedings of the 18th International System Dynamics Society Conference,* August 6-10, Bergen, Norway.

Hill, C.W.L. & Jones, G.R. (1998). *Strategic Management: An Integrated Approach.* Houghton Mifflin: Boston, MA.

Kampmann, C.E. (1996). Feedback loops gains and system behavior. In *Proceedings of the 12th International System Dynamics Society Conference,* July 21-25, Cambridge, MA.

Kanungo, S. (2003). Using system dynamics to operationalize process theory in information systems research. *Proceedings of the* 24th *International Conference on Information Systems,* 450-463.

Katsamakas, E. & Georgantzas, N. (2008). *Open source disruptive innovation strategies.* Working Paper, Fordham University, New York, NY.

Lyytinen, K. & Rose, G. (2003a). The disruptive nature of Information Technology innovations: the case of internet computing in systems development organizations. *MIS Quarterly,* 27(4), 557-595.

Lyytinen, K. & Rose, G. (2003b). Disruptive information system innovation: the case of internet computing. *Information Systems Journal,* 13, 301-330.

Meadows, D.H. (1989). System dynamics meets the press. *System Dynamics Review,* 5(1), 68-80.

Mojtahedzadeh, M.T. (1996). *A Path Taken: Computer-Assisted Heuristics for Understanding Dynamic Systems.* Ph.D. Dissertation. Rockefeller College of Public Affairs and Policy, SUNY: Albany NY.

Mojtahedzadeh. M.T., Andersen, D. & Richardson, G.P. (2004). Using *Digest®* to implement the pathway participation method for detecting influential system structure. *System Dynamics Review,* 20(1), 1-20.

Morecroft, J.D.W. (1985). Rationality in the analysis of behavioral simulation models, *Management Science,* 31, 900-916.

Oliva, R. (2004). Model structure analysis through graph theory: partition heuristics and feedback structure decomposition. *System Dynamics Review,* 20(4), 313-336.

Oliva, R. (1994). *A Vensim Module to Calculate Summary Statistics for Historical Fit.* MIT System Dynamics Group D-4584.

Oliva, R. & Mojtahedzadeh, M.T. (2004). Keep it simple: a dominance assessment of short feedback loops. In *Proceedings of the 22nd International System Dynamics Society Conference,* July 25-29, Keble College, Oxford University, Oxford UK.

Oliva, R., Sterman, J.D. & Giese, M. (2003). Limits to growth in the new economy: exploring the 'get big fast' strategy in e-commerce. *System Dynamics Review,* 19(2), 83-117.

Paich, M. & Sterman, J.D. (1993). Boom, bust and failures to learn in experimental markets. *Management Science,* 39(12), 1439-1458.

Raynor, M.E. (2007). *The Strategy Paradox.* Currency-Doubleday: New York, NY.

Repenning, N.P. (2002). A simulation-based approach to understanding the dynamics of innovation implementation. *Organization Science,* 13(2), 109-127.

Repenning, N.P. (2003). Selling system dynamics to (other) social scientists. *System Dynamics Review,* 19(4), 303-327.

Richardson, G.P. (1991). *Feedback Thought in Social Science and Systems Theory.* University of Pennsylvania Press: Philadelphia, PA.

Richardson, G.P. (1995). Loop polarity, loop prominence, and the concept of dominant polarity. *System Dynamics Review,* 11(1), 67-88.

Richmond, B. (1993). Systems thinking: critical thinking skills for the 1990s and beyond. *System Dynamics Review,* 9(2), 113-133.

Richmond. B. et al. (2006). *iThink® Software* (*version 9*). iSee Systems™: Lebanon NH.

Sastry, M.A. (1997). Problems and paradoxes in a model of punctuated organizational change. *Administrative Science Quarterly,* 42(2), 237-275.

Senge, P. et al.(1994). *The Fifth Discipline Fieldbook,* Currency-Doubleday: New York, NY.

Sterman, J.D. (1989). Modeling managerial behavior: misperceptions of feedback in a dynamic decision making experiment. *Management Science,* 35(3), 321-339.

Sterman, J.D. (1994). Beyond training wheels. In *The Fifth Discipline Fieldbook*, Senge P. et al, Currency-Doubleday: New York, NY.

Sterman, J.D. (2000). *Business Dynamics: Systems Thinking and Modeling for a Complex World.* Irwin McGraw-Hill: Boston, MA.

Sterman, J.D. (1984). Appropriate summary statistics for evaluating the historical fit of system dynamics models. *Dynamica,* 10(Winter), 51-66.

Theil, H. (1966). *Applied Economic Forecasting.* Elsevier Science (North Holland): New York, NY.

Tirole, J. (1988). *The Theory of Industrial Organization.* MIT Press: Cambridge, MA.

Tushman, M.L. & Anderson, P. (1986). Technological discontinuities and organizational environments. *Administrative Science Quarterly,* 31, 439-465.

Unsworth, K. (2001). Unpacking creativity. *Academy of Management Review,* 26, 289-297.

Veryzer, R.W. (1998). Discontinuous innovation and the new product development process. *Journal of Product Innovation Management,* 15(2), 136-150.

APPENDIX

Table 1. Hard-disk (HD) makers' population and users' jobs-to-do sector equations

Stock or Level (State) Variable ({·} = comments and/or units)	Equation #
HD Makers(t) = HD Makers(t - dt) + (enter - exit) * dt	(1)
INIT HD Makers = 18 {unit: firm}	(1.1)
Jobs To Do(t) = Jobs To Do(t - dt) + (new jobs – done jobs) * dt	(2)
INIT Jobs To Do = TIME / job articulation fraction {unit: job}	(2.1)
Flows or Rate Variables	
done jobs = jobs delivery delay effect * jobs sought {unit: job / year}	(3)
enter = ROUND (STEP (HD Makers * empirical growth fraction * annual shortage of jobs effect, TIME)) {unit: firm / year}	(4)
exit = ROUND (STEP (HD Makers * exit fraction, TIME)) {unit: firm / year}	(5)
new jobs = 1 - job articulation fraction * (Jobs To Do - done jobs) / TIME {unit: job / year}	(6)
Auxiliary Parameters and Converter Variables	
empirical growth fraction = 0.1385 {unit: 1 / year}	(7)
exit fraction = (1.048 - jobs delivery delay effect * annual shortage of jobs effect) {unit: 1 / year}	(8)
jobs delivery delay = Jobs To Do / jobs sought {unit: year}	(9)
jobs fraction = 0.103 {unit: 1 / year}	(10)
jobs sought per firm per year = 29.3 {unit: job / firm / year}	(11)
shortage of jobs ratio = done jobs / jobs sought {unit: unitless}	(12)
jobs sought = HD Makers * jobs sought per firm per year {unit: job / year}	(13)
year = 1 {Data time interval (i.e., unit: year)}	(14)
actual HD Makers = GRAPH(TIME {Christensen's (1992) HD Makers data}) (1973, 18.0), (1974, 20.0), (1975, 22.0), (1976, 24.0), (1977, 26.0), (1978, 28.0), (1979, 27.0), (1980, 26.0), (1981, 23.0), (1982, 20.0), (1983, 17.0), (1984, 14.0), (1985, 11.0), (1986, 10.4), (1987, 9.75), (1988, 9.12), (1989, 8.50), (1990, 7.88), (1991, 7.25), (1992, 6.62), (1993, 6.00)	(15)
annual shortage of jobs effect = GRAPH(shortage of jobs ratio / year {unit: 1 / year}) (0.00, 0.00), (0.1, 0.06), (0.2, 0.14), (0.3, 0.255), (0.4, 0.395), (0.5, 0.535), (0.6, 0.685), (0.7, 0.825), (0.8, 0.92), (0.9, 0.98), (1, 1.00)	(16)
jobs delivery delay effect = GRAPH(jobs fraction * jobs delivery delay {unit: unitless}) (0.00, 0.00), (0.1, 0.06), (0.2, 0.14), (0.3, 0.255), (0.4, 0.395), (0.5, 0.535), (0.6, 0.685), (0.7, 0.825), (0.8, 0.92), (0.9, 0.98), (1, 1.00)	(17)
job articulation fraction = GRAPH(TIME {unit: 1 / year}) (1973, 0.197), (1974, 0.221), (1975, 0.237), (1976, 0.259), (1977, 0.287), (1978, 0.325), (1979, 0.369), (1980, 0.416), (1981, 0.468), (1982, 0.527), (1984, 0.593), (1985, 0.667), (1986, 0.75), (1987, 0.844), (1988, 0.949), (1989, 1.07), (1990, 1.20), (1991, 1.35), (1992, 1.52), (1993, 1.71)	(18)

Table 2. Behavior reproduction testing sector equations

Stock or Level (State) Variable ({·} = comments and/or units)	Equation #
Est HD Makers(t) = Est HD Makers(t - dt) + (in - out) * dt; INIT Est HD Makers = 18 {unit: firm}	(19)
∑ee(t) = ∑ee(t - dt) + (add ee) * dt; INIT ∑ee = 0	(20)
∑x(t) = ∑x(t - dt) + (add x) * dt; INIT ∑x = 0 {Cumulative sum of the actual data}	(21)
∑xx(t) = ∑xx(t - dt) + (add xx) * dt; INIT ∑xx = 0 {Cumulative sum of the squared actual data}	(22)
∑xy(t) = ∑xy(t - dt) + (add xy) * dt; INIT ∑xy = 0 {Cumulative sum of the xy product}	(23)
∑y(t) = ∑y(t - dt) + (add y) * dt; INIT ∑y = 0 {Cumulative sum of the simulated data}	(24)
∑yy(t) = ∑yy(t - dt) + (add yy) * dt; INIT ∑yy = 0 {Cumulative sum of the squared simulated data}	(25)
n(t) = n(t - dt) + (add n) * dt; INIT n = 1e-9 {The current count n of data points}	(26)
Flows or Rate Variables	
add ee = e: residuals^2 / DT {Adds to the sum of squared errors between actual and simulated data}	(27)
add n = sample / DT {Increments n, i.e., adds one to the number of observations}	(28)
add x = x / DT {Adds to the cumulative sum of the actual data}	(29)
add xx = x^2 / DT {Adds to the sum of the squared actual data}	(30)
add xy = x * y / DT {Adds to the cumulative sum of the xy product of actual and simulated data}	(31)
add y = y / DT {Adds to the cumulative sum of the simulated data}	(32)
add yy = y^2 / DT {Adds to the cumulative sum of the squared simulated data}	(33)
in = STEP(2, 1973) - STEP(2, 1978) {unit: firm / year}	(34)
out = STEP(1, 1978) - STEP(1, 1980) + STEP(3, 1980) - STEP(3, 1985) + STEP(0.625, 1985) - STEP(0.625, 1993) {unit: firm / year}	(35)
Auxiliary Parameters and Converter Variables	
bias TIS = ((∑x / n) - (∑y / n))^2 / (1e-9 + MSE) {The unequal bias Theil inequality statistic (TIS) is the MSE fraction caused by unequal means of the actual and simulated data}	(36)
covariance TIS = (2 * s x * s y * (1 - r)) / (1e-9 + MSE) {The unequal covariance Theil inequality statistic (TIS) is the MSE fraction caused by imperfect correlation between actual and simulated data}	(37)
e: residuals = x − y {The difference between sampled actual and simulated data}	(38)
MSE = ∑ee / n {The mean squared error between actual and simulated data}	(39)
r = ((∑xy / n) - (∑x / n) * (∑y / n)) / (s x * s y + 1e-9) {The correlation between x and y}	(40)
RR = r^2 {The coefficient of determination R^2 is the square of the correlation coefficient}	(41)
s x = SQRT ((∑xx / n) - (∑x / n)^2) {The standard deviation of x}	(42)
s y = SQRT ((∑yy / n) - (∑y / n)^2) {The standard deviation of y}	(43)
sample = PULSE (DT, year 1973, year) * (STEP (1, year 1973) - STEP (1, year 1993 + DT / 2)) {Sterman (2000, Ch. 21 + CD) suggests sampling once a year between in order to compare actual and simulation data only where actual data exist}	(44)
variance TIS = (s x - s y)^2 / (1e-9 + MSE) {The unequal variance Theil inequality statistic (TIS) is the MSE fraction caused by the unequal variance of actual and simulated data}	(45)
x = sample * Est HD Makers {The actual data sampled}	(46)
y = sample * HD Makers {The simulated data sampled}	(47)
year 1973 = 1973 {The data start time}	(48)
Year 1993 = 1993 {The data end time}	(49)
Zero = 0 {This plots a horizontal line at the origin of the y axis in the time domain}	(50)

This work was previously published in Best Practices and Conceptual Innovations in Information Resources Management: Utilizing Technologies to Enable Global Progressions, edited by M. Khosrow-Pour, pp. 231-250, copyright 2009 by Information Science Reference (an imprint of IGI Global).

Chapter 17
Using a Systems Thinking Perspective to Construct and Apply an Evaluation Approach of Technology–Based Information Systems

Hajer Kefi
IUT Paris and University of Paris Dauphine, France

ABSTRACT

In this article, we use soft systems methodology and complexity modeling to build an evaluation approach of a data warehouse implemented in a leading European financial institution. This approach consists in building a theoretical model to be used as a purposeful observation lens, producing a clear picture of the problematic situation under study and aimed at providing knowledge to prescribe corrective actions.

INTRODUCTION

In this article, we discuss a research approach constructed and applied to evaluate the performance and the multiple impacts of a corporate data warehouse implemented in a financial institution. The first section examines the epistemological and the methodological underpinnings of our approach based upon soft systems methodology and systemic modeling. We will especially focus on

the reasons why we have chosen a systemic view to build an evaluation approach of an information technology (IT)-based information system (IS). We will argue that pragmatic issues arising from the characteristics of the empirical field under investigation and/or the researcher status within this field can lead to seek an alternative to the positivist paradigm on one side and the interpretive paradigm on the other side. In the second section, we discuss the theoretical development of our evaluation tool conceived as a structuring framework to investigate the field with special lenses, and also allowing the description of emergent and unpredictable events. The third section describes how this approach has been applied in an empirical research process conducted during a period of 17 months using multiple research techniques. Finally, results, limitations, implications, and recommendations for future research are presented.

SEEKING A METHODOLOGICAL APPROACH: TO MAKE SENSE, BUILD THE MODEL

Since the Delone and McLean (1992) quest for a dependent variable to assess technology-based information systems (IS/IT) success, the evaluation of these systems, in terms of their intrinsic performance (technical or task-oriented), and/or impacts on individuals, groups, organizations, and societies is still a hot issue, generating much interest among a wide range of researchers, in management science, economics, sociology, computer science, and so forth.

This issue illustrates, in our opinion, the divergent and nevertheless complementary perspectives and points of view that characterize the information systems field. We believe that this diversity does not contribute to enhancing the identity crisis within this discipline (Benbasat & Zmud, 2003; Galliers, 2003) but helps researchers and practitioners to develop multiple evaluation tools and frameworks that can satisfy a multiplicity of requirements: technical, financial, productivity-oriented, behavioural, and so forth. The first step for a researcher or a practitioner involved in an evaluation process is to define the perspective that will be adopted. Such a choice obviously depends on his or her intentions, interests, and theoretical and professional background. It also depends on the objectives of the study being conducted: theory development, theory testing (empirical studies),

practical recommendations, corrective actions prescription (action research), and so forth.

The research study discussed in this article is related to an evaluation process conducted by a doctoral student in management information systems, who has been mandated by the chief information officer of a leading European financial institution to assess the performance of a corporate data warehouse (DW) implemented within this firm and to prescribe corrective actions in order to promote success and avoid failure. To cope with this task, combining scientific rigor, pertinent observation, objective assessment, and corrective actions, a research approach built upon systems thinking is iteratively developed, applied, and adjusted over time (Churchman, 1979). Theoretical work is not a prerequisite to enter the research field. Empirical and theoretical tasks are combined to help the researcher give meaning to what he or she observes and experiences (Avison, Lau, Myers, & Nielson, 1999; Avison, Baskerville, & Myers, 2001). This is what Checkland and Scholes (1990) call an *Organized Purposeful Action* defined as a "deliberate, decided, willed action, whether by an individual or by a group" (p. 1).

The dilemma here is that to produce purposeful action, the researcher cannot ignore the knowledge already accumulated. He is also willing to produce "new" experience-based knowledge. Now, where might the knowledge to guide action be found? The temptation is great to try an affiliation to the positivist research tradition (Hirschheim, 1992). Whereas the omnipresent social dimension of the IS field and the human intentions embedded in organized purposeful action, the constructivist and interpretivist research strategy seem to be more appropriate (Galliers, 1992).

Pragmatically, we adopt a systems thinking epistemology positioning that we will apply using pluralistic research techniques (qualitative and quantitative). As argued by many authors (Alter, 2004; Checkland, 1999; LeMoigne, 1977; Mora, Gelman, Cervantes, Mejía, & Weitzenfeld, 2002), using "systems" thinking in the information "systems" research field is not evident at all. This is because the word "systems" does not refer to the same thing in these two areas of knowledge. While the "systematic" perspective is predominant in the IS field, particularly in systems development, it is the "systemic" or "systems" perspective that mostly characterizes systems thinking. In the systematic perspective, a system is a group of elements in interaction. Reasoning in a systematic manner implies identifying all the embedded elements of an object and studying all the interactions between them, in respect of the reductionism principle of the positivist paradigm (Popper, 1963). In the systemic approach related to the systems thinking paradigm and particularly to the general systems theory (Von Bertalanffy, 1968), a system is an abstract notion that means a whole, or *holon:* an emergent entity that is not simply the collection of its constituent elements. According to this view, a way of thinking, solving problems, and learning require the construction of a *model* that we can call "general system" to help restructure the real-world situation in an "investigable" or modeled manner.

This philosophy has been supported by Jean-Louis LeMoigne (1977), professor emeritus in France and the initiator of the European Program on Complexity Modeling and the association for complex thinking (www.mcxapc.org). According to LeMoigne, real-world situations are necessarily *complex*, that is, ill-structured and not recognizable directly and in a one and only "best" way. A purposeful "general system," or *holon,* is needed to make a recognition (between many others) of a given real-world situation.

Figure 1 below describes this mechanism. The general system, or holon, is the method that transforms a soft phenomenon in a well-structured object under study (the model). Rather than a transformation, LeMoigne prefers the term of "modeling." The representation system is a very important component of the modeling mechanism. It is a surrogate of the researcher and encompasses his or her intentions, research objectives, and view points. Accordingly, to develop knowledge, that is, a structured understanding of natural facts, it is not necessary for the observer to be neutral and to remain out of the subject under study to guarantee objectivity.

This approach produces a substantial theoretical contribution to the systemic thinking paradigm. Nevertheless, it is not, in our opinion, sufficiently explicit to be applied per se in our study. We have chosen to build upon soft systems methodology (Checkland, 1999; Checkland & Scholes, 1991) to define our research approach. In SSM, the adjective "soft" is linked to the systemic perspective, as opposed to "hard" systems related to the systematic perspective. It also advocates (as for complex or systemic modeling) the neces-

Figure 1. Systemic modeling lemoigne (1977, p.57)

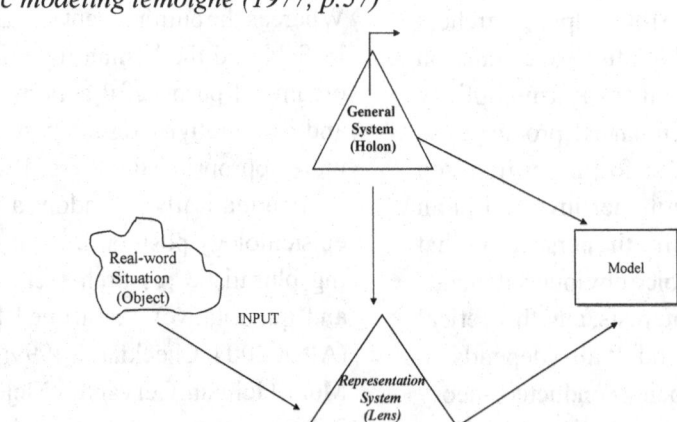

sary *feed-back* between real-world and systems thinking about real world. SSM and complexity modeling are neighboring and complementary research perspectives, yet they continue to ignore each other.

We have decided to adopt the systemic view (built upon SSM and systemic modeling) in our research approach. And accordingly, to attribute meaning to what we observe and experience in an investigated situation, we have to build *a model*. This model will be used as a diagnostic tool that aims to produce a clear picture of this situation and provides knowledge to prescribe corrective actions.

The core question we investigate is: "What are the critical elements that conduct and converge over time to the success (or the failure) of the DW implementation project within the organization and are there any solutions to prescribe in order to promote success and avoid failure?"

A longitudinal case study is conducted, based on the assumption that the adoption and use of the DW should be conceptualized as a form of organizational change and that such perspective allows us to anticipate, explain, and evaluate different consequences following the introduction of such a tool in the organization (Orlikowski, 1993). We argue that these consequences occur at multiple

levels: the IS/IT level, the individual level, and the organizational level. The focus is then to develop a context-based, process-oriented description and explanation of the phenomenon, rather than an objective and static description (Kannelis, Lycett, & Paul, 1998; Orlikowski, 1993).

This approach draws on Pettigrew's (1990) conceptualization of the organizational change study within the organization in a manner that considers both the content and context of change over time. Multiple research techniques will be used to draw a clear picture of the reality we are describing, from multiple viewpoints and using multiple data sources, including questionnaire, structured and semi-structured interviewing, documentation review, and observation.

THEORETICAL DEVELOPMENT

We consider three dimensions of what Markus and Robey (1988) call a good theory, that is, a "theory that guides research, which when applied, increases the likelihood that information technology will be employed with desirable consequences for users, organizations and other interest parties" (p. 583).

Figure 2. Represents the conventional seven-stage model of SSM.

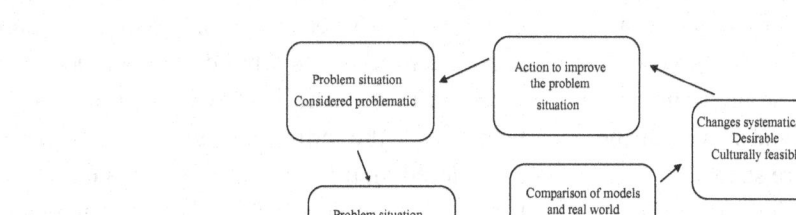

These dimensions are: causal agency, logical structure, and level of analysis.

The causal agency refers to the nature of causality relationships between the two major components of the theory: IS/IT and organization. Here we have chosen to break with the frameworks largely used in this kind of research area, such as the studies investigating the organizational impacts of IS/IT or those related to IS/IT development and implementation literature. These frameworks consist of causal and variance models that proceed to the identification of dependent and independent variables, and define the relationship between the technology and the organization, in terms of causal unidirectional relations, based on the deterministic assumptions of either a technological or an organizational imperative (Markus & Robey, 1988).

The framework proposed here is different from those identified above. It is based on a structurational perspective (DeSanctis & Poole, 1994; Giddens, 1987; Orlikowski, 1992, 1993, 2000; Swanson & Ramiller, 1997) that emphasizes the dual nature of technology, comprising a social and a technical interrelated component, and considers a bi-directional and recursive relation between technology and organization, via diverse attitudes and behaviors of human agents during task execution. Three main categories of agents are identified: the decision makers (or managers), the users, and the conceptors/developers. The institutional properties of the organization are considered in this framework to be: (1) the context that affects the ways in which these agents interact with technology; and (2) the field in which the impacts of these interactions are shaped.

We have to notice here that this structurational perspective belongs to the socio-technical tradition thinking in IS (Bostrom & Heinen, 1977) and that both of them are built upon the systems thinking paradigm (Checkland, 1999).

The logical structure concerns the hypothesized relationships between causes and effects, or outcomes, among the components of the theory and how they are shaped over time. That means whether causes are related to effects in an invariant, sufficient, and necessary relationship (variance model) or in a recipe of sufficient conditions occurring over time.

We argue that a process model for IS/IT evaluation is more appropriate to investigate over time the conditions of success or failure of an IS/IT project throughout its lifecycle, considering the content of the impacts it generates and the context in which these impacts occur.

Process-oriented IS/IT evaluation models supposing longitudinal studies are recommended in many studies, such as those of King and Rodriguez (1978), Hamilton and Chervany (1981a, 1981b) and Delone and McLean (1992, 2003).

The level of analysis refers to the entities about which the theory poses concepts and relationships. We adopt a multi-level perspective for IS/IT evaluation. These levels are considered over time, beginning with the (IS/IT) tool and his intrinsic characteristics (technical and social ones); then we consider the impacts it generates at the individual level (the users considered individually, then on the organization as a whole).

As such, our process-oriented and multi-level evaluation approach has some similarities with the Delone and McLean Model (1992). However, we propose an extension to this model by breaking with the simple unidirectional causes-effects relationships it advocates and that have been so criticized (Ballantine et al., 1998). The structurational perspective allows us to build an interactionist and recursive relationship between the three levels: the technology, the individual, and the organization considered, in this order over time, and the institutional context. Evaluating the performance and effectiveness of a technology-based information system within this framework requires a global understanding of the phenomenon incorporating the strategic and organizational context, and more specifically the context of development and use, and the interactions between actors during the system life cycle. We argue that the complexities related to all these aspects should not be ignored or underestimated.

Our theoretical model consists of two components:

- The first component is a process-oriented evaluation instrument of a technology-based information system that determines: (1) the perceived performance of this system by its users; (2) its individual impacts on users by identifying the resulting appropriation types; (3) its organizational impacts defined in terms of organizational changes.

- The second component of the model displays the context factors that affect the perceived performance of the IS/IT and its individual and organizational impacts. These factors are related to the organizational context, the strategic context, and the context of development and use.

According to the structurational model defined above, the two components are recursively interrelated (Figure 3).

Now we explicitly define the major components of our theoretical model and define their causal relationships.

The Strategic and Organizational Context

Organization size, corporate strategy, structure, culture, role of the IS/IT function in the organization, IT strategy, and role of the leadership in the decision-making concerning IS/IT are some of the most-cited factors related to the strategic and organizational context of IS/IT implementation projects. In most of the studies, they are considered in a contingency perspective in the evaluation approaches (Myers, Kappelman, & Prybutok, 1998; Saunders & Jones, 1992).

A deeper stream of research is based on an inductive perspective aimed at identifying, exclusively through data collected and analyzed in the field, the critical elements that shape the impacts associated with the implementation and use of IS/IT tools (Orlikowski, 1993).

We propose to begin our investigation of the strategic and organizational context by considering three aspects largely cited in the literature:

Figure 3. Research model of IS/IT evaluation

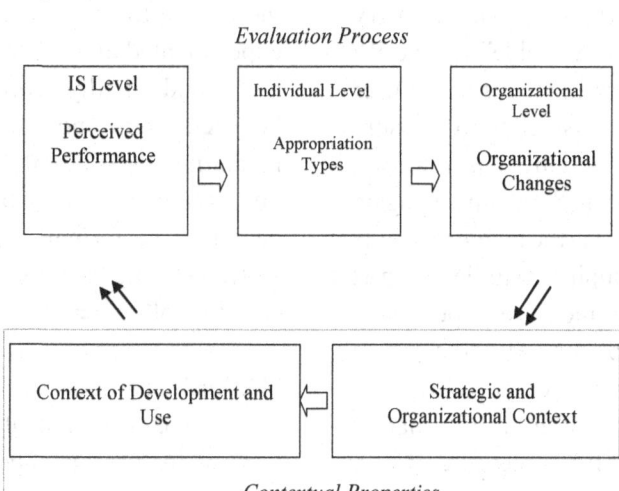

(1) corporate strategies, (2) structure and culture, and (3) role of IT function.

The Context of Development and Use

The approaches that focus on the context of development and use are based on theories of attitudes and behavior and try to identify the situational factors that lead to intentions to IS/IT utilization and acceptance. They argue that increased utilization and acceptance lead to positive performance impacts. Within these situational factors, Zmud (1979) focuses on the construct of individual differences; Ives and Olson (1984) and Kappelman (1995) on user involvement (Barki & Hartwick1989); on user participation, user involvement, and user attitude (Barki and Hartwick1994); and Goodhue and Thompson (1995) on the characteristics of the task and the individual and the concept of task-technology-fit.

In these studies, the subtleties between these approaching concepts are determined and their impacts on IS/IT performance studied either as independent variables or as mediating ones.

The IS/IT level: Perceived Performance Evaluation

This component of our model is related to the perceptual approaches of IS/IT evaluation that try to identify "the dependent variable" for the IS/IT effectiveness or success and focus on the user perception of utility and satisfaction to measure it. The so-called user satisfaction or user information satisfaction construct is the main surrogate measure for this variable. It is the theme of a wide range of theoretical and empirical studies, such as Franz and Robey's (1986) measure of perceived usefulness; Doll and Torzadeh's (1988) measure of end-user computing satisfaction; Bailey and Pearson's (1983) measure of user information satisfaction; and Davis' (1989) measure of perceived usefulness and perceived ease of use. And

more recently, it's been seen in the instruments developed by Straub, Limayem, and Karahanna-Evaristo (1995); Garrity and Sanders (1998); Ishman (1998); and so forth.

We draw on this stream of research to establish our first hypothesis:

H1: The IS/IT quality assessed in terms of task-support-fit, ease of use and interface quality, jointly with the degree of usage of this tool, determine its perceived impacts in terms of: global user satisfaction, impacts on productivity, decision-making impacts and quality of work life satisfaction.

The Impacts of Development and Use Context on the Perceived Performance of the IS/IT

The key issue here concerns the nature of the relationship between the factors related to this context and the determinants of the perceived performance of the IS/IT defined above.

Following Goodhue and Thompson (1995) and Barki and Hartwick (1994), we consider task characteristics and user participation in the development process to be the determinant variables of perceived IS/IT performance.

However, user involvement (a priori and a posteriori), user attitude toward IS/IT tools (mostly the degree to which they regard themselves as experimented and skilled in manipulating these tools) and, finally, collaboration relationships between users and conceptors/developers, are mediating variables that affect the way previous determinant variables shape the ultimate perceived performance (McKeen & Guimares, 1997; Saleem, 1996). Our second hypothesis is then the following:

H2: Task characteristics and user participation in the development process have an impact on the perceived performance of the IS/IT. This impact is moderated by mediating factors related to user involvement, perceived experience in IS/IT use and collaboration relationships between users and conceptors/developers.

The Impacts of the Strategic and Organizational Context on the Perceived Performance of the IS/IT

These impacts have not been deeply investigated in previous literature. We can, however, build upon some works (Franz & Robey, 1986; Lucas, 1973) that support the intuitively accepted hypothesis according to which:

H3: The strategic and organizational context contribute to shape the IS/IT development and use context and accordingly have an impact on the perceived performance of the IS/IT. The strategic and organizational impact can also directly affect that perceived performance.

The Individual Level: Identifying Appropriation Types

Different interaction schemes may occur between the users and their IS/IT tools. These are what we call appropriation types. According to DeSanctis and Poole (1994) and the adaptive structuration theory, the appropriation concept results from studying the structurational impacts of an IS/IT at the individual level, that is to say, situating the user as the analysis unit.

More precisely, the studies on the concepts of infusion, diffusion, and routinization (Cooper & Zmud, 1990; Saga & Zmud, 1996) adopt a process-oriented perspective to describe the human-machine subsystem in its ongoing and day-to-day practices. These result from a series of events: implementation, use, and acceptance.

Saga and Zmud (1996) establish a utilization taxonomy at the individual level—the extended use, the integrated use, and the emergent use—and

explain each of them using different determinant factors, such as the IS/IT function maturity or the leadership role. We have adopted this point of view to establish our fourth hypothesis:

H4: Multiple appropriation types at the individual level emerge over time. They are affected by the perceived performance of the IS/IT and by the prevailing context.

The Organizational Level: Investigating Changes

The approaches that focus on the organizational level to evaluate IS/IT are based on the assumption that the information system, its technological support, and the human being using it to perform his or her tasks is a subsystem within the organizational system and that the performance evaluation of this subsystem has to be assessed via its contribution to the achievement of the organizational objectives, mainly those related to organizational effectiveness (Chakraborty, 1994; Mirani & Lederer, 1998) and IT-enabled organizational change (Grover, Jeong, & Teng, 1995; Kettinger, Teng, & Guha, 1997). While most of the studies that investigate these aspects follow a causal, variance, and deterministic framework, those based on the structurational model identified above try to go further and argue that the implementation and use of IS/IT should be conceptualized as a form of organizational change (Orlikowski, 1993, 1996 ; Orlikowski & Hofman, 1997).

These studies offer a deeper framework to emphasize the criticality of the organizational context in shaping technology use in organizations and in describing the changes that have occurred among them recursively. Hence, they support hypotheses H5 and H6:

H5: Organizational impacts defined in terms of organizational changes: planned change, improvisational change, and/or emergent change, appear over time. They are affected by the different individual appropriation types and by the prevailing context.

H6: These organizational changes (H5) potentially produce incremental or radical changes among the pre-existing context.

THE CONDUCT OF THE RESEARCH PROCESS

A longitudinal case study was conducted in a multinational financial organization specialized in assets management. The IS/IT under study is related to a data warehouse implementation project conceived as a decision support system and aimed to be an organization-wide information system that supports all the activities of assets performance analysis and client reporting.

The study was conducted during a period of 17 months. It consisted of three main phases, which not only differed in time and purpose but also with respect to the research methodology and techniques used.

Phase 1: (five months)

- Purpose: Study of the strategic and organizational context and the context of development and use (including the DW project specificities).
- Research techniques: Qualitative data collection and analysis using documentary analysis (content analysis), observation, and interviewing of the three categories of actors (decision makers, IS/IT conceptors/developers, and users; 6 semi-structured interviews and 15 structured interviews were conducted)

Phase 2: (five months)

- Purpose: Constructing and applying an instrument that measures the DW performance and relates the performance criteria used to some specificities of the development and use context.
- Research techniques: Questionnaire construction and administration: A total of 101 usable questionnaires were collected and analyzed using statistical data analysis.

Phase 3: (seven months)

- Purpose: (1) Identifying appropriation types at the individual level, and relating them to the performance scores obtained in phase 2 and the characteristics of the user-machine characteristics during all the observation period; (2) describing the organizational changes over time, relating them to the identified appropriation types and the prevailing context.
- Research techniques: Qualitative data collection and analysis using documentary analysis, observation, interviewing (11 structured interviews), and participative intervention (in collaboration with the actors) that aimed to solve problems related to usage dysfunctions identified for each appropriation type.

ANALYSIS AND RESULTS

As stated above, distinct research phases were conducted. Data analysis performed in one phase can be interpreted and reused in the following phases. The results obtained in phase one concerning context properties are used and actualized (identifying the changes) during all the research process. The results of phase two are used and reinterpreted in phase three.

Phase 1: Context Properties

We have focused on the strategic and organizational context and identified the characteristics of the company concerning:

- **Corporate strategy:** Maintaining competitive advantage by applying the climate-intimacy model in all the targeted markets using standardized management processes built upon standardized IS/IT tools (among them consolidated data bases and the DW under study).
- **Strategy and culture:** A global structure built upon a network of local firms acting in their respective local markets. A multicultural and collaborative climate prevails that promotes mobility of the know how and competencies throughout the global firm
- **IS/IT function:** It is a global function where global and standardized IS/IT solutions (proprietary or externalized) are built and implemented, beginning by one pilot site (one of the local firms), then extended in the other sites. The DW project was built in the Paris site, then it began to be extended to the other sites.

Phase 2: DW Perceived Performance

The questionnaire has been adapted from previous literature (Baroudi & Orlikowski, 1988; Davis, 1989; Doll & Torzadeh, 1988; Franz & Robey, 1986; Goodhue & Thompson, 1995; Ishman, 1998; Kappelman, 1995; Myers et al, 1997; Saleem, 1996).

A pre-test has been realized in order to adapt the original items to our specific case. Then, all the adapted items were translated into French. The questionnaire has been established in two versions, French and English, and then administered in the two sites where the DW is functional, in Paris and London.

The 101 usable questionnaires obtained represent a response rate of 53.2 %.

We used the statistical data analysis package SPSS 10.0 to execute the following treatments:

- A descriptive data analysis
- A confirmative data analysis through a series of factor analyses
- A series of hierarchical regression analyses

All of them aim to test hypotheses H1 and H2.

Descriptive Data Analysis

It concerns two categories of variables:

- The performance criteria variables (dependent variables): global satisfaction, impacts on productivity, impacts on quality of work life satisfaction, impacts on decision making, usage (frequency/regularity and perceived dependence), task-technology-fit, ease of use, and interface quality.
- The context variables (independent variables), including user participation in the development process, characteristics of the human-machine subsystem (user-task), user involvement a priori (during the development process), user involvement a posteriori (during the use process), collaboration relationships between users and conceptors/developers, and finally user perceived experience in IS/IT.

We calculated aggregated and detailed performance scores for the first category of variables and, using frequency analyses performed upon the second category of variables, determined groups of users presenting the same context of development and use specificities.

Confirmative Data Analysis

- The factor analysis produced satisfactory results in terms of reliability of data: internal consistency (Alpha Cronbach's > 0.7 for most of the variables), content and convergent/discriminant validity. Twenty-four items over a total of 26 load to their respective factors (the variables they are assumed to measure in the theoretical model).

- The hierarchic regression we executed on the basis of the previous results aim to examine how the different categories of variables behave in terms of causal relationships. We consider that the ultimate dependent variables are the perceived impacts of the DW: global user satisfaction, impacts on productivity, decision making impacts and quality of work life satisfaction. We establish a regression equation for each of these variables where we test the determinant power of those variables related to task-support-fit, ease of use, interface quality, and degree of usage, introduced progressively (hypothesis H1). Then we test the determinant power of the context variables (hypothesis H2), also introduced progressively.

The results obtained strongly confirm the first hypothesis H1 ($p < 0.01$ and satisfactory correlations). Global user satisfaction, impacts on productivity, decision making impacts, and quality of work life satisfaction are effectively determined by the DW quality assessed in terms of task-support-fit, ease of use and interface quality, and by the degree of usage of this tool, in terms of frequency and regularity of use and perceived dependence of users.

Concerning the second hypothesis H2, evidence showed no significant determinant power of all the context variables for the impacts criteria related to global user satisfaction, impacts on productivity and impacts and quality of work life satisfaction; however, impacts on decision making

are effectively related to human-machine characteristics, user involvement, and their perceived experience in IS/IT.

Phase 3: Individual Appropriation Types and Organizational Impacts

We have to notice here that the third hypothesis H3 has not been tested using the previous questionnaire because it is not frequent to find items related to strategic and organizational context variables in satisfaction questionnaires in IS/IT usage addressed to a population of users belonging to the same organization. We argue that these context specificities have to be assessed and related to the results obtained during the evaluation process, using qualitative data collection techniques, such as observation and interviewing.

For this concern, we found that the role of IS/IT function in this firm as a global entity producing standardized solutions is the main explaining factor of a context of development and use with no significant participation of users in the development process, except for those who will use the DW in decision-making tasks.

Now, concerning the individual level of analysis, the results obtained through the questionnaire, interviewing of the actors (users, developers, and managers), and observation have converged to define four appropriation types that describe different interactions schemes between the DW and the users, depending on the role they played in the development process, their own characteristics, mainly their perceived experience in IS/IT, and their collaboration with developers during the development and use of this tool.

These appropriation types are defined along two axes:

- The utilization type: DW has been conceived as a decision support system. Two utilization categories exist: the primary utilization (decision making) and the secondary utilization (reporting and treatments of data to be used by the decision makers who do not have direct access to the DW).

- The perceived experience in IS/IT: That is, whether the users regard themselves as experienced and skilled users of such tools or not.

These results provide support to hypothesis H4.

We have also noticed that each appropriation type is characterized by homogenous performance scores and presents the same utilization problems related to data quality or task support misfit.

A collaborative working group including the researcher and the representatives of (1) the DW conceptors/developers team, (2) each group of users, (3) managers of users departments, and (4) managers of the IS/IT function, has been established in order to identify these problems, design solutions, deliver them, and provide all the support and assistance to implement them.

This collaborative group is a temporary structure (maintained during three months) and has not been institutionalized, whereas some of the changes it helped create can be defined as organizational changes: the DW hotline implementation within the IS/IT function, routinizing the collaborative use of the DW between the users departments, are some of the most significant changes and thus provide support for our fifth research hypothesis H5.

Regarding the observation period (only seven months for phase three), can we detect the recursive impact of the DW implementation process on the prevailing institutional context?

What we can say is that role of the IS/IT function as a global entity delivering standardized solutions to all the sites of the group has been reconsidered to adjust the implemented solutions to the real needs of the users. Thus, we can conclude that this recursive impact is also confirmed, and so is the final hypothesis H6.

IMPLICATIONS AND RECOMMENDATIONS FOR FUTURE RESEARCH

In this article, a theoretical model is developed as a process where three levels of evaluation are considered over time: the IS/IT level, by measuring the perceived quality of the application and of the information, the degree of use of that application, and its perceived impacts by users, in terms of satisfaction, for example; the individual level, by the identification of individual appropriation types; and the organizational level, by the study of the organizational changes. The whole process is integrated among the organizational and strategic context and the context of development and use.

This model has been applied in an in-depth field study conducted over 17 months and related to a DW implementation project in a financial institution. Multiple research techniques have been used, including documentary analysis, observation, cooperative intervention, interviews, and a questionnaire.

Results suggest that the perceived performance of the DW, jointly with the context factors related to the characteristics of the human-machine subsystem characteristics and the role of the IS/IT function in the organization, determine multiple appropriation types. To be effective, these involve a process of organizational change over time.

From a theoretical point of view, we argue that the data warehouse studied in this research process has been conceived as an IT artifact fully embedded and explicitly specified in time, place, discourse, and community, as recommended by Orlikowski and Iacono (2001). We have constructed a multifaced evaluation tool and have integrated it into an organizational context in a process-oriented approach. The systemic approach adopted here helped us go beyond the dichotomy positivist methodology *versus* interpretive methodology dichotomy in an attempt to integrate them in a pluralistic research perspective.

Finally, we suggest that this study can be reproduced in other research contexts, including multiple organizations and multiple sectors, in a comparative perspective, and/or to provide generalization and external validity to the results. It can also be more extended in time in order to investigate more deeply the recursive impacts of an IS/IT implementation project on its institutional context.

REFERENCES

Alter. (2004). Desperately seeking systems thinking in the information systems discipline. In *Proceedings of Twenty-Fifth International Conference on Information Systems*, 757-770.

Avison, D., Baskerville, R., & Myers, M. (2001). Controlling action research projects. *Information Technology & People, 14*(1), 28-45.

Avison, D., Lau, F., Myers, M., & Nielson, P.A. (1999). Action research. *Communications of the ACM, 42*(1), 94-97.

Bailey, E. J., & Pearson, S. W. (1983). Development of a tool for measuring and analyzing computer user satisfaction. *Management Science, 29*(5), 530-545.

Ballantine, J., Bonner, M., Levy, M., Martin, A., Munro, I., & Powell, P.L. (1998). Developing a 3-D model of information systems success. In E. J. Garrity & G. L. Sanders (Eds.), *Information system success measurement* (pp. 46-59). Hershey, PA: Idea Group Publishing.

Barki, H., & Hartwick, J. (1989). Rethinking the concept of user involvement. *MIS Quarterly*, (13), 53-63.

Barki, H. and Hartwick, J. (1994). Explaining the Role of User Participation in Information System Use. *Management Science, 40(4)*, pp. 440-465.

Baroudi, J.J. and Orlikowski, W.J. (1988). A short-form measure of user information satisfaction: a psychometric evaluation and notes on use. *Journal of Management Information Systems*; 4(4), pp.44 - 59

Benbasat, I., & Zmud, R. (2003). The identity crisis within the IS discipline: defining and communicating the discipline's core properties. *MIS Quarterly, 27*(2), 183-194.

Bostrom, R. and Heinen, S. (1977). MIS Problems and Failures: A Socio-Technical Perspective. Part I: The Causes. *MIS Quarterly, September,* 17-31

Chakraborty, R. (1994): Information systems and organizational success: A quantitative modeling approach. TDQM-94-08. Retrieved from http://web.mit.edu/tdqm/www/articles/94/94-08.html.

Checkland, P. (1999). *Systems thinking, systems practice*. John Wiley & Sons Editions.

Checkland, P., & Scholes, J. (1990). *Soft systems methodology in action*. John Wiley & Sons Editions.

Churchman, C.W. (1979). *The system approach and its enemies*. New York: Basic Books.

Cooper, R.B., & Zmud, R.W. (1990). Information technology implementation research: A technological diffusion approach. *Management Science, 36*(2), 123-139.

Davis, F.D. (1989). Perceived usefulness, perceived ease of use, and user acceptance of information technology. *MIS Quarterly, 13*(3), 319-340.

Delone, W. H., & McLean, E. R. (1992). Information systems success : The quest for the dependent variable. *Information Systems Research, 3*(1), 60-95.

Delone, W. H., & McLean, E. R. (2003). The Delone and McLean model of information systems Success: A ten year update. *Journal of Management Information Systems, 19*(4), 9-30.

DeSanctis, G., & Poole, M. S. (1994). Capturing complexity in advanced technology use: Adaptive structuration theory. *Organization Science, 5*(2), 121-146.

Doll, W. J., & Torzadeh, G. (1988). The measurement of end-user satisfaction. *MIS Quarterly, 12*(2), 259-274.

Foray, D., & Mairesse, J. (Ed.). (1999). Innovations et Performances. EDHESS Editions.

Franz, C. R., & Robey, D. (1986). Organizational context, user involvement, and the usefulness of information systems. *Decision Sciences, 17*(3), 329-356.

Galliers, R. (1992). Choosing information systems research approaches. In R. Galliers (Ed.), *Information systems research: Issues, methods and practical guidelines.* Oxford: Blackwell Scientific Publications.

Galliers, R. D. (2003). Change as crisis or growth? Toward a trans-disciplinary view of information systems as a field of study - a response to Benbasat and Zmud's call for returning to the IT artifact. *Journal of the Association for Information Systems, 4*(6), 337-351.

Garrity, E. J., & Sanders, G. L (1998). Introduction to information systems success measurement. In E. J. Garrity & G. L. Sanders (Eds.), *Information system success measurement* (1-11). Hershey, PA: Idea Group Publishing.

Giddens, A. (1987). La constitution de la société : éléments de la théorie de la structuration. Ed. PUF.

Goodhue, D. L., & Thompson, R. L. (1995). Task-technology fit and individual performance. *MIS Quarterly, 19*(2), 213-236.

Grover, V., Jeong, S. R., & Teng, J. T. C. (1995). The implementation of business process reengineering. *Journal of Management Information Systems, 12*(1), 109-129.

Hamilton, S., & Chervany, N. L. (1981a). Evaluating information system effectiveness – Part I: Comparing evaluation approaches. *MIS Quarterly, 5*(3), 55-69.

Hamilton, S., & Chervany, N. L. (1981b). Evaluating information system effectiveness – Part II: Comparing evaluator viewpoints. *MIS Quarterly, 5*(4), 79-86.

Hirschheim, R. A. (1992). Information systems epistemology: An historical perspective. In R. Galliers (Ed.), *Information systems research: Issues, methods and practical guidelines.* Oxford: Blackwell Scientific Publications.

Ishman, M. (1998). Measuring information success at the individual level. In E. J. Garrity & G. L. Sanders (Eds.), *Information system success measurement* (pp. 60 – 78). Hershey, PA: Idea Group Publishing.

Ives, B.. & Olson, M. H. (1984). User Involvement and MIS success: A review of research. *Management Science, 30*(5), 586-603.

Kannelis, P., Lycett, M., & Paul, R. J. (1998). An interpretive approach to the measurement of information systems success. In E. J. Garrity & G. L. Sanders (Eds.), *Information system success measurement* (pp. 133-151). Hershey, PA: Idea Group Publishing.

Kappelman, L. A. (1995). Measuring user involvement: A diffusion of innovation perspective. *Data Base, 26.*

Kettinger, W. J., Teng, J. T. C., & Guha, S. (1997). Business process change: A study of methodologies, techniques and tools. *MIS Quarterly, 21*(1), 55-88.

King, W. R., & Rodriguez, J.I. (1978). Evaluating management information systems. *MIS Quarterly, 2*(3), 43-51.

LeMoigne, J. L. (1977). La théorie du système général: Théorie de la modélisation. Editions PUF.

Lucas, H. C. (1973). A descriptive model of information systems in the context of the organization. *Data Base, 5*(2), 27-36.

Markus, M. L., & Robey, D. (1988). Information technology and organizational change: Casual structuring in theory and research. *Management Science, 34*(5), 583-598.

McKeen, J. D., & Guimares, T. (1997). Successful strategies for user participation in systems development. *Journal of Management Information Systems, 14*(2), 33-150.

Mirani, R., & Lederer, A.L. (1998): An instrument for assessing the organizational benefits of IS projects. *Decision Sciences, 29*(4), 803-823.

Mora, M., Gelman, O., Cervantes, F., Mejía, M., & Weitzenfeld, A. (2003). A systemic approach for the formalization of the information systems concept: why information systems are systems?. In J. J. Cano (Ed.), *Critical reflections on information systems: A systemic approach* (pp. 1-29). Hershey, PA: Idea Group Publishing

Myers, B. L., Kappelman, L.A., & Prybutok, V. R. (1998). A comprehensive model for assessing the quality and productivity of the information systems success: Toward a theory for information systems assessment. In E. J. Garrity & G. L. Sanders (Eds.), *Information system success measurement* (pp. 94-121).

Orlikowski, W. (1992). The duality of technology: Rethinking the concept of technology in organizations. *Organization Science, 3*(3), 398-427.

Orlikowski, W. J. (1993). CASE tools as organizational change: Investigating Incremental and Radical changes in systems development. *MIS Quarterly, 17*(3), 309-340.

Orlikowski, W. J. (1996). Improvising organizational transformation over time: A situated change perspective. *Information Systems Research, 7*(1), 63-92.

Orlikowski, W. J. (2000). using technology and constituting structures: A practice lens for studying technology in organizations. *Organizations Science, 11*(4), 404-428.

Orlikowski, W. J., & Hofman, D. (1997). An improvisional model of change management: The case of groupware technologies. *Sloan Management Review, 38*(2), 11-21.

Orlikowski, W. J., & Iacono, C.S. (2001). Research commentary: Desperately seeking the "IT" in IT research – A call to theorizing the IT artifact. *Information Systems Research, 12* (2), 121-134.

Pettigrew, A. M., (1990). Longitudinal field research on change: Theory and practice. *Organization Science, 1*(3), 267-292.

Popper, K. R. (1963). Conjectures and refutations. Routledge & Kegan Paul.

Saga, V. L., & Zmud, R.W. (1996). Introduction de logiciels de gestion dans des petites entreprises liées à une profession libérale. *Systèmes d'Information et Management, 1*(1), 51-73.

Saleem, N. (1996). An empirical test of the contingency approach to user participation information systems development. *Journal of Management Information Systems, 13*(1), 145-166.

Saunders, C. S., & Jones, J. W. (1992). Measuring performance of the information systems function. *Journal of Management Information Systems, 8*(4), 63-82.

Straub, D.W., Limayem M., & Karahanna-Evaristo, E. (1995). Measuring system usage: Implications for IS theory testing. *Management Science, 41*(8), 1328-1342.

Swanson, E. B., & Ramiller, N. C. (1997). The organizing vision in information systems innovation. *Organization Science, 8*(5), 458-474.

Von Bertalanffy, L. (1968). *General systems theory*. George Braziller, NY Editions.

Zmud, R. W. (1979). Individual differences and MIS success: A review of the empirical literature. *Management Science, 25*(10), 966-979.

This work was previously published in Information Resources Management Journal, Vol. 20, Issue 2, edited by M. Khosrow-Pour, pp. 108-121, copyright 2007 by IGI Publishing (an imprint of IGI Global).

Chapter 18
The Distribution of a Management Control System in an Organization

Alfonso A. Reyes
Universidad de los Andes, Colombia

ABSTRACT

This chapter is concerned with methodological issues. In particular, it addresses the question of how is it possible to align the design of management information systems with the structure of an organization. The method proposed is built upon the Cybersin method developed by Stafford Beer (1975) and Raul Espejo (1992). The chapter shows a way to intersect three complementary organizational fields: management information systems, management control systems, and organizational learning when studied from a systemic perspective; in this case from the point of view of management cybernetics (Beer 1959, 1979, 1981, 1985).

UNDERSTANDING CONTROL IN AN ORGANIZATIONAL CONTEXT

When Norbert Wiener defined cybernetics as the science of control and communication in the animal and the machine (Wiener 1948) he was using the Greek word κυβερνητηζ, or steersman, as his main inspiration. Indeed, he was recalling the ancient practice of steering a ship towards a previously agreed destination regardless of changing conditions of currents and winds. This simple idea of connecting communication (at that time used as a synonymous of information flow) and control by a continuous feedback process opened up a huge space of possibilities to explain physical, biological and social phenomena related to self-regulation (Heims 1991). This is the case, for instance, of a heater in a physical domain, or the homeostatic mechanism to regulate body temperature in mammals (Ashby 1956). In all these cases, however, it is important to notice that control is far from its naïve interpretation as a crude process of coercion, but instead it refers to self-regulation. This is the meaning of control used in this chapter.

Cybernetics has evolved in many branches since its early years (Espejo & Reyes 2000). One of these variations has focused on the study of communication and control processes in organizations; this is the topic of management cybernetics (Beer 1959, 1966, 1979) and is the conceptual underpinning of this chapter.

Given the close relation between information and control in self-regulating systems (as organizations) this chapter addresses the question of how information should be distributed across the structure of an organization in order to allow self-regulation to be effective. In order to achieve this, we would like to show a way of relating three organizational fields: management information systems, management control systems and organizational learning. This is done from a methodological point of view by describing a step-by-step method (although it is not intended to be linear) to build a network of homeostatic mechanisms. But before describing the method, it is important to clarify with more detail the meaning of control used herein.

In an organizational context controlling a system is a process intended to close the gap between the observed outcomes produced by the organization and the expectations previously agreed among relevant stakeholders. It is, therefore, a self-regulating process.

An organization, on the other hand, is understood in this context as a closed network of relationships constituted by the recurrent interplay of roles and resources in a daily basis. In other words, people in organizations play formally defined roles that underpin the working relations they carry out with other organizational members. When these relations allow them to create, regulate and produce the goods and services they want to offer, an organization with a particular identity emerges; a *human interaction system* (Espejo 1994). This is an operational way to distinguish between a group of people that meets regularly to do something (as fans that used to meet at football matches) and an organization (when those fans constitute a club).

There are different ways to describe what an organization is doing; one way is to make explicit the transformation process by which this organization is producing the goods or services it is offering. Figure 1 shows a simple representation of such description. Notice that this description is suitable not only for an organization as a whole (like an insurance company that transforms information into specific products) but also to any other organizational processes like those carry out by the human resource department of a bank or those constituting the quality system of a company.

Our concern is to model the self-regulating (or control) process of any organizational system that could be described as a transformation process.

Figure 1. A representation of a system-in-focus as a transformation process

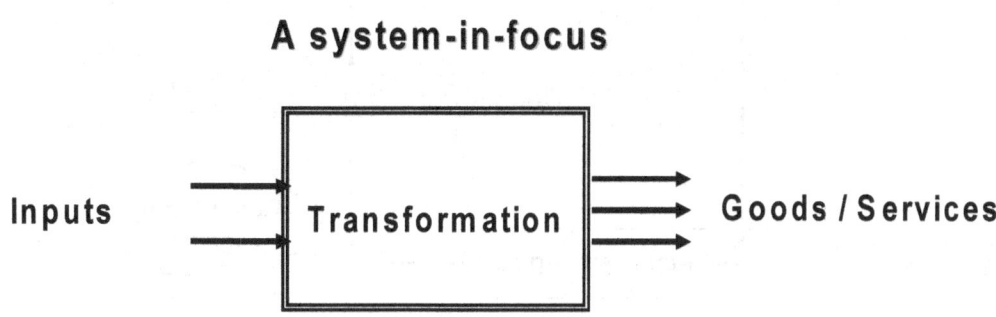

From now on we will call an instance of these processes a system-in-focus.

Figure 2 describes a self-regulating mechanism for a system-in-focus. This control cycle starts by observing the behavior of the system, that is, by looking at a set of indices that measure the critical success factors of its operation. If the state of these indices does not match a set of expected values, then the system is out of control. The manager (or any organizational member) that is responsible for its operation has to address the reasons of such an undesirable behavior and as a result of this inquiry design a set of strategic or operational actions to intervene on the system. Once this decision is agreed and carried out by relevant organizational agents the cycle starts over again.

Notice how this self-regulating mechanism allows us to explain the intertwined relation between several organizational fields. First, the need to define a set of critical success factors (CSF) for the system-in-focus and a corresponding set of indices to measure them, along with the regular reports needed to inform management about the behavior of the system are the basis of a management information system. Secondly, inquiring for the reasons underpinning an unexpected behavior of the system-in-focus, designing a strategic and operational action and directing its execution through the doing of other organizational members are at the core of a management control system. And finally, the control loop itself can be related with the continuous operation of four stages: *observing* the state of a system; *assessing* the mismatch with expected outcomes; *designing* a set of actions; and *implementing* them to close the loop. These four stages are characteristics of an individual learning loop usually known as the OADI learning model (Kim 1993). When the loop is designed in such a way that it operates for a particular role (instead of a particular individual) we are entering in the field of organizational learning (Argyris 1993); we will go back to this point later on.

So far we have shown a model that allows the conceptual integration of (management) information systems, (management) control systems and organizational learning; all of them related to self-regulation of a system-in-focus. The methodological problem now is how we can define an integrated set of these self-regulating mechanisms given a particular organization.

Figure 2. A general model for a self-regulating mechanism in an organizational system

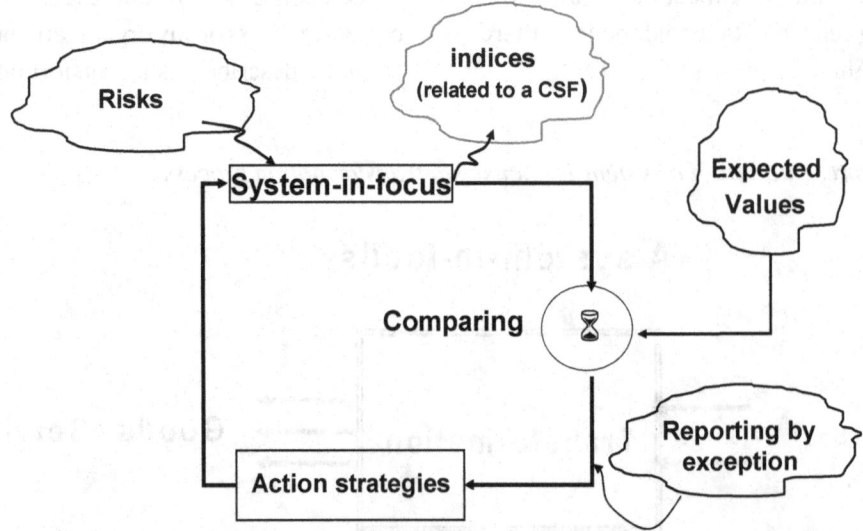

A METHOD TO DISTRIBUTE CONTROL IN AN ORGANIZATION

Our goal is to identify a complete set of control loops distributed across the structure of the organization. The steps presented here are based on the Viplan method (Espejo 1989; Espejo et al 1999).

Step 1: Naming a System-in-Focus

The first step of the method consists of identifying precisely the organizational system that will be the focus of control. This could be an organization as a whole, a strategic business unit of an organization, an area of such organization or a support process. In every case what is important is to name the system as a transformation process (see Figure 1). A canonic form to name this transformation is as follows: the system-in-focus S produces *X* by means of the activities *Y* with the purpose *Z*. In short, we are answering three main questions related to the system: ¿What is produced? ¿How is it produced? and ¿with what purpose is it produced?

Next follows the identification of the stakeholders of the system-in-focus. To do this notice that from the elements shown in Figure 1 it is possible to differentiate five stakeholders (see Figure 3): those who supply the inputs to the transformation (called *suppliers*); those carrying out the activities of the transformation (called generically *actors*); those who receive the goods/services of the transformation (usually called *clients*) and those responsible for the management of the transformation (normally called *owners*). It is also important to take into account the larger organizational context in which the system is operating; this consideration allows us to identify what are called the *interveners* of the system. They are stakeholders that although do not belong to the system-in-focus their organizational role may directly affect the system's transformation. This is, for instance, the case of both the competitors and the regulators of the system.

Viplan provides the mnemonic TASCOI to facilitate this process of naming the system-in-focus; certainly T(ransformation), A(ctors), S(uppliers), C(lients), O(wner), and I(nterveners).

Step 2: Unfolding of Complexity

Once the system-in-focus has been named what follows is to recognize the way this system orga-

Figure 3. Stakeholders of a system-in-focus

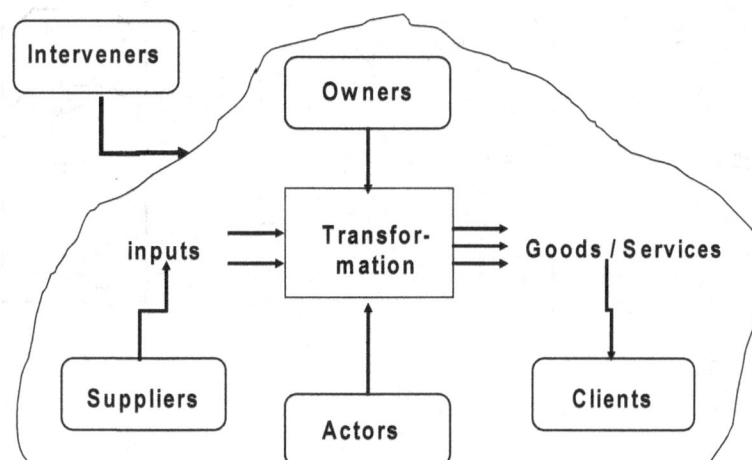

nizes its resources to carry out its transformation. There are four complexity drivers that normally guide an organization to distribute its activities: technology, geography, market segmentation and time (Espejo et al 1999).

The first one refers to the way the activities constituting the transformation are organized according to the technology selected. We can use different technologies to produce the same transformation. For instance, Banks at the beginning of the 20th century used a technology based on the use of books and manual calculations to deliver its services to clients. Nowadays Internet is the main technological driver to deliver these services. The roles, resources used and the way activities are carried out have changed dramatically over this period in the banking industry. Therefore, choosing the appropriate technology to produce a given transformation is crucial to organize the primary activities of an organization. The other way around, given a particular system-in-focus, it is always possible to describe the primary activities implied (and carried out) by the selected

technology. To describe these activities and their relations we used technological models (Espejo et al 1999).

A technological model is a macro flow diagram showing the activities needed to produce the transformation of the system-in-focus. Figure 4 shows an example for SATENA, a Colombian airline company.

The second complexity driver for structuring an organizational system refers to the distribution of activities in geographically diverse locations. Indeed, sometimes we need to take into account the best location of actors, suppliers and clients in order to organize activities of a transformation. For instance, if a company produces pavement-related products, it makes sense to have activities related to the production process near the quarries whereas its sales division will be near its clients. In the same way, and for economic reasons, some companies do prefer to have their manufacturing processes distributed in different countries (according to the cost of raw materials and salaries of the local work-force) while the assembly of

Figure 4. An example of a technological model for SATENA, a Colombian airline company

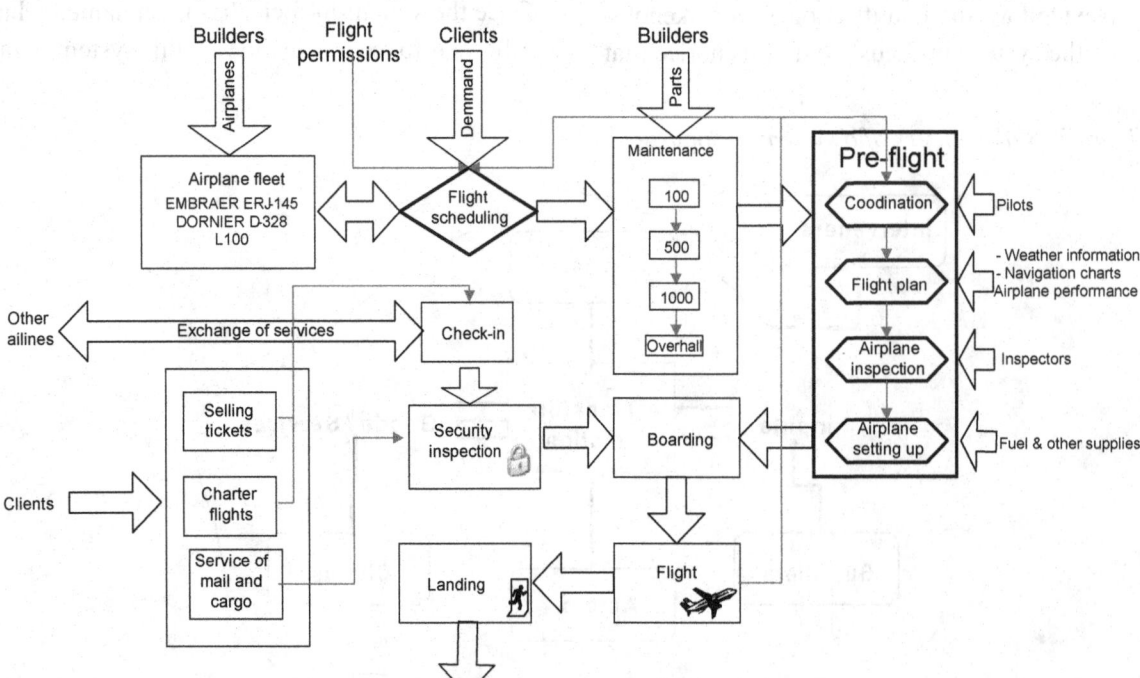

final products could be located in another country. Similarly, a multinational company may group its primary activities according to geographical criteria to distribute its products and services. In all these cases geographical models are used to describe this distribution of activities. Figure 5 shows an example for SATENA, we can see how their resources are distributed across the country and inside each particular city where it operates.

The third driver refers to the grouping of activities that are necessary to produce, in a differentiated way, the goods or services offered to a segmented market. Market segmentation is a good practice to increase market share for many products and services. In this case, each new product/service will respond to more specific needs of potential clients and, therefore, it is probable that the company has to incorporate a further specialization of activities in its production process (or into the design of customize services). This, in

turn, may affect the relations with suppliers (to get new raw material) and the relations with clients (to tackle the newly differentiated market).

Client-supplier models are helpful to describe the way a company groups its activities according to this segmentation strategy. Figure 6 shows an example of this kind of models for SATENA. Here we have that this company offers two main services: air transportation and plain maintenance to other companies. The first service, in turn, is divided into four sub-services: passenger transportation, packages delivery, charter flights, and planes renting. The model shows the relations between suppliers, services and clients taking into account this market segmentation.

Finally, the last driver refers to the need of differentiating activities according to time chunks. This is usually the case of an organization that has several turns to carry out its activities. For instance that happens when a company is using the same production line to produce different

Figure 5. An example of a geographical model for SATENA

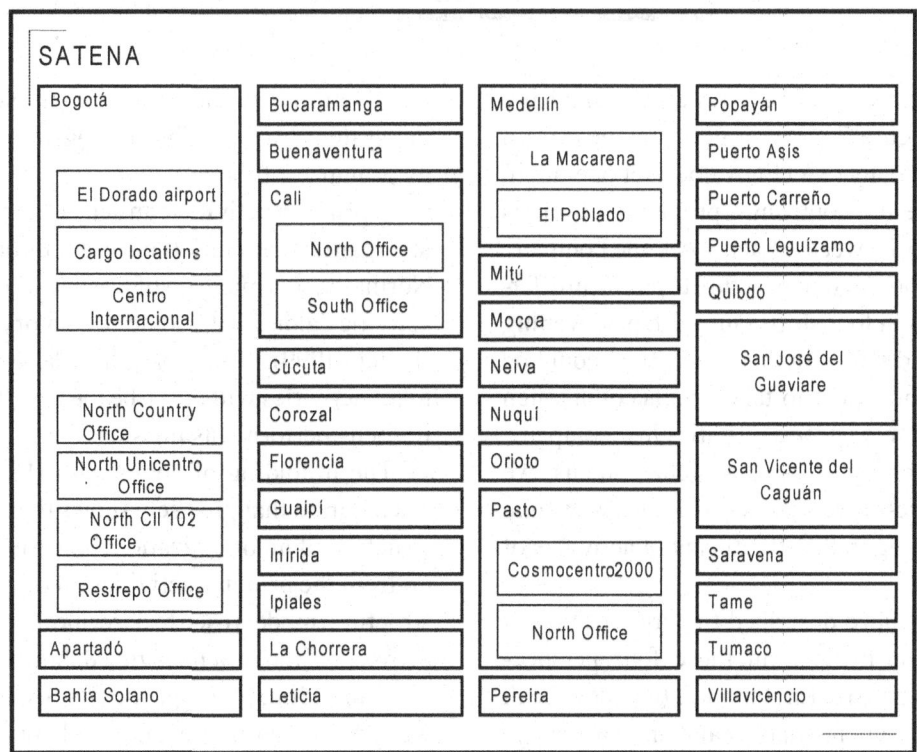

Figure 6. An example of a client-supplier model for SATENA

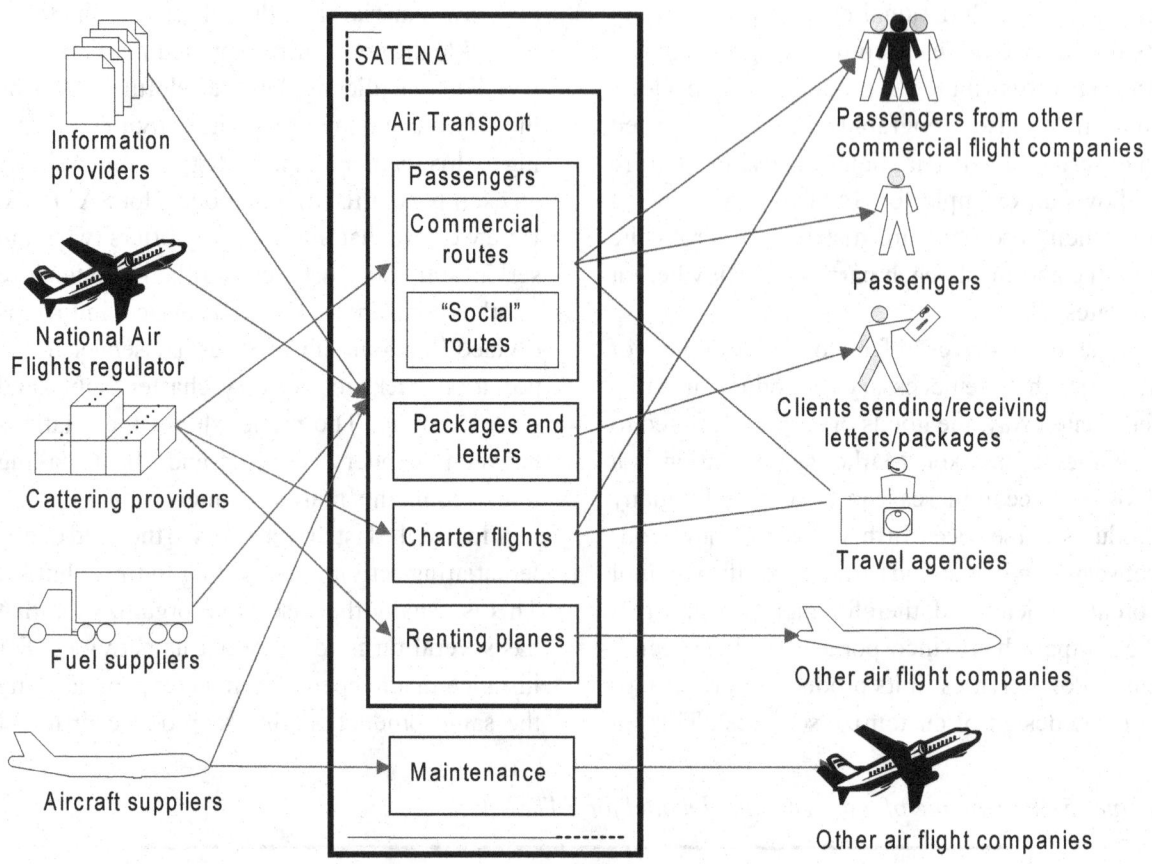

products following a cyclic time-pattern during a month. Figure 7 shows a good example from another company (SATENA does not operate in shifts). Here the company's production cycle is divided into 12 weeks. Each week the company uses its line production structure to produce different products in four shifts (am shift, pm, evening and weekend shift). In this sense, time could be an important aspect to take into account when designing how to group activities in a company. In a similar way, given a system-in-focus, we can use time-models to describe the way time is participating in the structuring of activities of the system.

Once we have described the organization of the system-in-focus from these four perspectives using the structural models (technological, geographical, client-supplier and time) we can go

on to summarize the organization's structure of the system by depicting the logical ordering of its primary activities.

A primary activity is an activity that produces an organization's task (i.e., its goods or services). Normally, it is made up of a set of sub-(primary) activities along with some regulatory functions (Espejo 1999). Therefore, the relation of all primary activities will describe the way the system-in-focus performs its mission.

The unfolding of complexity (Espejo 1999) is a diagram that precisely shows the operational structure of an organizational system. It is build up from the primary activities taken from all the structural models used to describe the grouping of activities of the system. Figure 8 shows an example of an unfolding of complexity for SATENA. Here we can see that the second level corresponds to

Figure 7. An instance of a Time-Model

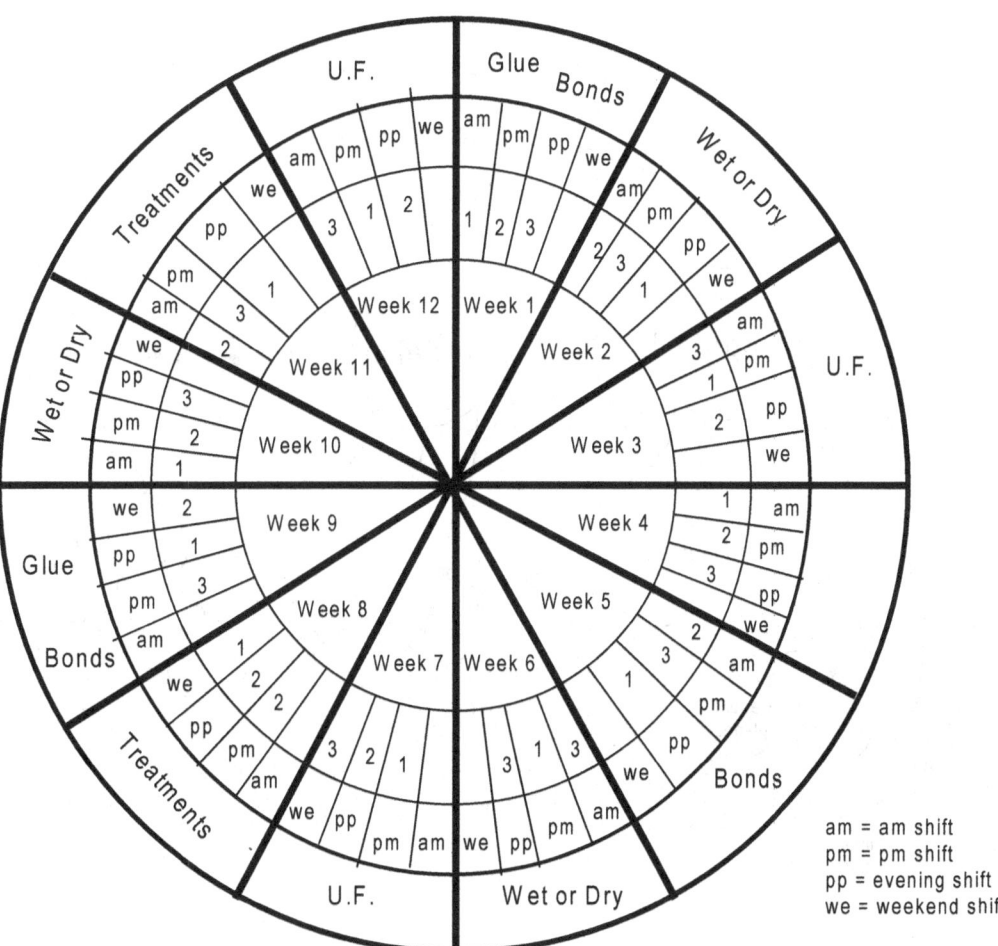

the geographical model whereas subsequent levels come from the client-supplier model indicating the way SATENA groups activities according to particular services. Notice that all services are not provided in all cities, for instance maintenance of planes to other companies is only provided in Bogotá. We can continue doing the unfolding by depicting in the last level activities taken from the technological model. However, for the sake of simplicity we are leaving aside this level in Figure 8.

The unfolding of complexity is a means to depict the *operational structure* of an organization. The operational structure refers to the set of interrelated primary activities that produces the organization goods/services. By contrast, the *supporting structure* is the set of interrelated activities that regulate the primary activities. This distinction is similar to one between missional processes and supporting processes normally used in ISO certification projects. This is a systemic way of describing the structure of an organization in which we can simultaneously see the organization as a whole (the first level of the unfolding of complexity) as well as all its primary activities organized in a cascade of subsumed logical levels. Notice that from one level to another there are no hierarchical relations. Indeed, related activities

Figure 8. An example of a complexity unfolding for SATENA

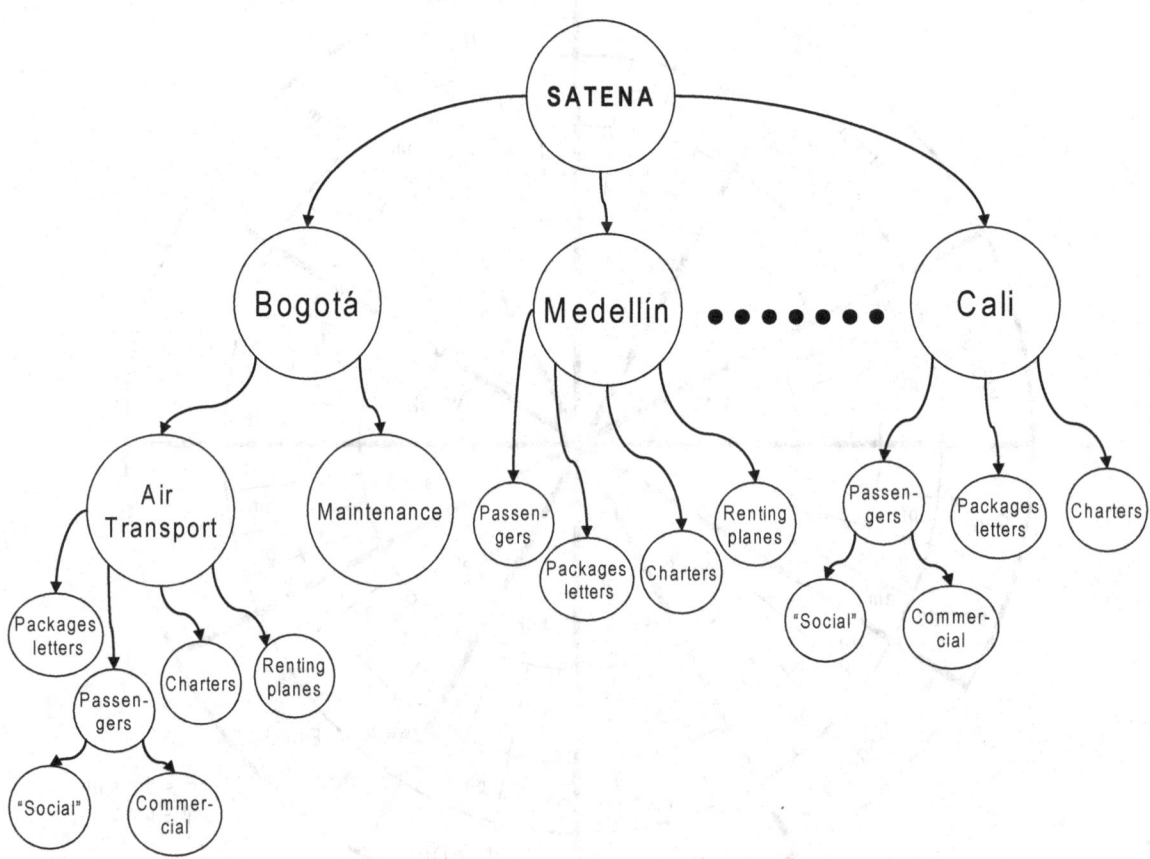

in level n+1 *constitute* the activity they are part of in level n. This is a radically different way of describing the structure of an organization that leaves aside the fragmented view offered by the traditional organization chart.

The unfolding of complexity shows an interrelated set of primary activities autonomously producing, at different structural levels, the organization's goods/services. Each primary activity acts as a "smaller" whole by itself. On the other hand, the organization chart is a functional description of a company that normally difficult a holistic view of the organization by its members.

Step 3: Defining Indices

As mentioned in the previous section, the unfolding of complexity of the system-in-focus shows all primary activities necessary to produce its goods/services. Each of these primary activities can be treated, in turn, as (sub)systems-in-focus at different structural levels and, therefore, they could be named in the same way as explained in step 1. In particular, each one is doing a transformation of inputs to produce specific outputs to clients. So each primary activity has a manager (or an organizational member) who is responsible for its effective performance. Surely the manager, tacitly or explicitly, has chosen a set of critical success factors (CSF) to focus his/her managerial role. Each CSF, in turn, should have associated one or more indices that measure it through time. Figure 9 shows the distribution of CSF and indices across primary activities of a system-in-focus. Of course it is quite possible that several of this CSF were the same for different primary activities.

Notice that this same figure shows the distribution of a managerial information system across this organization. Specifically, it shows what type of reports containing what kind of information (indices) should go to what relevant roles in the organization (in fact, the manager responsible for each particular primary activity). Notice that the higher it goes in the diagram the more aggregate the indices are until we reach the first level in which

we have indices referring to the performance of the system-in-focus as a whole.

This relationship between the level of aggregation of indices (or information in general) and the structural level of the manager getting the reports is what some authors call the alignment between information systems and organizational structure (Espejo 1993). A mismatch in this alignment implies a manager receiving either information too detailed (in respect to its managerial task) or too aggregated for its actual capabilities. The first one is the case of a manager that is concerned with so much detail that soon loses track of the holistic view of its primary activity and collapses under the pressure of too much information. The second one refers to those managers that are very well informed of such things for which they do

Figure 9. Distribution of indices (and CSF) across primary activities of an organization

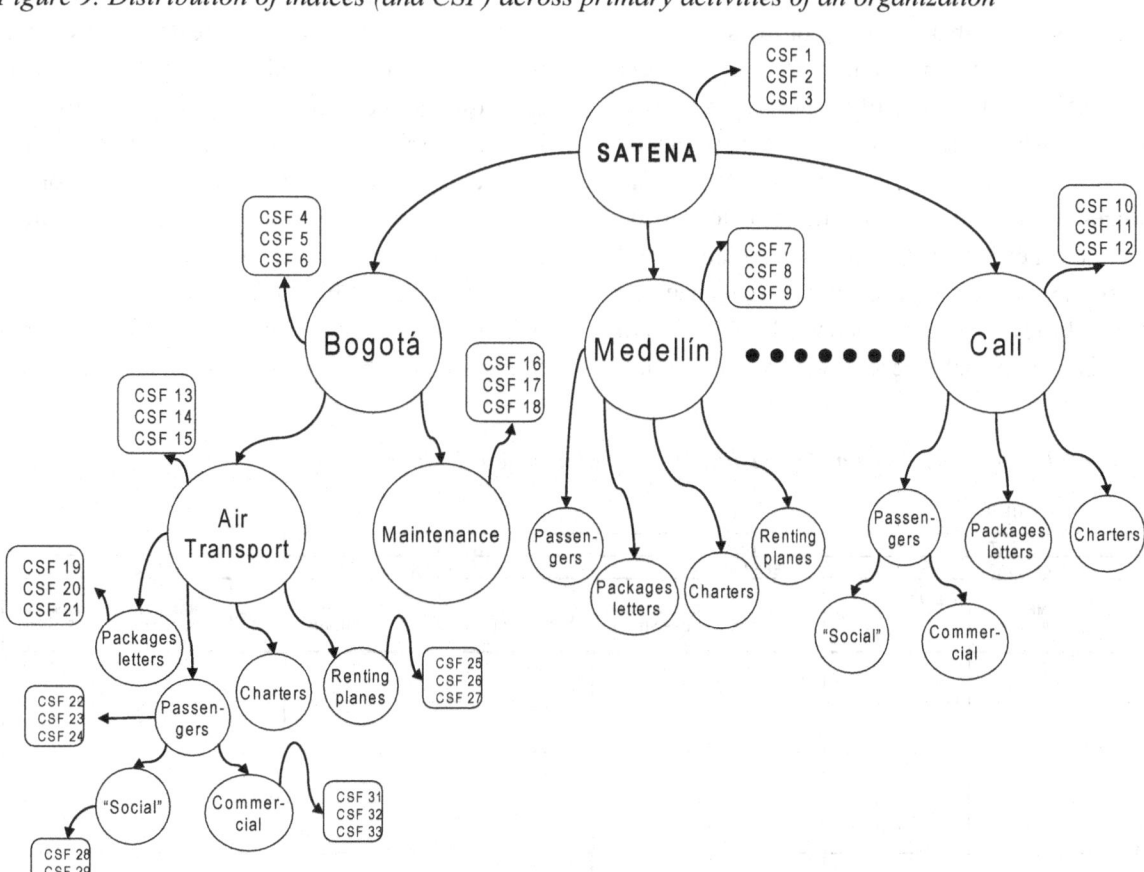

not have any capability to take decisions nor mobilize resources to do something about them. In both cases, the information (indices) received by the information system is irrelevant; therefore it make sense to say that the alignment between the two facilitates the effective management of the system-in-focus.

Nowadays many organizations have their own set of indices. So, and independently of its actual relevance, it is quite important to recognize such efforts from the start. Figure 10 shows a table in which it is possible to keep a record of all indices that have been built for the system-in-focus. This table classifies indices in three main categories: efficacy, efficiency and effectiveness, although it can be extended to allow other categories. For each indicator there is a code, a name and an operational definition. But before going any further, it is important to explain the way we understand this taxonomy of indices.

Let us recall that the name of the system-in-focus, using the mnemonic TASCOI, answers three main questions regarding the transformation process: ¿what is produced? ¿how is it produced? and ¿for what purpose it is produced? We say that indices that measure the first question belong to the category of efficacy; those measuring the second question relate to efficiency and those measuring the third one refer to effectiveness. Another way to put this is to say that efficacy measures the relation between what has been produced (or offered, in the case of a service) and what has been planned. Efficiency, in turns, measures the optimal use of resources needed to produce the goods/services of the system-in-focus. Effectiveness, on the other hand, measures to what extend the purpose of the transformation has been accomplished.

Notice that this taxonomy defined in such a way implies that all three categories form an orthogonal set. In other words, none of the indices in one category can be calculated as a function of the other two. In simple terms, it is possible to have a system-in-focus whose indices of efficacy and efficiency (at any moment in time) are high but its effectiveness is low; in the same way it is possible to have a state in which indices of efficiency and effectiveness are high but indices of efficacy are low; and also a state in which while indices of efficacy and effectiveness are high, the system is inefficient.

The following table (Figure 11) shows a distribution of indices used by the system-in-focus according to the primary activities of the unfolding of complexity of such system. Rows in the table correspond to the primary activities of the system-in-focus while columns refer to the indices registered before (using Table in Figure 10).

This table (Figure 11) can be used to show weaknesses of the actual managerial information system of the organization. In fact, an empty

Figure 10. A table to register the existing indices of a system-in-focus

SYSTEM-IN-FOCUS:

WHAT			HOW			WHAT FOR			OTHERS		
EFFICACY			EFFICIENCY			EFFECTIVENESS / IMPACT					
CODE	NAME OF INDICATOR	OPERATIONAL DEFINITION	CODE	NAME OF INDICATOR	OPERATIONAL DEFINITION	CODE	NAME OF INDICATOR	OPERATIONAL DEFINITION	CODE	NAME OF INDICATOR	DEFINICION OPERACIONAL
E.1			F.1			I.1			0.1		
E.2			F.2			I.2			0.2		
E.3			F.3			I.3			0.3		
E.4			F.4			I.4			0.4		
E.5			F.5			I.5			0.5		

Figure 11. A distribution of indices (classified by categories) among primary activities (SATENA)

	EFFICACY					EFFICIENCY					EFFECTIVENESS					OTHERS			
SATENA	E1	E2				F1	F2	F3			I1					O1			
Bogotá	E1						F2		F4										
Air transport		E3	E4							F5									
Packages and letters				E5						F5									
Passengers	E1	E3				F1													
"Social"																			
Commercial																			
Charters																			
Renting planes																			
Maintenance		E3				F1													
Medellín	E1						F2		F4										
Passengers	E1	E3				F1													
Packages and letters				E5						F5									
Charters																			
Renting planes																			
...																			
...																			
...																			
...																			
Cali	E1						F2		F4										
Passengers	E1	E3				F1													
"Social"																			
Commercial																			
Packages and letters				E5						F5									
Charters																			

third column (for a given primary activity) shows a lack of indices measuring the effectiveness of such primary activity (e.g. most primary activities in SATENA). In other words, the manager will not be aware of the impact of the task s(he) is responsible for in this activity. Similarly, we could point out other managerial problems if other columns are void for a giving primary activity. An empty row is, of course, an extreme case in which a manager is acting by feeling because it lacks any measure (i.e., information) that could guide his/her decisions. This is the case of air transportation of "social" and commercial passengers for SATENA.

Developing an appropriate management (distributed) control system should take care of these weaknesses just mentioned. This means defining new indices (if needed) to measure the efficacy, efficiency and effectiveness of each primary activity of the system-in-focus. Once Table in Figure 11 is filled, we will have the general specifications for a managerial information system (MIS) supporting the distributed control of the system-in-focus.

However, in order to specify more detailed requirements for the MIS, it is quite useful to fill in a form for each indicator (see Figure 12). This form will register, among others, the following information: a) Name of the indicator; b) Associated Primary Activity; c) CSF related; d) Type (efficacy, efficiency, effectiveness, other); e) Operational definition, that is the function of variables defining the indicator; f) Relation of variables, indicating for each one its unit, the level of aggregation, its frequency and source (the role responsible for providing or getting the information); g) Level of aggregation of the indicator; h) Goal (that reflects managerial expectations about the CSF associated with the indicator); i) Criteria for interpretation (that facilitates the way to ascribe meaning to the indicator); j) Context for interpretation (that includes other variables or indices that should be looked at in order to understand the behavior of the indicator in a given period); k) the role responsible to produce the indicator (to calculate and generate the report); l) the role responsible to interpret and use the indicator

Figure 12. A form to specify detailed information of indices

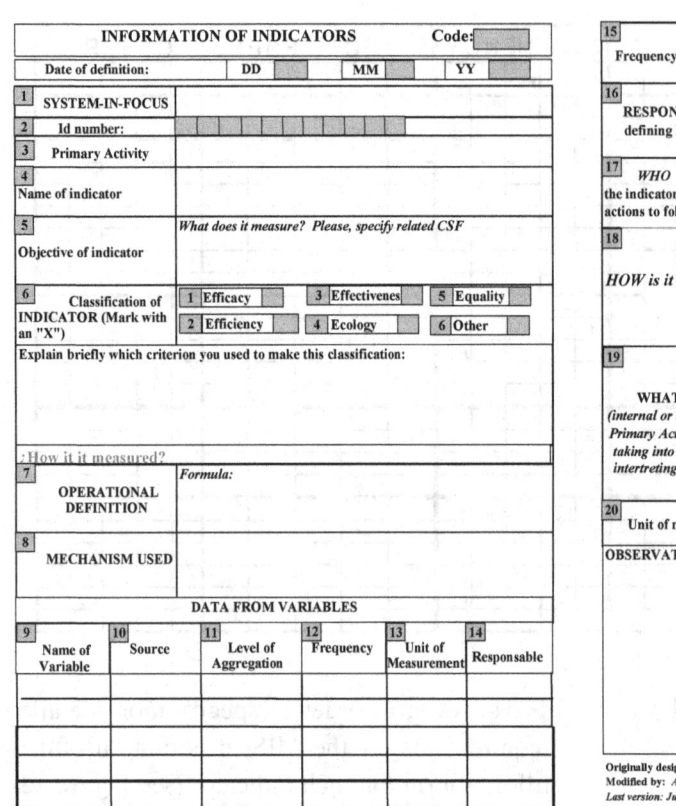

(that usually will be the role responsible for the management of the associated primary activity); and m) Date of definition (that allows a historical tracking of the indicator).

This record should be updated any time an indicator suffers a modification and, in fact, should be part of the MIS itself. This is so because indices are aligned to the strategy of the organization through the CSF of each primary activity. That means that a strategic change in the organization could imply a modification of management priorities. This, in turn, may produce a need to update CSF and, therefore, indices. In other words, as much as an organization is alive and subject to regular changes so should be its distributed control system. This explains the reason to keep track of indices through time.

Step 4: Calculating Indexes

Figure 9 shows how indices are distributed through primary activities of the system-in-focus. Each indicator is measuring at least one CSF which reflects management's priorities. These aspects could be of quite different nature: productivity, opportunity, costs/benefits, quality, market share, and so on. Each one, therefore, could have a different unit of measurement: time, quantity, money, percentage, etc. This lack of uniformity, along with the number of indices at any moment in time, could make complex the interpretation of reports produced by the MIS.

In order to reduce such complexity there is a useful method called Ciberfilter (Beer 1975; 1979). This method normalizes any indicator by defining a set of three indexes. The way it works is as follows. First, for each indicator we differentiate

322

three different kinds of values: its *actuality*, its *capability* and its *potentiality*.

The actuality of a given indicator is the value it gets regularly by the MIS; its capability is defined as the maximum value (or the minimum, depending on how the indicator was defined) that the indicator may achieve taking into account all the structural limitations of the corresponding primary activity. If the indicator is defined in such a way that the bigger its value the better (as in the case of measuring productivity or revenue) then capability is the *maximum value* it can get given current structural limitations; on the other hand, if the definition of the indicator implies that the smaller its value the better (as in the case of measuring costs or delays) then capability is the *minimum value* it can get given current structural limitations. Examples of these restrictions could be related to insufficient resources (people, money, etc.), obsolete technology, poor training, and so on. On the other hand, the potentiality of an indicator will be the maximum value that it can achieve (or the minimum) if we invest enough resources in reducing these structural restrictions. Notice that these two types of values are goals defined by recognizing the structural limitations of the primary activity under consideration.

Whereas the capability of a given indicator may be the result of management experience or the output of a benchmarking process, its potentiality is the outcome of a negotiation process. Indeed, the manager responsible for the performance of a primary activity, after recognizing that with its actual resources (people, technology, budget, etc,) s/he could get a (maximum) value for a performance indicator (i.e., its capability), s/he may set for a better goal for this indicator (i.e., its potentiality) as long as s/he can get enough resources to invest in reducing these limitations.

Secondly, once these values (or goals) are set, three indexes can be calculated for each indicator. These indexes are called: *achievement, latency* and *performance* (Beer 1975; 1979). Figure 13 shows the way these indexes are defined. Achievement is the ratio between actuality and capability; latency is the ratio between capability and potentiality; and performance is the ratio between actuality and potentiality (or the product between achievement and latency). Notice that we have to invert these ratios if the indicator is defined in such a way that the smaller its value the better. Indexes should never be greater than unity.

Notice that by definition all indexes are numbers between 0 and 1; in other words, they indicate percentages no matter what is the measurement unit of the corresponding indicator. This is exactly what we were looking for, that is a way to normalize all indicators; therefore, any time a manager gets a report from the MIS, s/he gets an index whose value is always a percentage. How could s/he interpret this value?

Figure 13. Three indexes for a given indicator (Beer 1979)

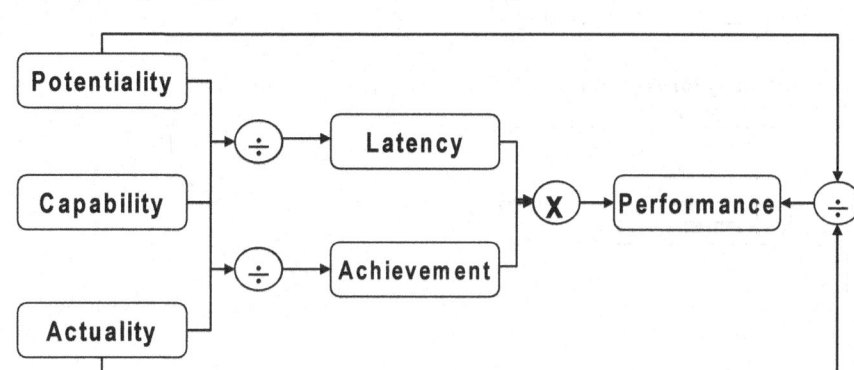

Considering that the maximum value that actuality could achieve is capability, then a low level of achievement indicates weaknesses in the management of current resources, whereas a low level of latency indicates that investment is not having an expected effect in reducing the structural limitations of the primary activity. Notice, on the other hand, that performance shows the balance of the other two indexes in the sense that it achieves its maximum value (that is 100%) if and only if simultaneously the indexes of achievement and latency are 100%. In other words, a low level of performance for a given indicator shows that either management of current resources is poor or that we are not investing in improving the primary activity. This is quite important because traditional MIS tends to concentrate in information from the past (measuring and reporting what happened) whereas cyberfilter additionally shows what is the impact of investment over the performance of primary activities. In this way management becomes more proactive.

Finally, managers should fix a range of accepted values for each index so that s/he will get reports from the MIS only by exception. This means that a report is produced only when a given index is falling out of the range previously defined. This setting of the expected values for each index and for each indicator will normally go through a process of tuning until it reaches some stability; this is part of the learning process of managers as part of their structural coupling with the MIS.

Step 5: Setting Control (Learning) Loops

So far, we have shown how indices that are produced by a MIS could be distributed across all primary activities of a system-in-focus. Moreover, each indicator is a way to measure a crucial aspect for the management of each primary activity (indeed, it is a critical success factor). Therefore, Figure 9 is also showing a distributed control system for an organization.

Figure 14. A management of risks as part of a control cycle

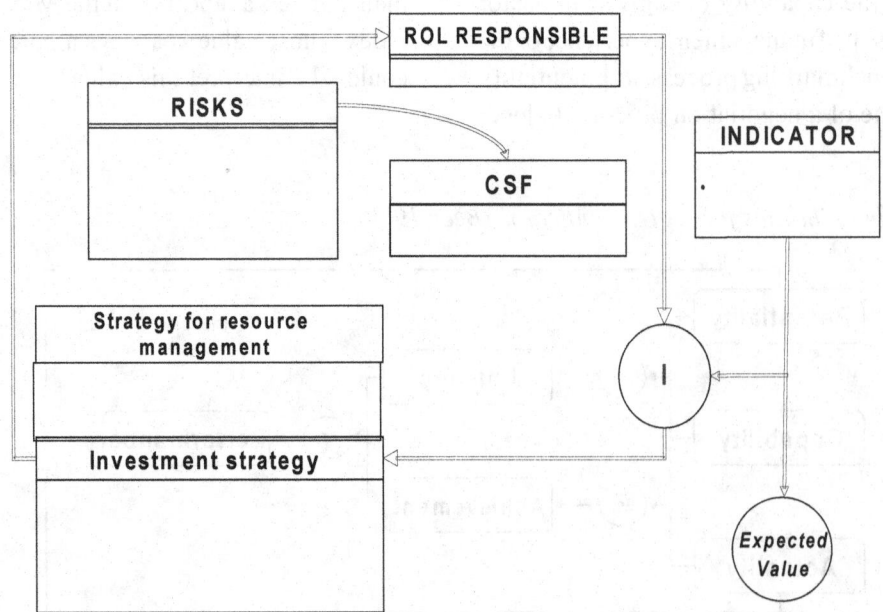

Figure 15. A distributed control system in an organization

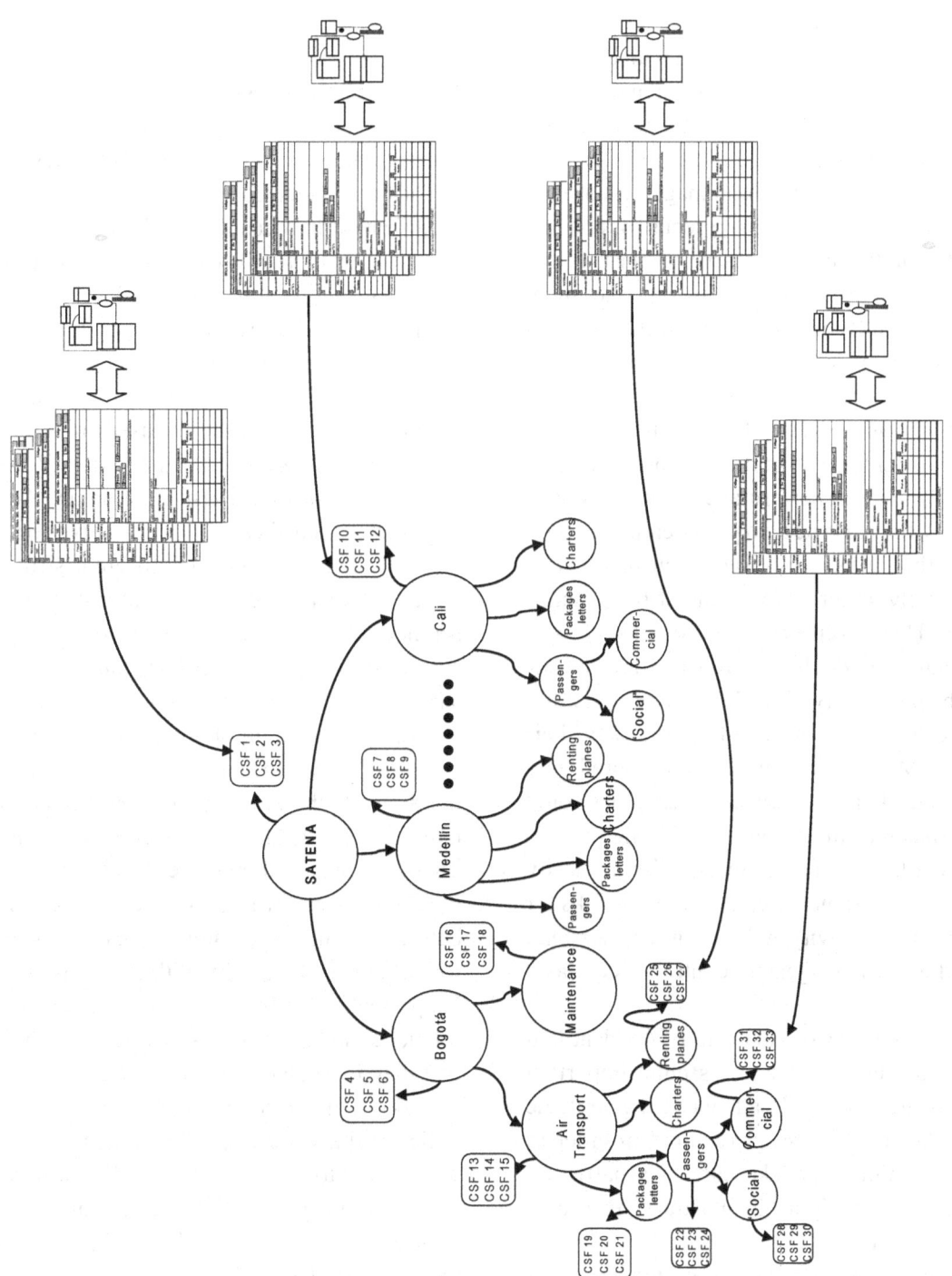

In fact, for each CSF we have a manager responsible to keep this aspect (of a primary activity) under control. In order to do this, as we saw in Figure 2, a manager has to enter into a learning loop in which s/he is able to observe (indices), assess (the reasons of any mismatch with expected values), design (i.e., choose a particular decision), and implement (i.e., transform this decision into effective action). If each one of the organizational members responsible for the management of each primary activity in a system-in-focus has both the capacity and the means to carry out these learning cycles as part of its managerial role, there is a good chance of having an effective control system of primary activities. Notice that in this case a distributed MIS is essential.

But managers not only have to take care (directly or indirectly) of the day-to-day aspects that may affect the performance of their primary activities, they also need to pay attention to those aspects that, although of rare occurrence, may dramatically affect the outcome of the primary activity. These events are usually called *risks*.

Managers not only have to learn how to estimate the probability of risk's occurrences (PRO) but also they have to be able to quantify their impact (IMP). If this impact is measured as a percentage, then the *relevance* of each risk could be calculated as the product (PRO * IMP). This exercise allows managers to establish a priority of risks. At the same time, to determine risks for each primary activity will produce a *risk-map* useful for the management control system as a whole.

Finally, for each one of the top risks identified in each primary activity, it is quite important that managers define in advanced the strategic action and the investment required in order to prevent, diminish or take care of a particular risk occurrence. Figure 14 illustrate this aspect of management.

Learning (i.e., controlling) cycles like these are associated with each CSF of each primary activity of a system-in-focus. This is precisely the way of distributing a control system in an organization. Figure 15 summarizes in a single picture the self-regulating mechanism that has been presented here.

FINAL REMARKS

We have shown a method to a step-by-step building of a distributed control system for a particular organization-in-focus. The method consists of five steps that are mutually interconnected, that is, the process is nonlinear but a cyclical one: outcomes of one step may affect previous steps.

First, we have to explicitly identify the organizational borders of our system-in-focus. The mnemonic TASCOI is the tool used to distinguish the transformation process of this system. Relevant stakeholders of this transformation are Actors, Suppliers, Clients, Owners and Interveners.

Secondly, we used four different structural models (technological, geographical, time and segmentation) in order to describe the way the system-in-focus organizes its primary activities. The whole (systemic) picture is represented by a diagram called an unfolding of complexity.

Thirdly, each primary activity, in the logical hierarchy of the unfolding of complexity, has a manager responsible for its effective performance. To assure this, each manager has to define a set of CSF relevant for its task. Then, it is crucial to define one or more indicators to measure each CSF for each primary activity of the system-in-focus. We presented three main types of indices: efficacy, efficiency and effectiveness defined in such a way that they form an orthogonal space.

Taking into account that we may have many indices with different measurement units, we need a way to normalize them. We showed this in step fourth by identifying three values for each indicator (actuality, capability and potentiality); the last two relate to the structural limitations of their corresponding primary activity. We then built three indexes (achievement, latency and performance) which values are inside the [0,1] range; so they could be interpreted as percentages. By definition a low achievement indicates poor management of actual resources in the primary activity; whereas a low latency points to an ineffective investment plan. The index of performance, in turn, by definition is balancing a short-term tactical management with a medium-term strategic management.

In step 5 we showed how the distribution of control for the system-in-focus means setting a learning loop (observe, asses, design and implement) for each CSF in every primary activity. Indexes are a fundamental part of these learning loops and so are the MIS that provide them. This relation among CSF and indices regarding primary activities shows the way to align a MIS with the organizational structure of the system-in-focus.

Finally, management should be aware not only of daily perturbations that may affect indices but also aware of risks whose occurrence may dramatically impinge upon the performance of CSF. Distributing these learning/control loops across primary activities shows a way to implement a distributed management control system in an organization.

The method has been applied extensively during the last three years in a regular postgraduate course on managerial control systems in the department of industrial engineering in the Universidad de los Andes. Students taking the course have to apply the method in a system-in-focus selected from the organization they work for. About twenty of these applications have actually being implemented.

REFERENCES

Argyris, C. (1993). *On organizational learning.* Cambridge, MA: Blackwell.

Beer, S. (1959). *Cybernetics and management.* London: The English University Press.

Ashby, R. (1956). *An introduction to cybernetics.* London: Chapman & Hall Ltd.

Beer, S. (1966). *Decision and control.* Chichester: Wiley.

Beer, S. (1979). *The heart of enterprise.* Chichester: Wiley.

Beer, S. (1981). *Brain of the firm.* Chichester: Wiley.

Beer, S. (1985). *Diagnosing the system for organisations.* Chichester: Wiley.

Espejo, R. (1989). A cybernetic method to study organisations. In Espejo, R. and Harnden, R. (eds), *The viable system model: interpretations and applications of Stafford Beer's VSM.* Chichester: Wiley.

Espejo, R. (1992). Cyberfilter: A management support system. In Holtham, C.(ed.), *Executive information systems and decision support.* London: Chapman.

Espejo, R. (1993). Strategy, structure and information management. *Journal of Information Systems,* (3), 17-31.

Espejo, R. (1994). ¿What is systemic thinking? *Systems Dynamics Review,* (10), Nos. 2-3 (Summer-Fall), 199-212.

Espejo R., Bowling, D., & Hoverstadt, P. (1999). The viable system model and the viplan software. *Kybernetes,* (28) Number 6/7, 661-678.

Espejo, R, and Reyes, A. (2000). Norbert Wiener. In Malcom Warner (ed), *The international encyclopedia of business and management: The handbook of management thinking.* London: Thompson Business Press.

Heims, S. (1991). *The cybernetics group.* Cambridge, MA: MIT Press.

Kim, D. (1993). The Link between individual and organizational learning. *Sloan Management Review*, Fall, 37-50.

Wiener, N. (1948). *Cybernetics: Or control and communication in the animal and the machine.* New York: Wiley.

Chapter 19
Making the Case
for Critical Realism:
Examining the Implementation of Automated Performance Management Systems

Phillip Dobson
Edith Cowan University, Australia

John Myles
Edith Cowan University, Australia

Paul Jackson
Edith Cowan University, Australia

ABSTRACT

This article seeks to address the dearth of practical examples of research in the area by proposing that critical realism be adopted as the underlying research philosophy for enterprise systems evaluation. We address some of the implications of adopting such an approach by discussing the evaluation and implementation of a number of automated performance measurement systems (APMS). Such systems are a recent evolution within the context of enterprise information systems. They collect operational data from integrated systems to generate values for key performance indicators, which are delivered directly to senior management. The creation and delivery of these data are fully automated, precluding manual intervention by middle or line management. Whilst these systems appear to be a logical progression in the exploitation of the available rich, real-time data, the statistics for APMS projects are disappointing. An understanding of the reasons is elusive and little researched. We describe how critical realism can provide a useful "underlabourer" for such research, by "clearing the ground a little ... removing some of the rubbish that lies in the way of knowledge" (Locke, 1894, p. 14). The implications of such an underlabouring role are investigated. Whilst the research is still underway, the article indicates how a critical realist foundation is assisting the research process.

INTRODUCTION

Many recent articles from within the information systems (IS) arena present an old-fashioned view of realism. For example, Iivari, Hirschheim, and Klein (1998) see classical realism as seeing "data as describing objective facts, information systems as consisting of technological structures ('hardware'), human beings as subject to causal laws (determinism), and organizations as relatively stable structures" (p. 172). Wilson (1999) sees the realist perspective as relying on "the availability of a set of formal constraints which have the characteristics of abstractness, generality, invariance across contexts." (p. 162)

Fitzgerald and Howcroft (1998) present a realist ontology as one of the foundational elements of positivism in discussing the polarity between hard and soft approaches in IS. Realism is placed alongside positivist and objectivist epistemologies and quantitative, confirmatory, deductive, laboratory-focussed, and nomothetic methodologies. Such a traditional view of realism is perhaps justified within the IS arena as it reflects the historical focus of its use; however, there now needs to be a greater recognition of the newer forms of realism—forms of realism that specifically address all of the positivist leanings emphasised by Fitzgerald and Howcroft (1998). A particular example of this newer form of realism is critical realism. This modern realist approach is primarily founded on the writings of the social sciences philosopher Bhaskar (1978, 1979, 1986, 1989, 1991) and is peculiarly European in its origins.

Critical realism is becoming influential in a range of disciplines including geography (Pratt, 1995), economics (Fleetwood 1999; Lawson, 1997), organization theory (Tsang & Kwan, 1999), accounting (Manicas, 1993), human geography (Sayer, 1985), nursing (Ryan & Porter, 1996; Wainwright, 1997), logistics and network theory (Aastrup 2002), and library science (Spasser, 2002). Critical realism has been proposed as a suit-able underlabourer for IS research (Dobson, 2001, 2002; Mingers, 2001, 2002), yet there have been few practical examples of its use in IS research. The application of critical realism within the IS field has been limited to date. Mutch (1999, 2000, 2002) has applied critical realist thinking in the examination of organizational use of information. In so doing, he comments how difficult it is to apply such a wide-ranging and sweeping philosophical position to day-to-day research issues. Mingers (2002) examines the implications of a critical realist approach, particularly in its support for pluralist research. Dobson (2001, 2002) argues for a closer integration of philosophical matters within IS research and suggests a critical realist approach has particular potential for IS research. Carlsson (2003) examines IS evaluation from a critical realist perspective. This article seeks to address the dearth of practical examples of critical realist use in IS by proposing the review of APMS implementation from such a perspective.

The Case Example

The Sarbanes-Oxley Act was introduced in 2002 to address high-profile accounting scandals in the U.S. The act requires that senior executives must advise stockholders immediately of any issues that are likely to affect company performance. This liability is personal and thus makes senior executives liable for the effectiveness and immediacy of their internal measurement systems and reporting. Similar legislation has been introduced in many other countries, including Australia, where the Corporations Act was implemented earlier in 2001. The development of effective performance reporting and management tools is one necessary consequence of the Sarbanes-Oxley Act and similar legislation. The resulting requirement for executives to have unimpeded, unmediated access to organizational data suggests that such tools require minimal or no human intervention in the analysis and collection of the data. This automated component in corporate performance management systems will lead to the growth of a new class of monitoring system distinct from traditional business intelligence (BI) and business activity monitoring (BAM) tools—these so-called automated performance management systems can be argued to ultimately rest on a lack of trust or confidence in traditional reporting tools and management structures. The research described in this article seeks to understand the issues involved in implementing such performance measurement systems and proposes the adoption of critical realism as a basic underlying philosophical grounding for the research.

Realist Review as a Foundational Platform

The lack of practical examples of critical realist use is perhaps not difficult to understand given the philosophy provides little real methodological guidance. Contemporary realist examination requires precision and contextualized detail, this contextualization being a necessary consequence of an underlying, ontologically bold philosophy (Outhwaite, 1987, p. 34). Along with most realist approaches, critical realism encompasses an external realism in its distinction between the world and our experience of it. This assumption necessarily implies that any knowledge gained of an external world must typically be provisional, fallible, incomplete, and extendable. As Stones (1996) suggests, realist methodologies and writings need to reflect a continual commitment to caution, scepticism, and reflexivity.

In contrast to traditional realist approaches, critical realism also suggests a so-called depth realism and argues for a stratified ontology. This concept suggests that reality is made up of three ontologically distinct realms—first, the empirical, that is, experience; second, the actual, that is, events (i.e., the actual objects of experience); and third, the transcendental, non-actual or deep, that is, structures, mechanisms, and associated powers. Critical realism argues that:

the world is composed not only of events and our experience or impression of them, but also of (irreducible) structures and mechanisms, powers and tendencies, etc. that, although not directly observable, nevertheless underlie actual events that we experience and govern or produce them. (Lawson, 1997, p. 8)

The deep structures and mechanisms that make up the world are the primary focus of such an ontological realism. The realist seeks a deep knowledge and understanding of a social situation. It argues against single concentration on observed events and requires an understanding of the deeper structures and mechanisms that often belie the surface event level observation.

Bhaskar (1979) presents fundamental difficulties with the way that prediction and falsification have been used in the open systems evident within the social arena. For the critical realist, a major issue with social investigation is the inability to create closure—the aim of "experiment" in the natural sciences. Bhaskar argues that this inability implies that theory cannot be used in a predictive manner and can **only** play an explanatory role in social investigations since:

*in the absence of spontaneously occurring, and given the impossibility of artificially creating, closed systems , the human sciences must confront the problem of the direct scientific study of phenomena that only manifest themselves in open systems—for which orthodox philosophy of science, with its tacit presupposition of closure, is literally useless. In particular it follows from this condition that criteria for the rational appraisal and development of theories in the social sciences, which are denied (in principle) decisive test situations, **cannot be predictive** and so must be **exclusively explanatory.** (Bhaskar, 1979, p. 27)*

As Mingers (2002) suggests, such an argument has specific ramifications with respect to the use of statistical reasoning to predict future results. Bhaskar (1979) argues that the primary measure of the "goodness" of a theory is in its explanatory power—from Bhaskar's perspective, predictive use of theories is not possible in open social systems and therefore predictive power cannot be a measure of goodness.

As Sayer (2000) suggests the target for realist research is not the determination of an "objective" or generalisable truth but the achievement of the best we can do at the time, that is, "practically adequate" explanations. This practical focus within critical realism sees knowledge as existing in a "historically specific, symbolically mediated and expressed, practice-dependent form" (Lawson 1997) that is potentially transformable as subsequent deeper knowledge is gained. The realist denies easy generalisability and requires a heavy focus on context.

Implications for APMS Examination

The APMS examined in this study were founded on large-scale data warehousing applications that form a part of various automated business (or corporate) performance measurement systems. All projects were based on SAP's business warehouse product, and the data warehouses sourced their data from SAP's R3 enterprise resource planning (ERP) systems as well as a myriad of other non-SAP production systems. The organisations ranged from a large government business enterprise to a mixture of global commodity companies.

The data warehousing systems had the common objective of producing automatic performance measurement management reporting via a mixture of Microsoft Excel spreadsheets and Web-based reports. The objective of the APMS was for performance measures to be presented directly to senior management in a form that precluded any manual manipulation. In most cases, this was achieved through implementing

new security/authorisation layers to protect the reporting document.

Most of the systems examined are languishing as implementation and process change management failed to get traction. Generally these systems have not become embedded within the various organizations as meaningful tools. They are generally used in an ad hoc fashion and are seen by some as just "expensive toys."

In contrast to the general failure, however, two of the APMS are in fact producing useful outcomes with over 60% of managers and information analysts using the tool throughout the business with production benefits being realised. A cursory examination of the different systems has not produced any easy explanation for the differences in implementation success. Given that such systems are expensive and difficult to produce, the organizations were understandably interested in determining the possible reasons for the patchy success.

This widely felt concern prompted a doctoral research study to be conducted by an experienced IS industry consultant. A discussion group involving two academics and the researcher was then formed to analyse and review the critical realist approach being utilised, resulting in this article. Figure 1 reflects the approach adopted in the research.

The research stages illustrated in are described below. Each stage number corresponds to the number in a circle on the figure.

1. A literature review was conducted based on the DeLone and McLean I/S Success Model (DeLone & McLean, 1992) by contrasting the DeLone and McLean Ten Year review (DeLone & McLean, 2002) and the Wixon and Watson Data Warehousing Success model (Wixom & Watson, 2001). A consolidated model was proposed based on the information systems literature. This literature review also concentrated on available operations management literature where

Figure 1. Research approach

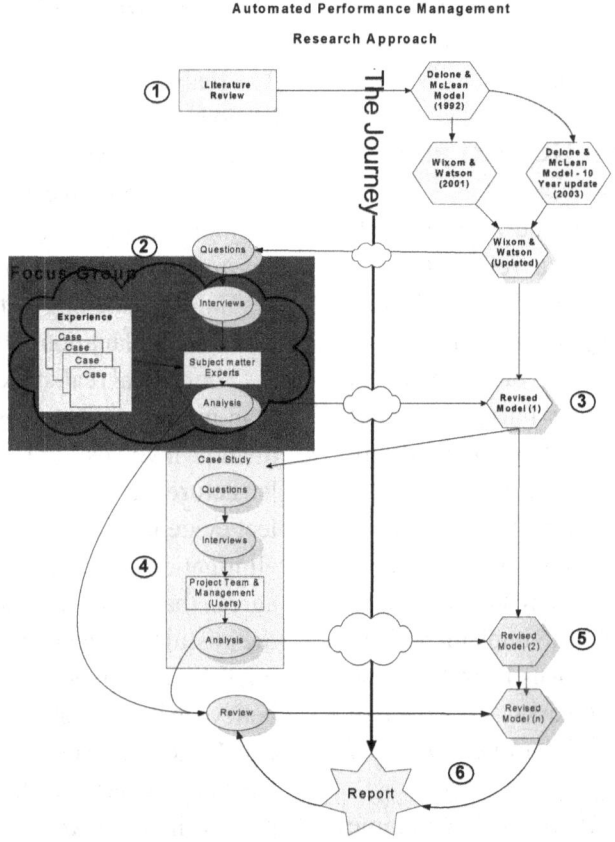

there have been a number of recent research publications. Through a process of review and consolidation by comparing and contrasting the different domains, a model for performance measurement system success was proposed.

2. This model was then used as the basis for defining a set of questions for semi-structured, qualitative interviews.

3. Once refined, the questions were used in a set of interviews utilising a focus group (Krueger, 1988). This focus group was composed of I/S industry experts, active in the performance measurement system area. Given the level of organizational interest in the perceived failure of the APMS, recruiting participants was not difficult. Against this data, the results were further analysed and

a revised model was produced (Model 1).

4. Model 1 is being tested against a case study (Yin, 1989) with further refinements to the model being made as required. This will result in an updated model (Model 2).

5. Through a number of reviews and case interviews, more refinements to the model will occur (Model 3 & 4).

6. A final model will be synthesised and is to be included in the doctoral thesis to be submitted for examination.

The approach is based upon continual comparison of the data collected in each stage with the developing model. Constant, iterative comparison of the data with the developed model and conceptual categories leads to a continuously refined explanatory model.

Throughout the study, critical realism provided a foundational platform for developing the research. The following realist elements were important in the study development:

The Realist Focus on Context and Setting

Pawson, Greenhalgh, Harvey, and Walshe (2004) describe realist review as "a relatively new strategy for synthesizing research which has an explanatory rather than a judgemental focus. It seeks to unpack the mechanisms of how complex programmes work (or why they fail) in particular contexts and settings" (p. 21). Such methods are becoming more prevalent in the analysis of the effectiveness of social programs. It is the contention of this paper that a similar approach can be effective in examining the heavily social and contextual nature of complex APMS implementation. Critical realist evaluation moves from the basic evaluative question—*what* works—to *what is it about this implementation that works for whom in what circumstances.*

In the context of the APMS research, it became evident that contextual issues were paramount in explaining the success and failure of the implementations. With the focus group interviews and individual case follow-up, the fundamental discussion is always around the particular circumstances of the implementation. This emphasis on context impacted the underlying research focus. The critical realist focus on retroductive prepositional-type questioning led to a contextual basis for the study seeking to answer "Under what conditions might APMS implementation prove successful?" rather than "What are the (predictive) critical success factors for an APMS implementation?" A simplistic critical success factors approach tends to deny the heavy contextuality and complexity of large-scale systems implementation.

Realist Emphasis on Explanation and Ex-Post Evaluation

The realist focus on explanation rather than prediction necessarily encourages an emphasis on ex-post evaluation. The realist would suggest that ex-ante or predictive evaluation is difficult given the highly complex nature of the implementation environment. Ex-post evaluations after the event are more in keeping with the underlying realist focus on explanation rather than prediction.

The critical realist focus on explanation rather than prediction suggests that the critical realist method involves "the postulation of a possible [structure or] mechanism, the attempt to collect evidence for or against its existence and the elimination of possible alternatives." The realist agrees that we have a good explanation when (1) the postulated mechanism is capable of explaining the phenomenon, (2) we have good reason to believe in its existence, (3) we cannot think of any equally good alternatives. (Outhwaite, 1987). Such an approach has specific impacts on the research process in that it argues for research heavily oriented toward confirming or denying theoretical proposals. For the realist, the initial explanatory focus may be on proposing (i.e., transcending or speculating) non-experienced and perhaps non-observable mechanisms and structures that may well be outside the domain of investigation. As Wad (2001, p. 2) argues:

If we take explanation to be the core purpose of science, critical realism seems to emphasise thinking instead of experiencing, and especially the process of abstraction from the domains of the actual and the empirical world to the transfactual mechanisms of the real world.

For the APMS study, the case examples were of previously implemented systems, and the focus was on confirming or denying a postulated model. The model developed from the focus group interviews is being further refined by examining an actual case study.

The Realist Need for an "Analytical Dualism"

The original Delone and McLean model (1992) of IS success in Figure 2 is realist in focus, as it emphasizes causal factors; however, the critical realist would have difficulty agreeing with the simplistic notion that organizational impacts are solely pre-determined by individual factors. The realist argues for a deeper multi-level analysis that recognizes that individual agency (micro) level impacts are only one of the components. Such an analysis ignores the duality of structure in that agency actions are both constrained and enabled by pre-existing structures.

Any research study founded on critical realism needs to reflect this duality of structure and agency. Archer (1995) proposes that such a duality is difficult to properly examine in social situations and therefore argues for an "analytical" or artificial dualism whereby structure (macro) and agency (micro) are artificially separated in order to properly examine their interaction. Hedström and Swedberg (1998) propose three basic mechanisms:

1. Situational mechanisms (macro-micro level)
2. Action-formation mechanisms (micro-micro level)
3. Transformational mechanisms (micro-macro level)

The typology implies that macro-level events or conditions affect the individual (step 1), the individual assimilates the impact of the macro-level events (step 2), and a number of individuals generate, through their actions and interactions, macro-level outcomes (step 3). Such a critical realist perspective on technology is presented by Smith (2005) when he suggests that:

technology introduces resources and ideas (causal mechanisms) that may enable workers to change their practices, but these practices are also constrained and enabled by the structures in which they are embedded ... Thus ... a researcher must try to understand how the generative mechanisms, introduced by the technology into a particular context of structural relations that pre-existed the intervention, provided the resources and ideas that resulted in changes (or not) to individual practices that then either transform or reproduce those original structural relations. (p. 16)

Such a representation highlights the historicity of information technology (IT) implementation and argues for a consideration of the environment prior to IT initiation. The framework also suggests

Figure 2. Delone and Mclean model (1992)

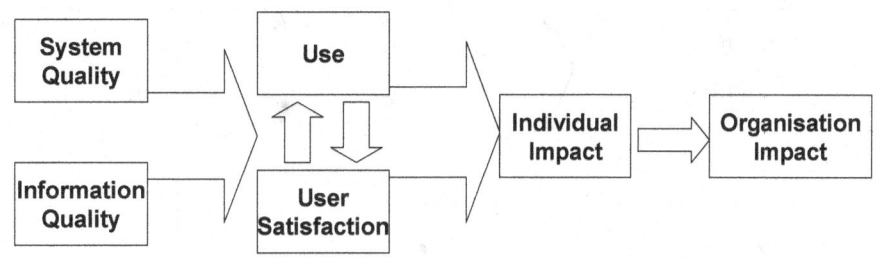

that any study of APMS implementation would need to view the implementation as fundamentally a change of pre-existing *social* practices.

The original Delone and McLean model emphasizes the micro-macro interaction when it suggests individual impacts aggregate to organizational impacts. However, from a realistic perspective, it has no recognition of the macro-micro and micro-micro level interactions.

The 2002 changes made to the original Delone and McLean model (see Figure 3) were the introduction of service quality and two dimensions, organisational and individual impact, being combined into one dimension called net benefits (Delone and McLean, 2002). From a realist perspective, this again moves the model further away from a realist position in that the organizational and individual impacts are conflated. Archer (1995) argues against such conflation when she suggests that "structure and agency can only be linked by examining the interplay between them over time, and that without the proper incorporation of time the problem of structure and agency can never be satisfactorily resolved" (p. 65). The static simplistic representation of Delone and McLean is inconsistent with such a view.

The models did, however, provide guidance as to the various categories that might be used in the grounded theory analysis.

An extension of Delone and McLean's original model developed by Wixom and Watson (2001) to model data warehousing success provided further depth to the analysis. The new model (Figure 4) helped to identify the various levels of analysis needed and associated impacts at each level. The increasing richness of the model suggests a more subtle and differentiated interaction between its elements and reduces the dependence upon a few "critical" success factors.

An Emphasis on the Social Nature of IT Implementation

The defining characteristic of APMS is that it is the automated communication of key performance indicators. As such, the implementation and operation of such a system can be highly political and sensitive. Performance measurement can be defined as the process of quantifying the efficiency and effectiveness of action and a performance measurement system as the set of metrics used to quantify both the efficiency and effectiveness of actions (Bourne, Neely, et al., 2003). The devel-

Figure 3. The reformulated I/S success model (DeLone & McLean, 2002, p. 9)

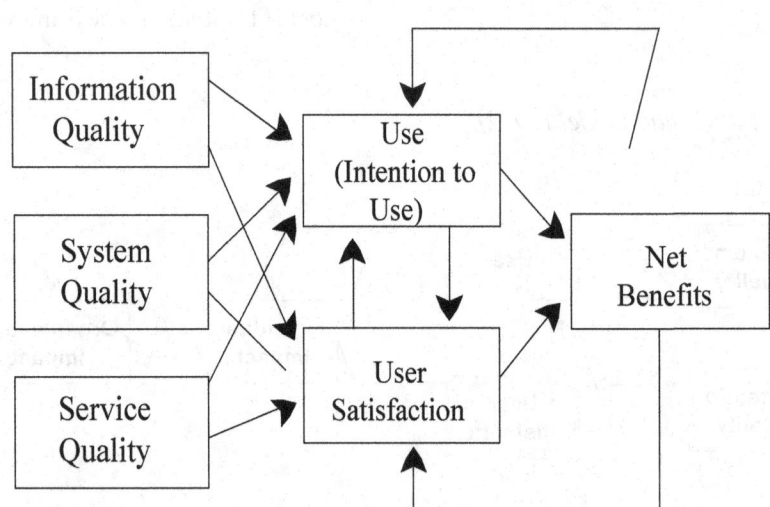

Figure 4. Research model for data ware housing success (Wixom & Watson, 2001)

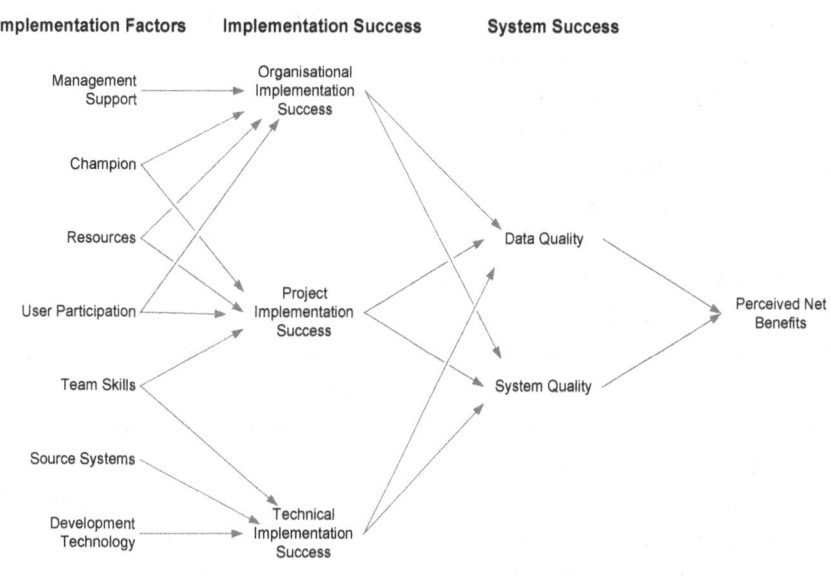

Research Model for Data Warehousing Success (Wixom & Watson, 2001)

opment of any performance management system must adhere to the following definite stages:

- define the performance to be measured
- determine and agree on appropriate performance metrics
- implement systems to monitor performance against these metrics
- implement systems to communicate these metrics to concerned stakeholders

Each such stage in the development of a performance management system can be expected to be personalised, potentially highly political, possibly controversial, and affect the acceptance of the final management system.

The final communication of performance figures is inherently social. As Pawson et al. (2005) suggest, this collection of performance figures is usually followed by public disclosure of underperforming sectors. Such a public disclosure ideally leads to "sanction instigation" whereby the broader organizational community act to "boycott, censure, reproach or control the underperform-

ing party." The final phase is termed "miscreant response" in which "failing parties are shamed, chastised, made contrite and so improve performance in order to be reintegrated." As Pawson et al. (2005) argues these social processes are all fallible and can all lead to unintended outcomes. The initial performance metric may be inappropriate or measuring the wrong problem, the dissemination may be inappropriate, public reactions may take the form of apathy or panic rather than reproach, thus leading to attempts to "resist, reject, ignore or actively discredit the official labelling." The potential for active resistance seems more likely given the automated nature of an APMS system. Automated communication may be seen to imply a lack of trust in intervening management structures and could lead to active resistance.

Organizational goals are set by management; high-level requirements are set by management, as are timelines, resources, and objectives. The design solution of APMS, its overarching principles and objectives, depend upon the ideologies, requirements, and principles of these decision makers. These principles are based upon a normative threat (the Sarbanes-Oxley legislation and similar such acts) as well as the drive to maximise productivity through control and early intervention. The ideology of industrialization, that increasing labor productivity is the foundation of increasing wealth and the improvement of social and economic conditions, also makes rational resistance difficult. The solution of APMS is therefore conservative, preserving the power *status quo* and serving the needs of those who need to control, measure, and manipulate. Here we can observe a structure of legitimated management and regulation interacting with the agency of individual and idiosyncratic leaders and subordinates. Critical realism allows that these structures have a causative function, derived from the ontological commitment of protagonists. These causal events may have elements that can be generalized, but their universality needs to be understood in the context of agency and individualism.

Conversely, where there is an emphasis on authority and control, this is antithetical to knowledge commitments and the hostages one gives to fortune, when one gives away knowledge. One of the complicating factors in systems design in particular, as indeed it is in any form of innovation, are the implications of change for participants involved in and stakeholders affected by the change. Innovation of any kind is knowledge intensive and controversial, "uncertain, fragile, political and imperialistic" (Kanter, 1996, p. 95). It crosses boundaries, redefines job descriptions and requires close communication. This leads inexorably to the fact that "Information systems development is also a political process in which various actors stand to gain or lose power as a result of design decisions" (Robey & Markus, 1984, p. 5).

New divisions of labour and requirements for cooperation, a transcendence of current work processes, will break down existing divisions of labour and require extensive cooperation. Particularly in organisations with command and control management paradigms and Fordist conceptions

of the structure of work, boundary spanning and the unimpeded flow of information will be perceived as a threat to those whose authority is based upon the existence of boundaries and fiefdoms. The adjustment and threat to power structures defined through knowledge is a high-risk area for projects whose focus and objective is to codify knowledge and ways of doing things and make them freely available. The case of APMS is particularly interesting because it is managers whose knowledge is being codified and commoditized and whose ability to intervene and massage production figures is being withdrawn. It is managers whose fiefs are becoming subject to a super-Panopticon, accessed by the CEO himself, who may ring up at 8 a.m. and complain about the previous day's poor production quality. The stance of critical realism can sensitise researchers not only to the collision of conflicting structures but also to the motivations of the protagonists who inhabit those structures and have careers to build or mortgages to pay.

People in organisations are usually aware of the importance of their knowledge to their position, status, and remuneration, and any reduction may well be met with lack of full cooperation. The implementation of APMS moves this to the next level. Martin (1988) states that "the major resource distribution by technological change is knowledge: groups with knowledge of the old system may lose control of knowledge under the new system." (p. 119) Scarbrough and Corbett (1992) assert that the higher the levels of autonomy and job specialisation, the greater the power of the job holder. If this is correct, then if these two parameters are reduced by technological change, it is more than likely that the change will be resisted at some stage of the technology change project: either in design, implementation, or use. This resistance is a denial of the legitimacy of the technological solution and may have nothing to do with whether the solution is "the best for the company" or even represents a best possible reorganisation of work processes. Critical realism recognises the role of individual agency in the withdrawal of support and legitimation for the normative and regulative structures implied by the "organisation as machine" metaphor in which APMS finds its validation. The *automated* aspect of an APMS has implications for the autonomy of the manager in that the APMS is intended to bypass the manager's intervention. The performance management aspect of the system has implications derived from surveillance and control and the concomitant power structures.

The diverse range of stakeholders, subordinate to the accountable managers, are line staff, whose actions have already been "informated" by the implementation of an operational information system. They are responsible for data entry (which must be timely and accurate for the APMS to succeed). There are the technical personnel who set up and maintain the APMS. They must understand the needs of the other "culture" and be competent in the execution of the technology. There appear to be quite different purposes and value orientations within these groups.

There is a requirement for a high degree of structure and order in the interaction between systems and the delivery of meaningful outcomes. The derivation of a few key numbers from highly complex ERP systems requires the correct functioning of many software and hardware systems and types of components, as well as standardised (highly "structurated") rules, processes, and meta data definitions.

The Ontological Depth of Critical Realism

In line with the recognition of continuing micro/macro interaction and the social implications of IT implementation, Carlsson (2003) proposes a multi-leveled investigation of the research situation. As Figure 5 indicates, the framework includes macro phenomena, like structural and institutional phenomena, as well as micro phenomena, like behaviour and interaction. The framework highlights the importance of wider macro-level issues on individual situated activity. As Carlsson suggests (2003, p. 13), the self and situated activity focus concentrates on "... the way individuals respond to particular features of their social environment and the typical situations associated with this environment" (Layder, 1993).

Critical realism is ontologically bold in the sense that it not only encompasses an external realism in its distinction between the world and our experience of it, but it also suggests a stratified ontology and a so-called depth realism in defining the objects that make up such a world. This concept suggests that reality is made up of three ontologically distinct realms—first, the empirical, that is, experience; second, the actual, that is,

Figure 5. A realist research map (Carlsson, 2003, p.13, adapted from Layder, 1993)

	Element	Focus
H I S T O R Y	CONTEXT	Macro social forms, e.g. gender, national culture, national economic situation
	SETTING	Immediate environment of social activity, e.g. organisation, department, team
	SITUATED ACTIVITY	Dynamics of face to face interaction
	SELF	Biographical experience and social involvements

events (i.e., the actual objects of experience); and third, the transcendental, non-actual, or deep, that is structures, mechanisms, and associated powers. The deep structures and mechanisms that make up the world are thus the primary focus of such an ontological realism, *events* as such not being the primary focus. An important element within critical realism is that these deep structures and mechanisms may, in fact, be only observable through their effects and thus a causal criterion for existence is accepted:

Observability may make us more confident about what we think exists, but existence itself is not dependent on it. In virtue of this, then, rather than rely purely upon a criterion of observability for making claims about what exists, realists accept a causal criterion too (Collier, 1994). According to this a plausible case for the existence of unobservable entities can be made by reference to observable effects which can only be explained as the products of such entities.... A crucial implication of this ontology is the recognition of the possibility that powers may exist unexercised, and hence ...the nature of the real objects present at a given time constrains and enables what can happen but does not pre-determine what will happen. (Sayer, 2000, p. 12)

Realist researchers need to be able to account for the underlying ontological richness they implicitly assume and also need to reflect the belief that any knowledge gains are typically provisional, fallible, incomplete, and extendable. Realist methodologies and writings thus must reflect a continual commitment to caution, scepticism, and reflexivity.

DISCUSSION

The focus group meetings with previous APMS project participants confirmed the importance of many of the factors identified in the various models. The study is still ongoing with the in-depth examination of the case study yet to be completed. In the case study, the organization had previously tried to implement automated performance management on at least five occasions with very little success. The final attempt was, however, successful in, that the system is being used to report meaningful data. One of the key aspects being identified is that the successful APMS appears to have a degree of sustainability that other systems did not have. According to Backström, van Eijnatten, & Kira (2002), a sustainable work system can be described as a work system that consciously strives toward simultaneous development at different levels: individual, group/firm, and region/society. The term "sustainability" is also referred to as corporate sustainability (Liyanage & Kumar, 2003) and may convey a difference in meaning to many, but generally it consists of external influences that are not commonly refereed to within the information systems discipline. They can include economy and technology, ecology and demography, and governance and equity.

The notion of timeliness also emerges as an underlying structure. It addresses how quickly, when, or by what date an enhancement or change can be applied to affect the automated performance reporting. The ability to react to a new measure within a reporting cycle is very important. Governments, external regulators, and other *ex machina* bodies do not necessarily wait for a business to be ready to report a particular measure. Sometimes these measures are driven internally due to a need to correct or enhance a particular business process.

From the ongoing study, it is becoming evident that external structures and the constraints and mandates they impose have severely affected APMS implementations. Such a conclusion is consistent with the critical realist view, in that it reveals the evident analytical duality in the way that agents are both constrained and enabled by pre-existing internal and external structures that they transform and reinforce through their ongoing actions.

CONCLUSION

APMS implementation is highly complex, socially and technologically. In a sense, such systems are the pinnacle of enterprise information systems, relying upon the technological success of base systems, the adequacy of their own technology, and the organisational coherence and commitment of a wide range of affected stakeholders. In line with Pawson et al. (2005, p. 22), the use of critical realism as an underlying philosophy for the APMS research appears to offer some particular benefits:

a. It has firm roots in the social sciences and allows one to identify and make salient the external, objectified, social structures that function as causal elements in the success and failure of implementation. Using this paradigm, one is allowed to explore in depth the social aspects of systems use and implementation;

b. It is grounded in the rigor of structured, analytical philosophy, and one can be reasonably confident in its reliability and consistency as a base paradigm for research development;

c. It is not a prescriptive method or formula for developing research but provides a logic of inquiry that is "inherently pluralist and flexible," embracing both "qualitative" and "quantitative," "formative" and "summative," "prospective" and "retrospective," perspectives – it suggests but does not prescribe which "rocks to look under;"

d. It seeks not to judge but to explain and is driven by the question "What works for whom in what circumstances and in what respects?" It supports the pragmatic realization, after many years of information systems failure, that "there is no silver bullet;"

e. It learns from (rather than "controls for") real-world phenomena such as diversity, change, idiosyncrasy, adaptation, cross-

contamination, and "programme failure" —its outcomes therefore make a good fit within the context of organisational learning and professional reflection;

f. It engages stakeholders systematically, as experienced but nevertheless fallible experts whose "insider" understanding of historical reasoning and action needs to be documented, formalised, reflected upon, and validated within complex, multi-level explanatory models.

Realist review does, however, have a number of limitations:

a. It is not an easy foundation on which to build in that it recognizes complexity in social research and requires a pluralist and innovative development process. It is an approach that requires experience, both in research and in subject matter. As Pawson et al. (2004) suggest, realist review is not for the novice.

b. The research generated cannot be taken to be reproducible and has therefore limited generalisability. Expressed differently, this is an honest recognition of the fact that social systems, while they contain real structures, are in fact open-ended and informed with individual agency and situational specificity.

c. Research based around critical realism cannot provide easy answers, as much as users or researchers would like this to be the case. Conclusions reached are always provisional, fallible, incomplete, and extendable and rely upon the reader to draw conclusions about transferability and reuse.

Perhaps the greatest benefit of adopting a critical realist underlabouring is the emphasis on deep understandings and context. The emphasis throughout the study has been to try and understand why particular APMS implementations succeed whereas others did not. The underlying contextual emphasis is always on "what works for whom in what circumstance."

REFERENCES

Aastrup, J. (2002). *Networks producing intermodal transport*. Unpublished doctoral dissertation, Copenhagen Business School.

Archer, M. (1995). *Realist social theory: The morphogenetic approach*. Cambridge: Cambridge University Press.

Backström, T., Eijnatten, F. M. van, & Kira, M. (2002). A complexity perspective on sustainable work systems. In P. Docherty, J. Forslin, & A. B. Shani, (Eds.), *Creating sustainable work systems: Emerging perspectives and practice*. London: Routledge.

Bhaskar, R. (1978). *A realist theory of science*. Sussex: Harvester Press.

Bhaskar, R. (1979). *The possibility of naturalism*. Hemel, Hempstead: Harvester Wheatsheaf.

Bhaskar, R. (1986). *Scientific realism and human emancipation*. Verso: London.

Bhaskar, R. (1989). *Reclaiming reality: A critical introduction to contemporary philosophy*. London: Verso.

Bhaskar, R. (1991). *Philosophy and the idea of freedom*. Oxford: Blackwell.

Bourne, M., Neely, A., et al. (2003). Implementing performance measurement systems: A literature review. *International Journal of Business Performance Management, 5*(1), 1-24.

Carlsson, S. (2003). Advancing information systems evaluation (research): A critical realist approach. *Electronic Journal of Information Systems Evaluation, 6*(2), 11-20.

Collier, A. (1994). *Critical realism: An introduction to the philosophy of Roy Bhaskar*. London: Verso.

DeLone, W. H., & McLean, E. R. (1992). Information systems success: The quest for the dependent variable. *Information Systems Research 3*(1), 60-95.

DeLone, W. H., & McLean, E. R. (2002). Information systems success revisited. *Proceedings of the 35th Hawaii International Conference on System Sciences*, Hawaii.

Dobson, P. (2001). The philosophy of critical realism -- An opportunity for information systems research. *Information Systems Frontiers*.

Dobson, P. (2002). Critical realism and IS research—Why bother with philosophy? *Information Research*, January. Retrieved from http://InformationR.net/ir/

Fitzgerald, B., & Howcroft, D. (1998). Towards dissolution of the IS research debate: From polarization to polarity. *Journal of Information Technology, 13*, 313-326.

Fleetwood, S. (Ed.) (1999). *Critical realism in economics: Development and debate*. London: Routledge.

Hedström, P., & Swedberg, R. (1998). Social mechanisms: An introductory essay. In P. Hedström & R. Swedberg (Eds.), *Social mechanisms: An analytical approach to social theory* (pp. 1-31). Cambridge University Press, New York.

Iivari, J., Hirschheim, R., & Klein, H. K. (1998). A paradigmatic analysis contrasting information systems development approaches and methodologies. *Information Systems Research, 9*(2), 164-193.

Kanter, R. M. (1996). When a thousand flowers bloom: Structural, collective, and social conditions for innovation in organizations. In P. S. Myers (Ed.), *Knowledge management and organizational design*. Newton, MA: Butterworth-Heinemann.

Krueger, R. A. (1988). *Focus groups: A practical guide for applied research.* Newbury Park, CA: Sage Publications.

Lawson, T. (1997). *Economics and reality.* Routledge: London.

Layder, D. (1993). *New strategies in social research: An introduction and guide.* Cambridge: Polity Press.

Liyanage, J. P., & Kumar, U. (2003). Towards a value-based view on operations and maintenance performance management. *Journal of Quality in Maintenance Engineering, 9*(4), 1355-2511.

Locke, J. (1894). An essay concerning human understanding. In A. C. Fraser (Ed.), *vol. 1.* Oxford: Clarendon.

Manicas, P. (1993). Accounting as a human science. *Accounting, Organizations and Society, 18,* 147- 161.

Martin, D. D., (1988). Technological Change and Manual Wiork. In D. Gallie (Ed.), *Employment in Britain* (102-127). Oxford: Basil Balckwell.

Mingers, J. (2001). Combining IS research methods: Towards a pluralist methodology. *Information Systems Research, 12*(3), 240-259.

Mingers, J. (2002). Realizing information systems: Critical realism as an underpinning philosophy for information systems. In *Proceedings Twenty-Third International Conference on Information Systems, 295-303.*

Mutch, A. (1999). Critical realism, managers and information. *British Journal of Management, 10,* 323-333.

Mutch, A. (2000). Managers and information: Agency and structure. *Information Systems Review, 1.*

Mutch, A. (2002). Actors and networks or agents and structures: Towards a realist view of information systems. *Organization, 9*(3), 477-496.

Outhwaite, W. (1987). *New philosophies of social science: Realism, hermeneutics, and critical theory.* New York: St. Martin's Press.

Pawson, R., Greenhalgh, T., Harvey G., & Walshe K. (2004). Realist synthesis: An introduction (RMP Methods Paper 2/2004). Manchester, ESRC Research Methods Programme, University of Manchester.

Pawson, R., Greenhalgh, T., Harvey, G., & Walshe K. (2005). Realist review – A new method of systematic review designed for complex policy interventions. *Journal of Health Services Research & Policy 10*(1), 21–34.

Pratt, A. (1995). Putting critical realism to work: The practical implications for geographical research. *Progress in Human Geography 19*(1), 61-74.

Robey, D., & Markus, M.L. (1984). Ritual in information systems design. *MIS Quarterly, 8,* 5-15

Ryan, S., & Porter, S. (1996). Breaking the boundaries between nursing and sociology: A critical realist ethnography of the theory-practice gap. *Journal of Advanced Nursing, 24,* 413-420.

Sayer, A. (1985). Realism in geography. In R. J. Johnston (Ed.), *The future of geography* (pp. 159-173). London: Methuen.

Sayer, A. (2000). *Realism and social science.* London: Sage Publications Ltd.

Sayer, R. A. (1992). *Method in social science: A realist approach.* Routledge: London.

Scarbrough, H., & Corbett, J. M. (1992). *Technology and organization: Power, meaning and design.* Routledge: London.

Smith, M. L. (2005). *Overcoming theory-practice inconsistencies: Critical realism and information systems research.* Unpublished manuscript, Department of Information Systems, London School of Economics and Political Science, working paper 134.

Spasser, M. A. (2002). Realist activity theory for digital library evaluation: Conceptual framework and case study. *Computer Supported Cooperative Work, 11*, 81-110.

Stones, R. (1996). *Sociological reasoning: Towards a past-modern sociology*. MacMillan.

Tsang, E., & Kwan, K., (1999). Replication and theory development in organizational science: A critical realist perspective, Academy of Management Review (24:4), 759-780

Wad, P. (2001, August 17-19). *Critical realism and comparative sociology*. Draft paper for the IACR conference.

Wainwright, S. P. (1997). A new paradigm for nursing: The potential of realism. *Journal of Advanced Nursing, 26*, 1262-1271

Wilson, F. (1999). Flogging a dead horse: The implications of epistemological relativism within information systems methodological practice. *European Journal of Information Systems, 8*(3), 161-169.

Wixom, B. H., & Watson, H. J. (2001). An empirical investigation of the factors affecting data warehousing success. *MIS Quarterly, 25*(1), 17-41.

Yin, R. K. (1989). *Case study research: Design and methods (vol. 5)*. Beverley Hills, CA: SAGE Publications Ltd.

This work was previously published in Information Resources Management Journal, Vol. 20, Issue 2, edited by M. Khosrow-Pour, pp. 138-152, copyright 2007 by IGI Publishing (an imprint of IGI Global).

Compilation of References

Aastrup, J. (2002). *Networks producing intermodal transport.* Unpublished doctoral dissertation, Copenhagen Business School.

Abdel-Hamid, T. and Madnick, S. (1989). Lessons learned from modeling the dynamics of software development. *Communications of the ACM, 32*(12), 1426-1455.

Abdel-Hamid, T. K., & Madnick, S. E. (1991). *Software project dynamics.* Englewood Cliffs, NJ: Prentice-Hall.

Ackoff, R. (1960). Systems, organizations and interdisciplinary research. *General System Yearbook, 5,* 1-8.

Ackoff, R. (1967). Management misinformation systems. *Management Science, 14*(4), 147-156.

Ackoff, R. (1971). Towards a system of systems concepts. *Management Science, 17*(11), 661-671.

Ackoff, R. (1973). Science in the systems age: Beyond IE, OR and MS. *Operations Research, 21*(3), 661-671.

Ackoff, R. (1973). Science in the Systems Age: Beyond IE, OR, and MS. *Operations Research, May-Jun,* 661-671.

Ackoff, R. (1981). *Creating the corporate future.* New York: John Wiley & Sons.

Ackoff, R. (1981). The art and science of mess management. *Interfaces, 11*(1), 20-26.

Ackoff, R. (1993, November). From mechanistic to social systems thinking. In *Proceedings of Systems Thinking Action Conference*, Cambridge, MA.

Ackoff, R. (1999). *Ackoff's best: his classic writings on management.* New York: Wiley.

Ackoff, R. (2006). Why few organizations adopt systems thinking. *Systems Research and Behavioral Science, 23*(5), 705-708. doi:10.1002/sres.791

Ackoff, R. L. (1978). *The art of problem solving.* New York: John Wiley and Sons.

Ackoff, R. L. (1999). *Ackoff's best - His classical writings on management.* John Wiley and Sons.

Ackoff, R., Gupta, S., & Minas, J. (1962). *Scientific method: Optimizing applied research decisions.* New York: Wiley.

Ackoff, R.L. (1971). Towards a system of systems concepts. *Management Science, Theory Series, 17*(11), 661-671.

Ackoff, R.L. (1974). *Redesigning the future: A systems approach to societal problems.* New York: John Wiley & Son.

Ackoff, R.L. (1989). From data to wisdom. *Journal of Applied Systems Analysis, 16,* 3-9.

Ackoff, R.L. (1993). Idealized design: Creative corporate visioning. *Omega, International Journal of Management Science, 21*(4), 401-410.

Ackroyd, S., & Fleetwood, S. (2000). *Realist perspectives on management and organisations.* London: Routledge.

Adam, F., & Fitzgerald, B. (2000). The status of the information systems field: Historical perspective and practical orientation. *Information Research, 5*(4), 1-16.

Agile Technology. (2004) http://www.agileallienceeurope.org.

Alberts, D., & Hayes, R. (2005). *Power to the edge: Command and control in the information age* (CCRP Publication Series). CCRP.

Aldrich, H. (1999). *Organizations evolving.* Thousand Oaks, CA: Sage.

Alfeld, L.E. and Graham, A. (1976). *Introduction to Urban Dynamics.* MIT Press, Cambridge MA. Reprinted by Productivity Press: Portland OR and currently available from Pegasus Communications: Waltham, MA.

Alter, S (2006). *The work system method: Connecting people, processes, and IT for business results.* Larkspur CA: Work System Press.

Alter, S. (1999). A general, yet useful theory of information systems. *Communications of the Association for Information Systems, 1*, 13.

Alter, S. (2001). Are the fundamental concepts of information systems mostly about work systems? *CAIS, 5*(11), 1-67.

Alter, S. (2002). The work system method for understanding information systems and information systems research. *Communications of the AIS, 9*(6), 90–104.

Alter, S. (2003). 18 reasons why IT-reliant work systems should replace the IT artifact as the core subject matter of the IS field. *Communications of the AIS, 12* (23), 365-394.

Alter, S. (2004). *Desperately seeking systems thinking in the information systems discipline.* Paper presented at ICIS 25, Washington, DC.

Alter, S. (2005). Architecture of Sysperanto - A model-based ontology of the IS field. *Communications of the AIS, 15* (1), 1-40.

Alter, S. (2006). *The work system method: connecting people, processes, and IT for business results.* Lakspur, CA: Work System Press.

Alter, S. (2007). Could the work system method embrace system concepts more fully? *Information Resources Management Journal, 20*(2), 33–43.

Alter, S. (2007). *Service responsibility tables: A new tool for analyzing and designing systems.* Paper presented at the 13th Americas Conference on Information Systems, Keystone, CO.

Alter. (2004). Desperately seeking systems thinking in the information systems discipline. In *Proceedings of Twenty-Fifth International Conference on Information Systems,* 757-770.

American Productivity & Quality Center (APQC). (1999). *Creating a knowledge-sharing culture.* Consortium Benchmarking Study -- Best-Practice Report.

Anderson, P. (1999). Complexity theory and organization science. *Organization Science, 10*(3), 216-232.

Andoh-Baidoo, F., White, E., & Kasper, G. (2004). Information systems' cumulative research tradition: A review of research activities and outputs using pro forma abstracts. In *Proceedings of the Tenth Americas Conference on Information Systems,* New York, NY (pp. 4195-4202).

Andriole, S., & Freeman, P. (1993, May). Software systems engineering: The case for a new discipline. *Software Engineering Journal,* pp. 165-179.

Anthony, S. and Christensen, C. (2004). Forging innovation from disruption. *Optimize,* (Aug) issue 24.

Archer, M. (1995). *Realist social theory: The morphogenetic approach.* Cambridge: Cambridge University Press.

Argyris, C. (1993). *On organizational learning.* Cambridge, MA: Blackwell.

Argyris, C., & Schon, D. A. (1978). *Organizational learning. A theory of action perspective.* Addison- Wesley.

Arnold, S., & Lawson, H. (2004). Viewing systems from a business management perspective: The ISO/IEC 15288 Standard. *Systems Engineering, 7*(3), 229-242.

Asbjornsen, O. A., & Hamann, R. J. (2000, May). Toward a unified systems engineering education. *IEEE TSMC Part C-Applications and Reviews, 30*(2), 175–182.

Ashby, R. (1956). *An introduction to cybernetics*. London: Chapman & Hall Ltd.

Atkinson, G. (2004). Common ground for institutional economics and system dynamics modeling. *System Dynamics Review, 20*(4), 275-286.

Atkinson, S., & Moffat, J. (2005). *The agile organization: From informal networks to complex effects and agility* (CCRP Information Age Transformation Series). CCRP.

Aurum, A., & Wohlin, C. (Eds.). (2005). *Engineering and managing software requirements*. Heidelberg: Springer.

Avison, D. (2000). *Multiview: An exploration in information systems development* (2nd ed). Alfred Waller Ltd.

Avison, D. E., & Fitzgerald, G. (2003). Where now for development methodologies? *Communications of the ACM, 46*(1), 79–82. doi:10.1145/602421.602423

Avison, D. E., & Wood-Harper, A. T. (1990). *Multiview: An exploration in information systems development*. Henley on Thames, UK: Alfred Waller (McGraw-Hill Publishing Company).

Avison, D., & Horton, J. (1992). *Evaluation of information systems* (Working Paper). Southampton: University of Southampton, Department of Accounting and Management Science.

Avison, D., Baskerville, R., & Myers, M. (2001). Controlling action research projects. *Information Technology & People, 14*(1), 28-45.

Avison, D., Lau, F., Myers, M., & Nielson, P.A. (1999). Action research. *Communications of the ACM, 42*(1), 94-97.

Avison, D., Wood-Harper, A. T., Vidgen, R. T., & Wood, J. R. G. (1998). A further exploration into information systems development: The evolution of Multiview2. *Information Technology and People, 11*(2), 124-139.

Avison, D., Wood-Harper, A., Vidgen, R., & Wood, J. (1998). A further exploration into information systems development: The evolution of multiview2. *Information Technology & People, 11*(2), 124-139.

Avison, D.E., & Wood-Harper, A.T. (1986). Multiview: An exploration in information systems development. *The Australian Computer Journal, 18*(4).

Axelrod, R. (1997). *The complexity of cooperation: Agent-based models of competition and collaboration*. New Jersey: Princeton University Press.

Axelrod, R., & Cohen, M. (1999). *Harnessing complexity: Organizational implications of a scientific frontier*. New York: The Free Press.

Backhouse, J., Hsu, C., & Silva, L. (2006). Circuits of power in creating *de jure* standards: Shaping an international information systems security standard. *MIS Quarterly, 30*, 413-438.

Backström, T., Eijnatten, F. M. van, & Kira, M. (2002). A complexity perspective on sustainable work systems. In P. Docherty, J. Forslin, & A. B. Shani, (Eds.), *Creating sustainable work systems: Emerging perspectives and practice*. London: Routledge.

Bacon, J., & Fitzgerald, B. (2001). A systemic framework for the field of information systems. *The DATA BASE for Advances in Information Systems, 32*(2), 46-67.

Bahill, A. T., & Gissing, B. (1998, November). Re-evaluating systems engineering concepts using systems thinking. *IEEE TSMC Part C-Applications and Reviews, 28*(4), 516–527.

Bailey, E. J., & Pearson, S. W. (1983). Development of a tool for measuring and analyzing computer user satisfaction. *Management Science, 29*(5), 530-545.

Bajaj, A., Batra, D., Hevner, A., Parsons, J., & Siau, K. (2005). Systems analysis and design: Should we be researching what we teach? *Communications of the AIS, 15*, 478–493.

Ball, K., & Wilson, D. (2000). Power, control and computer-based performance monitoring: A subjectivist approach to repertoires and resistance. *Organization Studies, 21*(3), pp. 539-565.

Ballantine, J., Bonner, M., Levy, M., Martin, A., Munro, I., & Powell, P.L. (1998). Developing a 3-D model of information systems success. In E. J. Garrity & G. L.

Sanders (Eds.), *Information system success measurement* (pp. 46-59). Hershey, PA: Idea Group Publishing.

Ballantine, J., Levy, M., Munro, I., & Powell, P. (2003). An ethical perspective on information systems evaluation. *International Journal of Agile Management Systems, 2*(3), 233-241.

Banathy, B.H. (1992). *Systems design of education: Concepts and principles systems for effective practice.* New York: Educational Technology Publications.

Banathy, B.H. (1996). *Designing social systems in a changing world.* New York: Plenum Press.

Banville, C., & Landry, M. (1989). Can the field of MIS be disciplined? *Communications of the ACM, 32*(1), 48-60.

Baresi, L, Di Nitto, & Ghezzi, C. (2006). Toward open-world software: issues and challenges. *IEEE Computer, 39*(10), 36-43.

Bariff, M., & Galbraith, J. R. (1978). Intraorganizational power considerations for designing information systems. *Accounting, organizations and society, 3*(1), 15-27.

Barkhi, R., & Sheetz, S. (2001). The state of theoretical diversity of information systems. *CAIS, 7*(6), 1-19.

Barki, H. and Hartwick, J. (1994). Explaining the Role of User Participation in Information System Use. *Management Science, 40(4),* pp. 440-465.

Barki, H., & Hartwick, J. (1989). Rethinking the concept of user involvement. *MIS Quarterly,* (13), 53-63.

Baroudi, J.J. and Orlikowski, W.J. (1988). A short-form measure of user information satisfaction: a psychometric evaluation and notes on use. *Journal of Management Information Systems*; 4(4), pp.44 - 59

Barton, J. (1999). Pragmatism, systems thinking and system dynamics. In *Proceedings of the 17th Conference of the System Dynamics Society.* Palermo: System Dynamics Society.

Barton, J., Emery, M., Flood, R. L., Selsky, J., & Wolstenholm, E. (2004). A maturing of systems thinking? Evidence from three perspectives. *Systemic*

Practice and Action Research, 17(1), 3–36. doi:10.1023/B:SPAA.0000013419.99623.f0

Barton, J., Emery, M., Flood, R.L., Selsky, J.W., & Wolstenholme, E. (2004). A maturing of systems thinking? Evidence from three perspectives. *Systems Practice and Action Research, 17*(1), 3-36.

Bar-Yam, Y. (1997). *Dynamics of complex systems.* Reading, MA: Perseus Press.

Bar-Yam, Y. (2002a). Complexity rising: From human beings to human civilization, a complexity profile. In *Encyclopedia of Life Support Systems (EOLSS).* Oxford, UK: UNESCO, EOLSS Publishers.

Bar-Yam, Y. (2002b). General features of complex systems. In *Encyclopedia of Life Support Systems (EOLSS).* Oxford, UK: UNESCO, EOLSS Publishers.

Bar-Yam, Y. (2003). When systems engineering fails -- toward complex systems engineering. In Proceedings of the *International Conference on Systems, Man & Cybernetics* (Vol. 2, pp. 2021- 2028). Piscataway, NJ: IEEE Press, 2021- 2028.

Bar-Yam, Y. (2003a). When systems engineering fails: Toward complex systems engineering. In *Proceedings of the International Conference on Systems, Man & Cybernetics* (pp. 2021-2028). Piscataway, NJ: IEEE Press.

Bar-Yam, Y. (2003b). Unifying principles in complex systems. In M.C. Roco & W.S. Bainbridge (Eds.), *Converging technology (NBIC) for improving human performance* (pp. 1-32). Kluwer.

Bar-Yam, Y. (2004). Multiscale variety in complex systems. *Complexity, 9,* 37-45.

Bar-Yam, Y. (2005). About engineering complex systems: multiscale analysis and evolutionary engineering. In S. Brueckner et al. (Eds.), *ESOA 2004,* Heidelberg, Germany (pp. 16-31). Springer-Verlag. Lecture Notes in Computer Science, 3464.

Bar-Yam, Y. et al. (2004). The characteristics and emerging behaviors of system of systems. In *NECSI: Complex Physical, Biological and Social Systems Project* (pp.

1-16). Cambridge, MA: New England Complex Systems Institute.

Baskerville, R., & Wood-Harper, A. T. (1998). Diversity in information systems action research methods. *European Journal of Information Systems, 7*(2), 90–107. doi:10.1057/palgrave.ejis.3000298

Bate, R. (1998, July-August). Do systems engineering? Who, me? *IEEE Software*, pp. 65-66.

Bateson, G. (1972). *Steps to an ecology of the mind.* New York: Ballantine.

Beachboard, J., & Beard, D. (2005). Innovation in information systems education-II: Enterprise IS management: A capstone course for undergraduate IS majors. *CAIS, 15*, 315-330.

Beauregard, O. C. (1961). *Sur l'equivalance entre information et entropie.* In J.G. Miller (Ed. And Trans.), *Living systems* (p. 42). New York: Mc-Graw-Hill.

Beckerman, L. (2000). Application of complex systems science to systems engineering. *Systems Engineering, 3*(2), 96-102.

Beer, S. (1959). *Cybernetics and management.* London: The English University Press.

Beer, S. (1966). *Decision and control.* Chichester: Wiley.

Beer, S. (1979). *The heart of enterprise.* Chichester: Wiley.

Beer, S. (1981). *Brain of the firm.* Chichester: Wiley.

Beer, S. (1985). *Diagnosing the system for organisations.* Chichester: Wiley.

Bell, S., & Wood-Harper, A. T. (1998). *Rapid information systems development.* Maidenhead: McGraw Hill.

Bell, S., & Wood-Harper, A. T. (2005). *How to set up information systems: A non-specialist guide to the multiview approach.* London: Earthscan publications.

Benbasat, I., & Zmud, R. (2003). The identity crisis within the IS discipline: defining and communicating the discipline's core properties. *MIS Quarterly, 27*(2), 183-194.

Benbya, H., & McKelvey, H. (2006). Toward a complexity theory of information systems development. *Information Technology & People, 19*(1), 12-34.

Bennetts, P. D. C., Wood-Harper, T., & Mills, S. (2000). A holistic approach to the management of information systems development. *Systemic Practice and Action Research, 13*(2), 189–206. doi:10.1023/A:1009594604515

Bernhauer, J., & Mahon, M. (1994). The ethics of Michel Foucault. In G. Gutting (Ed.), *The Cambridge Companion to Foucault* (pp. 141-158). Cambridge, UK: Cambridge University Press.

Bertalanffy, L. V. (1968). *General system theory.* New York: George Braziller.

Bertalanffy, L. von. (1950). An outline of general systems theory. *British Journal of the Philosophy of Science, 1*, 134-164 (reprinted in Bertalanffy (1968)).

Bertalanffy, L. von. (1951). The theory of open systems in physics and biology. *Science, 111*, 23-29.

Bertalanffy, L. von. (1968). *General systems theory – foundations, developments, applications.* New York: G. Brazillier.

Bertalanffy, L. von. (1972). The history and status of general systems theory. *Academy of Management Journal, December*, 407-426.

Best, S., & Kellner, D. (1991). *Postmodern theory: Critical interrogations.* New York: Guilford Press.

Bhaskar, R. (1975). *A realist theory of science.* Sussex: Harvester Press.

Bhaskar, R. (1975). *A realist theory of science.* Sussex: Harvester Press.

Bhaskar, R. (1979). *The possibility of naturalism.* Hemel, Hempstead: Harvester Wheatsheaf.

Bhaskar, R. (1986). *Scientific realism and human emancipation.* Verso: London.

Bhaskar, R. (1989). *Reclaiming reality: A critical introduction to contemporary philosophy.* London: Verso.

Bhaskar, R. (1991). *Philosophy and the idea of freedom.* Oxford: Blackwell.

Bhaskar, R. (1993). *Dialectic: The pulse of freedom.* London: Verso.

Bhaskar, R. (1994). *Plato etc.* London: Verso.

Blanchette, S. (2005). *U.S. Army acquisition – The program executive officer perspective,* (Special Report CMU/SEI-2005-SR-002). Pittsburgh, PA: Software Engineering Institute.

Bloomfield, B., & Coombs, R. (1992). Information technology, control and power: The centralization and decentralization debate revisited. *Journal of Management Studies, 29*(4), 459-484.

Bødker, S. (1996). Creating conditions for participation: conflicts and resources in systems development. *Human-Computer Interaction, 11,* 215–236. doi:10.1207/s15327051hci1103_2

Boehm, B. (2000, March). Unifying software engineering and systems engineering. *Computer,* pp. 114-116.

Boehm, B. (2006). Some future trends and implications for systems and software engineering processes. *Systems Engineering, 9*(1), 1-19.

Boehm, B., & Basili, V. (2005). *The CeBASE framework for strategic software development and evolution.* EDSER-3 Position Paper. Retrieved from http://sunset.usc.edu/csse/TECHRPTS/2001/usccse2001-503/usccse2001-503.pdf.

Boehm, B., & Lane J. (2007). Using the incremental commitment model to integrate system acquisition, systems engineering, and software engineering. *USC-CSSE-TR-2207* (Short Version in Cross Talk, October 2007, pp, 4-9)

Boehm, B., & Lane, J. (2006). 21st century processes for acquiring 21st century systems of systems. *CrossTalk, 19*(5), 4-9.

Boehm, B., & Turner, R. (2004). *Balancing agility and discipline - a guide for the perplexed.* Boston: Addison-Wesley.

Boehm, B., & Turner, R. (2005). Management challenges for implementing agile processes in traditional development organizations. *IEEE Software, 5,* 30–39. doi:10.1109/MS.2005.129

Boehm, B., Abt, C., Brown, A., Chulani, S., Clark, & et al. (2000). *Software cost estimation with COCOMO II.* Upper Saddle River, NJ: Prentice Hall.

Boehm, B., Valerdi, R., Lane, J., & Brown, A. (2005). COCOMO suite methodology and evolution. *CrossTalk, 18*(4), 20-25.

Boer, P. F. (1999). *The valuation of technology: business and financial issues in R&D.* Hoboken, NJ: Wiley.

Boer, P. F. (2002). *The real options solution: finding total value in a high-risk world.* Hoboken, NJ: Wiley.

Boer, P. F. (2004). *Technology valuation solutions.* Hoboken, NJ: Wiley.

Boland, R. (1985). Phenomenology: A preferred approach to research in information systems. In E. Mumford, R.A. Hirschheim, G. Fitzgerald, & A.T. Wood-Harper (Eds.), *Research methods in information systems.* North-Holland: Amsterdam.

Boland, R. (1991). Information systems use as a hermeneutic process. In H. Nissen, H. Klein & R. Hirschheim (Eds.), *Information systems research: Contemporary approaches and emergent traditions.* North-Holland: Amsterdam.

Boland, R.J., Tenkasi, R.V., & Te'eni D. (1994). Design information technology to support distributed cognition. *Organization Science, 5*(3), 456-475.

Bostrom, R. and Heinen, S. (1977). MIS Problems and Failures: A Socio-Technical Perspective. Part I: The Causes. *MIS Quarterly, September,* 17-31

Boulding, K. (1956). General systems theory – the skeleton of the science. *Management Science, 2*(3), 197-208.

Boulding, K. (1964). General systems as a point of view. In J. Mesarovic (Ed), *Views on general systems theory.* New York: John Wiley.

Bourne, M., Neely, A., et al. (2003). Implementing performance measurement systems: A literature review. *International Journal of Business Performance Management, 5*(1), 1-24.

Bower, J.L. and Christensen, C. (1995). Disruptive technologies: catching the wave. *Harvard Business Review* (Jan-Feb), 43-53.

Bradley, F. (1914). *Essays on truth and reality.* Oxford: Oxford University Press.

Braman, S. (1989). Defining information: An approach for policymakers. *Telecommunications Policy, 13*(1), 233-242.

Briscoe, B., Odlyzko, A., & Tilly, B. (2006). Metcalfe's Law is wrong. *IEEE Spectrum*, July, 26-31.

Brocklesby, J., & Cummings, S. (1996). Foucault plays Habermas: An alternative philosophical underpinning for critical systems thinking. *Journal of the Operational Research Society, 47*(6), 741-754.

Brooke, C. (2002). What does it mean to be "critical" in IS research? *Journal of Information Technology, 17*(2), 49-57.

Brooks, F. P. (1987). No silver bullet: Essence and accidents of software engineering. In *Proceedings of the IFIP Tenth World Computing Conference* (pp. 1069-1076).

Brown, D. E., & Scherer, W. T. (2000, May). A comparison of systems engineering programs in the United States. *IEEE TSMC C-Applications and Reviews, 30*(2), 204–212.

Brown, D., & Scherer, W. (2000). A comparison of systems engineering programs in the United States. *IEEE Transactions on Systems, Man and Cybernetics (Part C: Applications and Reviews), 30*(2), 204-212.

Bryman, A., & Bell, E. (2003). *Business research methods.* Oxford: OUP.

Buckland, M. (1991). *Information and information systems.* New York: Praeger Publishers.

Buckland, M.K. (1991). Information as thing. *Journal of American Society for Information Science, 42*(5), 351-360.

Buede, E. (2000). *The engineering design of systems* (Wiley Series in Systems Engineering). New York: Wiley.

Bühler, K. (1982). *Karl Bühler: Semiotic foundations of language theory.* New York: Plenum Press.

Bukowitz, W., & Williams, R. (1999). *The knowledge management fieldbook.* London: Financial Times Prentice Hall.

Bullock, A., & Tromby, S. (2000). *New Fontana dictionary of modern thought* (p. 443). London: Harper Collins.

Burks, A.W. (1998). Peirce's evolutionary pragmatic idealism. *Synthese, 106*, 323-372.

Burrell, G. (1988). Modernism, post modernism and organizational analysis: The contribution of Michel Foucault. *Organization Studies, 9*(2), 221-235.

Burrell, G., & Morgan, G. (1979). *Sociological paradigms and organisational analysis.* London: Heinemann.

Bustard, D. W., & Keenan, F. (2005). Strategies for systems analysis; Groundwork for process tailoring. *IEEEE, ECBC.*

Bustard, D. W., & Keenan, F. M. (2005). Strategies for systems analysis: groundwork for process tailoring. In *Proceedings of 12th Annual IEEE International Conference and Workshop on the Engineering of Computer Based Systems (ECBS 2005)* (pp. 357-362). Greenbelt, MD, USA, 3-8 April.

Button, G., & Dourish, P. (1996). Technomethodology: Paradoxes and possibilities. In *Proceedings of the SIGCHI Conference on Human Factors in Computing Systems: Common Ground,* Vancouver, BC, Canada (pp. 19-26). New York: ACM. Retrieved from http://www.acm.org/sigchi/cyhi96/proceedings/papers/Button/jpd_txt.htm

Callaos, N., & Callaos, B. (2002). Towards a systemic notion of information: Practical consequences. *Informing Science, 5*(1), 1-11.

Calvano, C., & John, P. (2004). Systems engineering in an age of complexity. *Systems Engineering Journal, 7*(1), 25-34.

Cannon, W. B. (1939). *Wisdom of the body.* New York: Norton

Carlock, P., & Fenton, R. (2001). System of systems (SoS) enterprise system engineering for information-intensive organizations. *Systems Engineering, 4*(4), 242-261.

Carlsson, S. (2003). Advancing information systems evaluation (research): A critical realist approach. *Electronic Journal of Information Systems Evaluation, 6*(2), 11-20.

Carlsson, S. (2003, June 16-21). Critical realism: A way forward in IS research. In *Proceedings of the ECIS 2003 Conference* Naples, Italy.

Carr, N. (2003). IT doesn't matter. *Harvard Business Review, 81*(5), 41-49.

Carvel, J. (2008). Bank bailout puts £12.7bn NHS computer project in jeopardy. *Guardian newspaper.* Wednesday 29 October 2008 (see also http://www.guardian.co.uk/society/2008/oct/29/nhs-health)

Caulkin, S. (2004, May 2). Why IT just doesn't compute: Public sector projects even more likely to fail than private. *The Observer* (p. 9).

Cavaye, A. L. M. (1995). User participation in system development revisited. *Information & Management, 28*(5), 311–323. doi:10.1016/0378-7206(94)00053-L

Chakraborty, R. (1994): Information systems and organizational success: A quantitative modeling approach. TDQM-94-08. Retrieved from http://web.mit.edu/tdqm/www/articles/94/94-08.html.

Champion, D., & Stowell, F. A. (2001). Pearl: A systems approach to demonstrating authenticity in information system design. *Journal of Information Technology, 16,* 3–12. doi:10.1080/02683960010028438

Champion, D., Stowell, F. A., & O'Callaghan, A. (2004). Client led information system creation (CLIC): Navigating the gap. *Information Systems Journal, 15*(3), 213–231. doi:10.1111/j.1365-2575.2005.00191.x

Champion, D., Stowell, F., & O'Callaghan, A. (2005). Client-led information system creation (clic): Navigating the gap. *Information Systems Journal, 15*(3), 213-231.

Chan, A., & Garrick, J. (2002). Organisation theory in turbulent times: The traces of Foucault's ethics. *Organization, 9*(4), 683-701.

Checkland, P. (1981). *Systems thinking, systems practice.* Chichester: John Wiley.

Checkland, P. (1981). *Systems thinking, systems practice.* London: John Wiley and Sons.

Checkland, P. (1983). O.R. and the systems movement: mappings and conflicts. *Journal of the Operational Research Society, 34*(8), pp. 661-675.

Checkland, P. (1990). Information systems and systems thinking: Time to unite? In P. Checkland & J. Scholes (Eds.), *Soft systems methodology in action* (pp. 303-315). Chichester, UK: John Wiley & Sons Ltd.

Checkland, P. (1999). *Systems thinking, systems practice* (Includes a 30-year retrospective). Chichester, UK: John Wiley.

Checkland, P. (2000). Soft systems methodology: A 30-year retrospective. In P. Checkland, *Systems thinking, systems practice* (pp. A1-A65). Chichester: Wiley.

Checkland, P. B., & Holwell, S. E. (1998). *Information, systems and information systems.* Chichester: Wiley and Sons.

Checkland, P., & Holwell, S. (1995). Information systems: What's the big idea?. *Systemist, 7*(1), 7-13.

Checkland, P., & Holwell, S. (1998). *Information, systems and information systems: Making sense of the field.* Chichester, UK: John Wiley and Sons.

Checkland, P., & Holwell, S. (1998). *Information, systems and information systems.* Chichester: Wiley.

Checkland, P., & Poulter, J. (2006). *Learning for action: A short definitive account of soft systems methodology and its use by practitioner, teachers and students.* Chichester: Wiley.

Checkland, P., & Scholes, J. (1990). *Soft systems methodology in action*. John Wiley & Sons Editions.

Checkland, P., & Scholes, P. (1990). *Soft systems methodology in action*. Chichester: John Wiley and Sons.

Checkland, P.B. (2000). Soft systems methodology: A thirty year retrospective. *Systems Research and Behavioural Science, 17*(1), S11-S58.

Chesbrough, C., & Spohrer, J. (2006). A research manifesto for services science. *Communications of the ACM, 49*(7), 35-40.

Christensen, C.M. & Raynor, M.E. (2003). *The Innovator's Solution: Creating and Sustaining Successful Growth*. Harvard Business School Press: Boston MA.

Christensen, C.M. (1992). *The Innovator's Challenge: Understanding the Influence of Market Environment on Processes of Technology Development in the Rigid Disk Drive Industry*. Ph.D. Dissertation, Harvard Business School: Boston, MA.

Christensen, C.M. (1997). *The Innovator's Dilemma: When New Technologies Cause Great Firms to Fail*. Harvard Business School Press: Cambridge, MA.

Christensen, C.M., Johnson, M. & Dann, J. (2002). Disrupt and prosper. *Optimize* (Nov), 41-48.

Christensen, C.M., Raynor, M.E. & Anthony, S.D. (2003). Six Keys to Creating New-Growth Businesses. *Harvard Management Update* (Jan).

Chujo, H., & Kijima, K. (2006). Soft systems approach to project-based education and its practice in a Japanese university. *Systems Research and Behavioral Science, 23*(1), 89–106. doi:10.1002/sres.709

Churchman, C. W. (1970). Operations research as a profession. *Management Science, 17*, b37-b53.

Churchman, C. W. (1971). *The design of inquiring systems: Basic concepts of systems and organisation* (p. 18). New York: Basic Books.

Churchman, C. W. (1979). *The systems approach and its enemies*. New York: Basic Books.

Churchman, C.W. (1971). *Design of inquiring systems*. New York: Basic Books.

Churchman, C.W. (1971). *The design of inquiring systems: Basic concepts of systems and organization*. New York: Basic Books.

Churchman, C.W. (1979). *The system approach and its enemies*. New York: Basic Books.

CIO UK Web site. (2007). Late IT projects equals lower profits. Retrieved July 10, 2007, from *http://www.cio.co.uk/concern/resources/news/index.cfm?articleid=1563*

CIO UK. (2007). Late IT projects equals lower profits. Retrieved from http://www.cio.co.uk/concern/resources/news/index.cfm?articleid=1563.

Clarke, S. (2001). *Information systems strategic management : An integrated approach*. London: Routledge.

Clarke, S., & Lehaney, B. (2000). Mixing methodologies for information systems development and strategy: A higher education case study. *Journal of the Operational Research Society, 51*, 542-566.

Cleary, D. (2005). Perspectives on complex-system engineering. *MITRE Systems Engineering Process Office Newsletter, 3*(2), 1-4.

Clippinger, J.H. (1999). *The biology of business: Decoding the natural laws of enterprise*. Jossey-Bass.

Cockburn, A. (2001). *Writing effective use cases*. Boston: Addison-Wesley.

Collier, A. (1994). *Critical realism: An introduction to the philosophy of Roy Bhaskar*. London: Verso.

Collins, T. (2007). Academic study finds that NHS IT programme is "hampered by financial deficits, poor communication and serious delays." *Computer Weekly*. Retrieved from http://www.computerweekly.com/blogs/tony_collins/2007/05/academic-study-finds-that-nhs-1.html

Complexity at large. (2007). *Complexity, 12*(3), 3-9.

Contu, A., & Willmott, H. (2005). You spin me round: The realist turn in organization and management studies. *Journal of Management Studies, 42*(8), 1645-1662.

Cooper, R.B., & Zmud, R.W. (1990). Information technology implementation research: A technological diffusion approach. *Management Science, 36*(2), 123-139.

Córdoba, J. R., & Robson, W. D. (2003). Making the evaluation of information systems insightful: Understanding the role of power-ethics strategies. *Electronic Journal of Information Systems Evaluation, 6*(2), 55-64.

Cornford, T., & Smithson, S. (2006). *Project research in information systems.* Basingstoke: Palgrave Macmillan.

Courtney, J.F. (2001). Decision making and knowledge management in inquiring organizations: Toward a new decision making paradigm for DSS. *Decision Support Systems, 31,* 17-38.

Courtney, J.F., Croasdell, D.T., & Paradice, D.B. (1998). Inquiring organizations. *Australian Journal of Information Systems, 6*(1), 3-15. Retrieved July 10, 2007, from http://www.bus.ucf.edu/jcourtney/FIS/fis.htm

Crnkovic, I., Land, R., & Sjögren, A. (2003). Is software engineering training enough for software engineers? In *Proceedings 16th International Conference on Software Engineering Education and Training.* Madrid, March 2003. IEEE.

Cross, M. (2005, October). Public sector IT failures. *Prospect,* 48-52.

Culnan, M., & Swason, B. (1986). Research in management information systems 1980-1984: Points of work and reference. *MIS Quarterly, 10*(3), 289-302.

Curtis, B., Krasner, H., & Iscoe, N. (1988). A field study of the software design process for large systems. *Communications of the ACM, 31*(11), 1268-1287.

Curtis, P., Phillips, M., & Weszka, J. (2002). CMMI – the evolution continues. *Systems Engineering Journal, 5*(1), 7-12.

Daily Telegraph. (2007, April 25). NHS staff block reforms, says ex-minister. Retrieved from http://www.telegraph.co.uk/news/main.jhtml?xml=/news/2007/04/24/nreforms24.xml

Darier, E. (1998). Time to be lazy: Work, the environment and modern subjectivities. *Time & Society, 7*(2), 193-208.

Darier, E. (1999). Foucault and the environment: An introduction. In E. Darier (Ed.), *Discourses of the environment* (pp. 1-33). Oxford: Blackwell.

Davenport, T. H. (2000). *Mission critical: Realizing the promise of enterprise systems.* Boston, MA: Harvard Business School Press.

Davis, F. D. (1989). Perceived usefulness, perceived ease of use, and user acceptance of information technology. *MIS Quarterly, 13*(3), 319-340.

Davis, G. (1974). *Management information systems: Conceptual foundations, structure and development.* New York: McGraw-Hill.

DeLone, W. H., & McLean, E. R. (1992). Information systems success: The quest for the dependent variable. *Information Systems Research 3*(1), 60-95.

DeLone, W. H., & McLean, E. R. (2002). Information systems success revisited. *Proceedings of the 35th Hawaii International Conference on System Sciences*, Hawaii.

Delone, W. H., & McLean, E. R. (2003). The Delone and McLean model of information systems Success: A ten year update. *Journal of Management Information Systems, 19*(4), 9-30.

Demirkan, H., & Goul, M. (2006). AMCIS 2006 panel: Towards the service oriented enterprise vision: Bridging industry and academics. *CAIS, 18,* 546-556.

Denning, P. J. (2005). Recentering computer science. *Communications of the ACM, 48*(11), 15–19. doi:10.1145/1096000.1096018

Denning, P. J., & Dunham, R. (2006). Innovation as language action. *Communications of the ACM, 49*(5), 47–52. doi:10.1145/1125944.1125974

Denning, P., Comer, D., Gries, D., Mulder, M., Tucker, A., Turner, J., & Young, P. (1989). Computing as discipline. *Communications of the ACM, 32*(1), 9-23.

Denno, P., & Feeney, A. (2002). Systems engineering foundations of software systems integration. In J.-M. Bruel & Z. Bellahsène (Eds.), *OOIS 2002 Workshops*, Heidelberg, Germany (pp. 245–259). Springer. Lecture Notes in Computer Science, 2426.

Department of Defense. (2007) *Defense acquisition guidebook*. Retrieved June 12, 2007 from https://akss.dau.mil/dag/TOC_GuideBook.asp?sNode=RandExp=Y,2004

DeSanctis, G., & Poole, M. S. (1994). Capturing complexity in advanced technology use: Adaptive structuration theory. *Organization Science, 5*(2), 121-146.

Developer (2007). *Developer*, an online magazine for software developers at http://www.developerdotstar.com/mag/categories/systems_software_series.html.

Dhillon, G. (2004). Dimensions of power and IS implementation. *Information & Management, 41*, 635-644.

Dietrich, Y., Floyd, C., & Klichewski, R. (2002). *Social thinking-software practice*. Boston: MIT Press.

Dobbin, T. J., & Bustard, D. W. (1994, August). Combining soft systems methodology and object-oriented analysis: the search for a good fit. In *Proceedings of the 2nd Information Systems Methodologies Conference*.

Dobson, P. (2001). The philosophy of critical realism -- An opportunity for information systems research. *Information Systems Frontiers*.

Dobson, P. (2001a). Longitudinal case research: A critical realist perspective. *Systemic Practice and Action Research, 14*(3), 283-296.

Dobson, P. (2002). Critical realism and information systems research: Why bother with philosophy? *Information Systems Research, 7*(2). Retrieved from http://InformationR.net/ir/7-2/paper124,html.

Dobson, P. (2003). The SoSM revisited – A critical realist perspective. In Cano, J. (Ed) *Critical reflections of information systems: A systemic approach* (pp. 122-135). Hershey, PA: Idea Group Publishing.

Doherty, N., & King, M. (2001). An investigation of the factors affecting the successful treatment of organisational issues in systems development. *European Journal of Information Systems, 10*, 147-160.

Doll, W. J., & Torzadeh, G. (1988). The measurement of end-user satisfaction. *MIS Quarterly, 12*(2), 259-274.

Doolin, B. (2004). Power and resistance in the implementation of a medical management information system. *Information Systems Journal, 14*(4), 343-362.

Drucker, P. (1999). *Management challenges for the 21st century*. New York: Harper Business.

Dugmore, J. (2006, May). Benchmarking provision of IT services. *ISO Focus*, pp. 48-51.

Dunleavy, P., & Margettes, H. (2004, September 1-5). *Government IT performance and the power of the IT industry: A cross national analysis*. Paper presented the Annual meeting of American Political Science Association, Chicago.

Dutta, A. and Roy, R. (2005). Offshore outsourcing: a dynamic causal model of counteracting forces. *Journal of Management Information Systems, 22*(2), 15-35.

Economides, N. (1996), The economics of networks. *International Journal of Industrial Organization, 14*(6), 673-699.

Economides, N. (2001). The Microsoft antitrust case. *Journal Of Industry, Competition And Trade: From Theory To Policy, 1*(1), 7-39.

Economides, N., & Himmelberg, C. (1994). Critical mass and network evolution in telecommunications. *Proceedings of Telecommunications Policy Research Conference, 1-25*, Retrieved from http://ww.stern.nyu.edu/networks/site.html.

Economides, N., & White, L., J. (1996). One-way networks, two-way networks, compatibility, and antitrust. In D. Gabel & D. Weiman (Eds.), *The regulation and pricing of access*. Kluwer Academic Press.

Editorial policy statement. (2006). *MIS Quarterly, 23*(1), iii.

Einzig, P. (1966). *Primitive money* (2nd ed.). New York: Pergamon Press.

Electronic Industries Alliance. (1999). EIA Standard 632: Processes for engineering a system.

Ellis, D., Allen, D., & Wilson, T. (1999). Information science and information systems: Conjunct subjects disjunct disciplines. *Journal of the American Society for Information Science, 50*(12), 1095-1107.

Emery, M. (2000). The current version of Emery's open systems theory. *Systems Practice and Action Research, 13*(5), 623-643.

Emes, M., Smith, A., & Cowper, D. (2001). Confronting an identity crisis: How to "brand" systems engineering. *Systems Engineering, 8*(2), 164-186.

EngineeringVillage. (2007). Retrieved March 26, 2007 from http://www.engineeringvillage2.org/controller/servlet/Controller?EISESSION=1_ebf0681117f03f83d b5dses2andCID=quickSearchanddatabase=1

Eom, S. (2000). The contribution of systems science to the development of the decision support systems subspecialties: an empirical investigation. *Systems Research and Behavioral Science, 17*, 117–134. doi:10.1002/(SICI)1099-1743(200003/04)17:2<117::AID-SRES288>3.0.CO;2-E

Epstein, M., & Axtell, R. (1996). *Growing artificial societies: Social science from the bottom up.* Washington D.C.: The Brookings Institution.

Espejo R., Bowling, D., & Hoverstadt, P. (1999). The viable system model and the viplan software. *Kybernetes,* (28) Number 6/7, 661-678.

Espejo, R, and Reyes, A. (2000). Norbert Wiener. In Malcom Warner (ed), *The international encyclopedia of business and management: The handbook of management thinking.* London: Thompson Business Press.

Espejo, R. (1989). A cybernetic method to study organisations. In Espejo, R. and Harnden, R. (eds), *The viable system model: interpretations and applications of Stafford Beer's VSM.* Chichester: Wiley.

Espejo, R. (1992). Cyberfilter: A management support system. In Holtham, C.(ed.), *Executive information systems and decision support.* London: Chapman.

Espejo, R. (1993). Strategy, structure and information management. *Journal of Information Systems,* (3), 17-31.

Espejo, R. (1994). ¿What is systemic thinking? *Systems Dynamics Review,* (10), Nos. 2-3 (Summer-Fall), 199-212.

Everaert, D.N. (2006). Peirce's semiotics. In L. Hebert (dir.), *Signo* [online]. Rimouski, Quebec. Retrieved November 21, 2006, from http://www.signosemio.com.

Ewusi-Mensah, K. (1997). Critical issues in abandoned information systems development projects. *Communications of the ACM, 40*(9), 74-80.

Farbey, B., Land, F., & Targett, D. (1999). Moving IS evaluation forward: Learning themes and research issues. *Journal of Strategic Information Systems, 8*(2), 189-207.

Farhoomand, A. (1987). Scientific progress of management information systems. *Database, Summer,* 48-57.

Farhoomand, A., & Drury, D. (1999). A historiographical examination of information systems. *CAIS, 1*(19), 1-27.

Farhoomand, A., & Drury, D. (2001). Diversity and scientific progress in the information systems discipline. *CAIS, 5*(12), 1-22.

Farr, J. V. (2008, October). *System life cycle costing: economic analysis, estimation, management.* Draft textbook.

Farr, J., & Buede, D. (2003). Systems engineering and engineering management: Keys to the efficient development of products and services. *Engineering Management Journal, 15*(3), 3-9.

Farrell, J., & Katz, M. (2001). Competition or predation? Schumpeterian rivalry in network markets. (Working Paper No. E01-306). University of California at Berkeley. Retrieved from http://129.3.20.41/eps/0201/0201003.pdf

Feingenbaum, D. (1968). The engineering and management of an effective system. *Management Science, 14*(12), 721-730.

Fenzl, N. (2005). Information and self organization of complex systems. In M. Petitjean (Ed.), *Proceedings of the Third Conference on the Foundations of Information Science* (pp. 1-11). Basel, Switzerland: Molecular Diversity Preservation International.

Finnegan, P., Galliers, R. D., & Powell, P. (2002). Planning electronic trading systems: Re-thinking IS practices via triple loop learning. In S. Wrycz (Ed.), *Proceedings of the Tenth European Conference on Information Systems* (pp. 252-261). Retrieved from http://www.csrc.lse.ac.uk/asp/aspecis/20020114.pdf

Fitzgerald, B., & Howcroft, D. (1998). Towards dissolution of the IS research debate: From polarization to polarity. *Journal of Information Technology, 13*, 313-326.

Fleetwood, S. (Ed.) (1999). *Critical realism in economics: Development and debate*. London: Routledge.

Fleetwood, S., & Ackroyd, S. (Eds.). (2004). *Critical realist applications in organisation and management studies*. London: Routledge.

Flood, R. L. (1990). *Liberating systems theory*. New York: Plenum Press.

Flood, R. L., & Carson, E. R. (1988). *Dealing with complexity: An introduction to the theory and application of systems science*. New York and London: Plenum Press.

Flood, R. L., & Jackson, M. C. (1991a). Total systems intervention: A practical face to critical systems thinking. *Systems Practice, 4*, 197-213.

Flood, R. L., & Jackson, M. C. (Eds.). (1991b). *Critical systems thinking: Directed readings*. Chichester: John Wiley and Sons.

Flood, R. L., & Romm, N. R. A. (1996). *Diversity management: Triple loop learning*. Chichester: Wiley.

Flood, R., & Jackson, M. (1991). *Creative problem solving: Total systems intervention*. New York: Wiley.

Flood, R., & Romm, N. (Eds). (1996) *Critical Systems Thinking*. New York: Plenum Press.

Flood, R.L. (1999). *Rethinking the fifth discipline: Learning within the unknowable*. London: Routledge.

Fontana, M. (2006). Simulation in economics: evidence on diffusion and communication. *Journal of Artificial Societies and Social Simulation,* 9(2) (http://jasss.soc.surrey.ac.uk/9/2/8.html).

Foray, D., & Mairesse, J. (Ed.). (1999). Innovations et Performances. EDHESS Editions.

Ford, D. (2003). Trust and knowledge management: The seeds of success. In *Handbook on Knowledge Management 1: Knowledge Matters* (pp. 553-575). Heidelberg: Springer-Verlag.

Forrester, J. (1958). Industrial dynamics – A major breakthrough for decision makers. *Harvard Business Review, 36*, 37-66.

Forrester, J. (1971). Counterintuitive Behavior of Social Systems. *Technology Review, 73* (3), January.

Forrester, J. (1991). *Systems dynamics and the lessons of 35 years*. (Tech. Rep. D-4224-4). Retrieved from http://sysdyn.mit.edu/sd-group/home.html

Forrester, J. W. (1994, Summer-Fall). System dynamics, systems thinking, and soft OR. *System Dynamics Review, 10*(2-3), 245–256. doi:10.1002/sdr.4260100211

Forrester, J.W. & Senge, P.M. (1980). Tests for building confidence in system dynamics models. In AA Legasto Jr, JW Forrester and JM Lyneis (Eds), *TIMS Studies in the Management Sciences, Vol. 14: System Dynamics*. North-Holland: New York, NY, pp. 209-228.

Forrester, J.W. (1961). *Industrial Dynamics*. MIT Press: Cambridge, MA.

Forrester, J.W. (2003). Dynamic models of economic systems and industrial organizations. *System Dynamics Review, 19*(4), 331-345.

Forrester, N. (1983). Eigenvalue analysis of dominant feedback loops. In *Plenary Session Papers Proceedings of the 1st International System Dynamics Society Conference*, Paris, France: 178-202.

Forsberg, K., & Mooz, H. (1992). The relationship of systems engineering to the project life cycle. *Engineering Management Journal, 4*(3), 36–43.

Fortune, J., & Peters, G. (2005). *Information systems, achieve success by avoiding failure*. Chichester: John Wiley & Sons.

Foucault, M. (1977). *The history of sexuality volume one: The will to knowledge* (Vol. 1). London: Penguin.

Foucault, M. (1980). *Power/knowledge: Selected interviews and other writings 1972-1977*. Brighton: Harvester Press.

Foucault, M. (1980a). Truth and power. In P. Rabinow (Ed.), *The Foucault reader: An introduction to Foucault's thought* (pp. 51-75). London: Penguin.

Foucault, M. (1980b). Two lectures. In C. Gordon (Ed.), *Power/knowledge: Selected interviews and other writings Michel Foucault* (pp. 78-108). New York: Harvester Wheatsheaf.

Foucault, M. (1982a). Afterword: The subject and power. In H. Dreyfus & P. Rabinow (Eds.), *Michel Foucault: Beyond structuralism and hermeneutics* (pp. 208-226). Brighton: The Harvester Press.

Foucault, M. (1982b). On the genealogy of ethics: An overview of work in progress. In P. Rabinow (Ed.), *The Foucault reader: An introduction to Foucault's thought* (pp. 340-372). London: Penguin.

Foucault, M. (1984a). The ethics of the concern of the self as a practice of freedom (R. e. a. Hurley, Trans.). In P. Rabinow (Ed.), *Michel Foucault: Ethics subjectivity and truth: Essential works of Foucault 1954-1984* (pp. 281-301). London: Penguin.

Foucault, M. (1984b). *The history of sexuality volume two: The use of pleasure*. London: Penguin.

Foucault, M. (1984c). What is enlightenment? (C. Porter, Trans.). In P. Rabinow (Ed.), *The Foucault reader: An introduction to Foucault's thought* (pp. 32-50). London: Penguin.

Foucault, M. (1988a). Truth, power, self: An interview with Michel Foucault. In L. Martin, H. Gutman & P. Hutton (Eds.), *Technologies of the self: An interview with Michel Foucault* (pp. 9-15). Amherst: University of Massachusetts Press.

Foucault, M. (1988b). What is enlightenment? In P. Rabinow (Ed.), *The Foucault reader* (pp. 32-50). London: Penguin.

Frank, M. (2001). Engineering systems thinking: A multifunctional definition. *Systemic Practice and Action Research, 14*(3), 361-379.

Franke, M. (2001). The engineering of complex systems for the future. *Engineering Management Journal, 13*(12), 25-32.

Franz, C. R., & Robey, D. (1986). Organizational context, user involvement, and the usefulness of information systems. *Decision Sciences, 17*(3), 329-356.

Frege, G. (1952). *Translations from the philosophical writings of Gottlob Frege* (P. Geach & M. Black, Trans.). Oxford: Blackwell.

Frickel, S., & Gross, N. (2005). A general theory of scientific/intellectual movements. *American Sociological Review, 70*, 204-232.

Fuggetta, A. (2000). Software process: A roadmap. In *Proceedings of the ICSE International Conference on Software Engineering*, Limerick, Ireland (pp. 25-34). ACM Digital Library.

Gadamer, H. (1975). *Truth and method*. New York: Seabury Press.

Galliers, R. (1992). Choosing information systems research approaches. In R. Galliers (Ed.), *Information systems research: Issues, methods and practical guidelines*. Oxford: Blackwell Scientific Publications.

Galliers, R. (2004). Change as crisis or growth? Toward a trans-disciplinary view of information systems as a field of study: A response to Benbasat and Zmud's call for returning to the IT artifact. *JAIS, 4*(7), 337-351.

Gao, X., & Li, Z. (2006, December). Business process modeling and analysis using UML and polychromatic sets. *Production Planning and Control, 17*(8), 780–791. doi:10.1080/09537280600875273

Garcia, S. (1998). Evolving improvement paradigms: Capability maturity models & ISO/IEC 15504 (PDTR). *Software Process Improvement and Practice, 3*(1), 1-11.

Gardner, C. (2000). *The valuation of information technology: A guide for strategy, development, valuation, and financial planning.* Hoboken, NJ: Wiley.

Garrity, E. J., & Sanders, G. L (1998). Introduction to information systems success measurement. In E. J. Garrity & G. L. Sanders (Eds.), *Information system success measurement* (1-11). Hershey, PA: Idea Group Publishing.

Gellner, E. (1993). What do we need now? Social anthropology and its new global context. *The Times Literary Supplement, 16*, July, 3-4.

Gelman O., & Negroe, G. (1982). Planning as a conduction process. *Engineering National Academy Review, 1*(4), 253-270.

Gelman, O., & Garcia, J. (1989). Formulation and axiomatization of the concept of general system. *Outlet IMPOS (Mexican Institute of Planning and Systems Operation), 19*(92), 1-81.

Gelman, O., Mora, M., Forgionne, G., & Cervantes, F. (2005). Information systems and systems theory. In M. Khosrow-Pour (Ed.), *Encyclopedia of information science and technology* (vol. 3, pp. 1491-1496). Hershey, PA: Idea Group.

Gelman, O., Mora, M., Forgionne, G., & Cervantes, F. (2005). M. Khosrow-Pour (Ed.) Information Systems and Systems Theory. In *Encyclopedia of Information Science and Technology* (1491-1496). Hershey, PA: IGR.

Georgantzas, N.C. & Katsamakas, E. (2008). *Disruptive service-innovation strategy.* Working Paper, Fordham University, New York, NY.

Georgantzas, N.C. & Ritchie-Dunham, J.L. (2003). Designing high-leverage strategies and tactics. *Human Systems Management, 22*(1), 217-227.

Georgescu-Roegen, N. (1971). *The entropy law and the economic process.* Cambridge, MA: Harvard University Press.

Gharajedaghi, J. (1999). *Systems Thinking: Managing Chaos and Complexity: A Platform for Designing Business Architecture.* Butterworth-Heinemann: Boston MA.

Giddens, A. (1987). La constitution de la société : éléments de la théorie de la structuration. Ed. PUF.

Gigch, van J., & Pipino, L. (1986). In search of a paradigm for the discipline of information systems. *Future Computer Systems, 1*(1), 71-97.

Gill, G. (1995). Early expert systems: Where are they now? *MIS Quarterly, 19*(1), 51-81.

Glass, R. (2001). *Computing failure.com.* Upper Saddle River, NJ: Prentice Hall.

Glass, R. (2005). Never the CS and IS Twain shall meet? *IEEE Software,* 120–119.

Glass, R. (1998). *Software runaways.* London: Prentice.

Glass, R., Armes, V., & Vessey, I. (2004). An analysis of research in computing disciplines. *Communications of the ACM, 47*(6), 89-94.

Glass, R., Vessey, I., & Ramesh, V. (2002). Research in software engineering: An analysis of literature. *Information & Software Technology, 44*(8), 491-506.

Goguen, J. (1996). Formality and informality in requirements engineering. In *Proceedings of the Second IEEE International Conference on Requirements Engineering* (pp. 102-108). IEEE Computer Society Press.

Goguen, J. A. (1997). Toward a social, ethical theory of information. In G.C. Bowker, S.L. Star, W. Turner, & L. Gasser (Eds.), *Social science, technical systems, and cooperative work: Beyond the great divide* (pp. 27-56). Mahwah, NJ: Lawrence Erlbaum Associates.

Gonçalves, P., Lerpattarapong, C. and Hines, J.H. (2000). Implementing formal model analysis. In *Proceedings of the 18th International System Dynamics Society Conference*, August 6-10, Bergen, Norway.

Gonzales, R. (2005). Developing the requirements discipline: Software vs. systems. *IEEE Software*, (March/April): 59–61. doi:10.1109/MS.2005.37

Goodhue, D. L., & Thompson, R. L. (1995). Task-technology fit and individual performance. *MIS Quarterly, 19*(2), 213-236.

Gopinath, C., & Sawyer, J. E. (1999). Exploring the learning from an enterprise simulation. *Journal of Management Development, 18*(5), 477–489. doi:10.1108/02621719910273596

Gorgone, J., Gray, P., Stohr, E., Valacich, J., & Wigand, R. (2006). MSIS 2006: Model curriculum and guidelines for graduate degree programs in information systems. *CAIS, 17*, 1-56.

Goul, M., Henderson, J., & Tonge, F. (1992). The emergence of artificial intelligence as a reference discipline for decision support systems. *Decision Sciences, 11*(2), 1273-1276.

Graham, I. (1998). *Requirements engineering and rapid development*. London: Addison Wesley.

Granovetter, M. (1973). The strength of weak ties. *American Journal of Sociology, 78*, 6.

Gray, L. (1996). ISO/IEC 12207 software lifecycle processes. *Crosstalk: The Journal of Defense Software Engineering, 8*, 1-11.

Gregor, S. (2006). The nature and theory of information systems. *MIS Quarterly, 30*(3), 611–642.

Gregory, A. (2000). Problematizing participation: A critical review of approaches to participation in evaluation theory. *Evaluation, 6*(2), 179-199.

Gregory, A., & Jackson, M. C. (1992). Evaluation methodologies: A system for use. *Journal of the Operational Research Society, 43*(1), 19-28.

Gregory, W. (1996). Dealing with diversity. In R. Flood & N. Romm (Eds.), *Critical Systems Thinking* (37-61). New York: Plenum Press.

Gregory, W. J. (1992). *Critical systems thinking and pluralism: A new constellation*. Unpublished doctoral dissertation, City University, London.

Grierson, P. (1977). *The origin of money*. London: The Athlone Press.

Groff, R. (2000). The truth of the matter: Roy Bhaskar's critical realism and the concept of alethic truth. *Philosophy of the Social Sciences, 30*(3), 407-435.

Grover, V., Jeong, S. R., & Teng, J. T. C. (1995). The implementation of business process reengineering. *Journal of Management Information Systems, 12*(1), 109-129.

Gruber, T. (1995). Novemer). Toward principles for the design of ontologies used for knowledge sharing. *International Journal of Human-Computer Studies, 43*(5-6), 907–928. doi:10.1006/ijhc.1995.1081

Guba, E. G., & Lincoln, Y. S. (1989). *Fourth generation evaluation*. Newbury Park, CA: Sage Publications.

Gunaratne, S. (2003). Thank you Newton, welcome Prigogine: Unthinking old paradigms and embracing new directions. Part 1: Theoretical distinctions. *Communications, 28*, 435-455.

Habermas, J. (1974). *Theory and practice*. London: Heinemann.

Habermas, J. (1978). *Knowledge and human interests* (2nd ed.). London: Heinemann.

Habermas, J. (1984). *The theory of communicative action: Vol. 1: Reason and the rationalization of society*. London: Heinemann.

Habermas, J. (1987). *The theory of communicative action: Vol. 2: Lifeworld and system: A critique of functionalist reason*. Oxford: Polity Press.

Habermas, J. (1993). On the pragmatic, the ethical, and the moral employments of practical reason. In J. Habermas (Ed.), *Justification and application* (pp. 1-17). Cambridge: Polity Press.

Habermas, J. (2003). *Truth and justification.* Cambridge: Polity Press.

Hall, D.J., & Paradice, D.B. (2005). Philosophical foundations for a learning-oriented knowledge management system for decision support. *Decision Support Systems, 39*(3), 445-461.

Hall, D.J., Paradice, D.B., & Courtney, J.F. (2003). Building a theoretical foundation for a learning-oriented knowledge management system. *Journal of Information Technology Theory and Applications, 5*(2), 63-89.

Halvorsen, C., & Conrado, R. (2000). A taxiomatic attempt at comparing SPI frameworks. In *Proceedings of the Norsk Informatikk Konferanse '2000 (NIK'2000),* Bodø (pp. 101-116).

Hamilton, S., & Chervany, N. L. (1981a). Evaluating information system effectiveness – Part I: Comparing evaluation approaches. *MIS Quarterly, 5*(3), 55-69.

Handfield, R.B., & Melnyk, S.A. (1998). The scientific theory-building process: A primer using the case of TQM. *Journal of Operations Management, 16,* 321-339.

Handy, C. (1976). *Understanding organizations.* Aylesbury: Penguin.

HBSP. (1998). *Harvard Business review on knowledge management.* Cambridge, MA: Harvard Business School Press.

Hecht, H. (1999, March). Systems engineering for software-intensive projects. In Proceedings of the *ASSET Conference,* Dallas, Texas (pp. 1-4).

Hedström, P., & Swedberg, R. (1998). Social mechanisms: An introductory essay. In P. Hedström & R. Swedberg (Eds.), *Social mechanisms: An analytical approach to social theory* (pp. 1-31). Cambridge University Press, New York.

Heims, S. (1991). *The cybernetics group.* Cambridge, MA: MIT Press.

Heng, M., & de Moor, A. (2003). From Habermas's communicative theory to practice on the Internet. *Information Systems Journal, 13,* 331-352.

Hevner, A. R., March, S. T., Park, J., & Ram, S. (2004). Design science in information systems research. *MIS Quarterly, 28*(1), 75–105.

Hevner, A., & March, S. (2003, November). The information systems research cycle. *Computer,* pp. 111-113.

Hevner, A., March, S., Park, J., & Ram, S. (2004). Design science in information systems research. *MIS Quarterly, 21*(8), 75-105.

Highsmith, J. A. (2000). *Adaptive software development: A collaborative, approach to managing complex systems.* Dorset House.

Hill, C.W.L. & Jones, G.R. (1998). *Strategic Management: An Integrated Approach.* Houghton Mifflin: Boston, MA.

Hirschheim, R. (1985), Information systems epistemology: A historical perspective. In E. Mumford, R. Hirschheim, G. Fitzgerald, & A. Wood-Harper (Eds.), *Research methods in information systems* (pp. 1199-1214). Amsterdam: Elsevier.

Hirschheim, R. A. (1992). Information systems epistemology: An historical perspective. In R. Galliers (Ed.), *Information systems research: Issues, methods and practical guidelines.* Oxford: Blackwell Scientific Publications.

Hirschheim, R., & Klein, H. (2003). Crisis in the IS field? A critical reflection on the state of the discipline. *JAIS, 4*(10), 237-293.

Hirschheim, R., & Smithson, S. (1999). Evaluation of information systems: A critical assessment. In L. Willcocks & S. Lester (Eds.), *Beyond the IT productivity paradox* (pp. 381-409). Chichester, UK: John Wiley and Sons.

Hirschheim, R., Klein, H. K., & Lyytinen, K. (1995). *Information systems development and data modelling: Conceptual and philosophical foundations.* Cambridge: The University Press.

Hirschheim, R.A., & Klien, H.K. (1989). Four paradigms of information system development. *Social aspects of computing, 32*(10), 1199-1214

Hitchins, D.K. (2003). *Advanced systems thinking, engineering and management*. London: Attach House.

HMSO. (1993). Applying soft systems methodology to an SSADM feasibility study. *Information systems engineering library* (series). London: HMSO.

Hoffer, J., George, J., & Valachi, J. (1996). *Modern systems analysis and design*. Menlo Park, CA: Benjamin/Cummings.

Hole, E., Verma, D., Jain, R., Vitale, V., & Popick, P. (2005). Development of the ibm.com interactive solution marketplace (ISM): A systems engineering case study. *Systems Engineering, 8*(1), 78-92.

Holland, J. (1995). *Hidden order: How adaptation builds complexity*. MA: Perseus Books Reading.

Holsapple, C.W. (Ed). (2003). *Handbook on knowledge management 2: Knowledge directions*. Heidelberg: Springer-Verlag.

Holsapple, C.W. (Ed.). (2003). *Handbook on knowledge management 1: Knowledge matters*. Heidelberg: Springer-Verlag.

Holwell, S. (2000). Soft systems methodology: Other voices. *Systemic Practice and Action Research, 13*(6), 773–797. doi:10.1023/A:1026479529130

Honour, E. (2004). Understanding the value of systems engineering. In *Proceedings of the INCOSE Conference* (pp. 1-16). INCOSE.

Horton, K. S. (2000). The exercise of power and information systems strategy: The need for a new perspective. *Proceedings of the 8th European Conference on Information Systems (ECIS)*, Vienna.

Horwich, P. (1991). *Truth*. Oxford: Blackwell.

Howcroft, D., & Trauth, E. (Eds.). (2005). *Handbook of critical information systems research: Theory and application*. London: Edward Elgar.

Huber, G. (2004). *The necessary nature of future firms*. Thousand Oaks, CA: Sage Publications.

Hughes, J., & Wood-Harper, T. (1999). Systems development as a research act. *Journal of Information Technology, 14*, 83–94. doi:10.1080/026839699344764

Humphrey, W. (1998, February). Three dimensions of process improvement. Part I: Process maturity. *Crosstalk: The Journal of Defense Software Engineering*, 1-7.

Hunt, S. (2005). For truth and realism in management research. *Journal of Management Inquiry, 14*(2), 127-138.

IbisSoft. (2007) *Environment that promotes IS research*. Retrieved from http://www.ibissoft.se/english/index.htm?frameset=research_frame.htm&itemframe=/english/about_isenvironment.htm.

IEEE. (2001). *SWEBOK: Guide to the software engineering body of knowledge*. Los Alamitos, CA: IEEE Computer Society Press.

Iivari, J., & Huisman, M. (2007). The relationship between organizational culture and the deployment of systems development methodologies. *MIS Quarterly, 31*(1), 35–58.

Iivari, J., Hirschheim, R., & Klein, H. (2004). Towards a distinctive body of knowledge for information systems experts: Coding ISD process knowledge in two IS journals. *Information Systems Journal, 14*, 313–342. doi:10.1111/j.1365-2575.2004.00177.x

Iivari, J., Hirschheim, R., & Klein, H. K. (1998). A paradigmatic analysis contrasting information systems development approaches and methodologies. *Information Systems Research, 9*(2), 164-193.

INCOSE (2004, June). *INCOSE Systems engineering handbook*. (Version 2A).

INCOSE (2007). Proposing a framework for a reference curriculum for a graduate program in systems engineering. *INCOSE Academic Council Report*. Version 2007-04-30.

INCOSE. (2004). *Systems engineering handbook*. Author.

Industry Report. (2005). Facility considerations for the data center version 2.1 (White paper). APC and PAN-DUIT Companies. Retrieved July 10, 2007, from *http://www.apc.com*

Introna, L. D. (1997). *Management, information and power: A narrative of the involved manager.* Basing-stoke: Macmillan.

Ipe, M. (2003). *The praxis of knowledge sharing in organizations: A case study.* Doctoral dissertation. University Microfilms.

Irani, Z. (2002). Information systems evaluation: Navigating through the problem domain. *Information & Management, 40*, 11-24.

Irani, Z., & Fitzgerald, G. (2002). Editorial. *Information Systems Journal, 12*(4), 263-269.

Irani, Z., & Love, P. E. (2001). Information systems evaluation: Past, present and future. *European Journal of Information Systems, 10*(4), 189-203.

Irani, Z., Love, P. E., Elliman, T., Jones, S., & Them-istocleus, M. (2005). Evaluating E-government: Learn-ing from the experiences of two UK local authorities. *Information Systems Journal, 15*(1), 61-82.

Ishman, M. (1998). Measuring information success at the individual level. In E. J. Garrity & G. L. Sanders (Eds.), *Information system success measurement* (pp. 60 – 78). Hershey, PA: Idea Group Publishing.

ISO. (1995). *ISO/IEC 12207: Information technology – software life cycle processes.* Geneva: ISO/IEC.

ISO. (2002). *ISO/IEC 15288: Information technology – systems life cycle processes.* Geneva: ISO/IEC.

ISO. (2005a). *ISO/IEC 20000-1: Information technology – service management. Part 1: Specification.* Geneva: ISO/IEC.

ISO. (2005b). *ISO/IEC 20000-1: Information technol-ogy – service management. Part 2: Code of practice.* Geneva: ISO/IEC.

ISO. (2006a). *ISO/IEC 15504-5 information technol-ogy – process assessment. Part 5: An exemplar process assessment model.* Geneva: ISO/IEC.

ISO. (2006b). *ISO 9000 and ISO 14000 in plain language.* Retrieved from www.iso.org.

ISO. (2007). *Quality management principles.* Retrieved from www.iso.org.

Ives, B., Hamilton, S., & Davis, G. (1980). A framework for research in computer-based management information systems. *Management Science, 26*(9), 910-934.

Ives, B.. & Olson, M. H. (1984). User Involvement and MIS success: A review of research. *Management Sci-ence, 30*(5), 586-603.

Ives, W., Torrey, B., & Gordon, C. (2000). Knowledge sharing is human behavior. In *Knowledge management: Classic and contemporary works* (pp. 99-129).

Jackson, M. (1991). *Systems methodology for the man-agement sciences.* New York: Plenum.

Jackson, M. (2000). *Systems approaches to management.* New York: Kluwer.

Jackson, M. A. (1983). *System development.* London: Prentice Hall

Jackson, M. C. (1982). The nature of soft systems think-ing: The work of Churchman, Ackoff and Checkland. *Journal of Applied Systems Analysis, 9*, 17-29.

Jackson, M. C. (1992). An integrated programme for critical thinking in information systems research. *In-formation Systems Journal, 2*, 83-95.

Jackson, M. C. (1999). Towards coherent pluralism in management science. *Journal of the Operational Re-search Society, 50*(1), 12-22.

Jackson, M. C. (2000). *Systems approaches to manage-ment.* London: Kluwer Academic/Plenum Publishers.

Jackson, M. C. (2003). *Creative holism: Systems thinking for managers.* Chichester, UK: John Wiley and Sons.

Jackson, M. C. (2003). *Systems thinking. Creative holism for managers.* Chichester: Wiley.

Jackson, M. C. (2006). Creative holism: A critical sys-tems approach to complex problem situations. *Systems Research and Behavioral Science, 23*(5), 647–657. doi:10.1002/sres.799

Jackson, M. C., & Keys, P. (1984). Towards a system of system methodologies. *Journal of the Operational Research Society, 35,* 473-486.

Jacucci, E., Hanseth, O., & Lyytinen, K. (2006). Introduction: Taking complexity seriously in IS research. *Information Technology & People, 19*(1), 5-11.

James, W. (1976). *The meaning of truth.* Cambridge, MA: Harvard University Press.

Jamshidi, M. (2005). System-of-systems engineering - A definition. *Proceedings of IEEE System*, Man, and Cybernetics (SMC) Conference. Retrieved January 29, 2005 from http://ieeesmc2005.unm.edu/SoSE_Defn.htm

Jan, T. S., & Tsai, F. L. (2002, January - February). A systems view of the evolution in information systems development. *Systems Research and Behavioral Science, 19*(1), 61–75. doi:10.1002/sres.441

Jasperson, J. S., Carte, T., Saunders, C. S., Butler, B. S., Croes, H. J. P., & Zheng, W. (2002). Power and information technology research: A metatriangulation review. *MIS Quarterly, 26*(4), 397-459.

Johnson, K., & Dindo, J. (1998). Expanding the focus of software process improvement to include systems engineering. *Crosstalk: The Journal of Defense Software Engineering*, 1-13.

Johnson, P., & Lancaster, A. (2000). *Swarm users guide.* Swarm Development Group.

Johnson, R., Kast, F., & Rosenzweig, J. (1964). Systems theory and management. *Management Science, 10*(2), 367-384.

Johnstone, D., & Tate, M. (2004). Bringing human information behaviour into information systems research: An application of systems modelling. *Information Research, 9*(4), 1-31.

Kagan. (1982). The dates of the earliest coins. *American Journal of Archaeology, 86*(3), 343-360.

Kampmann, C.E. (1996). Feedback loops gains and system behavior. In *Proceedings of the 12th International System Dynamics Society Conference*, July 21-25, Cambridge, MA.

Kannelis, P., Lycett, M., & Paul, R. J. (1998). An interpretive approach to the measurement of information systems success. In E. J. Garrity & G. L. Sanders (Eds.), *Information system success measurement* (pp. 133-151). Hershey, PA: Idea Group Publishing.

Kanter, R. M. (1996). When a thousand flowers bloom: Structural, collective, and social conditions for innovation in organizations. In P. S. Myers (Ed.), *Knowledge management and organizational design.* Newton, MA: Butterworth-Heinemann.

Kanungo, S. (2003). Using system dynamics to operationalize process theory in information systems research. *Proceedings of the 24th International Conference on Information Systems*, 450-463.

Kappelman, L. A. (1995). Measuring user involvement: A diffusion of innovation perspective. *Data Base, 26.*

Karakatsios, K.Z. (1990). *Casim's user's guide.* Nicosia, CA: Algorithmic Arts.

Katsamakas, E. & Georgantzas, N. (2008). *Open source disruptive innovation strategies.* Working Paper, Fordham University, New York, NY.

Kauffman, S. (1995). *At home in the universe: The search for laws of self-organization and complexity.* Oxford University Press.

Kaufman, J. J., & Woodhead, R. (2006). *Stimulating innovation in products and services.* Hoboken, NJ: Wiley.

Keating, C. et al. (2003). System of systems engineering. *Engineering Management Journal, 15*(3), 36-45.

Keating, C., Rogers, R., Unal, R., Dryer, D., Sousa-Poza, A., Safford, R., Peterson, W., & Rabadi, G. (2003). System of systems engineering. *Engineering Management, 15*(3), 36-45.

Kellner, M., Curtis, B., deMarco, T., Kishida, K., Schulemberg, M., & Tully, C. (1991). Non-technological issues in software engineering. In *Proceedings of the 13th ICSE International Conference on Software Engineering* (pp. 149-150). ACM Digital Library.

Kettinger, W. J., Teng, J. T. C., & Guha, S. (1997). Business process change: A study of methodologies, techniques and tools. *MIS Quarterly, 21*(1), 55-88.

Kim, D. (1993). The Link between individual and organizational learning. *Sloan Management Review*, Fall, 37-50.

Kim, R., & Kaplan, S. (2006). Interpreting socio-technical co-evolution: Applying complex adaptive systems to IS engagement. *Information Technology & People, 19*(1), 35-54.

King, W. R., & Rodriguez, J.I. (1978). Evaluating management information systems. *MIS Quarterly, 2*(3), 43-51.

Kitchenham, B., et al. (2002). Preliminary guidelines for empirical research in software engineering. *IEEE Transactions on Software Engineering, 28*(8), 721-734.

Klein, H. (2004). Seeking the new and the critical in critical realism: Deja vu? *Information and Organization, 14*(2), 123-144.

Klein, H. K., & Myers, M. D. (1999). A set of principles for conducting and evaluating interpretive field studies in information systems. *MIS Quarterly, 23*(1), 67–93. doi:10.2307/249410

Klein, H., & Huynh, M. (2004). The critical social theory of Jurgen Habermas and its implications for IS research. In J. Mingers & L. Willcocks (Eds.), *Social theory and philosophy for information systems* (pp. 157-237). Chichester: Wiley.

Klir, G. (1969). *An approach to general systems theory*. New York: Van Nostrand.

Kljajić, M. (1994). *Theory of system*. Kranj, Slovenia: Moderna organizacija.

Kljajić, M., Bernik, I., & Škraba, A. (2000). Simulation approach to decision assessment in enterprises. *Simulation, 75*, 199–210. doi:10.1177/003754970007500402

Koch, C. (2006, January). The ABCs of ERP, *CIO Magazine*.

Koizumi, T. (1993). *Interdependence and change in the global system*. Lanham, New York, London: University Press of America.

Kristensen, J.K., & Buhler, B. (2001, March). The hidden threat to e-government. Avoiding large government IT failures. *OECD Public Management Policy Brief, Policy Brief no. 8*.

Kroes, P., Franssen, M., van de Poel, I., & Ottens, M. (2006). Treating socio-technical systems as engineering systems: Some conceptual problems. *Systems Research and Behavioral Science, 23*(6), 803–814. doi:10.1002/sres.703

Kruchten, P. (2005). Casting software design in the Function-Behavior-Structure framework. *IEEE Software*, (March-April): 52–58. doi:10.1109/MS.2005.33

Krueger, R. A. (1988). *Focus groups: A practical guide for applied research*. Newbury Park, CA: Sage Publications.

Kuhn, T. (1970). *The structure of scientific revolutions*. Chicago: Chicago University Press.

Kuhn, T. S. (1970). *The structure of scientific revolutions* (2nd ed.). University of Chicago Press: Chicago.

Kwon, T., & Zmud, R. (1987). Unifying the fragmented models of information systems implementation. In J. Boland & R. Hirshheim (Eds.), *Critical issues in information systems research* (pp. 227-251). New York: John Wiley.

Land, F., & Kennedy-McGregor, M. (1987). Information and information systems: Concepts and perspectives. In R. Galliers (Ed.), *Information analysis: Selected readings* (pp. 63-91). Sydney: Wesely.

Lane, J. (2005). *System of systems lead system integrators: Where do they spend their time and what makes them more/less efficient*. (Tech. Rep. No. 2005-508.) University of Southern California Center for Systems and Software Engineering, Los Angeles, CA.

Lane, J. (2006). *COSOSIMO Parameter Definitions*. (Tech. Rep. No. 2006-606). University of Southern California Center for Systems and Software Engineering, Los Angeles, CA.

Lane, J. A., & Boehm, B. (2007). System-of-systems cost estimation: Analysis of lead system integrator engineering activities. *Information Resources Management Journal, 20*(2), 23–32.

Lane, J.-A., Petkov, D., & Mora, M. (2008). Software engineering and the systems approach: A conversation with Barry Boehm. *International Journal of Information Technologies and Systems Approach, 1*(2), 99–103.

Langefors, B. (1995). *Essays of infology* (B. Dahlbom, ed). Lund: Studentlitteratur.

Langenwalter, G. A. (2000). *Enterprise resource planning and beyond: Integrating your entire organization.* Boca Raton, FL: CRC Press, Taylor and Francis.

Lansing, S. (2006). *Perfect order: Recognizing complexity in Bali.* Princeton University Press.

Larses, O., and El-Khoury, J. (2005). *Review of Skyttner (2001)* in O. Larses and J. El-Khoury, "Views on General Systems Theory." Technical Report TRITA-MMK 2005:10, Royal Institute of Technology, Stockholm, Sweden. Retrieved June 30, 2006 on the World Wide Web: http://apps.md.kth.se/publication_item/web. phtml?ss_brand=MMKResearchPublications&department_id='Damek'

Larson, T., & Levine, J. (2005). Serching for Management Information Systems: Coherence and Change in the Discipline. *Information Systems Journal, 15*, pp. 357-381.

Lauer, T.W. (2001). Questions and information: Contrasting metaphors. *Information Systems Frontiers, 3(*1), 41–48.

Laware, J., Davis, B., & Peruisch, K. (2006). Systems thinking: A paradigm for professional development. *The International Journal of Modern Engineering, 6*(2).

Lawson, T. (1997). *Economics and reality.* Routledge: London.

Layder, D. (1993). *New strategies in social research: An introduction and guide.* Cambridge: Polity Press.

Lazanski, T. J., & Kljajić, M. (2006). Systems approach to complex systems modeling with special regards to tourism. *Kybernetes, 35*(7-8), 1048–1058. doi:10.1108/03684920610684779

Lazlo, E., & Krippner, S. (1998). In J.S. Jordan (Ed.), *Systems theories and a priori aspects of perception* (47-74). Amsterdam: Elsevier Science.

Lazlo, E., & Lazlo, A. (1997). The contribution of the systems sciences to the humanities. *Systems Research & Behavioral Science, 14*(1), 5-19.

Leavitt, H., & Whisler, T. (1958). Management in the 80's. *Harvard Business Review, 36*(6), 41-48.

Lee, A. (2000). Systems thinking, design science and paradigms: Heeding three lessons from the past to resolve three dilemmas in the present to direct a trajectory for future research in the information systems field. Retrieved July 10, 2007, from *http://www.people.vcu.edu/~aslee/ ICIM-keynote-2000/ICIM-keynote-2000.htm*

Lee, A. (2004). Thinking about social theory and philosophy for information systems. In J. Mingers & L. Willcocks (Eds.), *Social theory and philosophy for information systems* (pp. 1-26). Chichester: Wiley, Chichester.

Lee, B., & Miller, J. (2004). Multi-project software engineering analysis using systems thinking. *Software Process Improvement and Practice, 9*(3). doi:10.1002/spip.204

Legge, K. (1984). *Evaluating planned organizational change.* London: Academic Press.

LeMoigne, J. L. (1977). La théorie du système général: Théorie de la modélisation. Editions PUF.

Leonard, D., & Straus, S. (1998). Putting your company's whole brain to work. In *Harvard Business Review on Knowledge Management* (pp. 109-136). Boston: Harvard Business School Press.

Leontief, W., et al. (1953). *Studies in the structure of the American economy.* New York: Oxford University Press.

Lewis, P. (1994). *Information systems development: Systems thinking in the field of information systems.* London: Pitman Publishing.

Lewis, P. (1995). New challenges and directions for data analysis and modeling. In F. Stowell (Ed.), *Information systems provision: The contribution of soft systems methodology* (pp. 186-205). London: McGraw-Hill.

Lewis, P. J. (1995). Applying soft systems methodology to an SSADM feasibility study: CCTA. *Systems Practice, 8*(3), 337–340. doi:10.1007/BF02250482

Li, Z. B., & Xu, L. D. (2003, May). Polychromatic sets and its application in simulating complex objects and systems. *Computers & Operations Research, 30*(6), 851–860. doi:10.1016/S0305-0548(02)00038-2

Liang, Y., West, D., & Stowell, F. A. (1997). An interpretivist approach to IS definition using object modelling. *Information Systems Journal, 8*, 63–180.

Liao, S., Chang, J., Shih-chieh, C., & Chia-mei, K. (2004). Employee relationship and knowledge sharing: A case study of a Taiwanese finance and securities firm. *Knowledge Management Research & Practice, 2*, 24-34.

Liebowitz, S. J. (2002). *Rethinking the networked economy: The true forces driving the digital marketplace.* New York: Amacom Press.

Liebowitz, S. J., & Margolis, S. E. (1994). *Network externality: An uncommon tragedy. Journal of Economic Perspectives, 19*(2), 219-234.

Liebowitz, S. J., & Margolis, S. E. (2002). *Winners, losers & Microsoft.* Oakland, CA: The Independent Institute.

Linstone, H. A. (1984). *Multiple perspectives for decision making. Bridging the gap between analysis and action.* New York: North Holland.

Lissack, M.R. (1999). Complexity: The science, its vocabulary, and its relation to organizations. *Emergence, 1*(1), 110-126.

Lissack, M.R., & Roos, J. (2000). *The next common sense: The e-managers guide to mastering complexity.* London: Nicholas Brealey Publishing.

Liyanage, J. P., & Kumar, U. (2003). Towards a value-based view on operations and maintenance performance management. *Journal of Quality in Maintenance Engineering, 9*(4), 1355-2511.

Locke, J. (1894). An essay concerning human understanding. In A. C. Fraser (Ed.), *vol. 1.* Oxford: Clarendon.

Longshore Smith, M. (2006). Overcoming theory-practice inconsistencies: Critical realism and information systems research *Information and Organization, 16*(3), 191-211.

Lucas, H. C. (1973). A descriptive model of information systems in the context of the organization. *Data Base, 5*(2), 27-36.

Lukes, S. (1974). *Power: A radical view.* London: Macmillan.

Lynch, T., & Gregor, S. (2004). User participation in decision support systems development: Influencing system outcomes. *European Journal of Information Systems, 13*, 286–301. doi:10.1057/palgrave.ejis.3000512

Lyytinen, K. & Rose, G. (2003a). The disruptive nature of Information Technology innovations: the case of internet computing in systems development organizations. *MIS Quarterly, 27*(4), 557-595.

Lyytinen, K. & Rose, G. (2003b). Disruptive information system innovation: the case of internet computing. *Information Systems Journal, 13*, 301-330.

Lyytinen, K. (1988). Stakeholders, information system failures and soft systems methodology: An assessment. *Journal of Applied Systems Analysis, 15*, 61–81.

Lyytinen, K., & Hirschheim, R. (1987). Information systems failures - A survey and classification of the empirical literature. *Oxford Surveys in Information Technology, 4*, 257-309.

Lyytinen, K., & Robey, D. (1999). Leaving failure in information systems development. *Information Systems Journal, 9*, 85–101. doi:10.1046/j.1365-2575.1999.00051.x

Madachy, R. (1996). Modeling software processes with system dynamics: Current developments. In *Proceedings of the System Dynamics Conference* (pp. 1-23). Cambridge, MA: Systems Dynamic Society.

Madachy, R. J. (2006). Integrated modeling of business value and software processes. In *Unifying the software process spectrum* (LNCS 3840, pp. 389-402).

Madachy, R. J. (2008). *Software process dynamics.* Wiley. IEEE Press.

Madachy, R. J., Boehm, B., & Lane, J. A. (2007). Software lifecycle increment modeling for new hybrid processes. In *Software Process Improvement and Practice.* Wiley. Retreived from http://dx.doi.org/10.1002/spip.332

Magee, C., & de Weck, O. (2004). Complex system classification. In *Proceedings of the 14th Annual INCOSE International Symposium of the International Council on Systems Engineering* (pp. 1-18). INCOSE.

Maier, M. (1998). Architecting principles for systems-of-systems. *Systems Engineering, 1*(4), 267-284.

Manicas, P. (1993). Accounting as a human science. *Accounting, Organizations and Society, 18,* 147- 161.

Manthorpe, W. (1996). The emerging joint system of systems: A systems engineering challenge and opportunity for APL. *John Hopkins APL Technical Digest, 17*(3), 305-313.

March, S., & Smith, G. (1995). Design and natural science research on information technology. *Decision Support Systems, 15,* 251-266.

Markus, M. L. (2002). Power, politics and MIS implementation. In M. Myers & D. Avison (Eds.), *Qualitative research in information systems.* London: Sage.

Markus, M. L., & Robey, D. (1988). Information technology and organizational change: Casual structuring in theory and research. *Management Science, 34*(5), 583-598.

Markus, M.L. and Mao, J.Y. (2004). Participation in development and implementation – updating an old, tired concept for today's IS contexts. *Journal of the Association for Information Systems, 5* (11: 14).

Martin, D. D., (1988). Technological Change and Manual Wiork. In D. Gallie (Ed.), *Employment in Britain* (102-127). Oxford: Basil Balckwell.

Martin, J. N. (1996). *Systems engineering guidebook: A process for developing systems and products.* Boca Ration, FL: CRC Press.

Mason, R., & Mitroff, I. (1973). A program of research on MIS. *Management Science, 19*(5), 475-485.

Mason, R., & Mitroff, I. (1981). *Challenging strategic planning assumptions.* New York: Wiley.

Mathieu, R. G. (2002). Top-down approach to computing. *IEEE Computer, 35*(1), 138–139.

McBride, N. (2005). Chaos theory as a model for interpreting information systems in organizations. *Information Systems Journal, 15*(3), 233-254.

McKeen, J. D., & Guimares, T. (1997). Successful strategies for user participation in systems development. *Journal of Management Information Systems, 14*(2), 33-150.

McLeod, R. (1995). Systems theory and information resources management: integrating key concepts. *Information Resources Management Journal, 8*(2), 5–14.

MConnell, S. (1996). *Rapid development.* Microsoft Press Books.

Meadows, D.H. (1989). System dynamics meets the press. *System Dynamics Review, 5*(1), 68-80.

Meadwell, H. (1994). The foundations of Habermas's universal pragmatics. *Theory and Society, 23*(5), 711-727.

MEDIX. (2004). *Medix Survey for the BBC of Doctor's Views about the National Programme for IT.* Retrieved from www.medix-uk.com

Menezes, W. (2002, February). To CMMI or not to CMMI: Issues to think about. *Crosstalk: The Journal of Defense Software Engineering,* 1-3.

Merali, Y. (2005, July). Complexity science and conceptualisation in the Internet enabled world. Paper presented at the *21st Colloquium of the European Group for Organisational Studies,* Berlin, Germany.

Mertens, D. (1999). Inclusive evaluation: Implications of transformative theory for evaluation. *American Journal of Evaluation, 20*(1), 1-14.

Mesarović, M. D., & Takahara, Y. (1989). *Abstract systems theory.* Springer-Verlag.

Metcalfe, M. (2004, November). Generalisation: Learning across epistemologies]. *Forum: Qualitative Social Research, 6*(1), Art. 17. Retrieved March 10, 2007, from: http://www.qualitative-research.net/fqs-texte/1-05/05-1-17-e.htm.

Metcalfe, M., & Powell, P. (1995). Information: A perceiver-concerns perspective. *European Journal of Information Systems, 4*, 121-129.

Michigan Tech (n.d.). Retrieved December 11, 2008 from http://www.sse.mtu.edu/sse.html

Midgley, G. (1996). What is this thing called CST? In R. Flood & N. Romm (Eds.), *Critical systems thinking* (11-24). New York: Plenum Press.

Midgley, G. (1996). What is this thing called CST? In R. L. Flood & N. Romm (Eds.), *Critical Systems Thinking: Current Research and Practice* (pp. 11-24). New York: Plenum Press.

Midgley, G. (1997). Mixing methods: Developing systemic intervention. In J. Mingers & A. Gill (Eds.), *Multimethodology: The Theory and Practice of Combining Management Science Methodologies.* (pp. 249-290). Chichester, UK: John Wiley and Sons.

Midgley, G. (2000). *Systemic intervention: Philosophy, methodology and practice.* New York: Kluwer Academic/ Plenum.

Midgley, G. (2003). Science as systemic intervention: Some implications of systems thinking and complexity for the philosophy of science. *Systemic Practice and Action Research, 16*(2), 77-97.

Miller, J. G. (1978). *Living systems.* New York: McGraw-Hill.

Miller, W. L., & Morris, L. (1999). *Fourth generation R&D: Managing knowledge, technology and innovation.* Hoboken, NJ: Wiley.

Minar, N., Burkhar, R., Langton C., & Askemnazi, M. (1996). *The Swarm Simulation System: A toolkit for building multi-agent simulations.* Retrieved from http:// alumni.media.mit.edu/~nelson/research/ swarm/.

Mingers, J. (1980, April). Towards an appropriate social theory for applied systems thinking: Critical theory and soft systems methodology. *Journal of Applied Systems Analysis, 7*, 41-50.

Mingers, J. (1984). Subjectivism and soft systems methodology: A critique. *Journal of Applied Systems Analysis, 11*, 85-113.

Mingers, J. (1992). Technical, practical and critical OR: Past, present and future? In M. Alvesson & H. Willmott (Eds.), *Critical management studies* (pp. 90-112). London: Sage.

Mingers, J. (1995). Information and meaning: Foundations for an intersubjective account. *Information Systems Journal, 5*, 285-306.

Mingers, J. (1995). Using soft systems methodology in the design of information systems. In F. Stowell (Ed.), *Information systems provision: The contribution of soft systems methodology* (pp. 18-50). London: McGraw-Hill.

Mingers, J. (1996). An evaluation of theories of information with regard to the semantic and pragmatic aspects of information systems. *Systems Practice, 9*(3), 187–209.

Mingers, J. (1997). Towards critical pluralism. In J. Mingers & A. Gill (Eds.), *Multimethodology: Theory and practice of combining management science methodologies* (pp. 407-440). Chichester: Wiley.

Mingers, J. (2000). The contributions of critical realism as an underpinning philosophy for OR/MS and systems. *Journal of the Operational Research Society 51*,1256-1270.

Mingers, J. (2000). The contributions of critical realism as an underpinning philosophy for OR/MS and systems. *Journal of the Operational Research Society, 51*, 1256-1270.

Mingers, J. (2000). What is it to be critical? Teaching a critical approach to management undergraduates. *Management Learning, 31*(2), 219-237.

Mingers, J. (2001). Combining IS research methods: Towards a pluralist methodology. *Information Systems Research, 12*(3), 240-253.

Mingers, J. (2001). Multimethodology- mixing and matching methods. In J. Rosenhead & J. Mingers (Eds), *Rational analysis for a problematic world revisited.* Chichester: Wiley.

Mingers, J. (2002). Realizing information systems: Critical realism as an underpinning philosophy for information systems. In *Proceedings Twenty-Third International Conference on Information Systems, 295-303.*

Mingers, J. (2003a). A classification of the philosophical assumptions of management science methods. *Journal of the Operational Research Society, 54*(6), 559-570.

Mingers, J. (2003b). *Information, knowledge and truth: A polyvalent view* (Working Paper No. 77). Canterbury: Kent Business School.

Mingers, J. (2004). Realizing information systems: Critical realism as an underpinning philosophy for information systems. *Information and Organization, 14*(2), 87-103.

Mingers, J. (2004a). Paradigm wars: Ceasefire announced, who will set up the new administration? *Journal of Information Technology, 19,* 165-171.

Mingers, J. (2004b). Re-establishing the real: Critical realism and information systems research. In J. Mingers & L. Willcocks (Eds.), *Social theory and philosophical for information systems* (pp. 372-406). London: Wiley.

Mingers, J. (2005). 'More dangerous than an unanswered question is an unquestioned answer': A contribution to the Ulrich debate. *Journal of the Operational Research Society, 56*(4), 468-474.

Mingers, J. (2006). *Realising systems thinking: Knowledge and action in management science.* New York: Springer.

Mingers, J., & Brocklesby, J. (1997). Multimethodology: Towards a framework for mixing methodologies. *Omega, International Journal of Management Science, 25*(5), 489-509.

Mingers, J., & Gill, A. (1997). *Multimethodology: The theory and practice of combining management science methodologies.* Chichester, UK: John Wiley & Sons Ltd.

Mingers, J., & Gill, A. (Eds.). (1997). Multi methodology. Chichester: Wiley.

Mingers, J., & Gill, A. (Eds.). (1997). *Multimethodology: Theory and practice of combining management science methodologies.* Chichester: Wiley.

Minnich, H. (2002). EIA IS 731 compared to CMMI-SE/SW. *Systems Engineering Journal, 5*(1), 62-72.

Mirani, R., & Lederer, A.L. (1998): An instrument for assessing the organizational benefits of IS projects. *Decision Sciences, 29*(4), 803-823.

Mises, L. V. (1912). *The theory of money and credit.* (H. E. Batson, Trans.). Indianapolis: Liberty Fund.

Mitev, N., & Wilson, M. (2004). What we may learn from the social shaping of technology approach. In J. Mingers & L. Willcocks (Eds.), *Social theory and philosophy for information systems* (pp. 329-371). Chichester: Wiley.

Mitroff, I., & Linstone, H. (1993). *The unbounded mind.* New York: Oxford University Press.

Mojtahedzadeh, M.T. (1996). *A Path Taken: Computer-Assisted Heuristics for Understanding Dynamic Systems.* Ph.D. Dissertation. Rockefeller College of Public Affairs and Policy, SUNY: Albany NY.

Mojtahedzadeh. M.T., Andersen, D. & Richardson, G.P. (2004). Using *Digest®* to implement the pathway participation method for detecting influential system structure. *System Dynamics Review, 20*(1), 1-20.

Monarch, I. (2000). Information science and information systems: Converging or diverging? In *Proceedings of the 28th Annual Conference of the Canadian Association in Information Systems*, Alberta.

Monod, E. (2004). Einstein, Heisenberg, Kant: Methodological distinctions and conditions of possibility. *Information and Organization. , 14*(2), 105-121.

Mora, M., & Gelman, O. (2008). The case for conceptual research in information systems. Paper submitted to the 2008 International Conference on Information Resources Management (Conf-IRM), Niagara Falls, Ontario, Canada (pp. 1-10).

Mora, M., Cervantes, F., Gelman, O., Forgionne, G., & Cano, J. (2006b). The interaction of systems engineering, software engineering and information systems disciplines: A system-oriented view. In *Proceedings of the 18th International Conference on Systems Research, Informatics and Cybernetics*, Baden-Baden, Germany (pp. 1-5).

Mora, M., Forgionne, G., Cervantes, F., Garrido, L., Gupta, J., & Gelman, O. (2005). Toward a comprehensive framework for the design and evaluation of intelligent decision-making support systems (i-DMSS). *Journal of Decision Systems, 14*, 321–344. doi:10.3166/jds.14.321-344

Mora, M., Forgionne, G., Gupta, J., Cervantes, F., & Gelman, O. (2003). A framework to assess intelligent decision-making support systems, knowledge-based intelligent information and engineering systems (LNAI 2774 (PT 2), pp. 59-65).

Mora, M., Gelman, Cervantes, F., Mejia, M., & Weitzenfeld, A. (2003). A systemic approach for the formalization of the information system concept: Why information systems are systems. In J. Cano (Ed.), *Critical reflections of information systems: A systemic approach* (pp. 1-29). Hershey, PA: Idea Group.

Mora, M., Gelman, O., Cano, J., Cervantes, F. & Forgionne, G. (2006a). Theory of systems and information systems research frameworks. Proceedings from the *50th Annual Meeting of the International Society for the Systems Sciences*, paper 2006-282, (1-7), Somona State University, CA.

Mora, M., Gelman, O., Cervantes, F., Mejia, M., & Weitzenfeld, A. (2003). A systemic approach for the formalization of the information system concept: Why information systems are systems? In J. Cano (Ed.), *Critical reflections of information systems: A systemic approach* (pp. 1-29). Hershey, PA: Idea Group Publishing.

Mora, M., Gelman, O., Forgionne, G., & Cervantes, F. (2004, May 19-21). Integrating the soft and the hard systems approaches: A critical realism based methodology for studying soft systems dynamics (CRM-SSD). In *Proceedings of the 3rd. International Conference on Systems Thinking in Management (ICSTM 2004)*, Philadelphia, PA.

Mora, M., Gelman, O., Forgionne, G., & Cervantes, F. (in press). Information Systems: a systemic view. M. Khosrow-Pour (Ed.) In *Encyclopedia of information science and technology, 2nd ed.* Hershey, PA: IGR.

Mora, M., Gelman, O., Forgionne, G., Petkov, D., & Cano, J. (2007). Integrating the fragmented pieces of is research paradigms and frameworks: A systems approach. *Information Resources Management Journal, 20*(2), 1-22.

Mora, M., Gelman, O., Frank, M., Cervantes, F., & Forgionne, G. (2008). Toward an interdisciplinary engineering and management of complex IT-intensive organizational systems: A systems view. *International Journal of Information Technologies and the Systems Approach, 1*(1), 1-24.

Mora, M., Gelman, O., Macias, J. & Alvarez, F. (2007c). The management and engineering of IT-intensive systems: a systemic oriented view. Proceedings from the *2007 IRMA International Conference*, (1448-1453), Vancouver, BC, Canada: Idea Group.

Mora, M., Gelman, O., O'Connor, R., Alvarez, F., & Macías, J. (2007a). On models and standards of processes in SE, SwE and IT&S disciplines: Toward a comparative framework using the systems approach. In K. Dhanda & R. Hackney (Eds.), *Proceedings of the ISOneWorld 2007 Conference*, , Engaging Academia and Enterprise Agendas, Las Vegas, USA (pp. 49/1-18).

Mora, M., Gelman, O., O'Connor, R., Alvarez, F., & Macías, J. (2007b). A systemic model for the description and comparison of models and standards of processes in SE, SwE, and IT disciplines. In *E-Proceedings of the International Conference on Complex Systems 2007*, NECSI, Boston, MA, USA (pp. 1-8).

Mora, M., Gelman, O., O'Connor, R., Alvarez, F., & Macías, J. (2007c). An overview of models and standards of processes in SE, SwE, and IS disciplinesIn A. Cater-Steel (Ed.), *Information technology governance and service management: Frameworks and adaptations* (pp. 1-20). Hershey, PA: IGI Global. 1-20.

Mora, M., Gelman, O., O'Connor, R., Alvarez, F., & Macías, J. (2007b). On models and standards of processes in SE, SwE and IT&S disciplines: Toward a comparative framework using the systems approach. In K. Dhanda & R. Hackney (Eds.), *Proceedings of the ISOneWorld 2007 Conference: Engaging Academia and Enterprise Agendas*. Information Institute. Track in System Thinking/Systems Practice, Las Vegas, USA, April 11-13, 2007, paper-49, 1-18.

Morecroft, J.D.W. (1985). Rationality in the analysis of behavioral simulation models, *Management Science, 31*, 900-916.

Morris & Travis. (2001, February 16). Straw Calls Halt to £80m IT System. *The Times*.

Morris, C. (1938). Foundations of the theory of signs. In O. Neurath (Ed.), *International Encyclopedia of United Science, 1*(2). Chicago: University of Chicago Press.

Moser, I., & Law, J. (2006). Fluids or flows? Information and qualculation in medical practice. *Information Technology & People, 19*(1), 55-73.

Mouratidis, H., Giorgini, P., & Manson, G. (2003). Integrating security and systems engineering: Towards the modeling of secure information systems. *Advanced Information Systems Engineering* (LNCS 2681, pp. 63-78). Retrieved from http://www.sse.mtu.edu/sse.html

Mulder, M., et al. (1999). ISCC'99: An information systems-centric curriculum '99: program guidelines for educating the next generation of information systems specialists, in collaborating with industry. Retrieved July 10, 2007, from *http://www.iscc.unomaha.edu*

Mumford, E. (1995). *Effective requirements analysis and systems design: The ETHICS method*. Basingstoke: Macmillan.

Mumford, E., & Henshall, D. (1979). *A participative approach to computer system design*. London: Associated Business Press.

Mumford, E., Hirschheim, R., Fitzgerald, G., & Wood-Harper, T. (Eds.). (1985). *Research methods in information systems*. Amsterdam: North Holland.

Murray, S. (2003). *A quantitative examination to determine if knowledge sharing activities, given the appropriate richness lead to knowledge transfer, and if implementation factors influence the use of these knowledge sharing activities*. Doctoral dissertation. University Microfilms.

Mutch, A. (1999). Critical realism, managers and information. *British Journal of Management, 10*, 323-333.

Mutch, A. (1999). Information: A critical realist approach. In T. Wilson & D. Allen (Eds.), *Proceedings of the 2nd Information Seeking in Context Conference* (pp. 535-551). London: Taylor Graham.

Mutch, A. (2000). Managers and information: Agency and structure. *Information Systems Review, 1*.

Mutch, A. (2002). Actors and networks or agents and structures: Towards a realist view of information systems. *Organization, 9*(3), 477-496.

Mutch, A. (2005). Critical realism, agency and discourse: Moving the debate forward. *Organization, 12*(5), 781-786.

Myers, B. L., Kappelman, L.A., & Prybutok, V. R. (1998). A comprehensive model for assessing the quality and productivity of the information systems success: Toward a theory for information systems assessment. In E. J. Garrity & G. L. Sanders (Eds.), *Information system success measurement* (pp. 94-121).

Myers, M. (2004). The nature of hermeneutucs. In J. Mingers & L. Willcocks (Eds.), *Social theory and philosophy for information systems* (pp. 103-128). Chichester: Wiley.

Myers, M.D. (1994). A disaster for everyone to see: An interpretive analysis of a failed project. *Accounting, management and information technology, 4*(4), 185-210.

National Aeronautics and Space Administration (NASA). (1995, June). *Systems Engineering Handbook*. SP-610S.

Nauta, D. (1972). *The meaning of information*. Mouton: The Hague.

Nellhaus, T. (1998). Signs, social ontology, and critical realism. *Journal for the Theory of Social Behaviour, 28*(1), 1–24.

Nelson, H. (1994). The necessity of being "undisciplined" and "out-of-control:" Design action and systems thinking. *Performance Improvement Quarterly, 7*(3), 22-29.

Ngwenyama, O. K., & Lee, A. S. (1997). Communication richness in electronic mail: critical social theory and the contextuality of meaning. *MIS Quarterly, 21*(2), 145–167. doi:10.2307/249417

Niiniluoto, I. (2002). *Critical scientific realism.* Oxford: Oxford University Press.

Nolan, R., & Wetherbe J. (1980). Toward a comprehensive framework for MIS research. *MIS Quarterly, June,* 1, 1-20.

Nonaka, I., & Konno, N. (1998). The concept of Ba: Building a foundation for knowledge creation. *California Management Review (Special Issue on Knowledge and the Firm), 40*(3), 40-54.

Nonaka, I., & Takeuchi, H. (1995). *The knowledge-creating company: How Japanese companies create the dynamics of innovation.* New York: Oxford University Press.

Nunamaker, J., Chen, T., & Purdin, T. (1991). Systems development in information systems research. *Journal of Management Information Systems, 7*(3), 89-106.

O'Connor, J., & McDermott, I. (1997). *The art of systems thinking.* San Francisco: Thorsons.

O'Dell, C. (2008). *Web 2.0 and knowledge management.* American Productivity & Quality Center (APQC). Retrieved from http://www.apqc.org.

O'Neill, T. (1995). Implementation frailties of Guba and Lincoln's fourth generation evaluation theory. *Studies in Educational Evaluation, 21*(1), 5-21.

Odlyzko, A. (2001). Internet growth: Myth and reality, use and abuse. *Journal of Computer Resource management, 102,* Spring, pp. 23-27.

OECD. (2004). Highlights of the OECD information technology outlook 2004. Retrieved July 10, 2007, from *http://www.oecd.org*

OGC. (2007). *The official introduction to the ITIL service lifecycle.* London: TSO.

Oliga, J. (1996). *Power, ideology and control.* New York: Plenum.

Oliva, R. & Mojtahedzadeh, M.T. (2004). Keep it simple: a dominance assessment of short feedback loops. In *Proceedings of the 22nd International System Dynamics Society Conference,* July 25-29, Keble College, Oxford University, Oxford UK.

Oliva, R. (1994). *A Vensim Module to Calculate Summary Statistics for Historical Fit.* MIT System Dynamics Group D-4584.

Oliva, R. (2004). Model structure analysis through graph theory: partition heuristics and feedback structure decomposition. *System Dynamics Review, 20*(4), 313-336.

Oliva, R., Sterman, J.D. & Giese, M. (2003). Limits to growth in the new economy: exploring the 'get big fast' strategy in e-commerce. *System Dynamics Review, 19*(2), 83-117.

Orlikowski, W. (1992). The duality of technology: Rethinking the concept of technology in organizations. *Organization Science, 3*(3), 398-427.

Orlikowski, W. J. (1993). CASE tools as organizational change: Investigating Incremental and Radical changes in systems development. *MIS Quarterly, 17*(3), 309-340.

Orlikowski, W. J. (1996). Improvising organizational transformation over time: A situated change perspective. *Information Systems Research, 7*(1), 63-92.

Orlikowski, W.J. (2000). using technology and constituting structures: A practice lens for studying technology in organizations. *Organizations Science, 11*(4), 404-428.

Orlikowski, W. J., & Hofman, D. (1997). An improvisional model of change management: The case of groupware technologies. *Sloan Management Review, 38*(2), 11-21.

Orlikowski, W. J., & Iacono, C.S. (2001). Research commentary: Desperately seeking the "IT" in IT research – A call to theorizing the IT artifact. *Information Systems Research, 12* (2), 121-134.

Orlikowski, W., & Iacono, S. (2001). Desperately seeking the IT in IT research. *Information Systems Research, 7*(4), 400-408.

Ormerod, R. (1996). Information systems strategy development at Sainsbury's supermarket using "soft" OR. *Interfaces, 16*(1), 102-130.

Ormerod, R. (2005). Putting soft OR methods to work: the case of IS strategy development for the UK Parliament. *Journal of the Operational Research Society, 56*(12), 1379-1398.

Outhwaite, W. (1987). *New philosophies of social science: Realism, hermeneutics, and critical theory.* New York: St. Martin's Press.

Paich, M. & Sterman, J.D. (1993). Boom, bust and failures to learn in experimental markets. *Management Science, 39*(12), 1439-1458.

Parker, M. M., Benson, R., & Trainor, H. E. (1988). *Information economics: Linking business performance to information technology.* Englewood Cliffs, NJ: Prentice Hall.

Pather, S., & Remenyi, D. (2004). *Some of the philosophical issues underpinning research on information systems: From positivism to critical realism.* Paper presented at the SAICSIT 2004, Prague.

Paulk, M. (1995, January). How ISO 9001 compares with CMM. *IEEE Software*, 74-83.

Paulk, M. (1998). *ISO 12207, ISO 15504, SW-CMM v1.1, SW-CMM v2 Draft C Mapping 1.* SEI/CMU.

Paulk, M. (1999). Analyzing the conceptual relationship between ISO/IEC 15504 (software process assessment) and the capability maturity model for software. In *Proceedings of the 1999 International Conference on Software Quality*, Cambridge, MA (pp. 1-11).

Pawson, R., Greenhalgh, T., Harvey G., & Walshe K. (2004). Realist synthesis: An introduction (RMP Methods Paper 2/2004). Manchester, ESRC Research Methods Programme, University of Manchester.

Pawson, R., Greenhalgh, T., Harvey, G., & Walshe K. (2005). Realist review – A new method of systematic review designed for complex policy interventions. *Journal of Health Services Research & Policy 10*(1), 21–34.

Peirce, C. (1878, January). How to make our ideas clear. *Popular Science Monthly.*

Peirce, C. S. (1931). The essential Peirce: Selected philosophical writings. *The Peirce Edition Project 1998.* Indiana University Press.

Peirce, C.S. (1931-1958). *The collected papers of C.S. Peirce.* C. Hartshorne & P. Weiss (Eds.). (1931-1935). Vol. I-VI. A.W. Burks (Ed.). (1958). Vol. VII-VIII. Cambridge, MA: Harvard University Press.

Perkins, J. (2002). Education in process systems engineering: Past, present and future. *Computers & Chemical Engineering, 26*, 283–293. doi:10.1016/S0098-1354(01)00746-3

Petkov, D., & Petkova, O. (2008). The work system model as a tool for understanding the problem in an introductory IS Project. *Information Systems Education Journal, 6*(21).

Petkov, D., Misra, R., & Petkova, O. (2008). Some suggestions for further diffusion of work system method ideas in systems analysis and design. *CONISAR/ISECON Proceedings*, Phoenix, November.

Petkov, D., Petkova, O., Andrew, T., & Nepal, T. (2007). Mixing multiple criteria decision making with soft systems thinking techniques for decision support in complex situations. *Decision Support Systems, 43*, 1615–1629. doi:10.1016/j.dss.2006.03.006

Petkova, O., & Roode, J. D. (1999). An application of a framework for evaluation of factors affecting software development productivity in the context of a particular organizational environment. *South African Computing Journal, 24*, 26–32.

Pettigrew, A. M., (1990). Longitudinal field research on change: Theory and practice. *Organization Science, 1*(3), 267-292.

Pfleeger, S. L. (2008). *Software engineering theory and practice.* Upper Saddle River, NJ: Prentice Hall.

Piccoli, G., & Ives, B. (2005). IT-dependent strategic initiatives and sustained competitive advantage: A review and synthesis of the literature. *MIS Quarterly, 29*(4), 747-776.

Poper, K. (1973). *The logic of scientific discovery.* (Nolit, Belgrade, Trans.) (Original work published 1968, London, Hutchinson, 3rd Edition).

Popper, K. (1959). *The logic of scientific discovery.* London: Hutchinson.

Popper, K. (1969). *Conjectures and refutations.* London: Routledge and Kegan Paul.

Porra, J., Hirschheim, R., & Parks, M. (2005). The history of Texaco's corporate information function: A general systems theoretical interpretation. *MIS Quarterly, 29*(4), 721-746.

Porter, M., & Millar, V. (1985). How information gives you a competitive advantage. *Harvard Business Review, 63*(4), 149-160.

Pratt, A. (1995). Putting critical realism to work: The practical implications for geographical research. *Progress in Human Geography 19*(1), 61-74.

Pressman, J., & Wildavsky, A. (1973). *Implementation: How great expectations in Washington are dashed in Oakland.* Oakland, CA: University of California Press.

Pressman, R. (2009). *Software Engineering, A practitioner's approach* (7th ed.). New York: McGraw Hill.

Prewitt, E., & Cosgrove, L. (2006). *The state of the CIO 2006* (CIO Report pp. 1-8). Retrieved July 10, 2007, from *http://www.cio.com/state*

Prigogine, I. (1997). *The end of certainty: Time, chaos, and the new laws of nature.* New York: The Free Press.

Probert, S. (2004). Adorno: A critical theory for IS. In J. Mingers & L. Willcocks (Eds.), *Social theory and philosophy for information systems* (pp. 129-156). Chichester: Wiley.

Putnam, H. (1981). *Reason, truth, and history.* Cambridge: Cambridge University Press.

QAA. (2007). *Subject benchmark statements: Computing.* Quality Assurance Agency, UK. Retrieved from http://www.qaa.ac.uk/academicinfrastructure/benchmark/statements/

Quine, W. (1992). *Pursuit of truth.* Boston: Harvard University Press.

Ramsey, F. (1927). Facts and propositions. *Proceedings of the Aristotelian Society, 7.*

Rapoport, A. (1968). Systems Analysis: General systems theory. *International Encyclopedia of the Social Sciences, 14*, 452-458.

Rashid, A., Meder, D., Wiesenberger, J., & Behm, A. (2006). Visual requirement specification in end-user participation. *MERE workshop 2006.* IEEE. Retrieved from http://www.collabawue.de/cms/dokumente/Rashid_et_al_visual_requirement_specification_%20in_end-user_participation.pdf

Raynor, M.E. (2007). *The Strategy Paradox.* Currency-Doubleday: New York, NY.

Reed, M. (2001). Organization, trust and control: A realist analysis. *Organization Studies, 22*(2), 201-223.

Reed, M. (2005). Reflections on the "realist turn" in organization and management studies. *Journal of Management Studies, 42*(8), 1621-1644.

Reifer, D. (2003). Is the software engineering state of the practice getting closer to the state of the art? *IEEE Software, 20*(6), 78–83. doi:10.1109/MS.2003.1241370

Remenyi, D., & Sherwood-Smith, M. (1999). Maximise information systems value by continuous participative evaluation. *Logistics Information Management, 12*(1/2), 145-156.

Repenning, N.P. (2002). A simulation-based approach to understanding the dynamics of innovation implementation. *Organization Science,* 13(2), 109-127.

Repenning, N.P. (2003). Selling system dynamics to (other) social scientists. *System Dynamics Review,* 19(4), 303-327.

Richardson, G.P. (1991). *Feedback Thought in Social Science and Systems Theory.* University of Pennsylvania Press: Philadelphia, PA.

Richardson, G.P. (1995). Loop polarity, loop prominence, and the concept of dominant polarity. *System Dynamics Review,* 11(1), 67-88.

Richardson, S.M., Courtney, J.F., & Haynes, J.D. (2006). Theoretical principles for knowledge management systems design: Application to pediatric bipolar disorder. *Decision Support Systems, 42,* 1321-1337.

Richmond, B. (1993). Systems thinking: critical thinking skills for the 1990s and beyond. *System Dynamics Review,* 9(2), 113-133.

Richmond, B. (1994, July). System dynamics/systems thinking: Let's just get on with it. In *Proceedings of the 1994 International System Dynamics Conference,* Sterling, Scotland. Retrieved from http://www.hps-inc.com/st/paper.html

Richmond. B. et al. (2006). *iThink® Software (version 9).* iSee Systems™: Lebanon NH.

Rittel, H., & Weber, M. (1973). Dilemmas in a general theory of planning. *Policy Sciences, 4,* 155-169.

Rittel, H.W.J., & Webber, M.M. (1973). Dilemmas in a general theory of planning. *Policy Sciences, 4,* 155-169.

Robey, D., & Boudreau, M. (1999). Accounting for the contradictory organizational consequences of information technology: Theoretical directions and methodological implications. *Information Systems Research,* 10(2), 167-185.

Robey, D., & Markus, M.L. (1984). Ritual in information systems design. *MIS Quarterly, 8,* 5-15

Roedler, G. (2006). *ISO/IEC JTC1/SC7: Status and plans of alignment of ISO/IEC 15288 and ISO/IEC 12207.* Retrieved from www.15288.com.

Rorty, R. (1982). *Consequences of pragmatism.* Minnesota University Press.

Rosenhead, J. (1989). *Rational analysis for a problematic world.* West Sussex: John Wiley.

Rouse, W. B. (1992). *Strategies for innovation: Creating successful products, systems and organizations.* Hoboken, NJ: Wiley.

Rouse, W. B. (2001). *Essential challenges of strategic management.* Hoboken, NJ: Wiley.

Rouse, W. B. (2005). A theory of enterprise transformation. *Systems Engineering, 8*(4) 279-295.

Rouse, W. B. (Ed.). (2006). *Enterprise transformation: Understanding and enabling fundamental change.* Hoboken, NJ: Wiley.

Rouse, W.B. (2007). *Complex engineered, organizational and natural systems: Issues underlying the complexity of systems and fundamental research needed to address these issues* (Report submitted to the Engineering Directorate, National Science Foundation, Washington, DC).

Rowlinson, M., & Carter, C. (2002). Foucault and history in organization studies. *Organization, 9*(4), 527-547.

Russell, B. (1912). *The problems of philosophy.* Oxford: Oxford University Press.

Ryan, S., & Porter, S. (1996). Breaking the boundaries between nursing and sociology: A critical realist ethnography of the theory-practice gap. *Journal of Advanced Nursing, 24,* 413-420.

Saaty, T. (1990). *Multicriteria decision making - The analytic hierarchy process* (2nd ed.). Pittsburgh: RWS Publications.

Saga, V. L., & Zmud, R.W. (1996). Introduction de logiciels de gestion dans des petites entreprises liées à une profession libérale. *Systèmes d'Information et Management, 1*(1), 51-73.

Sage, A. (2000, May). System engineering education. *IEEE TSMC Part C-Applications and Reviews, 30*(2), 164–174.

Sage, A. P. (1995). *Systems management for information technology and software engineering.* Hoboken, NJ: John Wiley & Sons.

Sage, A. P. (2000). Transdisciplinarity perspectives in systems engineering and management. In M. A. Somerville & D. Rapport (Eds.) *Transdisciplinarity: Recreating integrated knowledge* (158-169), Oxford, U.K.: EOLSS Publishers Ltd.

Sage, A. P. (2005). *Systems of systems: Architecture based systems design and integration.* Keynote address. International Conference on Systems, Man and Cybernetics, Hawaii, USA.

Sage, A. P. (2006). The intellectual basis for and content of systems engineering. *INCOSE INSIGHT, 8*(2) 50-53.

Sage, A. P., & Rouse, W. B. (Eds.). (1999). *Handbook of systems engineering and management.* Hoboken, NJ: John Wiley and Sons.

Sage, A., & Cuppan, C. (2001). On the systems engineering and management of systems of systems and federations of systems. *Information, Knowledge, Systems Management, 2*, 325-345.

Saleem, N. (1996). An empirical test of the contingency approach to user participation information systems development. *Journal of Management Information Systems, 13*(1), 145-166.

Samaras, G. M., & Horst, R. L. (2005). A systems engineering perspective on the human-centered design of health information systems. *Journal of Biomedical Informatics, 38*, 61–74. doi:10.1016/j.jbi.2004.11.013

Sastry, M.A. (1997). Problems and paradoxes in a model of punctuated organizational change. *Administrative Science Quarterly, 42*(2), 237-275.

Saunders, C. S., & Jones, J. W. (1992). Measuring performance of the information systems function. *Journal of Management Information Systems, 8*(4), 63-82.

Savage, A., & Mingers, J. (1996). A framework for linking soft systems methodology (SSM) and Jackson system development (JSD). *Information Systems Journal, 6*, 109–129. doi:10.1111/j.1365-2575.1996.tb00008.x

Sayer, A. (1985). Realism in geography. In R. J. Johnston (Ed.), *The future of geography* (pp. 159-173). London: Methuen.

Sayer, A. (2000). *Realism and social science.* London: Sage Publications Ltd.

Sayer, R. A. (1992). *Method in social science: A realist approach.* Routledge: London.

Scarbrough, H., & Corbett, J. M. (1992). *Technology and organization: Power, meaning and design.* Routledge: London.

Schmandt-Besserat, D. (1992). *Before writing: From counting to cuneiform* (Vols. I-II). Austin, TX: University of Texas Press.

Schneider, M., & Somers, M. (2006). Organizations as complex adaptive systems: Implications of complexity theory for leadership research. *The Leadership Quarterly, 17*, 351-365.

SEI. (2003). *SEI annual report 2003.* Pittsburgh: Software Engineering Institute.

Senge, P. (1994). *The fifth discipline: The art and practice of the learning organization.* Doubleday.

Serafeimidis, V., & Smithson, S. (1999). Rethinking the approaches to information systems evaluation. *Logistics Information Management, 12*(1-2), 94-107.

Serafeimidis, V., & Smithson, S. (2003). Information systems evaluation as an organizational institution - Experience from a case study. *Information Systems Journal, 13*, 251-274.

Sewchurran, K., & Petkov, D. (2007). A systemic framework for business process modeling combining soft systems methodology and UML. *Information Resources Management Journal, 20*(3), 46–62.

Shannon, C. E. (1948). A mathematical theory of communication. *The Bell System Technical Journal, 27*, 379–423, 623–656.

Shapiro, C., & Varian, H. R. (1999). *Information rules – A strategic guide to the network economy.* Boston, MA: Harvard Business School Press.

Sheard, S., & Lake, J. (1998). Systems engineering and models compared. Retrieved from www.software.org.

Sheffield, J. (2004). The design of gss-enabled interventions: A Habermasian perspective. *Group Decision and Negotiation, 13*(5), 415-435.

Shenhar, A. (1994, February). Systems engineering management: A framework for the development of multidisciplinary discipline. *IEEE TSMC, 24*(2), 327–332.

Shenhar, A., & Bonen, Z. (1997). The new taxonomy of systems: Toward an adaptive systems engineering framework. *IEEE Transactions on Systems, Man and Cybernetics (Part A: Systems and Humans), 27*(2), 137-145.

Silver, M., Markus, M., & Beath, C. (1995) The Information Techonlogy Interation Model: a Foundation of the MBA Course. *MIS Quarterly, 19(3),* pp. 361-369.

Simon, H. (1967). *Model of man.* (5th printing) John Wiley and Sons.

Simon, H. A. (1962). The architecture of complexity. *Proceedings of the American Philosophical Society, 106,* 467–469.

Simon, H. A. (1969). *The sciences of the artificial.* Cambridge, MA: MIT Press.

Škraba, A., Kljajić, M., & Borštnar Kljajić, M. (2007, January). The role of information feedback in the management group decision-making process applying system dynamics models. *Group Decision and Negotiation, 16*(1), 77–95. doi:10.1007/s10726-006-9035-9

Škraba, A., Kljajić, M., & Leskovar, R. (2003). Group exploration of system dynamics models - Is there a place for a feedback loop in the decision process? *System Dynamics Review, 19*(3), 243–263. doi:10.1002/sdr.274

Skyttner, L. (2001). *General systems theory.* Singapore: World Scientific Publishing

Small, C. (2006). *An enterprise knowledge-sharing model: A complex adaptive systems perspective on improvement in knowledge sharing.* Doctoral dissertation, George Mason University. University Microfilms.

Small, C., & Sage, A. (2006). Knowledge management and knowledge sharing: A review. *Information, Knowledge, and Systems Management, 5*(3), 153-169.

Smith, M. L. (2005). *Overcoming theory-practice inconsistencies: Critical realism and information systems research.* Unpublished manuscript, Department of Information Systems, London School of Economics and Political Science, working paper 134.

Smithson, S., & Tsiavos, P. (2004). Re-constructing information systems evaluation. In C. Avgerou, C. Ciborra & F. Land (Eds.), *The social study of information and communication technology: Innovation, actors and contexts* (pp. 207-230). Oxford: Oxford University Press.

Software Engineering Institute (2001). *Capability maturity model integration* (CMMI) (Special report CMU/ SEI-2002-TR-001). Pittsburgh, PA: Software Engineering Institute.

Software Engineering Institute (SEI). (1996). *IDEAL^{SM}: A user's guide for software process improvement* (Handbook CMU/SEI-96-HB-001). Pittsburgh, PA: Software Engineering Institute, Carnegie Mellon University.

Software Engineering Institute. (2002). CMMI for systems engineering and software engineering. (CMU/ SEI-2002-TR-001). Retrieved from www.sei.edu.

Sommerville, I. (1998). Systems engineering for software engineers. *Annals of Software Engineering, 6,* 111-129.

Sommerville, I. (2007). *Software Engineering* (8th ed.). Harlow: Pearson.

Sowa, J., & Zachman, J. (1992). Extending and formalizing the framework for information systems architecture. *IBM Systems Journal, 31*(3), 590-616.

Sowa, J.F. (2000). *Knowledge representation – logical, philosophical and computational foundations.* Pacific Grove, CA: Brooks/Cole.

Spasser, M. A. (2002). Realist activity theory for digital library evaluation: Conceptual framework and case study. *Computer Supported Cooperative Work, 11*, 81-110.

Spohrer, J., Maglio, P., Bailey, J., & Gruhl, D. (2007, January). Steps toward a science of service systems. *IEEE Computer*, 71-77.

Stamper, R. (1997). Organisational semiotics. In J.M. Mingers & F.A.Stowell (Eds.), *Information systems: An emerging discipline?* (pp. 267-283). Maidenhead: McGraw Hill.

Standish Group International. (2003). *The Extreme CHAOS Report*. Retrieved from www.standish-group.com.

Sterman, J. (2001). Systems dynamic modeling: Tools for learning in a complex world. *California Management Review, 43*(4), 8-25.

Sterman, J.D. (1984). Appropriate summary statistics for evaluating the historical fit of system dynamics models. *Dynamica, 10*(Winter), 51-66.

Sterman, J.D. (1989). Modeling managerial behavior: misperceptions of feedback in a dynamic decision making experiment. *Management Science*, 35(3), 321-339.

Sterman, J.D. (1994). Beyond training wheels. In *The Fifth Discipline Fieldbook*, Senge P. et al, Currency-Doubleday: New York, NY.

Sterman, J.D. (2000). *Business Dynamics: Systems Thinking and Modeling for a Complex World*. Irwin McGraw-Hill: Boston, MA.

Stevens Institute of Technology (2007). *SDOE 625 - Class notes for systems design and operational effectiveness*.

Stevens Institute of Technology. (2008). *Systems Engineering Research Center*. Retrieved from http://www.stevens.edu/sercuarc/

Stones, R. (1996). *Sociological reasoning: Towards a past-modern sociology*. MacMillan.

Stowell, F. (1995). *Information systems provision: The contribution of soft systems methodology*. London: McGraw-Hill.

Stowell, F. A. (1985). Experiences with SSM and data analysis. *Information technology training* (pp. 48-50).

Stowell, F. A. (1991). Towards client-led development of information systems. *Journal of Information Systems, 1*, 173–189. doi:10.1111/j.1365-2575.1991.tb00035.x

Stowell, F. A., & Cooray, S. (2006) Client led information systems creation (CLIC) reality or fantasy? *15th International Conference on Information System Development Methods and Tools and Theory and Practice*, Budapest Hungary. August 30-September 1, 2006.

Stowell, F. A., & Cooray, S. (2006, August 9). Addressing IS failure - Could a "systems" approach be the answer? *Special Symposium on Information Systems Research and Systems Approach*. Baden Baden, Germany.

Stowell, F. A., & West, D. (1994). *Client-led design: A systemic approach to information systems definition*. Maidenhead: McGraw-Hill.

Stowell, F., & Mingers, J. (1997). Introduction. In J. Mingers, & F. Stowell (Eds.), *Information systems: An emerging discipline?* London: McGraw-Hill.

Stowell, F., & West, D. (1996). Systems thinking, information systems practice. In *Proc. 40th Conference of the Int. Society For Systems Sciences*.

Straub, D.W., Limayem M., & Karahanna-Evaristo, E. (1995). Measuring system usage: Implications for IS theory testing. *Management Science, 41*(8), 1328-1342.

Strawson, P. (1950). Truth. *Proceedings of the Aristotelian Society, 24*.

Šuštersič, O., Rajkovič, V., Leskovar, R., Bitenc, I., Bernik, M., & Rajkovič, U..(2002, May-June). An information system for community nursing. *Public Health Nursing (Boston, Mass.), 19*(3), 184–190. doi:10.1046/j.0737-1209.2002.19306.x

Swanson, E. B., & Ramiller, N. C. (1997). The organizing vision in information systems innovation. *Organization Science, 8*(5), 458-474.

Swanson, G. A. (1993). *Macro accounting and modern money supplies*. Westport, CT: Quorum Books.

Swanson, G. A. (1998). Governmental justice and the dispersion of societal decider subsystems through exchange economics. *Systems Research and Behavioral Science, 15*, 413–420. doi:10.1002/(SICI)1099-1743(1998090)15:5<413::AID-SRES268>3.0.CO;2-F

Swanson, G. A. (2006). A systems view of the environment of environmental accounting. *Advances in Environmental Accounting and Management, 3*, 169–193. doi:10.1016/S1479-3598(06)03006-8

Swanson, G. A., & Miller, J. G. (1989). *Measurement and interpretation: A living systems theory approach.* New York: Quorum Books.

Swanson, G. A., Bailey, K. D., & Miller, J. G. (1997). Entropy, social entropy, and money: A living systems theory perspective. *Systems Research and Behavioral Science, 14*, 45–65. doi:10.1002/(SICI)1099-1743(199701/02)14:1<45::AID-SRES151>3.0.CO;2-Y

Swarm Development Group. (2004). *Chris Langton, Glen Ropella.* Retrieved from http://wiki.swarm.org/.

SWEBOK. (2004). *Software engineering body of knowledge.* Defined by the IEEE CS and ACM, http://www.swebok.org/ironman/pdf/SWEBOK_Guide_2004.pdf

Sweeney, A., & Bustard, D. W. (1997). Software process improvement: making it happen in practice. *Software Quality Journal, 6*, 265–273. doi:10.1023/A:1018572321182

Symons, V., & Walsham, G. (1988). The evaluation of information systems: A critique. *Journal of Applied Systems Analysis, 15*, 119-132.

Tarski, A. (1944). The semantic conception of truth. *Philosophy and Phenomenological Research, 4*, 341-375.

Tashakkori, A., & Teddlie, C. (1998). *Mixed methodology: Combining qualitative and quantitative approaches.* London: Sage Publications.

Tashakkori, A., & Teddlie, C. (2003). *Handbook of mixed methods in social and behavioural research.* Thousand Oaks, CA: Sage.

Taylor, C. (1984). Foucault on freedom and truth. *Political Theory, 12*(2), 152-183.

Teboul, J. (2007). Service is front stage: Positioning services for value advantage. Paris: INSEAD Business Press.

Thayer, R. (1997). Software systems engineering: An engineering process. In R. Thayer & M. Dorfan (Eds.), *Software requirements engineering* (pp. 84-109). Los Alamitos, CA: IEEE Computer Society Press.

Thayer, R. (2002, April). Software systems engineering: A tutorial. *IEEE Computer*, pp. 68-73.

Theil, H. (1966). *Applied Economic Forecasting.* Elsevier Science (North Holland): New York, NY.

Thomé, B. (1993). *Systems engineering: Principles and practice of computer-based systems engineering.* Chichester: John Wiley and Sons.

Tirole, J. (1988). *The Theory of Industrial Organization.* MIT Press: Cambridge, MA.

Toulmin, S. (1964). *The uses of argument.* Cambridge: Cambridge University Press.

Tsang, E., & Kwan, K., (1999). Replication and theory development in organizational science: A critical realist perspective, Academy of Management Review (24:4), 759-780

Tuomi, I. (1999). Data is more than knowledge: Implications of the reversed knowledge hierarchy for knowledge management and knowledge memory. *Journal of Management Information Systems, 16*(3), 103-117.

Tushman, M.L. & Anderson, P. (1986). Technological discontinuities and organizational environments. *Administrative Science Quarterly, 31*, 439-465.

Ulrich, W. (1983). *Critical heuristics of social planning: A new approach to practical philosophy.* Berne: Haupt.

Ulrich, W. (1998). *Systems thinking as if people mattered: Critical systems thinking for citizens and managers.* Working paper No.23, Lincoln School of Management.

Ulrich, W. (2001). A philosophical staircase for information systems definition, design and development: A discursive approach to reflective practice in ISD (Part

1). *The Journal of Information Technology Theory and Application, 3*(3), 55–84.

Ulrich, W. (2003). Beyond methodology choice: critical systems thinking as critically systemic discourse. *Journal of the Operational Research Society, 54*(4), 325-342.

United States Air Force Scientific Advisory Board (2005). *Report on system-of-systems engineering for Air Force capability development.* (Public Release SAB-TR-05-04). Washington, DC: HQUSAF/SB.

United States Air Force. (1974, May 1) *Military standard - engineering management.* MIL-STD-499A.

Unsworth, K. (2001). Unpacking creativity. *Academy of Management Review, 26*, 289-297.

Valerdi, R (2005). *The constructive systems engineering cost model (COSYSMO).* Unpublished doctoral dissertation, University of Southern California, Los Angeles.

Valerdi, R., & Madachy, R. (2007). Impact and contributions of MBASE on software engineering graduate courses. *Journal of Systems and Software, 80*(8), 1185–1190. doi:10.1016/j.jss.2006.09.051

Valero-Silva, N. (1996). A Foucauldian reflection on critical systems thinking. In R. L. Flood & N. Romm (Eds.), *Critical systems thinking: Current research and practice.* (pp. 63-79.). London: Plenum.

van Bon, J., Pieper, M., & van deer Veen, A. (2006). *Foundations of IT service management, based in ITIL.* San Antonio, TX: Van Haren.

Varzi, A. (2004). Mereology. In E. N. Zalta (Ed.), *The Stanford Encyclopedia of Philosophy (Fall 2004 Edition).* Retrieved from http://plato.stanford.edu/archives/fall2004/entries/mereology/

Vega-Romero, R. (1999). *Care and social justice evaluation: A critical and pluralist approach.* Hull: University of Hull.

Veryzer, R.W. (1998). Discontinuous innovation and the new product development process. *Journal of Product Innovation Management, 15*(2), 136-150.

Vessey, I., Ramesh, V., & Glass, R. (2002). Research in information systems: An empirical study of diversity in the discipline and its journals. *Journal of Management Information Systems, 19*(2), 129-174.

Vickers, G. (1965). *The art of judgment.* London: Chapman & Hall.

Vickers, G. (1981). The poverty of problem solving. *Journal of Applied Systems Analysis, 8*, 15–22.

Vickers, G. (1983). *The art of judgement.* London: Harper Rowe.

Vickers, G. (Ed.). (1991). *Rethinking the Future.* New Brunswich: Transaction Publishers.

Vilet, V. (2000). *Software engineering: Principles and practices.* Wiley.

Vo, H. V., Chae, B., & Olson, D. L. (2006). Integrating systems thinking into IS education. *Systems Research and Behavioral Science, 23*(1), 107–122. doi:10.1002/sres.720

Vo, H. V., Paradice, D., & Courtney, J. (2001). Problem formulation in inquiring organizations: A multiple perspectives approach. In *Proceedings of the Seventh Americas Conference on Information Systems*, Boston, MA.

Von Bertalanffy, L. (1968). *General systems theory.* George Braziller, NY Editions.

von Bertalanffy, L. (1972, December). The history and status of general systems theory. *Academy of Management Journal*, pp. 407-426.

Wad, P. (2001, August 17-19). *Critical realism and comparative sociology.* Draft paper for the IACR conference.

Wainwright, S. P. (1997). A new paradigm for nursing: The potential of realism. *Journal of Advanced Nursing, 26*, 1262-1271

Wallace, L., & Keil, M. (2004). Software project risks and their effect on outcomes. *Communications of the ACM, 47*(4), 68-73.

Walsham, G. (1991). Organisational metaphors and information systems research. *European Journal of Information Systems, 1*, 83. doi:10.1057/ejis.1991.16

Walsham, G. (1993). *Interpreting information systems in organisations*. Chichester, UK: John Wiley and Sons.

Walsham, G. (1997). Actor-network theory and IS research: Current status and future prospects. In A. Lee, J. Liebenau, & J. DeGross (Eds.), *Information systems and qualitative research* (pp. 466-480). London: Chapman Hall.

Walsham, G. (1999). Interpretive evaluation design for information systems. In L. Willcocks & S. Lester (Eds.), *Beyond the IT productivity paradox* (pp. 363-380). Chichester, UK: John Wiley and Sons.

Walsham, G., & Waema, T. (1994). Information systems strategy and implementation: A case study of a building society. *ACM Transactions on Information Systems, 12*(2), 150-173.

Wand, Y., & Weber, R. (1990). An ontological model of an information system. *IEEE Transactions on Software Engineering, 16*(11), 1282-1292.

Wand, Y., & Woo, C. (1991). An approach to formalizing organizational open systems concepts. *ACM SIGOIS Bulletin, 12*(2-3), 141-146. Retrieved from www.acm.org.

Watts, D. (2003). Six degrees: The science of a connected age. New York: W.W. Norton & Co.

Weber, R. (1987). Toward a theory of artifacts: A paradigmatic basis for information systems research. *Journal of Information Systems, 2*, 3-19.

Webster. (1986). *Webster's ninth new collegiate dictionary*. Springfield, MA: Merriam-Webster, Inc.

Weick, K., & Sutcliffe, K. (2001). *Managing the unexpected: Assuring high performance in an age of complexity*. Jossey-Bass Publishers.

Weinberg, G. (1992). *Quality Software Management, Volume 1: Systems Thinking*. Dorset House Publishing, New York

Weinberg, G. (2007). a site for books, articles and courses at http://www.geraldmweinberg.com/

Weiss, C. (1970). The politicization of evaluation research. *Journal of Social Issues, 26*(4), 57-68.

Weiss, C. (1998). Have we learned anything new about the use of evaluation? *American Journal of Evaluation, 19*(1), 21-34.

Weiss, P. (1971). *Hierarchically organized systems in theory and practice*. New York: Hafner.

Weitemeyer, M. (1962). *Some aspects of hiring of workers in the Sippar region at the time of Hammurabi*. Copenhagen: Munksgaard.

Wendorff, P. (2002). Systems Thinking In Extreme Programming. In *Proceedings of the Tenth European Conference on Information Systems*, 203-207, available at http://www.csrc.lse.ac.uk/asp/aspecis/20020124.pdf

West, D. (1991). *Towards a subjective knowledge elicitation methodology for the development of expert systems*. Unpublished PhD thesis, University of Portsmouth.

West, D., & Stowell, F. A. (2000). Models and diagrams and their importance to information systems analysis and design. In D.W. Bustard, P. Kawalek, & M.T. Norris (Eds.), *Systems modelling for business process improvement* (pp. 295-31). London: Artech.

West, D., Liang, Y., & Stowell, F. A. (1996). Identifying, selecting and specifying objects in object-oriented analysis: An interpretivist approach. In J.M.Ward (Ed.), *Proceedings of the 1st UKAIS Conference*. Cranfield University.

Wheatley, M. (1992, 2006). *Leadership and the new science: Discovering order in an age of chaos*. Berrett-Koehler Publishers.

White, L., & Tacket, A. (1996). The End of Theory? *Omega, 24*(1), 47–56. doi:10.1016/0305-0483(95)00048-8

Wiebe, R., Compton, D., & Garvey, D. (2006, June). A system dynamics treatment of the essential tension between C2 and self-synchronization. In *Proceedings of the International Conference on Complex Systems*,

New England Complex Systems Institute, Boston, Massachusetts.

Wiener, N. (1948). *Cybernetics: Or control and communication in the animal and the machine.* New York: Wiley.

Wiener, N. (1948). *Cybernetics.* John Wiley and Sons.

Wikgren, M. (2005). Critical realism as a philosophy and social theory in information science? *Journal of Documentation, 61*(1), 11-22.

Willcocks, L. (2004). Foucault, power/knowledge and information systems: Reconstructing the present. In J. Mingers & L. Willcocks (Eds.), *Social theory and philosophy for information systems.* Chichester: Wiley.

Willmott, H. (2005). Theorising contemporary control: Some post-structuralist responses to some critical realist questions. *Organization, 12*(5), 747-780.

Wilson, B. (1984). *Systems: Concepts, methodologies, and applications.* Chichester, UK: John Wiley and Sons.

Wilson, B. (2002). *Soft systems methodology: Conceptual model and its contribution.* Chichester, UK: John Wiley and Sons.

Wilson, F. (1999). Flogging a dead horse: The implications of epistemological relativism within information systems methodological practice. *European Journal of Information Systems, 8*(3), 161-169.

Winter, M. C., Brown, D. H., & Checkland, P. B. (1995). A role for soft systems methodology in information systems development. *European Journal of Information Systems, 4,* 130–142. doi:10.1057/ejis.1995.17

Wittgenstein, L. (1974). *Tractatus logico-philosophicus* (Rev. ed.). London: Routledge and Kegan Paul.

Wixom, B. H., & Watson, H. J. (2001). An empirical investigation of the factors affecting data warehousing success. *MIS Quarterly, 25*(1), 17-41.

Wood-Harper, A. T. (1985). Research methods in information systems: Using action research. In E. Mumford, R., Hirschheim, G. Fitzgerald & A.T. Wood-Harper (Eds.), *Research methods in information systems.* Amsterdam: Elsevier.

Work, B. (1997). Some reflections on information systems curricular. In J.M. Mingers & F.A.Stowell (Eds), *Information systems: An emerging discipline?* (pp. 329-360). Maidenhead: McGraw Hill.

WOS EXPANDED (2007). Retrieved Match 26, 2007 from http://wos.izum.si/CIW.cgi

Wray-Bliss, E. (2003). Research subjects/research subjections: Exploring the ethics and politics of critical research. *Organization, 10*(2), 307-325.

Wright, R. (1998, October). Process standards and capability models for engineering software-intensive systems. *Crosstalk: The Journal of Defense Software Engineering,* 1-10.

Wynn, E., & Graves, S. (2007). *Tracking the virtual organization* (Working paper). Intel Corporation.

SSADM (1990, July). *SSADM Version 4, Reference Manuals.* Oxford: NCC, Blackwell.

Xu, L. D. (2000). The contribution of systems science to information systems research. *Systems Research and Behavioral Science, 17*(2), 105–116. doi:10.1002/(SICI)1099-1743(200003/04)17:2<105::AID-SRES287>3.0.CO;2-M

Xu, L. D. (2000, March-April). The contribution of systems science to information systems research. *Systems Research and Behavioral Science, 17*(2), 105–116. doi:10.1002/(SICI)1099-1743(200003/04)17:2<105::AID-SRES287>3.0.CO;2-M

Yin, R. K. (1989). *Case study research: Design and methods (vol. 5).* Beverley Hills, CA: SAGE Publications Ltd.

Zachman, J. (1987). A framework for information systems architecture. *IBM Systems Journal, 26*(3), 276-292.

Zhou, L., Burgoon, J., & Twitchell, D. (2004). A comparison of classification methods for predicting deception in computer-mediated communication. *Journal of Management Information Systems, 20*(4), 139-165.

Zhu, Z. (2006). Complementarism versus pluralism: are they different and does it matter? *Systems Research and Behavioral Science, 23*(6), 757–770. doi:10.1002/sres.706

Zhu, Z. C. (2000, March-April). WSR: A systems approach for information systems development. *Systems Research and Behavioral Science, 17*(2), 183–203. doi:10.1002/(SICI)1099-1743(200003/04)17:2<183::AID-SRES293>3.0.CO;2-B

Zmud, R. W. (1979). Individual differences and MIS success: A review of the empirical literature. *Management Science, 25*(10), 966-979.

Zuboff, S. (1988). *In the age of the smart machine: The future of work and power*. New York: Basic Books.

About the Contributors

David Paradice is Sprint Professor of MIS and Chair of the MIS Department at Florida State University. He has worked as a programmer analyst and consultant and was a member of a team that built one of the first interactive DSS used in the electric utility industry. Dr. Paradice has published numerous articles focusing on the use of computer-based systems in support of managerial problem formulation. His publications appear in *Journal of MIS, IEEE Transactions on Systems, Man & Cybernetics, Decision Sciences, Communications of the ACM, Decision Support Systems, Annals of Operations Research, Journal of Business Ethics,* and other journals. His research also appears as chapters in several books. He is active in the Association of Information Systems Special Interest Group on Data, Knowledge, and Decision Support Systems and the International Federation on Information Processing TC8 Working Group 8.3 on Decision Support Systems. Dr. Paradice has served on several corporate advisory boards. He is co-Editor-in-Chief with Professor Manuel Mora (Autonomous University of Aguascalientes, Mexico) of the *International Journal of Information Technology and the Systems Approach.*

***

Denis Edgar-Nevill is Head of the Department of Computing at Canterbury Christ Church University. He is a senior member of the Editorial Review Board of *IJITSA*. His publications span a number of IS, SE and CS disciplines; software quality, systems thinking, software engineering. His more recent research in the application of IS in Computer Forensics has led to his election as Chair of the UK national British Computer Society Cybercrime Forensics Specialist Group

John V. Farr is Associate Dean for Academics and Professor of Systems Engineering and Engineering Management in the School of Systems and Enterprises at Stevens Institute of Technology. He was the founding Director of the Department of Systems Engineering and Engineering Management at Stevens, which he led from 2000 to 2007. He taught at the United States Military Academy at West Point from 1992 to 2000, and achieved the rank of Professor of Engineering Management. Dr. Farr was one of the first civilian professors in engineering at the Academy. He is a past president and Fellow of the American Society for Engineering Management, a Fellow of the American Society of Civil Engineers, and a member of the Army Science Board and Air Force Studies Board of the National Academies. He is a former editor of the *Journal of Management in Engineering* and the founder of the *Engineering Management Practice Periodical.* He has authored over 100 technical publications including several textbooks. He is a registered Civil Engineer in New York and Mississippi, and holds an undergraduate degree from Mississippi State University, a master's from Purdue University, and a PhD in Civil Engineering from the University of Michigan.

Miroljub Kljajić graduated from the Faculty of Electrical Engineering, University of Ljubljana, Slovenia in 1970 where also he obtained his M.Sc. in 1972 and his D.Sc. degree in 1974. Since 1976 he has been employed by the Faculty of Organizational Sciences in Kranj in the field of System Theory, Cybernetics, Computer Simulation, and Decision Theory. He obtained the rank of Professor in 1986. His main research field is methods of modeling and simulation of complex systems. He is author of two books on the Theory of Systems and Discrete Event Simulation. He published over 200 scientific articles from which 27 in JCR. He is the recipient of four awards for the best papers in international conferences and received three recognitions for achievements in the field of decision support systems. He was a member of the research team that received the Prize of the Boris Kidrič Fund in 1976 and the Prize for Innovations and Improvements in 1984. For his successes in engineering pedagogical work he received the Golden Medal of the University of Maribor in 1989.

Raymond Madachy is an Associate Professor in the Systems Engineering Department at the Naval Postgraduate School. He has over 80 publications including the book Software Process Dynamics. His research interests include modeling and simulation of processes for architecting and engineering of complex software-intensive systems; economic analysis and value-based engineering of software-intensive systems; integrating systems engineering and software engineering disciplines; systems and software measurement, process improvement, and quality; quantitative methods for systems risk management; and integrating empirical-based research with process simulation.

Rory O'Connor is a Senior Lecturer in Software Engineering at Dublin City University and a researcher with Lero, the Irish Software Engineering Research Centre. He received a PhD in Computer Science from City University (London) and has previously held research positions at both the National Centre for Software Engineering and the Centre for Teaching Computing and has also worked as a software engineer and consultant for several Irish and European technology organizations. His research interests are centered on the processes whereby software intensive systems are designed, implemented and managed.

D. Petkov is professor and coordinator of BIS at ECSU, USA. He is a Deputy Editor (USA) of Systems Research and Behavioral Science, Senior Area Editor for Software Engineering and the Systems Approach of IJITSA and editorial board member of several other journals. His research has appeared in the *Journal of Systems and Software, Decision Support Systems, IRMJ, IJITSA, Telecommunications Policy, JITTA, Intl. J. on Technology Management, Kybernetes* and elsewhere.

Frank Stowell is Professor of Systems and Information Systems at the University of Portsmouth. He has a PhD in Organisational Change and his research centers around Participative design of Information Systems. He has been co-chair of a number of research council funded projects notably the Systems Practice for Managing Complexity project, (http://www.spmc.org.uk/) which has developed into a self sustaining network. He is past President of the UK Academy of Information Systems and the UK Systems Society (http://www.ukss.org.uk/) and he presently occupies the chair of the Council of Information Systems Professors. He has supervised a number of research projects including modeling complex decision making in mental health care. He has published papers and texts in the field and presented papers at a number of international conferences in Europe and the United States. Prior to his academic career he was employed by central government as a consultant within the Management

Systems Development Group and has experience of defining and developing IT supported management information systems.

G. A. Swanson is professor and past chair of the Department of Accounting and Finance at Tennessee Technological University. Dr. Swanson holds a BS from Lee University, an M.A.C.T. from the University of Tennessee, Knoxville, and the PhD in accounting from Georgia State University. He is a Tennessee CPA, ret. and founder and first president of the Tennessee Society of Accounting Educators. He has served the International Society for the Systems Sciences as SIG chair, president elect, president, past president and VP for administration among service to other professional organizations. His more than 100 scholarly publications include books such as *Measurement and Interpretation in Accounting: A Living Systems Approach* with James Grier Miller and articles in such journals as *The Accounting Review, Systems Research and Behavioral Science* (SR&BS), and the *International Journal of Social Economics*. Professor Swanson serves on various editorial boards and has recently guest-edited a special issue of SR&BS honoring James Grier and Jessie Miller who are among the founders of the systems movement.

Index